Student Solutions Manual

Physics for Scientists and Engineers
Foundations and Connections

Volume 1

Eric Mandell
Bowling Green State University

Brian Utter
James Madison University

Debora M. Katz
United States Naval Academy

CENGAGE Learning

Australia • Brazil • Mexico • Singapore • United Kingdom • United States

ISBN: 978-0-534-46676-3

Cengage Learning
20 Channel Center Street
Boston, MA 02210
USA

Cengage Learning is a leading provider of customized learning solutions with office locations around the globe, including Singapore, the United Kingdom, Australia, Mexico, Brazil, and Japan. Locate your local office at: **www.cengage.com/global**.

Cengage Learning products are represented in Canada by Nelson Education, Ltd.

To learn more about Cengage Learning Solutions, visit **www.cengage.com**.

Purchase any of our products at your local college store or at our preferred online store **www.cengagebrain.com**.

Printed in the United States of America
Print Number: 01 Print Year: 2015

Preface

This *Student Solutions Manual* has been written to accompany the textbook *Physics for Scientists and Engineers: Foundations and Connections, First Edition, Volume I*, by Debora M. Katz, and provides solutions to selected end-of-chapter problems. It is hoped that providing detailed, stepped-out solutions for the problems selected will allow students further practice in methods of problem solving.

Every textbook chapter has a matching chapter in this book and very often reference is made to specific equations or figures in the textbook. Solutions are given for approximately 20 problems in each textbook chapter. Problems were selected to illustrate important concepts in each chapter. The solutions follow the two-column worked example format—INTERPRET and ANTICIPATE, SOLVE, and CHECK and THINK—presented in the text.

An important note concerning significant figures: When the statement of a problem gives data, we should only keep the number of significant figures in our final answer as dictated by the number of significant figures in the data and the rules for dealing with those figures in calculations (see Chapter 1, pages 10–11). The last digit is uncertain; it can, for example, depend on the precision of the values assumed for physical constants and properties. When a calculation involves several steps, we carry out intermediate steps to many digits, but we write down only three. We "round off" only at the end of any chain of calculations, never anywhere in the middle.

ACKNOWLEDGMENTS

We take this opportunity to thank everyone who contributed to this edition of the *Student Solutions Manual to accompany Physics for Scientists and Engineers: Foundations and Connections, First Edition, Volume 1*.

Special thanks for managing and directing this project go to Senior Product Manager Charles Hartford, Product Manager Rebecca Berardy Schwartz, Content Developers Susan Dust Pashos and Ed Dodd, Associate Content Developer Brandi Kirksey, and Product Assistant Brendan Killion.

Our appreciation goes our reviewers, Douglas Sherman and James D. Olsen. Their careful reading of the manuscript and checking the accuracy of the problem solutions contributed immensely to the quality of the final product. Any errors remaining in the manual are the responsibility of the authors.

Finally, we express our appreciation to our families for their inspiration, patience, and encouragement, during the production of this manual.

We sincerely hope that this *Student Solutions Manual* will be useful to you in reviewing the material presented in the text, and in improving your ability to solve problems and score well on exams. We welcome any comments or suggestions that could help improve the content of this study guide in future editions; and we wish you success in your study of physics.

Eric Mandell
Bowling Green, OH

Brian Utter
Harrisonburg, VA

Debora M. Katz
Annapolis, MD

Table of Contents

1

Getting Started

3. (N) The age of the Earth is 4.5 billion years. What is the age of the Earth in the appropriate SI units?

<table>
<tr><td colspan="2">INTERPRET and ANTICIPATE
Given the age of the Earth, we can multiply by conversion factors to express the answer in the SI units of seconds.</td></tr>
<tr><td>SOLVE
We first write the number in scientific notation given that a billion is also 10^9.</td><td>$t = 4.5 \times 10^9$ years</td></tr>
<tr><td>We can now use the fact that 1 year= 365 days, 1 day = 24 hours, and 1 hour = 3600 seconds. Given that the age is an estimate known to two significant figures, 365 days in a year is sufficiently accurate for this estimate.</td><td>$t = 4.5 \times 10^9 \text{ years}\left(\frac{365 \text{ days}}{1 \text{ year}}\right)\left(\frac{24 \text{ hr}}{1 \text{ day}}\right)\left(\frac{3600 \text{ s}}{1 \text{ hr}}\right)$
$t = \boxed{1.4 \times 10^{17} \text{ s}}$</td></tr>
<tr><td colspan="2">CHECK and THINK
As expected, the estimated age of the Earth is a very large number of seconds.</td></tr>
</table>

8. (N) In Jules Verne's novel, *Twenty Thousand Leagues Under the Sea*, Captain Nemo and his passengers undergo many adventures as they travel the Earth's oceans.
a. If 1.00 league equals 3.500 km, find the depth in meters to which the crew traveled if they actually went 2.000×10^4 leagues below the ocean surface.

INTERPRET and ANTICIPATE
The problem statement gives us a conversion factor between leagues and kilometers, 1.00 league = 3.500 km. After converting to kilometers, we will then need to convert to meters using the fact that 1 km = 1000 m. We expect the numeric answer to be much larger than 20,000, since each league is equivalent to a large number of meters (a smaller unit).

SOLVE We first apply the given conversion factor in order to express the distance in kilometers.	$(2.000\times10^4 \text{ leagues})\left(\dfrac{3.500 \text{ km}}{1.00 \text{ league}}\right) = 7.00\times10^4 \text{ km}$
Now, we convert this result from kilometers to meters	$(7.00\times10^4 \text{ km})\cdot\left(\dfrac{1000 \text{ m}}{1 \text{ km}}\right) = \boxed{7.00\times10^7 \text{ m}}$
CHECK and THINK As expected, the numeric answer (70,000,000) is much larger than 20,000, since we are expressing the same distance using a smaller unit of measure. Given that this is greater than the radius of the Earth, one might ask how the passengers could have possibly traveled to a depth of 20,000 leagues under the sea!	

b. Find the difference between your answer to part (a) and the radius of the Earth, 6.38×10^6 m. (Incidentally, author Jules Verne meant that the total distance traveled, and not the depth, was 20,000 leagues.)

INTERPRET and ANTICIPATE We can see that our answer to part (a) is less precise than the above expression for the radius of the Earth. Our answer to part (a) will dictate the level of precision in the difference between these two quantities. We also expect the answer to be on the same order of magnitude as our answer to part (a).	
SOLVE Subtract the radius of the Earth from the answer in part (a). For addition and subtraction, the number with the fewest digits after the decimal place determines the precision of the final answer.	$7.00\times10^7 \text{ m} - 6.38\times10^6 \text{ m} =$ $7.00\times10^7 \text{ m} - 0.638\times10^7 \text{ m} = \boxed{6.36\times10^7 \text{ m}}$
CHECK and THINK The difference between these numbers is bigger than the radius of the Earth! There is no way for anyone to travel to such a depth on the Earth given the dimensions of the Earth itself. As suggested, it was not the author's intention that the number be interpreted as a depth. Even as a total distance traveled under the ocean, it would be rather extreme (more than ten times the radius of the Earth and well beyond the Earth's circumference)!	

10. (E, N) A popular unit of measure in the ancient world was the cubit (approximately the distance between the elbow and the end of the middle finger, when outstretched). Estimate the length of one cubit in centimeters. Given that there are about 1.609×10^3 m in 1 mile, how many cubits are there in 1 mile?

INTERPRET and ANTICIPATE

Suppose that you estimate the distance between your elbow and the end of your outstretched middle finger to be about 46 cm. You are then able to create a conversion factor for centimeters to cubits, given the information in the problem statement. We already have a relationship between centimeters and meters. We are then able to convert the number of meters in one mile to a number of cubits.

SOLVE We begin by first finding the number of centimeters in 1.609×10^3 meters (one mile).	$1 \text{ mile} = 1.609\times10^3 \text{ m}$ $1 \text{ mile} = 1.609\times10^3 \text{ m}\left(\frac{100 \text{ cm}}{1 \text{ m}}\right)$ $1 \text{ mile} = 1.609\times10^5 \text{ cm}$
Now, convert the number of centimeters to cubits, using your estimate.	$1 \text{ mile} = 1.609\times10^5 \text{ cm}$ $1 \text{ mile} = 1.609\times10^5 \text{ cm}\left(\frac{1 \text{ cubit}}{46 \text{ cm}}\right)$ $1 \text{ mile} = \boxed{3.5\times10^3 \text{ cubits}}$

CHECK and THINK

Approximately 3500 people would have to line up their forearms in order to cover the distance of one mile! This sounds like a plausible estimate.

11. (N) CASE STUDY On planet Betatron, mass is measured in bloobits and length in bots. You are the Earth representative on the interplanetary commission for unit conversions and find that 1 kg = 0.23 bloobits and 1 m = 1.41 bots. Express the density of a raisin (2×10^3 kg/m^3) in Betatron units.

INTERPRET and ANTICIPATE

First, let's predict whether we expect a larger or smaller number in Betatron units. Since a bloobit represents more mass than a kg, the density measured using bloobits should be a smaller number than the density measured using kg. In addition, 1 bot is smaller than 1 m, so 1 bot^3 is a smaller volume than 1 m^3. So, at the same density, the amount of mass in 1 bot^3 is less than the amount of mass in 1 m^3. Both factors will result in the numerical value of the density being less in Betatron units compared to SI units. We can convert

mass and length between SI units and Betatron units by "multiplying by 1" repeatedly using the facts that 0.23 bloobits = 1 kg and 1 meter = 1.41 bots.	
SOLVE First convert density from kg/m^3 to bloobits/m^3:	$\rho = 2\times10^3 \frac{\text{kg}}{\text{m}^3}\left(\frac{0.23\text{ bloobits}}{1\text{ kg}}\right)$ $\rho = 0.46\times10^3 \frac{\text{bloobits}}{\text{m}^3}$
Now convert from bloobits/m^3 to bloobits/bot^3:	$\rho = 0.46\times10^3 \frac{\text{bloobits}}{\text{m}^3}\left(\frac{1\text{ m}}{1.41\text{ bot}}\right)^3$ $\rho = \frac{0.46\times10^3}{(1.41)^3}\frac{\text{bloobits}}{\text{bot}^3}$ $\rho = 0.16\times10^3 \frac{\text{bloobits}}{\text{bot}^3} \approx \boxed{2\times10^2 \frac{\text{bloobits}}{\text{bot}^3}}$
CHECK and THINK This result makes sense. The numerical value of the density is less in Betatron units (around 200) than SI units (2000). Since the density of a raisin in SI units was given to one significant figure, we estimate the density in Betatron units to one significant figure as well. While calculating along the way though, additional significant figures were carried in order to reduce rounding errors.	

14. (C) As part of a biology field trip, you have taken an equal-arm balance (Fig. P1.14) to the beach. Your plan was to measure the masses of various mollusks, but you forgot to bring along your set of standard gram masses. You notice that the beach is full of pebbles. Although there are variations in color, texture, and shape, you wonder whether you can somehow use the pebbles as a standard mass set. Develop a procedure for assembling a standard mass set from the pebbles on the beach. Describe your procedure step by step so that someone else could follow it.

Figure P1.14

Even though the pebbles do not look the same or have the exact same size and shape, they can be used as a standard mass set. The important property for objects in a standard mass set is to have masses that allow us to reliably determine an unknown mass. One possibility is to select a set of pebbles in which each pebble balances with every other

pebble, such that every pebble will have the same mass. We could then determine an unknown mass to the nearest pebble.

A possible procedure is as follows: First, make sure the equal arm balance is balanced with nothing in either pan. If necessary, drape a bit of seaweed on one side or the other. Second, pick out a pebble and place it in one pan of the balance. This will be our standard pebble. Third, select a new pebble from the beach, and place it in the vacant pan. If the system balances, then remove the new pebble and put it in your pocket. It is now a member of the standard mass set. If the system does not balance, then take the new pebble and throw it back on the beach. Repeat the second and third steps. In this manner, each pebble that is a candidate for the standard mass set will be compared against a single, "reference" pebble. As more and more pebbles are tested in this way, you will accumulate a pocket full of pebbles that all balance with each other. This set of pebbles is a standard mass set.

To use the standard pebble set to measure the mass of a mollusk, simply place the mollusk on one side of the balance and place pebbles from the standard mass set one at a time on the other side until balance is achieved and the mass is determined in "pebbles."

This process provides a systematic way to quantify the masses of objects. It is important to hang on to the reference pebble, and make sure every investigator has access to it. The process for defining a scale for mass in the SI system is similar to that described above. In fact, the "standard pebble" in the SI system is a platinum-iridium cylinder that is kept in a vault in a laboratory in Paris… we call it the kilogram!

21. (A) In subsequent chapters, two different types of a physical quantity called *energy* will be defined. The kinetic energy K of an object is $K = \frac{1}{2}mv^2$. The gravitational potential energy U associated with an object can sometimes be expressed as $U = mgy$. In these expressions, m stands for mass, g is the acceleration with dimensions of length per time squared, v is a quantity called speed that has the dimensions of length per time, and y is a distance with units of length. Show that the two different expressions for energy are consistent in that they have the same dimensions.

INTERPRET and ANTICIPATE

The dimensions for each expression for energy can be analyzed and reduced down to some combination of units of length, mass, and time. It is expected that each expression should result in the same combination of units.

SOLVE First, we consider the expression for kinetic energy, replacing the quantities, *m* and *v*, with their dimensions. We reduce these units down to the simplest combination of fundamental units *length*, *mass*, and *time*.	$[\![K]\!]=\frac{1}{2}[\![m]\!][\![v]\!]^2$ $[\![K]\!]=\text{M}\left(\frac{\text{L}}{\text{T}}\right)^2$ $[\![K]\!]=\text{M}\frac{\text{L}^2}{\text{T}^2}$
We now consider the expression for gravitational potential energy, replacing the quantities, *m*, *g*, and *y*, with their dimensions. We again reduce these units down to the simplest combination of fundamental units *length*, *mass*, and *time*.	$[\![U]\!]=[\![m]\!][\![g]\!][\![y]\!]$ $[\![U]\!]=\text{M}\frac{\text{L}}{\text{T}^2}\text{L}$ $[\![U]\!]=\text{M}\frac{\text{L}^2}{\text{T}^2}$

CHECK and THINK

The dimensions of each expression for energy produce a quantity with the same units. While the formulas refer to two different types of energy, dimensionally they are equivalent and both refer to the same type of quantity, energy. We can expect to find some physical connection between these quantities in future chapters.

33. (N) Calculate the result for the following operations using the correct number of significant figures.

$$3.07670-10.988+\frac{\left(\frac{5.4423\times10^6}{4.008\times10^3}\right)}{\left(1.0093\times10^5-9.98\times10^4\right)}$$

INTERPRET and ANTICIPATE

In order to determine a final answer, we must follow the proper order of operations. We also should not limit the number of significant figures when calculating during intermediate steps, before arriving at a final answer. This can be taken into account by retaining a few extra significant figures at each intermediate step. Given that the order of operations will dictate that addition and subtraction are the last steps to occur, we expect that the number of digits to the right of the decimal in the final answer will be determined by the number with the smallest number of digits after the decimal point.

SOLVE For multiplication and division, the result should have the same number of significant figures as the value with the least number of significant	

figures. For addition and subtraction, the answer should be written to the same precision (e.g. number of digits to the right of the decimal point) as the least precise number. It is a good idea to keep extra significant figures for intermediate steps of calculations to avoid rounding errors and then determine the correct number of significant figures for the final value.	
First, we determine the values of the quantities in parentheses. We can keep extra significant figures for the intermediate steps to avoid rounding errors, but we express the results using significant figures for clarity. The ratio has four significant figures (limited by the denominator, which has four significant figures). The difference has only two significant figures ($9.98 \times 10^4 = 99800$ is known only to the hundreds place, therefore the answer can only be known to the hundreds place, $1.1 \times 10^3 = 1100$).	$\left(\frac{5.4423\times10^6}{4.008\times10^3}\right)=1.358\times10^3$ $\left(1.0093\times10^5-9.98\times10^4\right)=1.1\times10^3$
We are then able to find the value of the total fraction using these two results. Since the denominator is known to two significant figures, the resulting quantity is written to two significant figures.	$\frac{\left(\frac{5.4423\times10^6}{4.008\times10^3}\right)}{\left(1.0093\times10^5-9.98\times10^4\right)}=\frac{1.358\times10^3}{1.1\times10^3}$ $\frac{\left(\frac{5.4423\times10^6}{4.008\times10^3}\right)}{\left(1.0093\times10^5-9.98\times10^4\right)}=1.2$
We are left with three numbers, which need to be added and subtracted. For addition and subtraction, the final answer should be written to the same precision as the least precise value, that is, with	$3.07670-10.988+1.2=\boxed{-6.7}$

the *fewest number of digits to the right* of the decimal.	

CHECK and THINK
We must pay attention to the order of operations when performing a string of calculations that arrive at a final answer. It is often best to keep all the figures in your calculator and round in the last step. As a general rule, the number of significant figures in the final answer is determined by the least precise number used in the calculation, but we need to consider multiplication and division differently than addition and subtraction.

35. (N) Complete the following calculations and report your answer using scientific notation, the correct number of significant figures, and SI units.
a. Model the Earth as a sphere with a radius of 6378.1 km. Find its volume.

INTERPRET and ANTICIPATE
We can calculate the volume of a sphere using the formula $V = \frac{4}{3}\pi R^3$. The value of π can be determined to arbitrary precision, so we expect that the number of digits in the radius will determine the precision of the final answer.

SOLVE First, we convert the radius to meters, using the fact that 1 km = 1000 m, writing the answer in scientific notation. The value has five significant figures.	$6378.1 \text{ km}\left(\frac{1000 \text{ m}}{1 \text{ km}}\right) = 6.3781 \times 10^6 \text{ m}$
Using the equation for the volume of the sphere, we calculate an answer. The number 4/3 is exact as is *π*. The final answer calculated by multiplying or dividing values, will have the same number of significant figures as the least precise value used in the calculation. In this case, the final answer has five significant figures, the same as the radius.	$V = \frac{4}{3}\pi R^3 = \frac{4}{3}\pi\left(6.3781 \times 10^6 \text{ m}\right)^3$ $V = \boxed{1.0868 \times 10^{21} \text{ m}^3}$

CHECK and THINK
We calculated the volume of the sphere by multiplying known values. The final answer has five significant figures, the same as the least precise value used in the calculation, the value of the radius.

b. The mass of the Earth is 5.98×10^{24} kg. Find the density of the Earth.

<table>
<tr><td colspan="2">INTERPRET and ANTICIPATE
We will calculate the density using the given mass and the volume from part (a). Since the mass is known to a lower precision, its value will determine the precision of the final answer.</td></tr>
<tr><td>SOLVE
We use Equation 1.1 to determine the density of the Earth. The final answer, for an answer calculated by multiplying or dividing values, will have the same number of significant figures as the least precise value used in the calculation. In this case, the final answer has three significant figures, the same as the mass.</td><td>$\rho = \frac{m}{V} = \frac{5.98 \times 10^{24}\ \text{kg}}{1.0868 \times 10^{21}\ \text{m}^3}$
$\rho = \boxed{5.50 \times 10^3\ \frac{\text{kg}}{\text{m}^3}}$</td></tr>
<tr><td colspan="2">CHECK and THINK
We calculated the density of the Earth by dividing known values. The final answer has three significant figures, the same as the least precise value used in the calculation, the value of the mass.</td></tr>
</table>

45. (E) Estimate the number of living cells in a tiger.

<table>
<tr><td colspan="2">INTERPRET and ANTICIPATE
Answers may vary. In order to estimate the number of cells in a living tiger, it would be helpful to estimate the volume of a tiger and the volume of a cell within a tiger. The ratio of these quantities could be used to estimate the number of cells within the tiger. Of course, we expect this to be a very large number.</td></tr>
<tr><td>SOLVE
Here is one method, though your specific method and estimated numbers may be different: Model the head of the tiger as a small box and the body of the tiger as a larger rectangular box. For the head, we estimate that the volume is the volume of a box with side length</td><td>$V_{\text{head}} = (0.3\ \text{m})^3 = 0.027\ \text{m}^3$
$V_{\text{body}} = (1.5\ \text{m}) \times (0.50\ \text{m}) \times (0.75\ \text{m}) = 0.56\ \text{m}^3$
$V_{\text{tiger}} = 0.027\ \text{m}^3 + 0.56\ \text{m}^3 = 0.59\ \text{m}^3$</td></tr>
</table>

30 cm, or 0.3 m. The body of the tiger might be estimated to have dimensions of 1.5 m × 0.50 m × 0.75 m. Our estimate for the volume of the tiger is simply the sum of these numbers. As usual, we are free to use more significant figures in our intermediate calculation steps, but at the end will write an approximate answer given the roughness of the estimation.	
We now need to estimate the volume of a cell. In Table 1.4, we see an estimate for the size of a living cell is 10 μm. If this is the side length of a cube that represents the cell, then an estimate for the volume of a cell is the volume of the cube.	$V_{\text{cell}} = (10\ \mu\text{m}) \times (10\ \mu\text{m}) \times (10\ \mu\text{m}) \times \left(\frac{1\ \text{m}}{1 \times 10^6\ \mu\text{m}} \right)^3$ $V_{\text{cell}} = 1 \times 10^{-15}\ \text{m}^3$
Now, we estimate the number of cells in the tiger, *N*, as the ratio of the volume of the tiger and the volume of one cell.	$N = \frac{V_{\text{tiger}}}{V_{\text{cell}}} = \frac{0.59\ \text{m}^3}{1 \times 10^{-15}\ \text{m}^3} = 6 \times 10^{14}\ \text{cells} \approx \boxed{10^{14}\ \text{cells}}$

CHECK and THINK
Our estimate is on the order of 10^{14} cells. While only a rough estimate, this is a very large number of cells as expected.

48. (E) In 2011, artist Hans-Peter Feldmann covered the walls of a gallery at the New York Guggenheim Museum with 100,000 one dollar bills (Fig. P1.48). Approximately how much would it cost you to wallpaper your room in one-dollar bills, assuming the bills do not overlap? Consider the cost of the bills alone, not other supplies or labor costs.

Figure P1.48

INTERPRET and ANTICIPATE Answers may vary. In order to make this estimate, we start by estimating the area of the walls of a room. Then, we can use the area of a dollar bill as shown in Table 1.3, 100 cm^2. We take the ratio of the area of the walls divided by the area of a dollar to find out how many dollars we need.	
SOLVE If we assume the room is 12 ft × 9 ft × 8 ft, where 8 ft is the height, then we can determine the total area of the four walls.	$A = 2\times(12 \text{ ft}\times 8 \text{ ft}) + 2\times(9 \text{ ft}\times 8 \text{ ft})$
Using the fact that there are 30.48 cm in 1 ft, we can convert this to square centimeters.	$A = \left[2\times(12 \text{ ft}\times 8 \text{ ft}) + 2\times(9 \text{ ft}\times 8 \text{ ft})\right]\times\left(\frac{30.48 \text{ cm}}{1 \text{ ft}}\right)^2$ $A = 3\times10^5 \text{ cm}^2$
Finally, we divide the area of the walls by the area for one bill to determine the total number of bills needed.	$N = \frac{3\times10^5 \text{ cm}^2}{100 \text{ cm}^2} = 3000$ $\text{cost} \approx \boxed{\$3000}$
CHECK and THINK We would need \$3000 in dollar bills to wallpaper the room. This is a fairly large number, as we might expect.	

51. (N) A 350-seat rectangular concert hall has a width of 60.0 ft, length of 81.0 ft, and height of 26.0 ft. The density of air is 0.0755 lb/ft^2.

a. What is the volume of the concert hall in cubic meters?

INTERPRET and ANTICIPATE Using the dimensions of the room provided, we can calculate the volume of this rectangular space. We expect the final answer to have the same number of significant figures as the distances used in the calculation.	
SOLVE The volume of the room can found assuming the room is a rectangular box. For intermediate steps, we keep a few extra significant figures to avoid rounding errors.	$V = l \times w \times h = (60.0\text{ ft})(81.0\text{ ft})(26.0\text{ ft})$ $V = 1.2636 \times 10^5\text{ ft}^3$
We now convert the volume to meters, using 1 ft = 0.3048 m. The final answer when multiplying or dividing values, will have the same number of significant figures as the least precise value used in the calculation. In this case, the final answer has three significant figures, the same as each of the distances used in the calculation.	$V = 1.2636 \times 10^5\text{ ft}^3 \left(\frac{0.3048\text{ m}}{1\text{ ft}}\right)^3 = 3.5781 \times 10^3\text{ m}^3$ $V = \boxed{3.58 \times 10^3\text{ m}^3}$
CHECK and THINK The volume of the rectangular room was found and the answer of a few thousand cubic meters sounds reasonable. The final answer, found by multiplication, has three significant figures matching the number of significant figures of the distances used in the calculation.	

b. What is the weight of the air in the concert hall in newtons?

INTERPRET and ANTICIPATE Using the density provided and the volume calculated in part (a), we can determine the weight of the air in the room. Each value has three significant figures, so a result found by multiplying them will also have three significant figures.	
SOLVE Using an expression like Equation 1.1, we can express the weight of air in the room in	$W = \rho V$ $W = (0.0755\text{ lbs/ft}^3)(1.26 \times 10^5\text{ ft}^3) = 9.54 \times 10^3\text{ lbs}$

terms of the volume of the room and the weight density, which is given as a weight (in pounds) per cubic foot. (We are really calculating a weight in this case and not a mass.) The volume was calculated in part (a). The final answer has three significant figures, the same as the least precise value used in the calculation. (In this case, they both have three significant figures.)	
Now, we can convert the weight to newtons using 1 lb = 4.448 N.	$9.54\times10^3\ \text{lbs}\left(\dfrac{4.448\ \text{N}}{1\ \text{lb}}\right)=\boxed{4.24\times10^4\ \text{N}}$

CHECK and THINK

The weight was found using the density and volume provided. The final answer, found through multiplication, has three significant figures since the values used in the calculation are known to three significant figures.

55. (A) Two different expressions for finding the magnitude of a quantity called force F, are $F = ma$ and $F = \dfrac{p_f - p_i}{t_f - t_i}$. The acceleration a has dimensions of length per time squared. Momentum p is equal to mv, where v is speed with the dimensions of length per time. The subscripts f and i refer to the final and initial values of those quantities, respectively. Also, m stands for mass and t stands for time. Show that the two different expressions for force are consistent in that they result in a quantity with the same dimensions.

INTERPRET and ANTICIPATE

The dimensions for each expression for force can be analyzed and reduced down to some combination of units of length, mass, and time. It is expected that each expression for force should result in the same combination of fundamental units.

SOLVE	
First, we consider the expression $F = ma$, replacing the quantities, m	

and a, with their dimensions. We reduce these units down to the simplest combination of length, mass, and time.	$[\![F]\!]=[\![m]\!][\![a]\!]$ $[\![F]\!]=\boxed{\mathrm{M}\left(\mathrm{L}/\mathrm{T}^2\right)}$
We now consider the second expression, first substituting the quantities mv for the quantity p.	$F=\dfrac{p_f-p_i}{t_f-t_i}=\dfrac{mv_f-mv_i}{t_f-t_i}=\dfrac{m\left(v_f-v_i\right)}{t_f-t_i}$
We now replace the quantities, m, v and t, with their dimensions, reducing these units down to the simplest combination of length, mass, and time.	$[\![F]\!]=\dfrac{[\![m]\!][\![\left(v_f-v_i\right)]\!]}{[\![t_f-t_i]\!]}$ $[\![F]\!]=\dfrac{\mathrm{M}\left(\mathrm{L}/\mathrm{T}\right)}{\mathrm{T}}$ $[\![F]\!]=\boxed{\mathrm{M}\left(\mathrm{L}/\mathrm{T}^2\right)}$

CHECK and THINK

The dimensions of each expression for force are equivalent. Since we are told that both of these expressions are valid for finding the magnitude of a force, this is clue foreshadowing the fact that there is more than one way to look at a concept such as force in physics.

57. (N) During a visit to New York City, Lil decides to estimate the height of the Empire State Building (Fig. P1.57). She measures the angle θ of elevation of the spire atop the building as 20.0°. After walking 9.0×10^2 ft closer to the iconic building, she finds the angle to be 25.0°. Use Lil's data to estimate the height h of the Empire State Building.

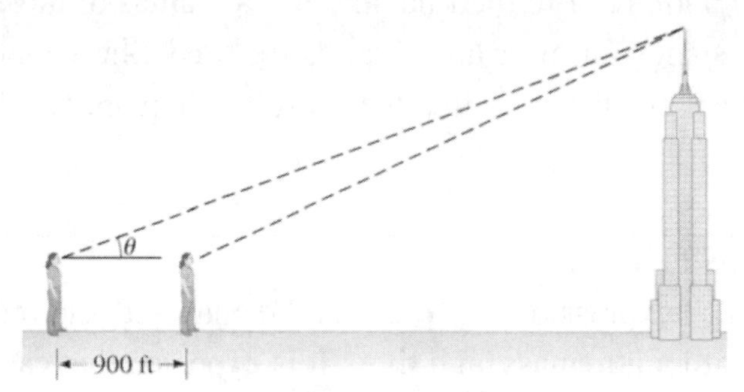

Figure P1.57

INTERPRET and ANTICIPATE

Knowing Lil's change in position and the angle to the spire allows us to draw triangles relating the angles to unknown quantities for both cases. We will need to write trigonometric relationships involving the angles and attempt to solve for the unknown height.

SOLVE

First, we sketch the situation for both positions. Triangles are formed with the given angles, the unknown height h, and an unknown initial distance d.

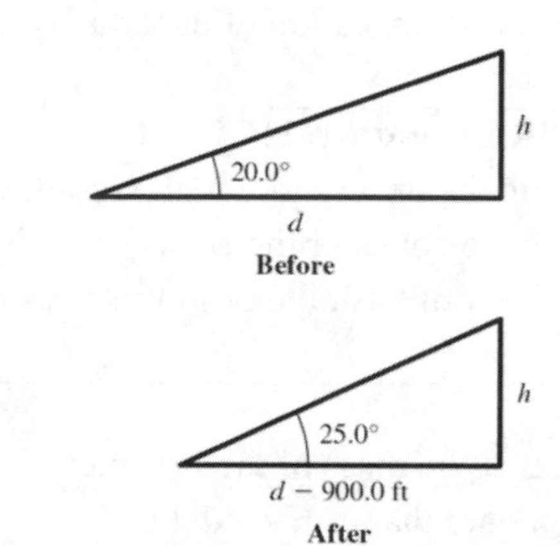

Figure P1.57ANS

Using this figure, the tangents of the angles are related to the ratio of the labeled sides.

$$\tan 20.0^\circ = \frac{h}{d}$$

$$\tan 25.0^\circ = \frac{h}{d - 900.0 \text{ ft}}$$

We have two equations and two unknowns. In order to eliminate d and solve for h, we solve each equation for d and set them equal to each other. The final answer has two significant figures, which is the precision of the 9.0×10^2 ft used in the final quotient.

$$d = \frac{h}{\tan 20.0^\circ}$$

$$d = \frac{h}{\tan 25.0^\circ} + 900 \text{ ft}$$

$$\frac{h}{\tan 20.0^\circ} = \frac{h}{\tan 25.0^\circ} + 900 \text{ ft}$$

$$2.7474\ h = 2.1445\ h + 900 \text{ ft}$$

$$0.603\ h = 900 \text{ ft}$$

$$h = \frac{900 \text{ ft}}{0.603} = 1500 \text{ ft} = \boxed{1.5 \times 10^3 \text{ ft}}$$

CHECK and THINK

We are able to estimate the height using the information given by sketching both situations and using trigonometry. The resulting height of 1.5×10^3 feet is around a third of a mile. This is quite tall, but seems possible for a skyscraper.

59. (N) You are part of a team in an engineering class that is working up a scale model of a new design for a life vest. You have been asked to find the mass of a piece of foam that will be used for flotation. Because the piece is too bulky to fit on your balance, you break it into two parts. You measure the mass of the first part as 128.3 ± 0.3 g, and the second part as 77.0 ± 0.3 g.

a. What are the maximum and minimum values for the total mass you might reasonably report? *Hint*: Propagation of uncertainty is described in Appendix A.

INTERPRET and ANTICIPATE To estimate the uncertainty in the result, we can apply a "worst-case scenario analysis." This means to consider the errors in the individual measurements to have all "conspired" to make the result as high or low as possible.	
SOLVE Suppose that the mass of the first piece is actually higher than expected. For instance, given the uncertainty of 0.3 g, it is reasonable to believe that the actual mass could be 0.3 g above the best estimate.	128.3 + 0.3 g = 128.6 g.
Similarly, the second piece could have a mass at the high end of the expected range.	77.0 + 0.3 = 77.3 g
Using these, we can calculate the total mass of foam.	128.6 g + 77.3 g = **205.9 g**
A similar worst case scenario allows us to estimate a lower bound for the mass of foam if each piece had an actual mass 0.3 g below the best estimate:	(128.3 – 0.3 g) + (77.0 – 0.3 g) = **204.7 g**
CHECK and THINK Given the uncertainty in the masses, we imagine that the actual mass is at the high or low end of the estimated range. This gives us an expectation that the actual mass is within the range 204.7 g – 205.9 g.	

b. What is the best estimate for the total mass of the foam? *Hint*: Propagation of uncertainty is described in Appendix A.

INTERPRET and ANTICIPATE	
When summing two measured values to calculate a new result, the best estimate for the result is simply the sum of the two measured values.	
SOLVE The best estimate for the total mass of foam is the sum of the two measured values:	$128.3\text{ g} + 77.0\text{ g} = 205.3\text{ g}$
CHECK and THINK As expected, the best estimate is within the range from part (a). These results could also be reported as: $M = \boxed{205.3 \pm 0.6\text{ g}}$ Note that when calculated the sum of two measured values, as we have done above, the uncertainties in the individual measurements sum to give a larger uncertainty in the final result.	

60. (A) Model the human body as three cylinders: a large one for the torso and two small ones for the legs. The height of each cylinder is half the height h of the person. The radius of the large cylinder is r, and the radii of the smaller cylinders are $r/2$. The body has a mass m. Assume the average density of the body is about the same as water, $\rho = 1000$ kg/m^3, and the radius of each cylinder is small compared with its height. Show that the body's surface area in meters squared is roughly given by $A = 0.129m^{0.5}h^{0.5}$, where m is in kilograms and h is in meters. *Note*: An empirical fit produces a more accurate estimate of the body's surface area: $A = 0.202m^{0.425}h^{0.725}$.

INTERPRET and ANTICIPATE
The total surface area of the three cylinders depends on their radii. We can eliminate this from our expression by finding the total volume of the cylinders (in terms of the radius), assuming the density is the same as water, and using the equation for density.

SOLVE A sketch helps us visualize the problem.	**Figure P1.60ANS**
The volume of any cylinder is area of its base times its height. There are three cylinders so we add their volumes to get the total volume.	$V = \pi r^2 \dfrac{h}{2} + 2\left(\pi\left(\dfrac{r}{2}\right)^2 \dfrac{h}{2}\right)$ $V = \dfrac{3}{4}\pi r^2 h \qquad (1)$
The average density is mass divided by volume (Eq. 1.1).	$\rho = \dfrac{m}{V} = \dfrac{4m}{3\pi r^2 h}$
The surface area of any cylinder is its circumference times its height plus the area of its base. We assume the area of the base is small and so we ignore it. We add to get the total area.	$A = 2\pi r \dfrac{h}{2} + 2\left[2\pi\left(\dfrac{r}{2}\right)\dfrac{h}{2}\right]$ $A = 2\pi r h \qquad (2)$
We can isolate r from Equation (1) and then substitute it into Equation (2).	$r = \sqrt{\dfrac{4m}{3\pi\rho h}}$ $A = 2\pi\sqrt{\dfrac{4m}{3\pi\rho h}}\,h$
We now use the approximation that the body's density is roughly the same as water.	$A = 2\pi\sqrt{\dfrac{4}{3\pi\left(1000 \text{ kg/m}^3\right)}}\,m^{0.5}h^{0.5}$ $A = 0.129 m^{0.5} h^{0.5}$

CHECK and THINK

Of course, we feel confident because we found the expected results. As a further check, let's find the SI units of the constant (0.129) and of A.

$$2\pi\sqrt{\frac{4}{3\pi\left(1000\ \text{kg/m}^3\right)}} = 0.129\ \frac{\text{m}^{3/2}}{\text{kg}^{1/2}}$$

$$[\![A]\!] = \frac{\text{m}^{3/2}}{\text{kg}^{1/2}}\left(\text{kg}\right)^{1/2}\left(\text{m}\right)^{1/2} = \text{m}^2 \text{ as expected.}$$

2

One-Dimensional Motion

6. In the traditional Hansel and Gretel fable, the children drop crumbs of bread on the ground to mark their path through the woods. Unfortunately, the crumbs are eaten by birds, and the children cannot find their way home. In this modern-day problem, the children use a device that releases a drop of food dye once per minute. As long as it does not rain, they can find their way home. As an extra bonus, they make a motion diagram as shown in Figure P2.6.

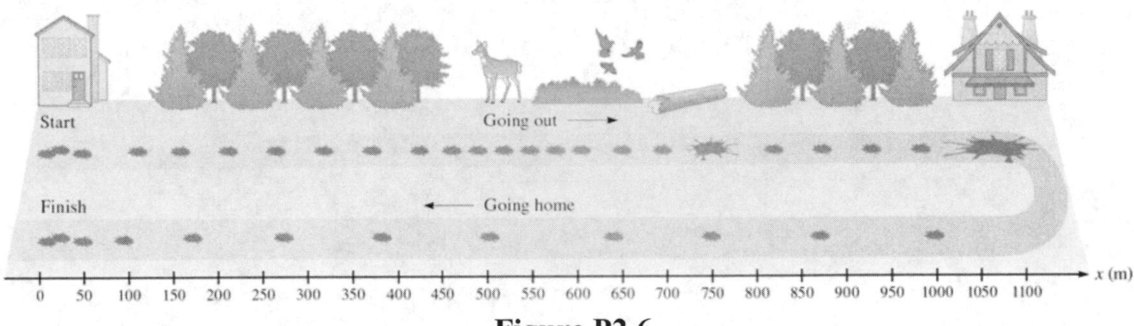

Figure P2.6

a. (C) Describe the motion of the children in words.

In a motion diagram, the dots are spaced at equal time intervals. Therefore, drops that are far apart occur for faster speeds when the position changes significantly in the given time interval and drops land on top of each other when the children are at rest. Based on this, they walk slowly away from home, quickly through the woods, and slowly past the deer. They rest on the fallen log and walk quickly through woods before resting at the ginger bread house. They run home, slowing down only once they are again near home.

b. (G) Using the coordinate system in Figure P2.6, make a position-versus-time plot. Note any ambiguities you encounter.

Using our interpretation of the motion diagram in part (a), we can plot the position versus time. The times when they are at rest are ambiguous, as we cannot easily determine how long they were at rest if the drops land on top of each other.

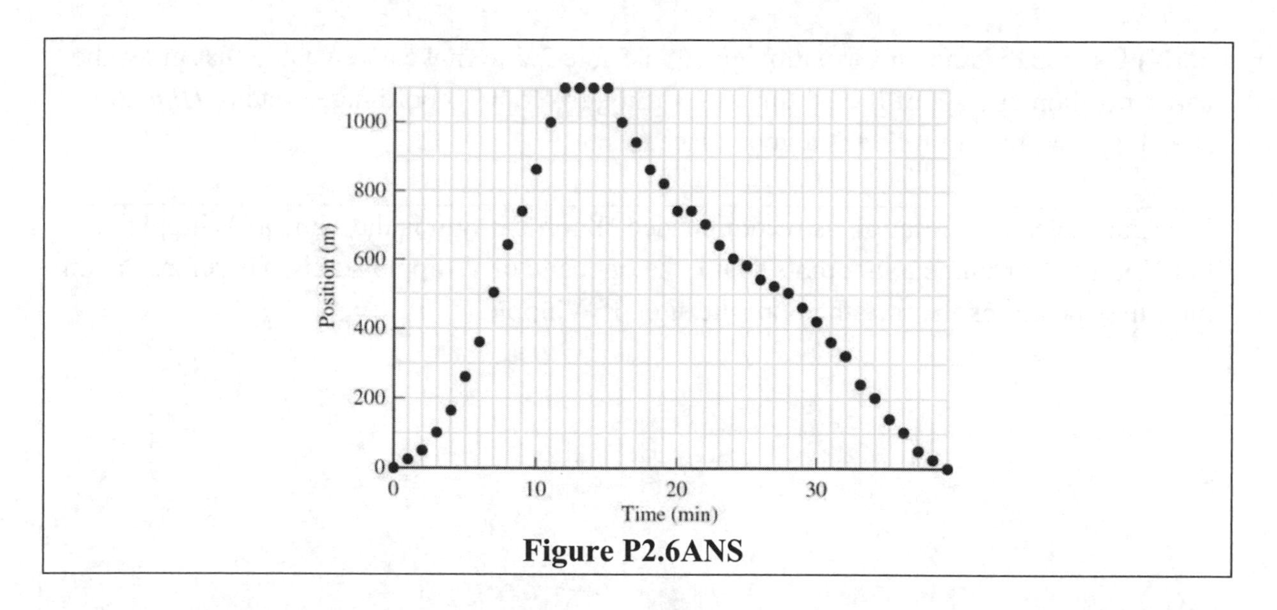

Figure P2.6ANS

c. (C) Is your position-versus-time graph consistent with your description in part (a)? Explain.

The plot is consistent with the description above. You should be able to follow along on the plot and match each phrase from part (a) to convince yourself that they are consistent.

10. CASE STUDY As shown in Figure 2.9, Whipple chose a coordinate system that was different from Crall's. There are many possible coordinate systems that can be used to analyze the motion of the cart. For example, a third student (Yoon) chose to place the origin of her coordinate system at position A (the car's initial position) and use a southward-pointing y axis as her positive axis.

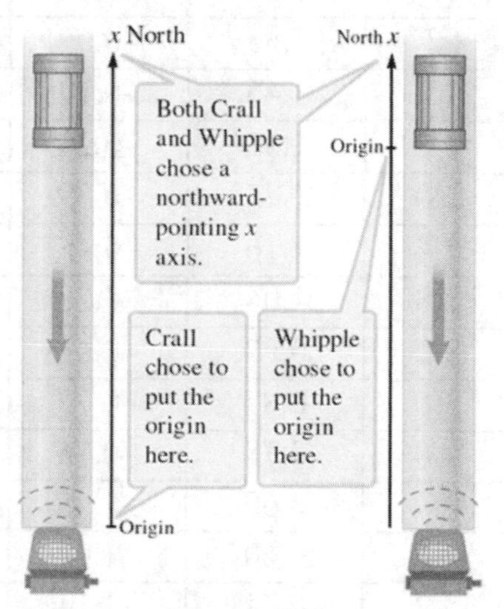

Figure 2.9

a. (N) Use the distance information given in Table 2.2 to make a new table that gives the cart's position at each of the 16 times according to Yoon's coordinate system. *Hint*: A sketch similar to Figure 2.9 is a good place to start.

We first sketch the situation as recommended. Yoon's origin is the same as Whipple's, but Yoon's coordinate axis points south, the opposite of Whipple's axis. Therefore, Yoon measures positions that are the opposite sign of Whipple.

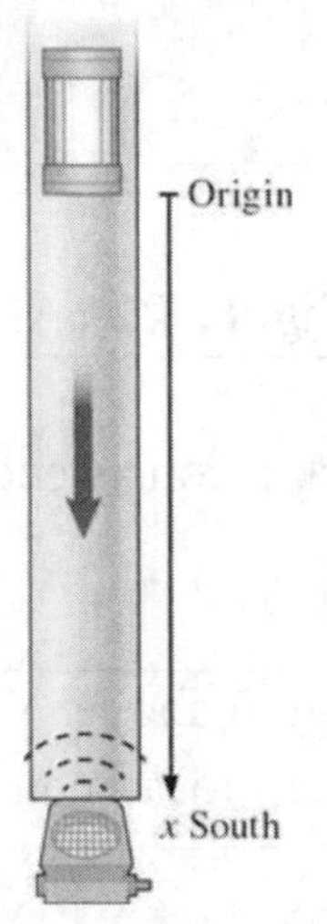

Figure P2.10aANS

Label	t (s)	Crall's x (m)	Whipple's x (m)	Yoon's x (m)
A	0.0	1.64	0	0
B	0.4	1.64	0	0
C	0.8	1.55	–0.09	0.09
D	1.2	1.45	–0.19	0.19
E	1.6	1.35	–0.29	0.29
F	2.0	1.25	–0.39	0.39
G	2.4	1.16	–0.48	0.48
H	2.8	1.06	–0.58	0.58
I	3.2	0.97	–0.67	0.67
J	3.6	0.87	–0.77	0.77
K	4.0	0.78	–0.86	0.86
L	4.4	0.69	–0.95	0.95
M	4.8	0.60	–1.04	1.04
N	5.2	0.51	–1.13	1.13
O	5.6	0.42	–1.22	1.22
P	6.0	0.33	–1.31	1.31

b. (G) Use the data in your table to make a new position-versus-time graph. *Hint:* CHECK and THINK about your graph by comparing it with Figure 2.6.

The data from part (a) can now be plotted to create a position versus time graph. The line starts at the same point as Whipple's since their axes both used the initial position of the cart as the origin, but the slope has the opposite sign since the positive direction of Yoon's axis is opposite that of Whipple's.

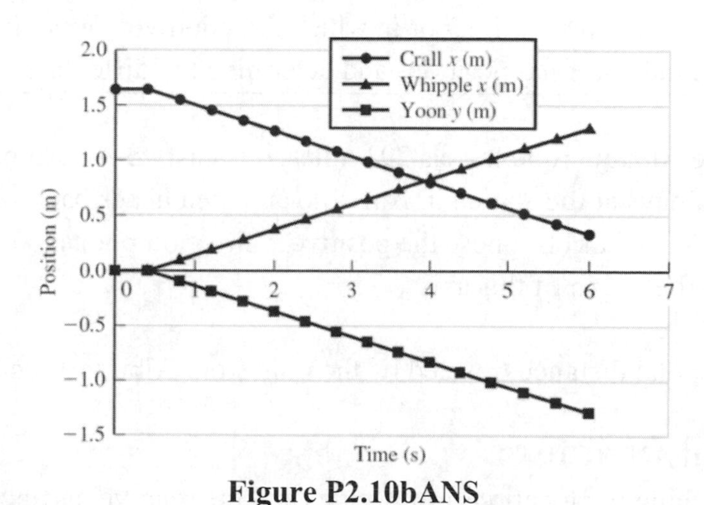

Figure P2.10bANS

12. (G) A particle moves from position *A* to position *C* as shown in Figure P2.3. Two different coordinates systems have been chosen. Write expressions for the particle's displacement from *A* to *C* using **a.** the top coordinate system, and **b.** the bottom coordinate system.

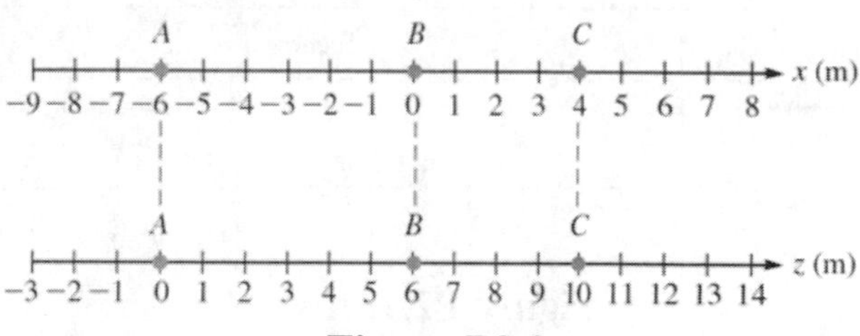

Figure P2.3

In each case, we need to find the coordinates of the two points using the specified coordinate system and then subtract them to find the displacement. For the top coordinate system,

$$\Delta\vec{x} = \vec{x}_C - \vec{x}_A = \left[4\hat{i} - (-6\hat{i})\right]\text{m} = \boxed{10\hat{i}\ \text{m}}$$

$$\Delta\vec{z} = \vec{z}_C - \vec{z}_A = \left[10\hat{\imath} - (0)\right]\text{m} = \boxed{10\hat{\imath}\ \text{m}}$$

c. (G) Compare your results. Do they make sense? Explain.

Yes, this makes sense. The displacement is independent of the coordinate system. We would expect that a question like "How far did the object move?" should have a single correct answer (or at least if the object is moving along a straight line!). It shouldn't matter where you lay down your ruler or in which direction you orient it, as long as you can measure the initial and final positions and determine the difference.

15. A train leaving Albuquerque travels 293 miles, due east, to Amarillo. The train spends a couple of days at the station in Amarillo and then heads back west 107 miles where it stops in Tucumcari. Suppose the positive x direction points to the east and Albuquerque is at the origin of this axis.

a. (N) What is the total distance traveled by the train from Albuquerque to Tucumcari?

INTERPRET and ANTICIPATE

We begin by sketching the locations of the cities and the relative distances between them. The total distance traveled by the train is the length of the whole path traversed by the train. For each leg of the journey, it does not matter in what direction the train travels, just how far, when determining distance traveled. The answer should be a positive number, expressed in meters.

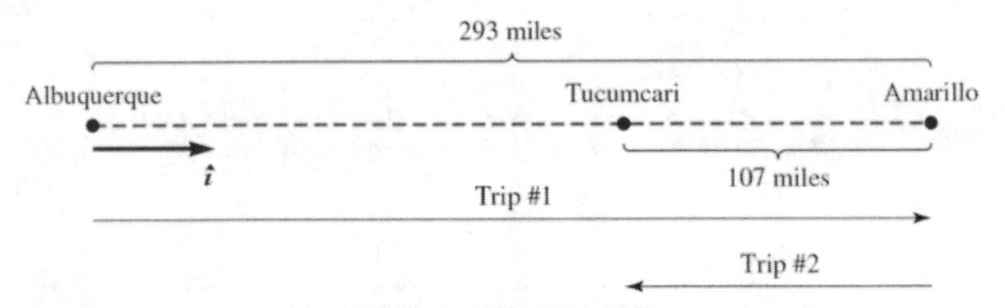

Figure P2.15ANS

SOLVE	
The train undergoes two displacements with different magnitudes. We first convert those magnitudes from miles into meters, using the fact that there are 1609 meters in one mile, retaining a few extra digits of precision because we	$d_1 = 293\ \text{miles}$ $d_1 = 293\ \text{miles} \times \left(\frac{1609\ \text{m}}{1\ \text{mile}}\right) = 4.71437 \times 10^5\ \text{m}$

have not finished the calculation process. The final answer should have three digits given the measure of miles with three significant digits in the problem statement.	$d_2 = 107 \text{ miles}$ $d_2 = 107 \text{ miles} \times \left(\frac{1609 \text{ m}}{1 \text{ mile}}\right) = 1.72163 \times 10^5 \text{ m}$
We then add the distances, or magnitudes of each displacement, to determine the total distance traveled.	$d_{tot} = d_1 + d_2 = 6.44 \times 10^5 \text{ m}$
CHECK and THINK This little back and forth journey for the train is approximately equal to a tenth of the radius of the Earth ($R_\oplus = 6.36 \times 10^6 \text{ m}$).	

b. (N) What is the displacement of the train for the entire journey? Give both answers in appropriate SI units.

INTERPRET and ANTICIPATE Referring to our sketch, the train undergoes two different trips, one to the east and one to the west. However, for the displacement, we need only consider the initial and final position of the train. Albuquerque is located at the origin of the coordinate system, so we can determine the position of Tucumcari by finding the distance between the two cities. Since the train never moves west of the starting point in Albuquerque, we expect that the net displacement points in the positive x direction, to the east.	
SOLVE Subtract the distance between Tucumcari and Amarillo from the distance between Albuquerque and Amarillo, to get the distance from Albuquerque to Tucumcari, labeled x_f.	$x_f = d_1 - d_2 = 4.71437 \times 10^5 \text{ m} - 1.72163 \times 10^5 \text{ m}$ $x_f = 2.99 \times 10^5 \text{ m}$
We can then write the vector.	$\vec{x}_f = 2.99 \times 10^5 \hat{i} \text{ m}$
According to Equation 2.1, the displacement is then equal to the difference between the final and initial positions.	$\Delta\vec{x} = \vec{x}_f - \vec{x}_i = 2.99 \times 10^5 \hat{i} \text{ m} - 0\hat{i} \text{ m}$ $\Delta\vec{x} = 2.99 \times 10^5 \hat{i} \text{ m}$

CHECK and THINK
The displacement does point in the positive x direction, as expected.

18. A particle is attached to a vertical spring. The particle is pulled down and released, and then it oscillates up and down. Using an upward-pointing y axis, the position of the particle is given by

$$\vec{y} = \left(y_0 \cos \frac{2\pi t}{T} \right) \hat{j}$$

Both y_0 and T are constants in time. The amplitude (y_0) is usually given in meters, and the period (T) is usually in seconds.

a. (G) Draw a sketch of this situation. Include the y axis.

The y axis points upwards and at $t = 0$ s, the particle is at $+y_0$.

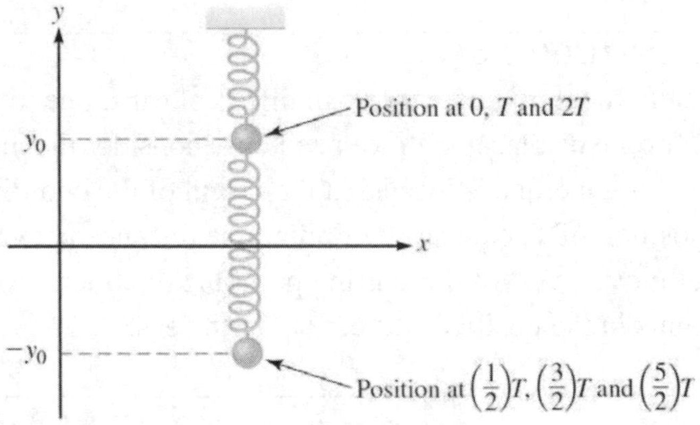

Figure P2.18ANS

b. (N) What is the position of the particle when $t = 0$, $t = T/2$, $t = T$, $t = 3T/2$, $t = 2T$, and $t = 5T/2$? Add these positions with clear labels to your sketch.

INTERPRET and ANTICIPATE
Since we have an expression for the position as a function of time, we can plug in the values of the time and solve for the position. Since the period is the time for one complete revolution, we expect that whatever the position is at $t = 0$, the particle has oscillated for one complete cycle and returns to the same position at $t = T$, and then returns to the same position after the second cycle at $t = 2T$. We would also expect that whatever the position is at $t = T/2$, the particle has oscillated and returns to the same position at $t = 3T/2$, and then returns to the same position after the second cycle $t = 5T/2$.

SOLVE We plug in each time into the expression for position to find the position at 0, *T*/2, *T*, 3*T*/2, 2*T*, and 5*T*/2.	$\vec{y}(t=0)=\left(y_0 \cos 0\right)\hat{j}=y_0\hat{j}$ $\vec{y}(t=\frac{T}{2})=\left(y_0 \cos \frac{\pi T}{T}\right)\hat{j}=-y_0\hat{j}$ $\vec{y}(t=T)=\left(y_0 \cos \frac{2\pi T}{T}\right)\hat{j}=y_0\hat{j}$ $\vec{y}(t=\frac{3T}{2})=\left(y_0 \cos \frac{3\pi T}{T}\right)\hat{j}=-y_0\hat{j}$ $\vec{y}(t=2T)=\left(y_0 \cos \frac{4\pi T}{T}\right)\hat{j}=y_0\hat{j}$ $\vec{y}(t=\frac{5T}{2})=\left(y_0 \cos \frac{5\pi T}{T}\right)\hat{j}=-y_0\hat{j}$

CHECK and THINK

As expected, the position is the same in each case. Since each point in time is separated by multiples of a period, the particle has traversed a whole number of oscillations and returned to the initial position.

21. During a relay race, you run the first leg of the race, a distance of 2.0×10^2 m to the north, in 22.23 s. You then run the same distance back to the south in 24.15 s in the second leg of the race. Suppose the positive *y* axis points to the north.

a. (N) What is your average velocity for the first leg of the relay race?

INTERPRET and ANTICIPATE

We can write the displacement vector for the first leg of the journey, and then find the average velocity for the first leg of the journey. The answer should be in m/s and the vector should point in the positive *y* direction since you are traveling to the north in the first leg of the journey.

SOLVE The displacement for the first leg of the journey can be constructed knowing the distance traveled to the north, given that the positive *y* axis points to the north.	$\Delta\vec{y}=2.0\times10^2\,\hat{j}$ m

Using Equation 2.2, we find the average velocity for the first leg of the journey	$\vec{v}_{av,y} = \frac{\Delta\vec{y}}{\Delta t} = \frac{2.0\times10^2\,\hat{j}\text{ m}}{22.23\text{ s}} = 9.0\,\hat{j}\text{ m/s}$
CHECK and THINK The average velocity does point in the positive y direction.	

b. (N) What is your average velocity for the entire race?

INTERPRET and ANTICIPATE In this case, you finish at the starting position, which means that there is no displacement for the entire race. This means that the average velocity for the race is 0 m/s.	
SOLVE We can first write the displacement for he entire race, and calculate the total time for the race.	$\Delta\vec{y} = 0\,\hat{j}\text{ m}$ $t_{total} = t_1 + t_2 = 22.23\text{ s} + 24.15\text{ s} = 46.38\text{ s}$
Equation 2.2 then allows us to calculate the average velocity.	$\vec{v}_{av,y} = \frac{\Delta\vec{y}}{\Delta t} = \frac{0\,\hat{j}\text{ m}}{46.38\text{ s}} = 0\,\hat{j}\text{ m/s}$
CHECK and THINK The average velocity for the entire race is 0 m/s, since the initial and final positions are the same and the total displacement is zero.	

24. Light can be described as a wave or as a particle known as a photon. The speed of light c is 3.00×10^8 m/s.

a. (N) Sirius is the brightest star in the night sky, 8.18×10^{16} m from the Earth. Find the time it takes a photon to reach us from Sirius. Give your answer in years.

INTERPRET and ANTICIPATE We know that the photon travels at the speed of light, which relates the distance traveled by the photon per unit time. Using the speed and distance, we can calculate the time.	
SOLVE Using the definition of speed, the time equals the distance divided by the speed of the photon.	$\Delta t = \frac{\Delta x}{c} = \frac{8.18\times10^{16}\text{ m}}{3.00\times10^{8}\,\frac{\text{m}}{\text{s}}} = 2.73\times10^{8}\text{ s}$

Using the fact that there are 3.16×10^7 seconds in a year, we can convert this to years.	$\Delta t = 2.73\times10^8\text{s}\cdot\left(\dfrac{1\text{ year}}{3.16\times10^7\text{s}}\right) = 8.64\text{ years}$
CHECK and THINK Given the speed and distance, we can determine the time for the photon to reach the Earth. Sirius is *very* far away, so even at traveling at the speed of light, the light takes nearly nine years to reach the Earth!	

b. (N) A light-year (ly) is the distance that light travels in 1 year. How far from the Earth is Sirius in light-years?

INTERPRET and ANTICIPATE With the definition of a light year, we can find the distance light travels in a year and use this to convert the distance from Sirius to light-years.	
SOLVE A light year is the distance light travels in one year.	$\Delta x_{light-year} = c\Delta t = \left(3.00\times10^8\dfrac{\text{m}}{\text{s}}\right)\left(3.16\times10^7\text{s}\right)$ $\Delta x_{light-year} = 9.48\times10^{15}\text{m}$
We can use this conversion factor to find the distance from Sirius in light-years.	$8.18\times10^{16}\text{m}\cdot\left(\dfrac{1\text{ light-year}}{9.48\times10^{15}\text{m}}\right) = 8.64\text{ light-years}$
CHECK and THINK In retrospect, we should expect the same numerical value as in part (a), where we found out how many years the light traveled to cover this distance. This is equivalent to measuring the distance in light-years.	

28. When you hear a noise, you usually know the direction from which it came even if you cannot see the source. This ability is partly because you have hearing in two ears. Imagine a noise from a source that is directly to your right. The sound reaches your right ear before it reaches your left ear. Your brain interprets this extra travel time (Δt) to your left ear and identifies the source as being directly to your right. In this simple model, the extra travel time is maximal for a source located directly to your right or left ($\Delta t = \Delta t_{max}$). A source directly behind or in front of you has equal travel time to each ear, so $\Delta t = 0$. Sources at other locations have intermediate extra travel times ($0 \le \Delta t \le \Delta t_{max}$). Assume a source is directly to your right.

a. (E) If the speed of sound in air at room temperature is $v_s = 343$ m/s, find Δt_{max}.

The distance between a person's ears is around 20 cm, or 0.2 m. The maximum extra time can be found using the speed of sound to find the time needed for the sound to travel the extra distance between the person's ears.

$$\Delta t_{max} = \frac{d}{v} = \frac{0.2\text{ m}}{343\ \frac{\text{m}}{\text{s}}} \approx 6\times10^{-4}\text{ s} = 0.6\text{ ms}$$

b. (E) Find Δt_{max} if instead you and the source are in seawater at the same temperature, where $v_s = 1531$ m/s.

The higher sound speed in water leads to a smaller time for sound to travel from one ear to the other.

$$\Delta t_{max} = \frac{d}{v} = \frac{0.2\text{ m}}{1531\ \frac{\text{m}}{\text{s}}} \approx 1.3\times10^{-4}\text{ s} = 0.13\text{ ms}$$

c. (C) Why is it difficult to locate the source of a noise when you are under water?

Since the time difference is about five times larger in air, it is easier for your brain to distinguish the time difference in air. In water, the time difference would be much smaller and you would perceive the sound as being nearly in front or behind you.

30. (A) The instantaneous speed of a particle moving along one straight line is $v(t) = ate^{-5t}$, where the speed v is measured in meters per second, the time t is measured in seconds, and the magnitude of the constant a is measured in meters per second squared. What is its maximum speed, expressed as a multiple of a?

INTERPRET and ANTICIPATE
This problem can be solved using the fact that the maximum value of a function will occur when its slope, or derivative, is zero. Speed is the magnitude of velocity, so we expect the answer to be greater than zero and have units of m/s.

SOLVE	
We first take the derivative of the speed with respect to time. Since the variable t occurs in two places in the expression for speed, we use the chain rule to	$\frac{dv(t)}{dt} = a\left[\left(\frac{d}{dt}t\right)\left(e^{-5t}\right) + t\left(\frac{d}{dt}e^{-5t}\right)\right]$

calculate the derivative. We then factor the result to express it in simplest form.	$\frac{dv(t)}{dt} = a\left[(1)(e^{-5t}) + (t)(-5e^{-5t})\right]$ $\frac{dv(t)}{dt} = a\left[e^{-5t} - 5te^{-5t}\right]$ $\frac{dv(t)}{dt} = ae^{-5t}(1-5t)$
To find the maximum speed, we now set the derivative of the speed equal to zero and solve for t, the time at which the maximum speed occurs.	$ae^{-5t}(1-5t) = 0$ $1 - 5t = 0$ $5t = 1$ $t = 0.2$ s
Finally, we substitute this time into the speed function to calculate the maximum speed.	$v_{max} = v(t = 0.2 \text{ s}) = a(0.2)e^{-5(0.2)}$ $v_{max} = 0.0736a$

CHECK and THINK

The derivative of the function was used to find the time at which the maximum occurs, which in turn was used to find the maximum value of the function.

32. (C) An object initially traveling in the positive x direction undergoes a change in velocity so that, after a finite amount of time passes, it ends up traveling in the negative x direction. Sketch and describe the slope of the position-versus-time graph for this object's motion.

The slope of a position-versus-time graph at any point in time is equal to the instantaneous velocity of the object in motion. Since the object is initially moving in the positive x direction, the slope of the position versus time graph will initially have a positive value. The object turns around and finishes with a negative velocity. At some point between the beginning and the end, the velocity must have been 0 m/s for the value of the velocity to cross over from positive to negative values. This means the slope of the position-versus-time graph must start off positive and decrease until it reaches 0 m/s. The slope must then become negative for it to travel in the negative x direction.

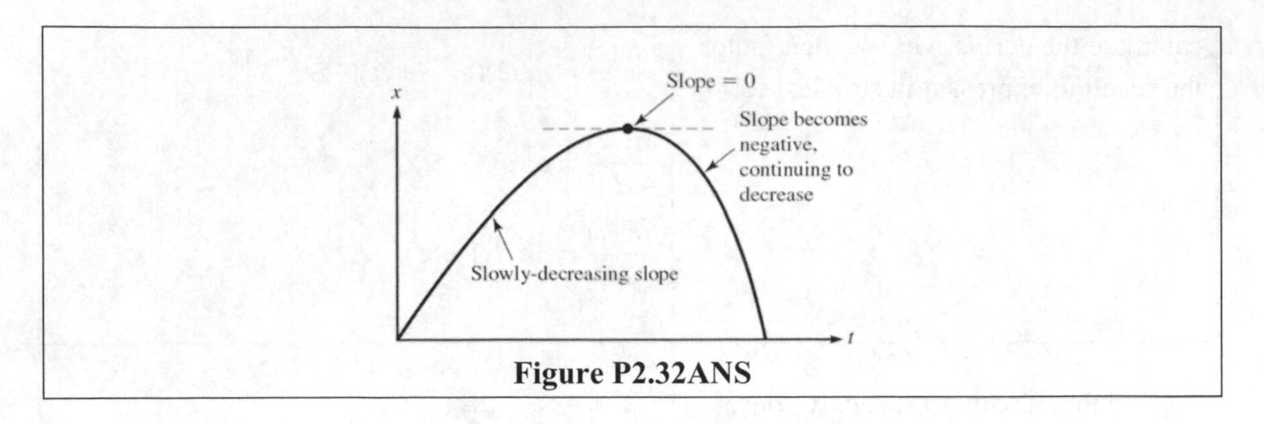

Figure P2.32ANS

34. A particle's position is given by $z(t) = -(7.50\,\text{m/s}^2)t^2$ for $t \geq 0$.

a. (A) Find an expression for the particle's velocity as a function of time.

The velocity is the derivative of the position with respect to time.

$$v_z(t) = \frac{dz}{dt} = -(15.0\,\text{m/s}^2)t$$

b. (C) Is the particle speeding up, slowing down, or maintaining a constant speed?

The particle is speeding up. The velocity is increasing in magnitude (though in the negative direction). The speed is the magnitude of the velocity, which is increasing linearly in time.

c. (N) What are the particle's position, velocity, and speed at $t = 6.50$ min?

INTERPRET and ANTICIPATE

We now have algebraic expressions for the position and velocity as functions of time. We should be able to plug in the time in order to determine the values.

SOLVE	
While we do need to "plug in the time," we also have to be careful about units. The units in the original function are "m/s²," so we need a time measured in seconds in order for the units to cancel to meters rather than something like $\frac{\text{m}\cdot\text{min}^2}{\text{s}^2}$. First, we convert the time to seconds.	$t = 6.50\text{ min}\left(\frac{60\text{ s}}{1\text{ min}}\right) = 390\text{ s}$

We can now plug this into our expressions for position and velocity. The speed is the magnitude of the velocity.	$z(390\text{ s}) = -(7.50\text{m/s}^2)(390\text{ s})^2 = 1.14\times10^6\text{m}$ $v_z(390\text{ s}) = -(15.0\text{m/s}^2)(390\text{ s}) = -5.85\times10^3\text{m/s}$ $\text{speed} = \lvert v_z(390\text{ s})\rvert = 5.85\times10^3\text{m/s}$

CHECK and THINK
With the expressions for position and velocity, and being careful to use consistent units, we are able to determine numerical values. As expected from part (b), the velocity is a negative value while the speed (magnitude of velocity) is positive.

43. In general, when an object moves through a medium such as water or air, the medium affects the object's motion. If nothing else (like gravity or a motor) acts to counter the effect of the medium, the object will decelerate. Above a certain speed threshold, the speed of the object is given by

$$v_x(t) = \frac{v_0}{bt + C}$$

along an arbitrary x axis. Both b and C are constants in time. The units of b are s^{-1}, and C is unitless. Finally, v_0 is a constant with units of meters per second.

a. (C) How do you know that the direction of the acceleration is opposite that of the velocity?

As time increases, the velocity decreases because t appears in the denominator. The object is slowing down, therefore the acceleration must be opposite the velocity.

b. (A) Show that the magnitude of its acceleration is given by $a_x = (b/v_0)v_x^2$.

The acceleration is found by taking the derivative of the velocity function. The velocity function can then be substituted back into the result to find the desired form for the acceleration.

$$a_x(t) = \left|\frac{dv_x}{dt}\right| = \frac{bv_0}{(bt + C)^2} = \frac{bv_x^2}{v_0}$$

c. (C) Is the acceleration constant? Can the equations in Table 2.4 be applied to this scenario?

No and no. The acceleration varies according to the equation in part (b). It is decreasing in time.

d. (C) How can a submarine move through the water at constant speed?

A motor must be used to produce a force that cancels the drag force from the water, thus producing zero acceleration and constant velocity.

46. In Example 2.6, we considered a simple model for a rocket launched from the surface of the Earth. A better expression for the rocket's position measured from the center of the Earth is given by

$$\vec{y}(t) = \left(R_{\oplus}^{3/2} + 3\sqrt{\frac{g}{2}} R_{\oplus} t \right)^{2/3} \hat{j}$$

where $R_{\oplus}$ is the radius of the Earth (6.38×10^6 m) and g is the constant acceleration of an object in free fall near the Earth's surface (9.81 m/s^2).

a. (A) Derive expressions for $\vec{v}_y(t)$ and $\vec{a}_y(t)$.

The velocity can be found by taking the time derivative of the position. The acceleration is equal to the derivative of the velocity with respect to time.

$$\vec{v}(t) = \frac{d\vec{y}}{dt} = \frac{2}{3}\left(R_{\oplus}^{3/2} + 3\sqrt{\frac{g}{2}} R_{\oplus} t \right)^{-1/3} \left(3\sqrt{\frac{g}{2}} R_{\oplus} \right) \hat{j}$$

$$\vec{v}(t) = \frac{d\vec{y}}{dt} = 2\sqrt{\frac{g}{2}} R_{\oplus} \left(R_{\oplus}^{3/2} + 3\sqrt{\frac{g}{2}} R_{\oplus} t \right)^{-1/3} \hat{j}$$

$$\vec{a}(t) = \frac{d\vec{v}}{dt} = -\frac{1}{3}\left(2\sqrt{\frac{g}{2}} R_{\oplus} \right)\left(R_{\oplus}^{3/2} + 3\sqrt{\frac{g}{2}} R_{\oplus} t \right)^{-4/3} \left(3\sqrt{\frac{g}{2}} R_{\oplus} \right) \hat{j}$$

$$\vec{a}(t) = -\left(g R_{\oplus}^2 \right)\left(R_{\oplus}^{3/2} + 3\sqrt{\frac{g}{2}} R_{\oplus} t \right)^{-4/3} \hat{j}$$

b. (G) Plot $y(t)$, $v_y(t)$, and $a_y(t)$. (A spreadsheet program would be helpful.)

The functions from part (a) can now be plotted.

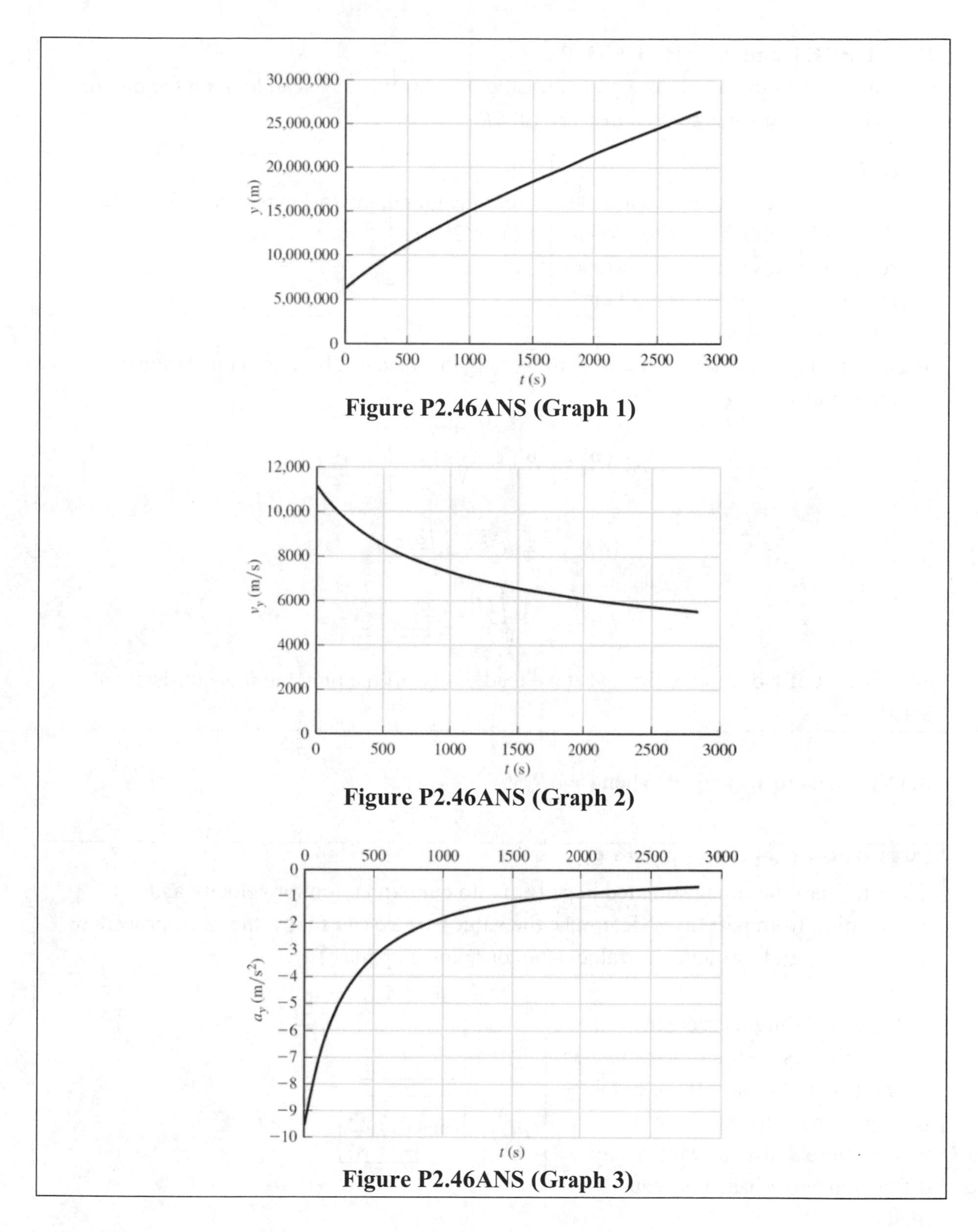

Figure P2.46ANS (Graph 1)

Figure P2.46ANS (Graph 2)

Figure P2.46ANS (Graph 3)

c. (N) When will the rocket be at $y = 4R_{\oplus}$?

INTERPRET and ANTICIPATE

One of the plots in part (b) is a plot of position versus time. We can look on the plot to find when the rocket reaches a distance of $4R_\oplus$.

SOLVE

The radius of the earth is about 6400 km. We look at our plot to see when the rocket reaches a distance of four times this value, around 25,000 km.

From the graph, the rocket reaches 25,000 km around 2600 s.

CHECK and THINK

We can actually check this precisely by putting this distance back into our original equation and solving for time:

$$4R_\oplus = \left(R_\oplus^{3/2} + 3\sqrt{\frac{g}{2}}R_\oplus t\right)^{2/3}$$

$$\left(4R_\oplus\right)^{3/2} = R_\oplus^{3/2} + 3\sqrt{\frac{g}{2}}R_\oplus t$$

$$t = \frac{4^{3/2} - 1}{3}\sqrt{\frac{2R_\oplus}{g}} = \boxed{2661\ \text{s}}$$

This is a lot of work to confirm what we read easily off the plot, but it is consistent at least!

d. (N) What are $\vec{v}_y$ and $\vec{a}_y$ when $y = 4R_\oplus$?

INTERPRET and ANTICIPATE

We can insert the time estimated in part (c) into our expression for velocity and acceleration from part (a) to determine the values, or we can follow the same procedure as in part (c) and estimate the values from our plots in part (b).

SOLVE

Either calculating the velocity and acceleration using the time estimated in part (c) or reading off the graph from part (b), we determine the values. Note that your values may differ somewhat based on your method.

$$\vec{v}(t) = \boxed{5590\hat{j}\ \frac{\text{m}}{\text{s}}}$$

$$\vec{a}(t) = \boxed{-0.613\hat{j}\ \frac{\text{m}}{\text{s}^2}}$$

CHECK and THINK

At a distance of $4R_\oplus$, the acceleration is approximately $1/16^{\text{th}}$ the value at the surface of the earth, consistent with what we expect for the inverse square law.

54. CASE STUDY: Crall and Whipple attached a fan to a cart placed on a level track and then released the cart. They made a position-versus-time graph (Fig. P2.54) and fit a curve to these data such that

$$x = 0.036\text{ m} + (0.0080\text{ m/s})t + (0.10\text{ m/s}^2)t^2$$

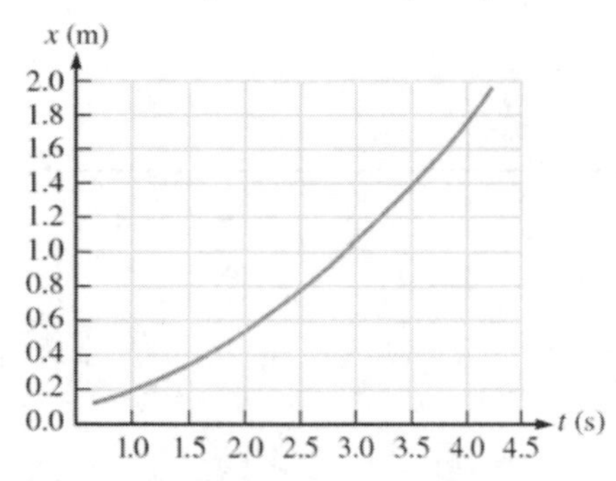

Figure P2.54

a. (A, G) Find and graph the velocity as a function of time.

The velocity is found by taking the derivative of the position function and the plotted versus time.

$$v_x = \frac{dx}{dt} = \boxed{0.0080\text{ m/s} + (0.20\text{ m/s}^2)t}$$

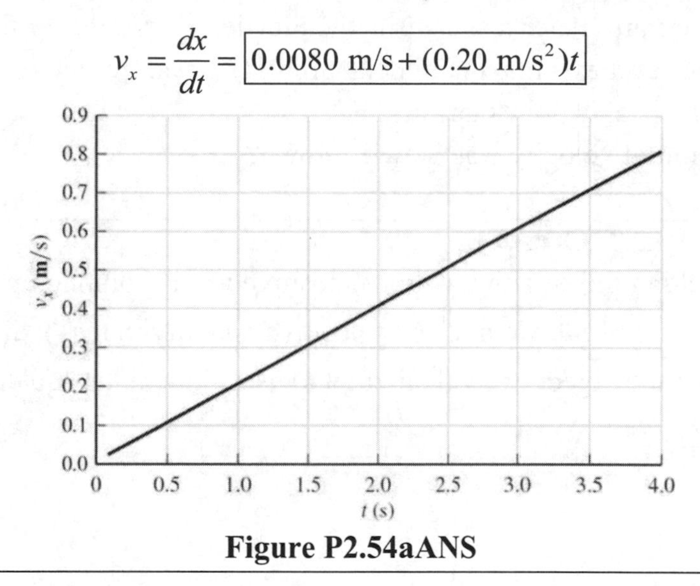

Figure P2.54aANS

b. (C) What is the shape of the velocity-versus-time graph? What do you expect the acceleration-versus-time plot to look like? Explain.

The plot of velocity is a straight line (and the equation for the velocity has the familiar form for a line equation "$y = mx + b$"). Since the acceleration is the slope of the velocity, it will be constant.

c. (A, G) Find and graph the acceleration as a function of time.

The acceleration is the derivative of velocity.

$$a_x = \frac{dv_x}{dt} = 0.20\ \text{m/s}^2$$

It is indeed constant as we predicted in part (b) and appears as a flat line when plotted versus time.

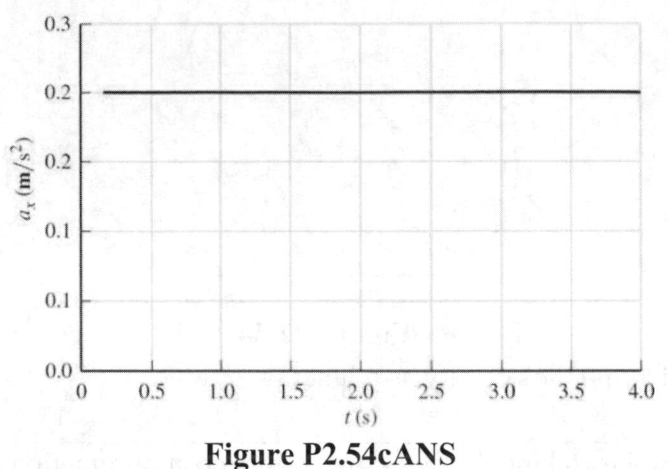

Figure P2.54cANS

58. In a cartoon program, Peter tosses his baby, Stewie, up into the air to keep the child entertained. Stewie reaches a maximum height of 0.873 m above the release point. Suppose the positive y axis points upward.

a. (N) With what initial velocity was Stewie thrown?

INTERPRET and ANTICIPATE

We begin by sketching the situation with the positive y axis pointing upward. We expect that the initial velocity of Stewie must be a positive quantity in the $\hat{j}$ direction. Knowing the displacement of the object, we will attempt to use a kinematic equation to solve for the unknown initial velocity.

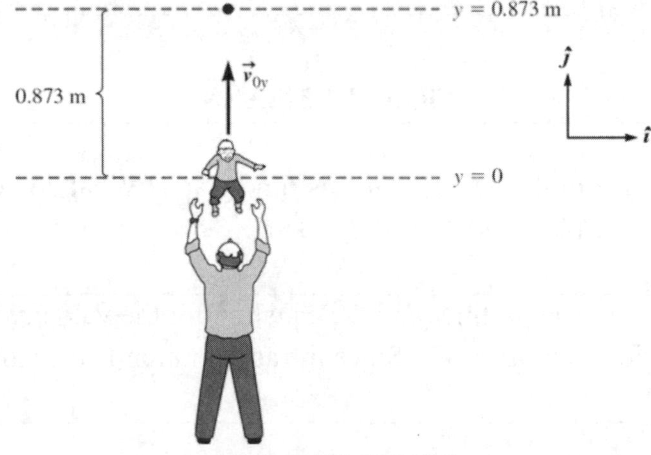

Figure P2.58ANS

SOLVE First, we collect and determine information on the five kinematic variables. At the top, the baby's velocity is momentarily zero.	$\Delta\vec{y} = 0.873\hat{j}$ m $\vec{a}_y = -g\hat{j} = -9.81\hat{j}$ m/s^2 $\vec{v}_y = 0$ $\vec{v}_{0y}$? t not needed
Equation 2.13 is the best choice given the variables we want to relate.	$v_y^2 = v_{y0}^2 + 2a_y\Delta y$
Solve this equation for the initial velocity.	$v_{y0}^2 = v_y^2 - 2a_y\Delta y$ $v_{y0} = \pm\sqrt{v_y^2 - 2a_y\Delta y}$
We choose the positive square root since the initial velocity was directed upward in what we called the positive y direction. Then, substitute the values and solve.	$v_{y0} = \sqrt{(0\text{ m/s})^2 - 2(-9.81\text{ m/s}^2)(0.873\text{ m})}$ $v_{y0} = 4.14$ m/s
Lastly, write the initial velocity as a vector.	$\vec{v}_{y0} = 4.14\hat{j}$ m/s

CHECK and THINK
As expected, the initial velocity must have been directed upward. The kinematic equations for constant acceleration describe the motion of an object as a function of time, as long as the acceleration remains unchanged.

b. (N) How much time did it take Stewie to reach the peak height?

INTERPRET and ANTICIPATE
Given the displacement, the fact that Stewie's speed is 0 m/s the instant he reaches the peak, and having found the initial velocity of Stewie, we look to the kinematic equations again to solve for the unknown time. Since the initial speed is less than the magnitude of the acceleration (which can be thought of as a change in velocity per second), we expect that it will take less than one second for the object to reach the peak and have a speed of 0 m/s.

SOLVE First, we collect and determine information on the five kinematic	

variables. At the top, the baby's velocity is momentarily zero.	$\Delta\vec{y} = 0.873\hat{j}$ m $\vec{a}_y = -g\hat{j} = -9.81\hat{j}$ m/s^2 $\vec{v}_y = 0$ $\vec{v}_{y0} = 4.14\hat{i}$ m/s t?
Equation 2.9 will allow us to make use of what we know and solve for the time.	$v_y = v_{y0} + a_y t$
We now solve this equation for time.	$t = \dfrac{v_y - v_{y0}}{a_y}$
Now, substitute the values and solve.	$t = \dfrac{0 \text{ m/s} - 4.14 \text{ m/s}}{-9.81 \text{ m/s}^2}$ $t = 0.422$ s
CHECK and THINK As expected, the time to reach the peak is less than one second.	

65. A sounding rocket, launched vertically upward with an initial speed of 75.0 m/s, accelerates away from the launch pad at 5.50 m/s^2. The rocket exhausts its fuel, and its engine shuts down at an altitude of 1.20 km, after which it falls freely under the influence of gravity.

a. (N) How long is the rocket in the air?

INTERPRET and ANTICIPATE
We can treat this rocket's flight as three separate stages: In the first, it is accelerating from its initial speed during the ascent at a constant acceleration until the engines are shut off. In the second, the rocket accelerates due to gravity until it reaches the top of its trajectory. In the third, it falls from rest at a constant acceleration due to gravity back the same total distance to Earth. We will use kinematic equations to calculate the times for each of these stages and add them together.

SOLVE We start by listing the kinematic quantities for stage 1. The rocket starts upwards at 75 m/s, continuing to accelerate at 5.5 m/s for a distance of 1200 m. We wish to find the time.	$v_i = 75.0$ m/s $v_f =$ unknown $a = 5.50$ m/s^2 $d = 1200$ m t ?
We can now use kinematic equation. One possibility is to use Equation 2.11, which relates all of the variables, and then solve the quadratic equation for time. Another method, which we use in this problem, is to use Equation 2.13 to find the final velocity and then use Equation 2.9 to solve for the time. This allows us to calculate all the kinematic variables and answer all three parts of the problem.	$v_f^2 = v_i^2 + 2a\Delta y$ $v_f = \sqrt{v_i^2 + 2a\Delta y}$ $v_f = \sqrt{(75.0 \text{ m/s})^2 + 2(5.50 \text{ m/s}^2)(1200 \text{ m})}$ $v_f = 137$ m/s $v_f = v_i + at$ $t_1 = \dfrac{v_f - v_i}{a} = \dfrac{137 \text{ m/s} - 75.0 \text{ m/s}}{5.50 \text{ m/s}^2} = 11.3 \text{ s}$
For stage 2, we again start by writing the kinematic variables. The rocket starts stage 2 at the final speed reached in stage 1 and falls with a *downward* acceleration due to gravity and reaches its peak position when the final velocity is zero at the top of the arc.	$v_i = 137$ m/s $v_f =$ unknown $a = -9.81$ m/s^2 $d = -1200$ m t ?
Using Equation 2.9, we can determine the time for stage 2. We can also use Equation 2.13 to find the distance traveled in this stage.	$v_f = v_i + at$ $t_2 = \dfrac{v_f - v_i}{a} = \dfrac{0 - 137 \text{ m/s}}{-9.80 \text{ m/s}^2} = 14.0 \text{ s}$ $v_f^2 = v_i^2 + 2a\Delta y = 0$ $\Delta y_2 = \dfrac{v_i^2}{2g} = \dfrac{(137 \text{ m/s})^2}{2(9.80 \text{ m/s}^2)} = 958 \text{ m}$

At the beginning of stage 3, the rocket is at an altitude of 1200 m + 958 m = 2158 m. The rocket is now in freefall with v_i = 0, a downward acceleration due to gravity, and a downward displacement back to Earth.	$v_i = 0$ m/s $v_f =$ unknown $a = -9.81$ m/s^2 $d = -2158$ m t ?
We can now use Eq. 2.13 to determine the final speed when the rocket hits the ground and Eq. 2.9 to determine the time in the air.	$v_f^2 = v_i^2 + 2a\Delta y$ $v_f = \sqrt{v_i^2 - 2g\Delta y}$ $v_f = \sqrt{(0)^2 - 2(9.80 \text{ m/s}^2)(-2158 \text{ m})}$ $v_f = 206$ m/s $v_f = v_i + at$ $t_3 = \dfrac{v_f - v_i}{a} = \dfrac{-206 \text{ m/s} - 0}{-9.80 \text{ m/s}^2} = 21.0 \text{ s}$
Given the time for each stage, we can now determine the total time of flight.	$t = t_1 + t_2 + t_3$ $t = 11.3 \text{ s} + 14.0 \text{ s} + 21.0 \text{ s} = \boxed{46.3 \text{ s}}$
CHECK and THINK The time for each stage has been calculated and sounds like a reasonable amount of time for a rocket to fly into the sky and back down to Earth.	

b. (N) What is the maximum altitude reached by the rocket?

INTERPRET and ANTICIPATE The hard work has been done in part (a) already. The maximum altitude must be higher than 1200 m when the rocket's engines were shut off.	
SOLVE Using the information from part (a), we add the distances from stages 1 and 2 to get the maximum altitude.	d_{max} = 1200 m + 958 m = 2160 m = 2.16 km
CHECK and THINK As expected the maximum altitude after stage 2 is larger than 1200 m.	

c. (N) What is the velocity of the rocket just before it strikes the ground?

INTERPRET and ANTICIPATE	
The final velocity is the velocity at the end of stage 3, calculated in part (a). If the rocket started upward at 75 m/s with only the acceleration due to gravity, it would reach a peack height and accelerate back to Earth reaching a speed of 75 m/s when it lands (ignoring air resistance of course). Since the rocket accelerates as it travels upwards for the first 1200 m, it goes even higher and should return to earth at a speed larger than 75 m/s.	
SOLVE Using part (a), we note the final speed at the end of stage 3.	$v_f = 206$ m/s, downward
CHECK and THINK The final speed is larger than 75 m/s as we expected.	

67. (N) While strolling downtown on a Saturday afternoon, you stumble across an old car show. As you are walking along an alley toward a main street, you glimpse a particularly stylish Alpha Romeo pass by. Tall buildings on either side of the alley obscure your view, so you see the car only as it passes between the buildings. Thinking back to your physics class, you realize that you can calculate the car's acceleration. You estimate the width of the alleyway between the two buildings to be 4 m. The car was in view for 0.5 s. You also heard the engine rev when the car started from a red light, so you know the Alpha Romeo started from rest 2 s before you first saw it. Find the magnitude of its acceleration.

INTERPRET AND ANTICIPATE
In the absence of more detailed data, the simplest approach is to model the motion of the Alpha Romeo as constant acceleration starting from rest. There are two significant intervals in this problem: (i) from the time at which the car starts from rest (lets call this time $t = 0$ s) to the time at which the car first comes into view, as it begins to cross the alley, and (ii) the interval during which the car crosses the alley. Note that the final time for the first interval is the initial time for the second. To assist in organizing the known and unknown variables, we define the following quantities: v_1: speed of the car as it crosses into the alley, v_2: speed of the car as it leaves the alley, d: distance between the traffic light and alley, and a_x: the constant acceleration of the car

Our goal is to solve for the unknown acceleration of the car, so the answer will have the form $a_x =$ ___ m/s^2

SOLVE First, we make a simple sketch including a coordinate system.	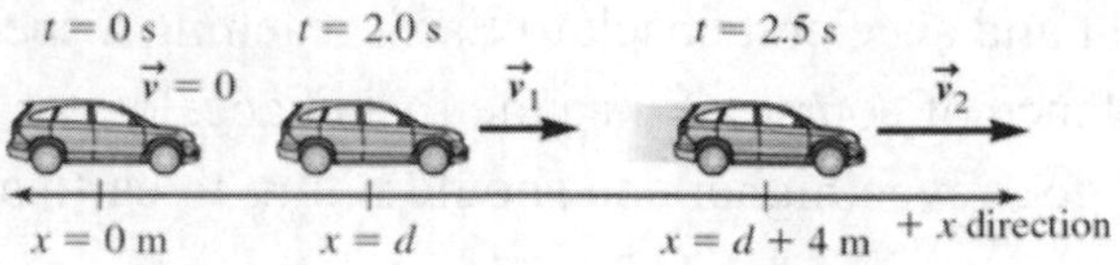 $t = 0$ s, $\vec{v} = 0$; $t = 2.0$ s, $\vec{v}_1$; $t = 2.5$ s, $\vec{v}_2$; $x = 0$ m; $x = d$; $x = d + 4$ m; + x direction **Figure P2.67ANS**
We list the known and unknown variables for the first interval…	First interval: $\Delta x = d$ (unknown) $v_{0x} = 0$ $v_x = v_1$ a_x unknown $t = 2$ s
… and then for the second interval.	Second interval: $\Delta x = 4$ m $v_{0x} = v_1$ $v_x = v_2$ a_x unknown $t = 0.5$ s
Our goal is to calculate a_x. We can't get there directly, but we use kinematic equations to relate variables. The acceleration should appear in our relationships, but we also notice that the speed v_1 appears for both intervals. We may be able to relate the acceleration, v_1, and known quantities to get two equations to solve for these two unknowns. For instance, we can apply Equation 2.9 to the first interval to obtain an expression containing two unknowns: a_x and v_1.	$v_x = v_{0x} + a_x \Delta t$ $v_1 = a_x (2\text{s})$ (1)

Now, applying Equation 2.10 to the second interval, we write another expression containing the two unknowns a_x and v_1.	$\Delta x = v_{0x}(\Delta t) + \frac{1}{2} a_x t^2$ $4\text{ m} = v_1(0.5\text{ s}) + \frac{1}{2} a_x (0.5\text{ s})^2$ $4\text{ m} = v_1(0.5\text{ s}) + (0.13\text{ s}^2)a_x$ (2)
We now have two equations with two unknowns, and can solve for a_x. Use Eq. 1 to substitute for v_1 in Eq. 2.	$4\text{ m} = v_1(0.5\text{ s}) + (0.13\text{ s}^2)a_x$ $4\text{ m} = (2\text{ s})(a_x)(0.5\text{ s}) + (0.13\text{ s}^2)a_x = (1.13\text{ s}^2)a_x$ $a_x = 4\text{ m}/(1.13\text{ s}^2) = 3.5\text{ m/s}^2$

CHECK AND THINK

We've estimated the acceleration of the Alpha Romeo. The units have come out correctly, and the value seems reasonable for a sports car. Using the approximation that 1 m/s is about 2 mph, this acceleration is about 7 mph per second, which would take the Alpha Romeo from 0 to 60 mph in about 9 seconds. The car can likely accelerate faster than this, so the driver probably didn't floor it!

77. The motion of a spacecraft in the outer solar system is described by the equation $x = 4.00t^2 - 3.00t + 5.00$, where x is in astronomical units (AU) and t is in years.

a. (N) What is the average speed of the spacecraft between $t = 1.00$ yr and $t = 3.00$ yr?

INTERPRET and ANTICIPATE

The average speed during an interval is the distance traveled divided by the time. Given the expression for distance and the time interval, we can determine the average speed in AU/yr.

SOLVE

In order to calculate average speed, we need the distance traveled and the interval of time. We first calculate the position at 1.00 and 3.00 years using the expression given.	$x = 4.00t^2 - 3.00t + 5.00$ $x\,(t = 1.00\text{ yr}) = 4.00(1.00)^2 - 3.00(1.00) + 5.00$ $x\,(t = 1.00\text{ yr}) = 6.00\text{ AU}$ $x\,(t = 3.00\text{ yr}) = 4.00(3.00)^2 - 3.00(3.00) + 5.00$ $x\,(t = 3.00\text{ yr}) = 32.0\text{ AU}$
The average speed is the distance traveled divided by the time interval.	$v_{\text{avg}} = \frac{x_f - x_i}{t_f - t_i} = \frac{32.0\text{ AU} - (6.00\text{ AU})}{(3.00 - 1.00)\text{ years}}$ $v_{\text{avg}} = 13.0\text{ AU/yr}$

CHECK and THINK
Given an expression for position, we calculated the distance traveled during the time interval specified, allowing us to calculate average speed.

b. (N) What is the instantaneous speed of the spacecraft at $t = 1.00$ yr and at $t = 3.00$ yr?

INTERPRET and ANTICIPATE	
The instantaneous speed of the spacecraft is the rate of change of its position at an instant. Mathematically, velocity is the time derivative of the position function.	
SOLVE The instantaneous velocity of the spacecraft is given by derivative of the position function versus time.	$v = \frac{dx}{dt} = \frac{d}{dt}(4.00t^2 - 3.00t + 5.00)$ $v = 8.00t - 3.00$
Plugging in $t = 1.00$ s and $t = 3.00$ s into $v = 8.00t - 3.00$ gives the velocity at each time.	$v(t = 1.00 \text{ yr}) = 8.00(1.00) - 3.00 = 5.00 \text{ AU/yr}$ $v(t = 3.00 \text{ yr}) = 8.00(3.00) - 3.00 = 21.0 \text{ AU/yr}$
CHECK and THINK With an equation for the position of the spacecraft as a function of time, we are able to take the derivative to find an expression for the velocity.	

c. (N) For what time t is the speed of the spacecraft zero?

INTERPRET and ANTICIPATE	
From part (b), we have a formula for velocity. Setting this function equal to zero, we can determine the time at which the velocity is zero.	
SOLVE To find the time at which $v = 0$, we set $v = 8.00t - 3.00 = 0$ and solve for t.	$t = 3.00/8.00 = \boxed{0.375 \text{ years}}$

As a final check, we plot the position and velocity versus time. The position initially decreases before beginning to rise, corresponding to an initial negative velocity which switches to an increasing positive velocity. The values for part (b) can be seen on the plot. In addition, around 0.375 years, the position graph is at a minimum and the velocity is zero.	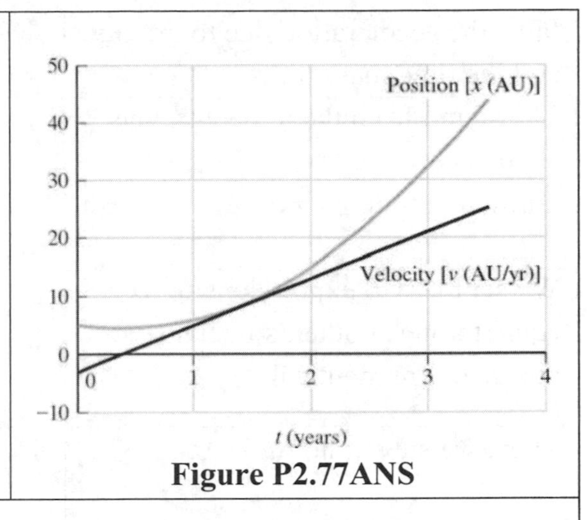 Figure P2.77ANS

CHECK and THINK
Setting the velocity function equal to zero, we find the time at which the velocity is instantaneously zero. Though we're not asked to plot the functions, sketching them can convince us that our results are consistent with a plot of the position and velocity functions.

80. Trying to determine its depth, a rock climber drops a pebble into a chasm and hears the pebble strike the ground 3.20 s later.

a. (N) If the speed of sound in air is 343 m/s at the rock climber's location, what is the depth of the chasm?

INTERPRET and ANTICIPATE	
The pebble is in free fall for a certain amount of time until it reaches the ground. The sound travels the same distance back for the climber to hear it. We need to determine expressions for the time for each of these two parts, which must add up to 3.20 s.	
SOLVE The pebble falls a distance d into the chasm in a time interval Δt_1 and the sound of the impact travels upward the same distance d in a time interval Δt_2 before the rock climber hears it. The total time interval is $\Delta t = \Delta t_1 + \Delta t_2 = 3.20$ s.	$\Delta t = \Delta t_1 + \Delta t_2 = 3.20$ s.
The pebble is in free fall during the time interval Δt_1 and has $v_i = 0$, so we can relate the distance that it	$d = \frac{1}{2}g(\Delta t_1)^2$

falls, the acceleration due to gravity, and the time that it falls.	
The sound from the impact travels at constant speed (v_s = 343 m/s) during the time interval Δt_2.	$d = v_s \Delta t_2$
We set the two expressions for d equal to one another, since the two distances are identical.	$\frac{1}{2}g(\Delta t_1)^2 = v_s \Delta t_2$
Then, we substitute $\Delta t_1 = \Delta t - \Delta t_2$.	$\frac{1}{2}g(\Delta t - \Delta t_2)^2 = v_s \Delta t_2$
We can rearrange this and recognize it as a quadratic equation.	$(\Delta t_2)^2 - 2\left(\Delta t + \frac{v_s}{g}\right)^2 \Delta t_2 + (\Delta t)^2 = 0$
Plugging in numbers, we have a quadratic equation in terms of the time interval Δt_2.	$(\Delta t_2)^2 - 2\left(3.20 \text{ s} + \frac{343 \text{ m/s}}{9.80 \text{ m/s}^2}\right)\Delta t_2 + (3.20 \text{ s})^2 = 0$ $(\Delta t_2)^2 - (76.4)\Delta t_2 + 10.24 = 0$
We can solve the quadratic equation, using the resulting time to calculate the distance traveled by the sound. That is, the height of the chasm.	$\Delta t_2 = \frac{76.4 \pm \sqrt{(-76.4)^2 - 4(1)(10.24)}}{2(1)}$ $\Delta t_2 = 0.134 \text{ s}$ $d = v_s \Delta t_2 = (343 \text{ m/s})(0.134 \text{ s}) = 46.1 \text{ m}$

CHECK and THINK

By writing expressions for the free fall of the pebble and the time for the sound to travel the same distance to the top of the chasm, we were able to determine the time for the sound to travel up the chasm and therefore its depth. A distance of 46 m, a four or five story building, sounds reasonable for the height of a chasm given the time interval of 3.2 s. (The reasonableness of the answer can only be 'checked' relative to the time interval, as a reasonable depth for a random chasm is unknown a priori.)

b. (N) What is the percentage of error that would result from assuming the speed of sound is infinite?

INTERPRET and ANTICIPATE	
Assuming the speed of sound is infinite would be to assume that the pebble was in free fall the entire time and the sound instantaneously made it back up the chasm. If the pebble was in free fall longer, we expect to determine a larger height of the chasm before the pebble hits the ground.	
SOLVE We imagine the same steps as part (a), but only include the free fall of the pebble.	$d=\frac{1}{2}g(\Delta t)^2=\frac{1}{2}(9.80\text{ m/s}^2)(3.20\text{ s})^2=50.2\text{ m}$
To calculate the error, we find the difference in distances divided by the actual distance, which we take to be the value from part (a), where we don't assume the sound speed is infinite.	$\frac{50.2-46.1}{46.1}=0.0889=\boxed{8.89\%}$
CHECK and THINK As expected, by ignoring the time that it takes for the sound to return to the climber, we overestimate the height of the chasm. This overestimate corresponds to about 9% error.	

88. (N) In Example 2.12, two circus performers rehearse a trick in which a ball and a dart collide. We found the height and time of the collision graphically. Return to that example, and find height and time by simultaneously solving the equations for the ball and the dart.

INTERPRET and ANTICIPATE	
With expressions for the positions of the ball and the dart, we can determine a time for which both objects are at the same position. In Example 2.12, we determined that this occurs at a position of 5.6 m, so we expect to find the same value (or at least a close value since our two methods might result in slightly different estimates).	
SOLVE The two simultaneous equations for the ball and the dart are listed here.	$y_{Bf}=\left(8.0-\frac{9.81}{2}t^2\right)\text{ m}$ $y_{Df}=\left(11.5t-\frac{9.81}{2}t^2\right)\text{ m}$

We want the position and time to be the same at the collision, or $y_{Bf} = y_{Df}$.	$\left(8.0 - \frac{9.81}{2}t^2\right) = \left(11.5t - \frac{9.81}{2}t^2\right)$
This equation can be solved to determine the time for which the two objects are at the same position.	$11.5t = 8.0$ $t = 0.70 \text{ s}$
This time can be inserted into either position equation to determine the position at this time. (Since we derived this assuming that both objects are at the same position, using either position equation will result in the same time.)	$y_{Bf} = \left(8.0 - \frac{9.81}{2}(0.70)^2\right) \text{ m} = \boxed{5.6 \text{ m}}$

CHECK and THINK

This agrees with what was determined graphically! We expect that both methods should result in the correct answer.

3

Vectors

4. (G) Vectors $\vec{A}$ and $\vec{B}$ have the same nonzero magnitude ($A = B$), where $\vec{C} = \vec{A} + \vec{B}$ and $\vec{D} = \vec{A} - \vec{B}$. If the magnitudes of $\vec{C}$ and $\vec{D}$ are the same, how are the directions of $\vec{A}$ and $\vec{B}$ related? *Hint*: Sketch these vectors.

The easiest approach is to sketch two vectors of equal length and think about their sum ($\vec{C}$) and difference ($\vec{D}$). We can start by imagining a couple simple cases:

If $\vec{A}$ and $\vec{B}$ (which are the same magnitude) are in the same direction, then when added ($\vec{C}$), the resultant would be twice as long as each of them. When subtracted ($\vec{D}$), the resultant would be zero. $\vec{C}$ and $\vec{D}$ are not equal in magnitude, which is what we're trying to accomplish.

By similar reasoning, if $\vec{A}$ and $\vec{B}$ are in opposite directions, then when subtracted, the resultant would be twice as long as either of them. When added, the resultant would be zero. Again, this does not satisfy our requirement.

If the directions of $\vec{A}$ and $\vec{B}$ are perpendicular, the results look much more promising, as shown in the figure. When added or subtracted, the resultant has the same magnitude. So, $\vec{A}$ and $\vec{B}$ must be perpendicular.

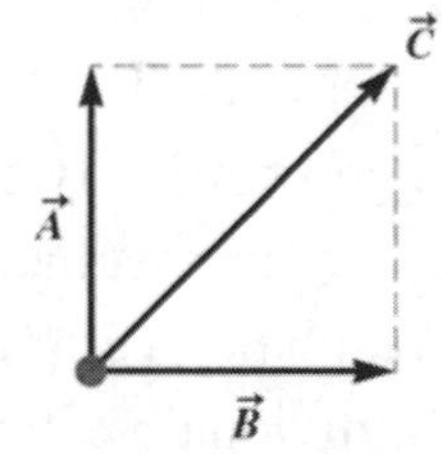

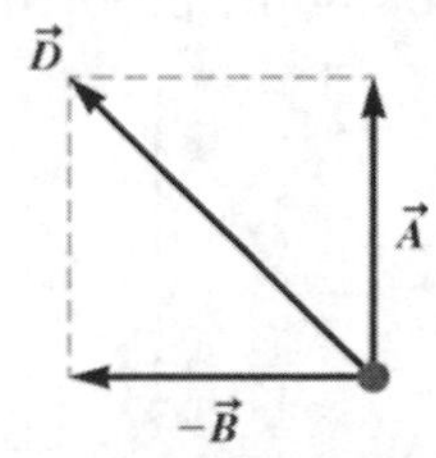

Figure P3.4ANS

You might wonder if we can solve this problem mathematically. In fact, it is possible, but gets messy quickly. Basically, given the two vectors $A_x\hat{i} + A_y\hat{j}$ and $B_x\hat{i} + B_y\hat{j}$, along with their sum $(A_x + B_x)\hat{i} + (A_y + B_y)\hat{j}$ and difference $(A_x - B_x)\hat{i} + (A_y - B_y)\hat{j}$, we want the original vectors to have the same length:

$$\sqrt{A_x^2 + A_y^2} = \sqrt{B_x^2 + B_y^2}$$

And, we want the sum and difference vectors to have the same length:

$$(A_x + B_x)^2 + (A_y + B_y)^2 = (A_x - B_x)^2 + (A_y - B_y)^2$$

This does not look fun! You might be able to convince yourself that the vectors in the figure, $\vec{A} = \hat{j}$ and $\vec{B} = \hat{i}$ (that is, $(A_x, A_y) = (0,1)$ and $(B_x, B_y) = (1, 0)$) actually do work, but you'll probably convince yourself that a graphical approach is much easier in the process!

6. (G) Figure P3.6 shows three vectors. Copy this figure on to your own paper and find $\vec{R} = \vec{F}_1 + \vec{F}_2 + \vec{F}_3$ geometrically.

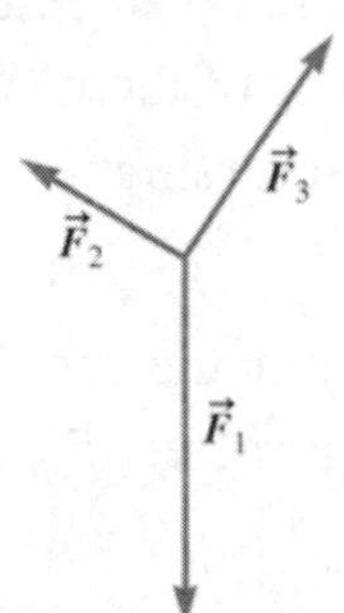

Figure P3.6

Taking the three vectors, we add them geometrically by placing the tail of each vector on the head of the previous vector as shown. The resultant vector is the total displacement using this procedure. In this case, the resultant vector is small, with approximately zero length compared to the lengths of the individual vectors. The precise length and direction is difficult to determine when completing the problem by hand.

Figure P3.6ANS

Chapter 3 – Vectors

11. (A) Three vectors $\vec{A}$, $\vec{B}$, and $\vec{C}$ are related to one another such that $2\vec{A}+3\vec{B}=18\vec{C}$. A fourth vector $\vec{D}$ exists such that $6\vec{C}+\vec{D}=\vec{A}$. Determine an expression for $\vec{D}$ in terms of the vectors $\vec{A}$ and $\vec{B}$.

INTERPRET and ANTICIPATE

Given the relationships between the four vectors, we can solve for the vector $\vec{C}$ in terms of $\vec{A}$ and $\vec{B}$, using the first expression, and substitute this into the second relationship to solve for $\vec{D}$.

SOLVE	
Beginning with the first expression, we solve for the vector $\vec{C}$.	$2\vec{A}+3\vec{B}=18\vec{C}$ $\frac{2}{18}\vec{A}+\frac{3}{18}\vec{B}=\vec{C}$ or, equivalently, $\vec{C}=\frac{1}{9}\vec{A}+\frac{1}{6}\vec{B}$
We now substitute this result into the second expression and solve for $\vec{D}$.	$6\vec{C}+\vec{D}=\vec{A}$ $6\left(\frac{1}{9}\vec{A}+\frac{1}{6}\vec{B}\right)+\vec{D}=\vec{A}$ $\frac{2}{3}\vec{A}+\vec{B}+\vec{D}=\vec{A}$ $\vec{D}=\vec{A}-\frac{2}{3}\vec{A}-\vec{B}$ $\vec{D}=\boxed{\frac{1}{3}\vec{A}-\vec{B}}$

CHECK and THINK

The algebraic manipulation of vectors is similar to the algebraic manipulation of scalar variables in some ways, including the associative and commutative law.

16. Vector $\vec{A}$ has a magnitude of 4.50 m and makes an angle of 64.0° with the positive x axis. Vector $\vec{B}$, with a magnitude equal to that of vector $\vec{A}$, points along the negative y axis (Figure P3.16).

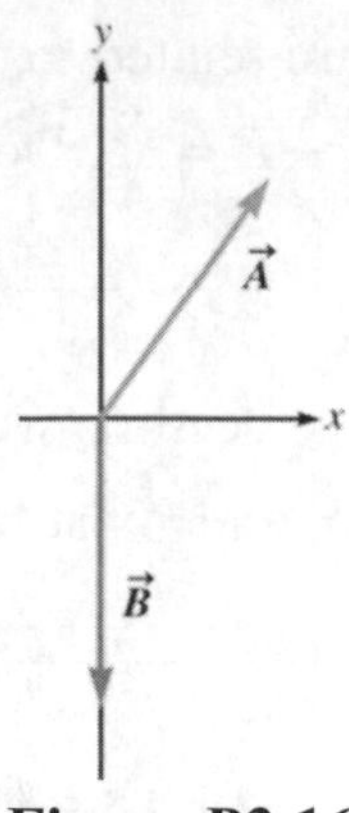

Figure P3.16

a. (G) Graphically find the magnitude and direction of the resultant vector $\vec{A}+\vec{B}$.

To add two vectors graphically, we place the tail of the second vector at the head of the first vector as shown. The resultant vector is the total displacement (from the beginning of the first vector to the end of the second vector) using this procedure.

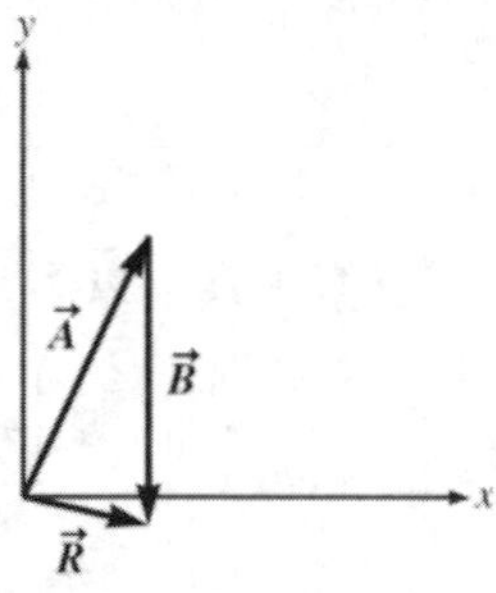

Figure P3.16aANS

b. (G) Graphically find the magnitude and direction of the resultant vector $\vec{A}-\vec{B}$.

We use the same procedure as in part (a) by interpreting $\vec{A}-\vec{B}$ as $\vec{A}+\left(-\vec{B}\right)$. That is, we flip the second vector to point in the opposite direction and then add it to the first vector.

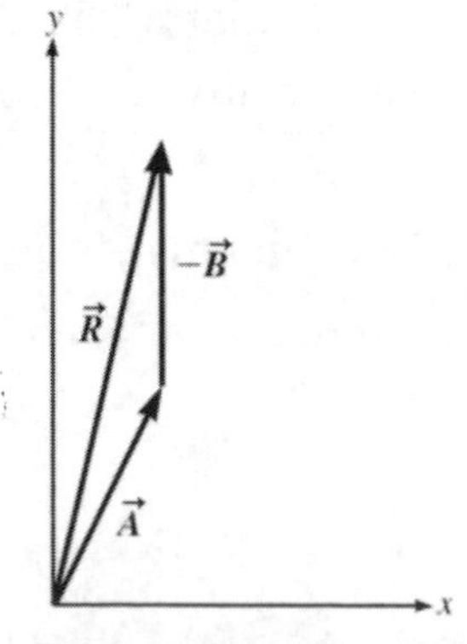

Figure P3.16bANS

c. (G) Graphically find the magnitude and direction of the resultant vector $\vec{B}-\vec{A}$.

Following the same plan as part (b), we flip the direction of vector $\vec{A}$ and then add the flipped vector to vector $\vec{B}$.

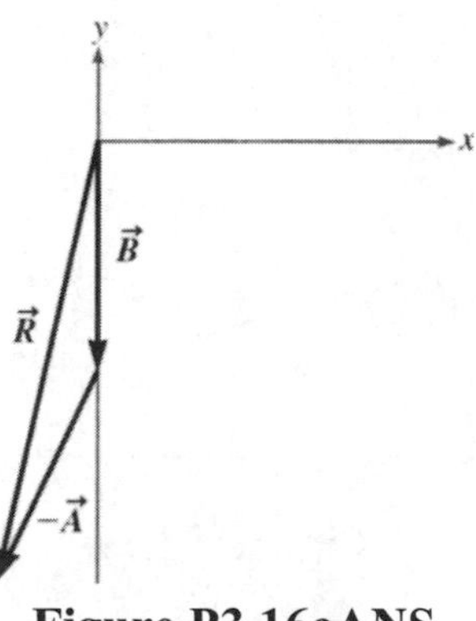

Figure P3.16cANS

d. (G) Graphically find the magnitude and direction of the resultant vector $2\vec{A}-\vec{B}$.

In this case, we first draw a vector that is twice the length of vector $\vec{A}$ (but still pointing in the same direction) and add it to a second vector that is the same length as $\vec{B}$ but pointing in the opposite direction.

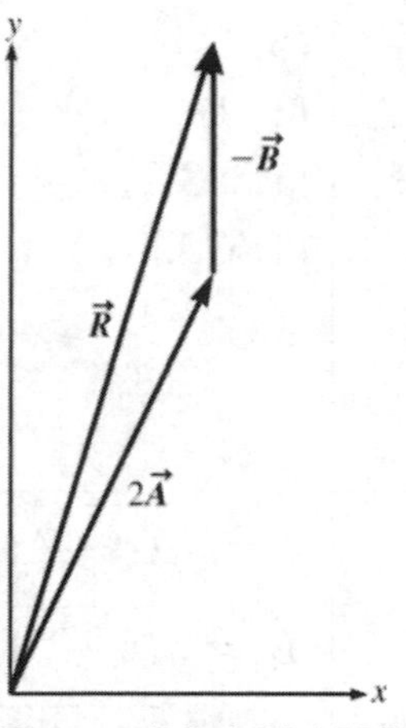

Figure P3.16dANS

19. (N) A museum curator is setting up a new display. The show has a particularly large painting that is 6.6 m long. The floor plan for the gallery is shown in Figure P3.19. Can the curator display this large painting in this gallery? If so, on which walls can it hang?

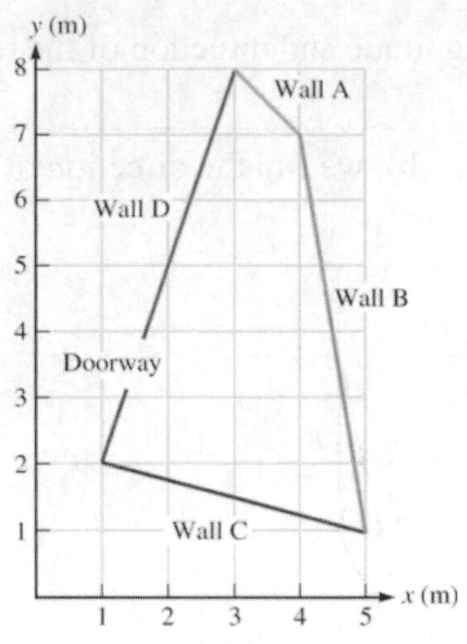

Figure P3.19

INTERPRET and ANTICIPATE Estimating the coordinates of the corners of the room, we can determine the length of each wall as the distance between two points. As the museum curator, we're hoping to find a wall that's at least 6.6 m long!	
SOLVE Estimating the positions of the corners of the room, starting from the point where wall D meets wall A and moving clockwise, we estimate distances in meters. There is also a doorway that we need to avoid, which spans the last two points.	(3, 8) (4, 7) (5, 1) (1, 2) (1.3, 3.2) (1.6, 3.8)
We can now estimate the length of each wall. The length of a line can be found given the x and y components of the distance between the endpoints using the Pythagorean theorem. Unfortunately, none of the walls is long enough to hang the 6.6 m painting!	$L=\sqrt{\Delta x^2+\Delta y^2}$ $L_A=\sqrt{(4-2.4)^2+(8-7)^2}=1.9\text{ m}$ $L_B=\sqrt{(4.2-4)^2+(7-1.2)^2}=5.8\text{ m}$ $L_C=\sqrt{(4.2-0.2)^2+(2.2-1.2)^2}=4.1\text{ m}$ $L_{D1}=\sqrt{(0.8-0.2)^2+(3-2.2)^2}=1.0\text{ m}$ $L_{D2}=\sqrt{(2.4-1)^2+(8-4)^2}=4.2\text{ m}$
CHECK and THINK It turns out that hanging the painting will be more difficult than calculating the lengths of	

the walls! After estimating the positions of the corners of the room and the position of the door, finding the length of the wall is pretty straightforward. We knew it might be a tight fit to begin with, so we're not *too* surprised that it doesn't fit.

20. A football is thrown along an arc. The ball is released at a height of 6.54 ft (1.99 m) above the ground and, after traveling a horizontal distance of 87.8 ft (26.8 m), is caught by a receiver at a point that is 4.22 ft (1.29 m) above the ground.
a. (G) Sketch the throw from the point of release to the moment the ball is caught (as viewed by an observer from the sideline), placing an x–y coordinate system in the picture with the origin at ground level, directly below where the ball is released.

The sketch should be a curve that begins above the ground and terminates at a point above the ground, but lower than the initial point.

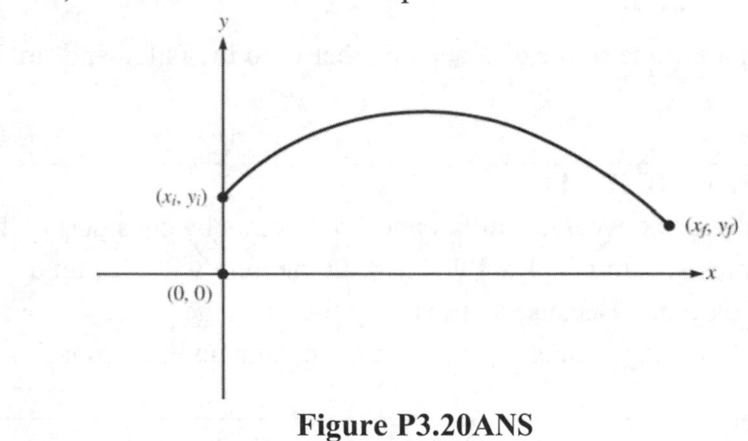

Figure P3.20ANS

b. (N) Determine the coordinates of the ball at the point it is released and where it is caught.

INTERPRET and ANTICIPATE
The choice of the origin in part (a) allows us to define coordinates for the initial and final position of the ball during its flight. All coordinate information should be converted to meters.

SOLVE First, convert all of the coordinate information to meters.	$y_i = 6.54\text{ ft}\times\left(\dfrac{0.3048\text{ m}}{1\text{ ft}}\right) = 1.99\text{ m}$ $y_f = 4.22\text{ ft}\times\left(\dfrac{0.3048\text{ m}}{1\text{ ft}}\right) = 1.29\text{ m}$ $x_i = 0$

	$x_f = 87.8\text{ ft} \times \left(\dfrac{0.3048\text{ m}}{1\text{ ft}}\right) = 26.8\text{ m}$
The coordinates of the initial and final points are then easily written.	$(x_i, y_i) = \boxed{(0, 1.99\text{ m})}$ $(x_f, y_f) = \boxed{(26.8\text{ m}, 1.29\text{ m})}$

CHECK and THINK
The coordinate values depend on the chosen coordinate system. A different choice would change the coordinate values for both points. The initial position makes sense since the point is directly above the origin, as in our sketch, and it makes sense that both coordinate values for the final position are positive.

c. (N) What is the magnitude of the displacement between the initial and final points in the ball's flight?

INTERPRET and ANTICIPATE
We can find the distance between the initial and final points by considering the change in the x and y coordinates separately. The Pythagorean theorem will then let us find the distance between the points. Because there is very little change in the y coordinate, we expect the distance to be approximately equal to the change in the x coordinate.

SOLVE	
The change in the horizontal position of the ball is found by comparing the difference in the x coordinates of the initial and final points.	$\Delta x = x_f - x_i$ $\Delta x = 26.8\text{ m} - 0 = 26.8\text{ m}$
The change in the vertical position of the ball is found by comparing the difference in the y coordinates of the initial and final points.	$\Delta y = y_f - y_i$ $\Delta y = 1.29\text{ m} - 1.99\text{ m} = -0.70\text{ m}$
The displacement of the ball could then be written.	$\Delta\vec{r} = \left(26.8\hat{\imath} - 0.70\hat{\jmath}\right)\text{ m}$

The Pythagorean theorem allows us to find the distance between the initial and final points.	$\ell = \sqrt{(\Delta x)^2 + (\Delta y)^2}$ $\ell = \sqrt{(26.8\text{ m})^2 + (-0.70\text{ m})^2}$ $\ell = \boxed{26.8\text{ m}}$
CHECK and THINK Though there is a difference in the y coordinate of the ball between the initial and final points, it is very small and does not appreciably affect the distance between the two points in this case, due to the large difference in the x coordinates. In other words, the distance is pretty much 26.8 m between the initial and final points as long as the y displacement isn't very large. This might not be the case if we determined, say, the distance between the initial point and when the ball reaches its maximum height.	

23. A truck driver delivering office supplies downtown travels 5.00 blocks north, 3.00 blocks east, and finally 2.00 blocks south.
a. (N) What is the magnitude (in blocks) and the direction of the driver's displacement?

INTERPRET and ANTICIPATE In words, we are given the magnitude and direction for three vectors. To find the total displacement, we can add the three vectors. We can then find the magnitude and direction of this resultant vector.	
SOLVE The driver's net east-west displacement is 0 + 3.00 + 0 = 3.00 blocks east, and his north-south displacement is 5.00 + 0 – 2.00 = 3.00 blocks north. We can picture this as the x and y displacements on a standard Cartesian coordinate system.	$\Delta x = 3.00$ blocks $\Delta y = 3.00$ blocks
The magnitude of the displacement can now be calculated using the Pythagorean theorem.	$R = \sqrt{\Delta x^2 + \Delta y^2} = \sqrt{(3.00)^2 + (3.00)^2}$ $R = \boxed{4.24\text{ blocks}}$
The components of this vector can also be used to determine the angle of this vector with respect to the x	$\theta = \tan^{-1}\left(\frac{3.00}{3.00}\right) = \boxed{45.0^\circ}$

axis by using the tangent function.	

CHECK and THINK
The driver's total displacement is 4.24 blocks at 45 degrees north of east. Since the driver has traveled east and more north than south, we would expect an answer that is in the northeast quadrant.

b. (N) What is the total distance traveled (in blocks) by the truck driver?

INTERPRET and ANTICIPATE
The distance traveled is irrespective of the direction, therefore it must be equal to or larger than the total displacement. In this case, since the motion is not entirely along the same direction, the total distance must be larger than the net displacement.

SOLVE The total distance traveled is simply the sum of the individual distances for each part of the trip.	$5.00 + 3.00 + 2.00 = \boxed{10.0 \text{ blocks}}$

CHECK and THINK
As expected this is larger than the magnitude of the displacement.

28. (N) A vector's x component is twice its y component. What is the acute angle between the two vector components?

INTERPRET and ANTICIPATE
Given the relationship between the components of the vector, we can use the inverse tangent to find the angle between the vector and the x-axis. Assuming the components are positive values, we expect the result to be in the first quadrant.

SOLVE
We first sketch a vector in which the x component is twice the y component.

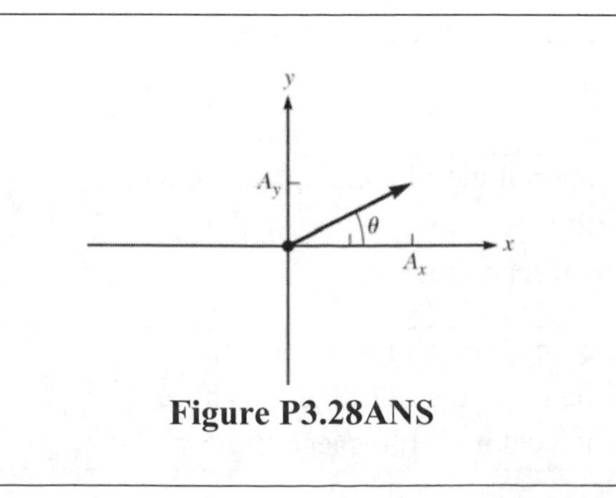

Figure P3.28ANS

We can specify the symbols and values we'll use in the solution. The equation shown is equivalent to the sentence, "Its horizontal component is twice its vertical component."	$A_x = 2A_y$
According to Equation 3.14, the angle is found by using the inverse tangent.	$\theta = \tan^{-1}\left(\frac{A_y}{A_x}\right) = \tan^{-1}\left(\frac{A_y}{2A_y}\right) = \tan^{-1}(0.5)$ $\theta = \boxed{27°}$

CHECK and THINK
The angle is less than 45 degrees, which means the vector is closer to the *x*-axis than to the *y*-axis. This answer is consistent with our drawing and is in the first quadrant as expected.

30. (G) Vector $\vec{B}$ has a magnitude of 19.45. Its direction measured counterclockwise from the x axis is 127.3°. Draw this vector on a coordinate system.

The vector can be drawn using the information given. The precise components were calculated in Problem 29. This angle, between 90 and 180 degrees, is in the second quadrant. Therefore, we expect the x component to be negative and the y component to be positive. Using the geometry specified:

$$B_x = B\cos\theta = 19.45\cos(127.3°) = -11.79$$
$$B_y = B\cos\theta = 19.45\sin(127.3°) = 15.47$$

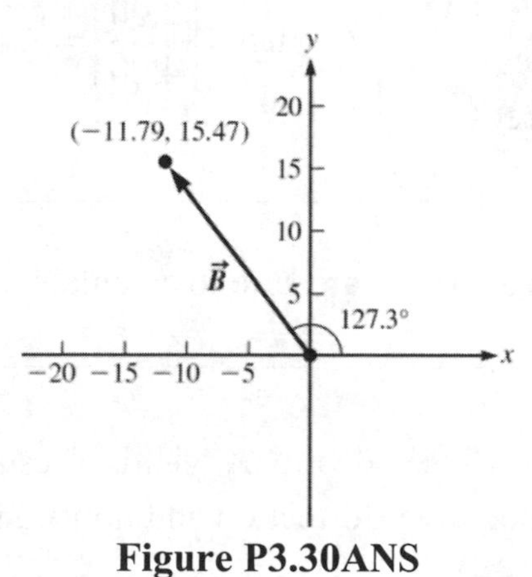

Figure P3.30ANS

33. Vector $\vec{A}$ has scalar components of 4.00 units in the negative x direction and 2.00 units in the negative y direction.

a. (N) Write the vector $\vec{A}$ in component form.

INTERPRET and ANTICIPATE The scalar components are simply the values of each of the components of the vector, so we can write the vector down directly with this information.	
SOLVE Using the given scalar components, we write the corresponding vector.	$A_x = -4.00, \quad A_y = -2.00$ $\vec{A} = \boxed{-4.00\,\hat{\imath} - 2.00\,\hat{\jmath}}$
CHECK and THINK This one is a straightforward substitution.	

b. (N) What are the magnitude and direction of $\vec{A}$?

INTERPRET and ANTICIPATE With the two components of a vector, the magnitude and direction can easily be found.	
SOLVE The magnitude of a vector is found according to Equation 3.12.	$\lvert\vec{A}\rvert = A = \sqrt{A_x^2 + A_y^2}$ $\lvert\vec{A}\rvert = \sqrt{(-4.00)^2 + (-2.00)^2} = \boxed{4.47}$
The vector is in the third quadrant and the tangent of the angle is related to the components of the vector based on Equation 3.14. The angle is found to be 27 degrees below the –x axis or 207 degrees counter-clockwise from the +x axis.	$\theta = \tan^{-1}\left(\dfrac{\lvert A_y\rvert}{\lvert A_x\rvert}\right)$ $\theta = \tan^{-1}\left(\dfrac{2.00}{4.00}\right) = \boxed{26.6^\circ \text{ below the } -x \text{ axis}}$
CHECK and THINK With the components of the vector, we are able to calculate both the magnitude and direction of the vector.	

c. (N) Find the vector $\vec{B}$ that when added to $\vec{A}$ yields a resultant vector with an x component 6.00 units in the negative x direction and no y component.

<table>
<tr><td colspan="2">INTERPRET and ANTICIPATE
We would like to add two vectors, with one being known, to produce another (known) vector. We can set up equations corresponding to this requirement and solve for the unknown values.</td></tr>
<tr><td>SOLVE
We can first write the vector $\vec{A}$ from part (a). Let's call the resultant vector $\vec{R}$ and express that as a vector as well.</td><td>$\vec{A} = -4.00\,\hat{i} - 2.00\,\hat{j}$
$\vec{R} = -6.00\,\hat{i} + 0\,\hat{j}$</td></tr>
<tr><td>We now write an equation that "the resultant is the sum of vectors $\vec{A}$ and $\vec{B}$" and solve for the unknown vector $\vec{B}$.</td><td>$\vec{R} = \vec{A} + \vec{B}$
$\vec{B} = \vec{R} - \vec{A}$
$\vec{B} = -6.00\,\hat{i} - \left(-4.00\,\hat{i} - 2.00\,\hat{j}\right)$
$\vec{B} = \boxed{-2.00\,\hat{i} + 2.00\,\hat{j}}$</td></tr>
<tr><td colspan="2">CHECK and THINK
Using the vectors provided, we are able to translate the required relationship into vector addition. We can double-check the result by adding the two vectors to confirm that the correct resultant is produced:
$$\vec{A} + \vec{B} = \left(-4.00\,\hat{i} - 2.00\,\hat{j}\right) + \left(-2.00\,\hat{i} + 2.00\,\hat{j}\right) = -6.00\,\hat{i}$$</td></tr>
</table>

34. A soccer player starts at one end of the field and runs along a straight line making an angle θ with the goal line. She is able to move d meters closer to the opposing goal line before going out of bounds.

a. (G) Make a sketch of this situation that includes the soccer player's displacement vector $\Delta\vec{r}$. Label d and θ on your sketch.

The quantity d corresponds to the change in the downfield position of the soccer player. The angle will relate this distance to the distance from the starting point to the side of the field where the player goes out of bounds and to the total displacement of the player ($\Delta\vec{r}$). It sounds like we'll be drawing a triangle and using some trigonometry!

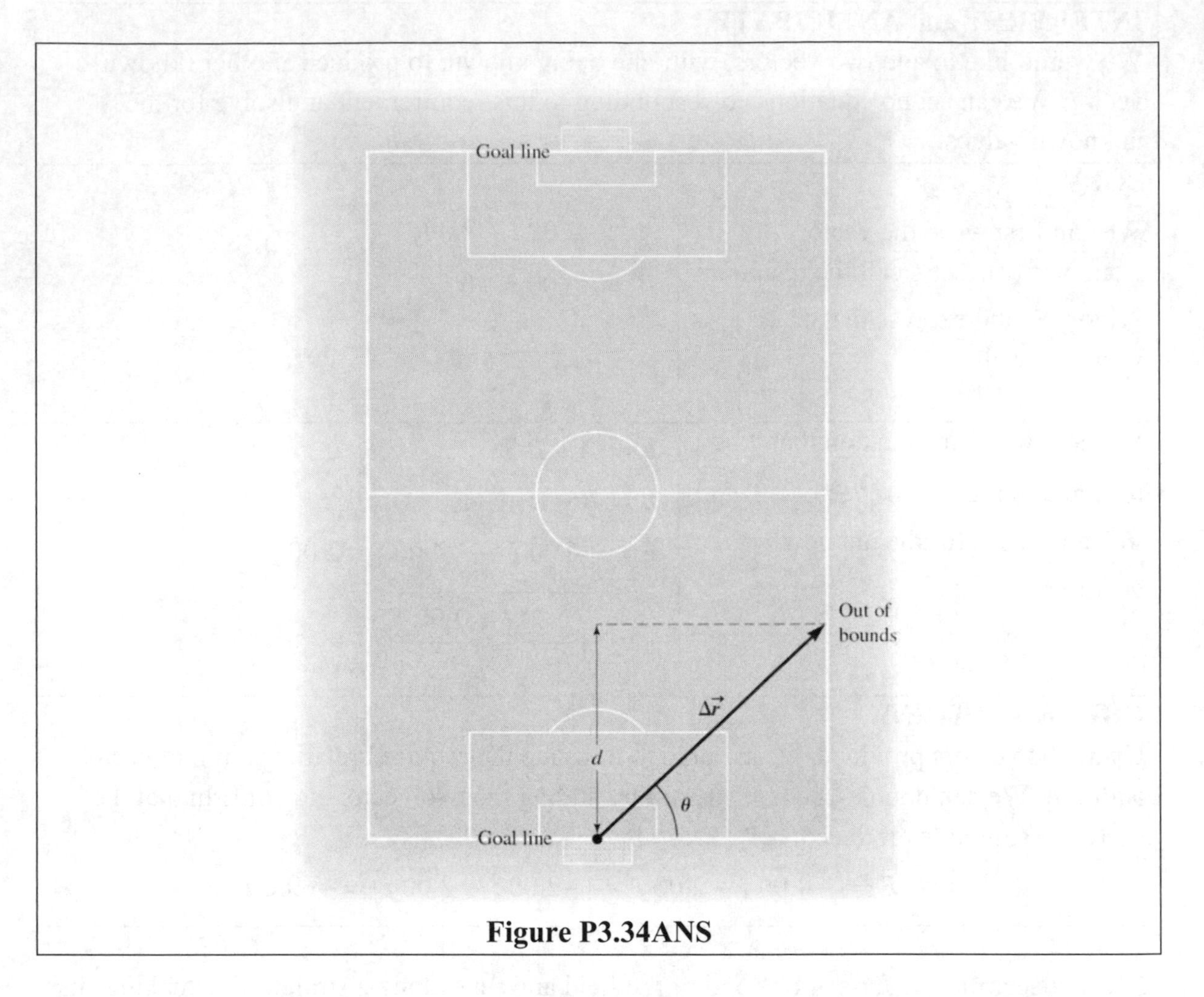

Figure P3.34ANS

b. (A) Write an expression for the distance the soccer player ran in terms of θ and *d*.

Now, using the figure in part (a), we can relate the variables using trigonometry. The distance the soccer player ran is given by the magnitude of her displacement vector, marked $\Delta\vec{r}$. The sine function relates *d* and $|\Delta\vec{r}|$. Solving for $|\Delta\vec{r}|$ gives the distance travelled in terms of the given quantities.

$$\sin\theta = \frac{d}{|\Delta\vec{r}|}$$

$$|\Delta\vec{r}| = \boxed{\frac{d}{\sin\theta}}$$

39. Vector $\vec{A}$ has an *x* component of 15.0 m, a *y* component of −6.00 m, and a *z* component of −3.00 m.

a. (N) Write the vector $\vec{A}$ in component form.

b. (N) Find the vector $\vec{B}$, pointing in the same direction as vector $\vec{A}$, but having one-third its length.

c. (N) Find the vector $\vec{C}$, pointing in the opposite direction of vector $\vec{A}$, but with three times its length.

INTERPRET and ANTICIPATE

Given the components of vector $\vec{A}$, we can write the vector in component form. A smaller or larger vector can easily be created by multiplying a known vector by a number that is smaller or larger than 1. In addition, a positive multiplier results in a vector pointing in the same direction, while a negative multiplier results in a vector pointing in the opposite direction.

SOLVE	
a. With the components provided, we can write the vector in component form.	$\vec{A} = \boxed{(15.0\text{ m})\,\hat{\imath} - (6.00\text{ m})\,\hat{\jmath} - (3.00\text{ m})\,\hat{k}}$
b. To create a vector that is in the same direction with one-third the size, we need to multiply by a positive number with the value 1/3.	$\vec{B} = \frac{1}{3}\vec{A}$ $\vec{B} = \frac{1}{3}\left[(15.0\text{ m})\,\hat{\imath} - (6.00\text{ m})\,\hat{\jmath} - (3.00\text{ m})\,\hat{k}\right]$ $\vec{B} = \boxed{(5.00\text{ m})\hat{\imath} - (2.00\text{ m})\hat{\jmath} - (1.00\text{ m})\hat{k}}$
c. To create a vector that is in the opposite direction with three times the size, we need to multiply by a negative number with the value 3.	$\vec{C} = -3\vec{A}$ $\vec{C} = -3\left[(15.0\text{ m})\,\hat{\imath} - (6.00\text{ m})\,\hat{\jmath} - (3.00\text{ m})\,\hat{k}\right]$ $\vec{C} = \boxed{(-45.0\text{ m})\hat{\imath} + (18.0\text{ m})\hat{\jmath} + (9.00\text{ m})\hat{k}}$

CHECK and THINK

Once we have the vector written in component form, we can see that the vector in part (b) is composed of components of 1/3 the magnitude while the vector in part (c) has components of 3 times the magnitude, but with opposite sign. This is what we were expecting.

42. The same vectors that are shown in Figure P3.6 are shown in Figure P3.42. The magnitudes are $F_1 = 1.90f$, $F_2 = f$, and $F_3 = 1.4f$, where f is a constant.

a. (N) Use the coordinate system shown in Figure P3.42 to find $\vec{R} = \vec{F}_1 + \vec{F}_2 + \vec{F}_3$ in component form in terms of f.

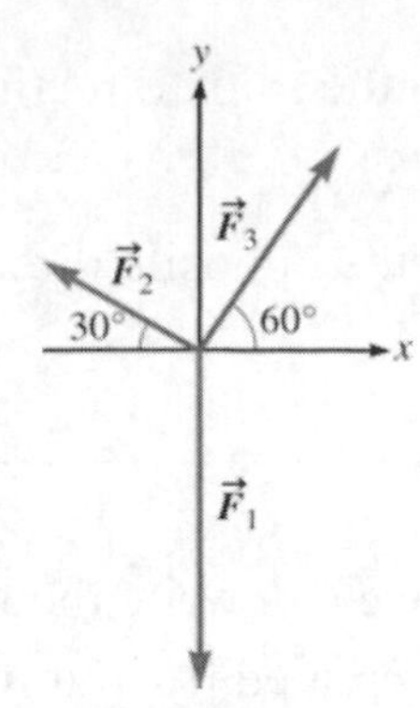

Figure P3.42

<table>
<tr><td colspan="2">INTERPRET and ANTICIPATE
We're given three vectors and asked to find their sum. With particular magnitudes (in terms of f) and angles, we can find the x and y components of each vector and add them together.</td></tr>
<tr><td>SOLVE
We write the three vectors in component form.</td><td>$\vec{F}_1 = -1.90 f\,\hat{j}$
$\vec{F}_2 = -f\cos 30^\circ\hat{i} + f\sin 30^\circ\,\hat{j} = -0.866 f\,\hat{i} + 0.500 f\,\hat{j}$
$\vec{F}_3 = 1.4 f\cos 60^\circ\hat{i} + 1.4 f\sin 60^\circ\,\hat{j} = 0.700 f\,\hat{i} + 1.21 f\,\hat{j}$</td></tr>
<tr><td>We can now add the x and y components of each vector.</td><td>$\vec{R} = (-0.866 f + 0.700 f)\hat{i} + (-1.90 f + 0.500 f + 1.21 f)\hat{j}$
$\vec{R} = \boxed{-0.166 f\,\hat{i} - 0.188 f\,\hat{j}}$</td></tr>
<tr><td colspan="2">CHECK and THINK
The result is consistent with Problem 6, in which we're asked to add the three vectors graphically, as discussed in part (c). Adding vectors graphically is a good check for consistency for numerical problems.</td></tr>
</table>

b. (N) If $R_x = 0.33$, what is R_y ?

<table>
<tr><td colspan="2">INTERPRET and ANTICIPATE
From part (a), we have an expression for R_x in terms of f, which means we can calculate f and then determine R_y.</td></tr>
<tr><td>SOLVE
Using the x components, we can determine the factor f.</td><td>$R_x = -0.166 f = 0.33$
$f = -1.99$</td></tr>
</table>

Knowing f, we can now calculate the y component.	$R_y = -0.188f = \boxed{0.373}$
CHECK and THINK With one component of the resultant vector given, we were able to determine the factor f and then the other component.	

c. (C) Check your result by comparing to your answer to that of Problem 6.

The vector is in the fourth quadrant and the magnitude is small compared to the lengths of the individual vectors, which is consistent with the results of Problem 6.

45. (N) A vector $\vec{A} = \left(5.20\hat{\boldsymbol{i}} - 3.70\hat{\boldsymbol{j}}\right)$ m and a vector $\vec{\boldsymbol{B}} = \left(1.04\hat{\boldsymbol{i}} + B_y\hat{\boldsymbol{j}}\right)$ m are related by a scalar quantity such that $\vec{A} = s\vec{\boldsymbol{B}}$. Determine the value of the y component of the vector $\vec{\boldsymbol{B}}$.

INTERPRET and ANTICIPATE Since the vectors are related by a scalar quantity, the separate components of the vectors must each be related by this factor. We can use the x component information to determine, the scalar quantity, s. The unknown y component can then be found.	
SOLVE We can use the relationship provided and substitute expressions for both vectors.	$\vec{A} = s\vec{\boldsymbol{B}}$ $\left(5.20\hat{\boldsymbol{i}} - 3.70\hat{\boldsymbol{j}}\right)\text{ m} = s\left(1.04\hat{\boldsymbol{i}} + B_y\hat{\boldsymbol{j}}\right)\text{ m}$ $\left(5.20\hat{\boldsymbol{i}} - 3.70\hat{\boldsymbol{j}}\right)\text{ m} = \left(s1.04\hat{\boldsymbol{i}} + sB_y\hat{\boldsymbol{j}}\right)\text{ m}$
Given the above equality, we can treat each component separately. Equating the x components from both sides of the equation, allows us to find s.	$5.20\text{ m} = s(1.04\text{ m})$ $s = 5.00$
Now, we equate the y components and solve for the unknown component.	$-3.70\text{ m} = s(B_y) = 5.00(B_y)$ $B_y = \boxed{-0.740\text{ m}}$

CHECK and THINK
When two vectors equal each other, their components must also be equal. Both of these vectors must also point in the same direction since they are related by a positive scalar quantity.

47. (N) Consider the vectors $\vec{A} = 16.5\hat{i} - 33.0\hat{j}$ and $\vec{B} = -2.00\hat{i} + 3.00\hat{j} - 4.00\hat{k}$. Find $\vec{R} = \vec{A} - \vec{B}$ in component form.

INTERPRET and ANTICIPATE	
This is a relatively straightforward vector subtraction problem. To subtract two vectors, we subtract each component of the two vectors.	
SOLVE Taking the equation for the resulting vector, we simply substitute the two vectors and subtract each component to find the components of the resultant vector.	$\vec{R} = \vec{A} - \vec{B}$ $\vec{R} = \left(-2.00\hat{i} + 3.00\hat{j} - 4.00\hat{k}\right) - \left(16.5\hat{i} - 33.0\hat{j}\right)$ $\vec{R} = \boxed{-18.5\hat{i} + 36.0\hat{j} - 4.00\hat{k}}$
CHECK and THINK We don't have a basis to expect a particular answer, but we've subtracted each component separately to calculate the resultant vector.	

51. (N) The resultant vector $\vec{R} = 2\vec{A} - \vec{B} - 2\vec{C}$ has zero magnitude. Vector $\vec{A}$ has an x component of 4.60 m and a y component of −12.1 m, and vector $\vec{B}$ has an x component of −3.00 m and a y component of −4.00 m. What are the x and y components of vector $\vec{C}$? (All of these are two-dimensional vectors.)

INTERPRET and ANTICIPATE	
We are given a relationship between four vectors in which three of them are specified. We should be able to solve for the remaining (unknown) vector.	
SOLVE We start by translating the descriptions of the known vectors into mathematical expressions.	$\vec{A} = 4.60\hat{i} - 12.1\hat{j}$ $\vec{B} = -3.00\hat{i} - 4.00\hat{j}$ $\vec{R} = 0$

We can now rearrange the expression and substitute our known vectors to solve for the unknown vector.	$2\vec{A}-\vec{B}-2\vec{C}=0$ $2\vec{C}=2\vec{A}-\vec{B}$ $\vec{C}=\vec{A}-\frac{1}{2}\vec{B}=(4.60\hat{i}-12.1\hat{j})-\frac{1}{2}(-3.00\hat{i}-4.00\hat{j})$ $\vec{C}=6.10\hat{i}-10.1\hat{j}$ or $C_x=\boxed{6.10\text{ m}}$ and $C_y=\boxed{-10.1\text{ m}}$

CHECK and THINK
As expected, even with multiple vectors, if all are specified except for one unknown vector, we are able to solve for the unknown vector.

55. (N) Two birds begin next to each other and then fly through the air at the same elevation above level ground at 22.5 m/s. One flies northeast, and the other flies northwest. After flying for 10.5 s, what is the distance between them? Ignore the curvature of the Earth.

INTERPRET and ANTICIPATE
Place the coordinate system so the two birds begin at the origin and sketch their trajectories, as shown in the figure below. Since the angles between their velocities and due north are the same (45 degrees away from north), the y components of their final positions will be the same. That is, they travel upwards at the same rate. Therefore, we only need to subtract the x components of their final positions to find the distance between the birds at any point in time.

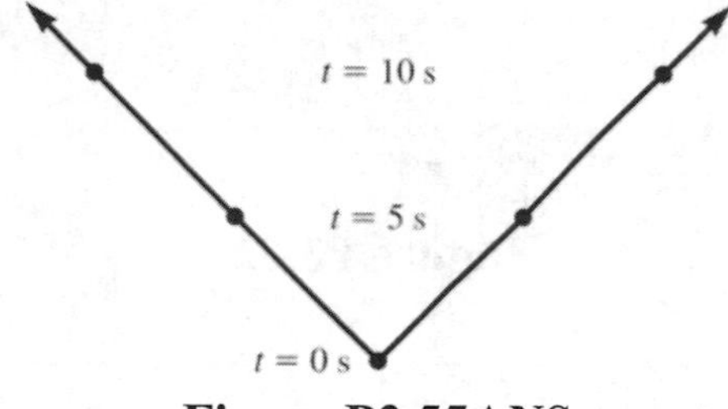

Figure P3.55ANS

SOLVE First, let's specify the symbols and values we'll use in the solution.	t = time = 10.5 s v = speed = $22.5\ \frac{\text{m}}{\text{s}}$ x_1 = x component of final position of bird flying northeast x_2 = x component of final position of bird flying northwest

Each flies a distance vt at an angle of forty-five degrees above the x-axis.	$x_1 = +vt\cos 45^\circ$ $x_2 = -vt\cos 45^\circ$
The final distance between the birds is equal to the difference in the x components.	$d = x_1 - x_2 = 2vt\cos 45^\circ$
We now insert numerical values.	$d = 2\left(22.5\dfrac{\text{m}}{\text{s}}\right)(10.5\ \text{s})\dfrac{\sqrt{2}}{2}$ $d = \boxed{334\ \text{m}}$
CHECK and THINK Given the magnitude of the speed, this seems like a reasonable distance between the birds after 10 seconds.	

62. (N) A glider aircraft initially traveling due west at 85.0 km/h encounters a sudden gust of 35.0 km/h winds directed toward the northeast (Fig. P3.62). What are the speed and direction of the glider relative to the ground during the wind gust? (The velocity of the glider with respect to the ground is the velocity of the gilder with respect to the wind plus the velocity of the wind with respect to the ground.)

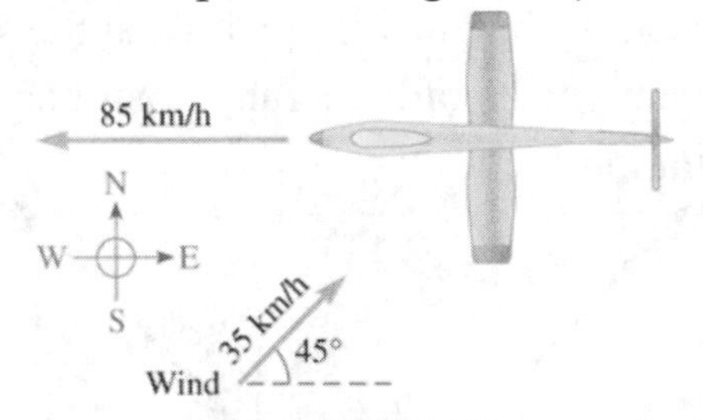

Figure P3.62

INTERPRET and ANTICIPATE The velocity of the glider given is its velocity with respect to the air. The velocity of the glider with respect to the ground is given by how fast the air moves relative to the ground (the wind speed that is given) plus how fast the glider moves relative to the air (the glider speed that is given). This eastern wind on the western-bound plane is a head wind that will tend to slow the plane down relative to the ground.	
SOLVE We first write the velocity of the glider relative to the air.	$\vec{v}_i = (-85.0\ \text{km/h})\hat{\imath}$

We can also express the velocity of the wind by calculating the components based on the given magnitude and direction.	$\vec{v}_{\text{wind}} = [(35.0 \text{ km/h})\cos 45°]\hat{\imath} + [(35.0 \text{ km/h})\sin 45°]\hat{\jmath}$ $\vec{v}_{\text{wind}} = (24.7 \text{ km/h})\hat{\imath} + (24.7 \text{ km/h})\hat{\jmath}$
The resultant velocity vector relative to the ground can now be calculated as described.	$\vec{v} = \vec{v}_i + \vec{v}_{\text{wind}}$ $\vec{v} = (-85.0 \text{ km/h})\hat{\imath} + (24.7 \text{ km/h})\hat{\imath} + (24.7 \text{ km/h})\hat{\jmath}$ $\vec{v} = (-60.3 \text{ km/h})\hat{\imath} + (24.7 \text{ km/h})\hat{\jmath}$
We can now easily determine the magnitude and direction of this velocity vector using Equations 3.12 and 3.14.	$\lvert\vec{v}\rvert = \sqrt{v_x^2 + v_y^2} = \sqrt{(-60.3)^2 + (24.7)^2} = 65.1 \text{ km/h}$ $\theta = \tan^{-1}\left(\frac{24.7}{-60.3}\right) = -22.3°$ Therefore, $\lvert\vec{v}\rvert = \boxed{65.1 \text{ km/h at } 22.3° \text{ north of west}}$

CHECK and THINK

As expected, with the headwind, the plane is traveling at a speed that is lower than the 85 km/h that it travels in still air. Its velocity also has a northern component since the wind blows it somewhat northward, in addition to slowing its westward progress. These effects can often be observed in east versus west flights that travel across the US through the Jetstream.

66. (A) Using the rules of vector addition, prove that $A = \sqrt{A_x^2 + A_y^2 + A_z^2}$ represents the magnitude of a three-dimensional vector.

We start with the three-dimensional vector $\vec{A} = A_x\hat{\imath} + A_y\hat{\jmath} + A_z\hat{k}$. Begin by considering the vector, $\vec{A}'$, that would be represented by the sum of only the x and y vector components, and its magnitude.

$$\vec{A}' = A_x\hat{\imath} + A_y\hat{\jmath}$$

$$A' = \sqrt{A_x^2 + A_y^2}$$

Note that the vector, $\vec{A}'$, lies in the xy plane and that the remaining vector component points in the z direction, perpendicular to the vector, $\vec{A}'$. We could imagine a new coordinate system that preserves the z-axis, but orients the positive x-axis along the direction of the vector, $\vec{A}'$.

$$\vec{A} = \left(\sqrt{A_x^2 + A_y^2}\right)\hat{i} + A_z\hat{k}$$

The magnitude of this vector is then found using the Pythagorean theorem since these two vector components are mutually perpendicular.

$$A = \sqrt{\left(\sqrt{A_x^2 + A_y^2}\right)^2 + A_z^2}$$

$$A = \sqrt{A_x^2 + A_y^2 + A_z^2}$$

This is the magnitude of a three-dimensional vector.

73. (N) A function is given as $f(x) = 3x$, where $x > 0$. Plot this function. What is the angle between the graph of this function and the positive x axis?

INTERPRET and ANTICIPATE

The graph of the function is a straight line in the first quadrant, which crosses the origin. By looking at the equation of the function, we can tell the slope of the line is equal to three. Using the slope, we can find the angle with respect to the x axis.

SOLVE

A plot of the line is shown in the figure.

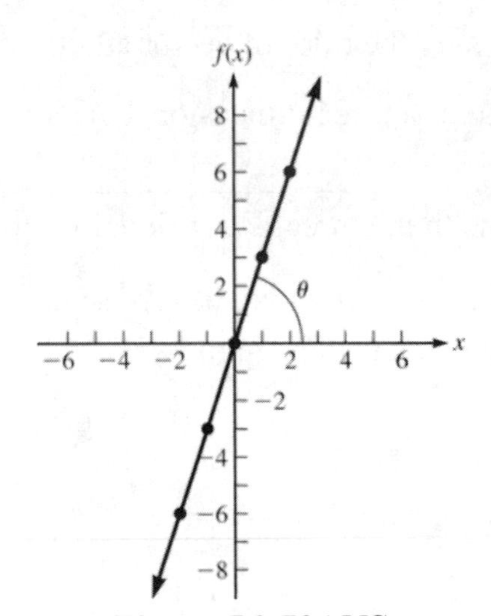

Figure P3.73ANS

The slope can be found as the derivative of the line or computing the "rise over run" using a couple points on the line.	$\text{slope} = \frac{dy}{dx} = 3$ or $\text{slope} = \frac{\text{rise}}{\text{run}} = \frac{f(x_2) - f(x_1)}{x_2 - x_1}$ $x_2 = 1, f(x_2) = 3$ $x_1 = 0, f(x_1) = 0$ $\text{slope} = \frac{3-0}{1-0} = 3$
We could also draw a triangle using points (0,0) and (1,3) and use the components to determine the angle with respect to the x axis. The angle in the first quadrant is found by using the inverse tangent.	$\theta = \tan^{-1}\left(\frac{A_y}{A_x}\right) = \tan^{-1}(3)$ $\theta = \boxed{72^\circ}$

CHECK and THINK

The angle is greater than 45 degrees, which means the graph of the function is closer to the y-axis than to the x-axis. This answer seems reasonable based on our figure.

4

Two- and Three-Dimensional Motion

6. A ball hangs from a string. The string is kept taut as the ball is displaced to one side and released. The ball swings freely back and forth. This is an example of a simple pendulum.

a. (G) Use the data in the accompanying table to create a motion diagram for this ball.

t (s)	x (cm)	y (cm)
0	30.90	4.89
1	29.44	4.43
2	25.14	3.21
3	18.36	1.70
4	9.69	0.47
5	0.00	0.00
6	−9.69	0.47
7	−18.36	1.70
8	−25.14	3.21
9	−29.44	4.43
10	−30.90	4.89
11	−29.44	4.43
12	−25.14	3.21
13	−18.36	1.70
14	−9.69	0.47
15	0.00	0.00
16	9.69	0.47
17	18.36	1.70
18	25.14	3.21
19	29.44	4.43
20	30.90	4.89

A motion diagram shows the position of an object at equal time intervals. The data table lists the position of the pendulum in an xy plane at equal time intervals (every second), so we are able to plot these points as y versus x.

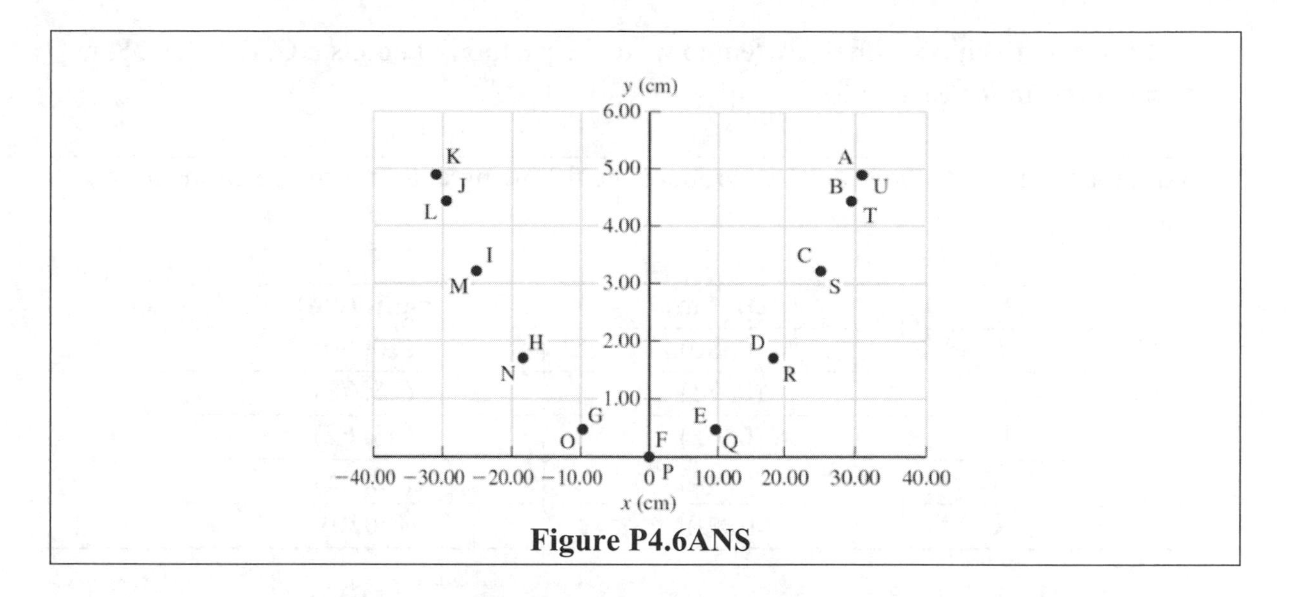

Figure P4.6ANS

b. (C) Does the ball maintain a constant speed? If not, where does it speed up or slow down?

The ball does not maintain a constant speed. Since the points plotted are at equal time intervals, the speed is fastest when the ball has traveled the furthest distance during the time interval. That is, the ball moves fastest where the points are furthest apart and slowest where the points are close together. For example, the fastest motion occurs at the bottom of the path, such as between points F and P.

8. Figure P4.8 shows the motion diagram of two balls, one on the left and one on the right. Each ball starts at a point labeled *i*. The ball on the left is released and falls straight down. At the same time, the ball on the right is launched horizontally and follows the path shown.

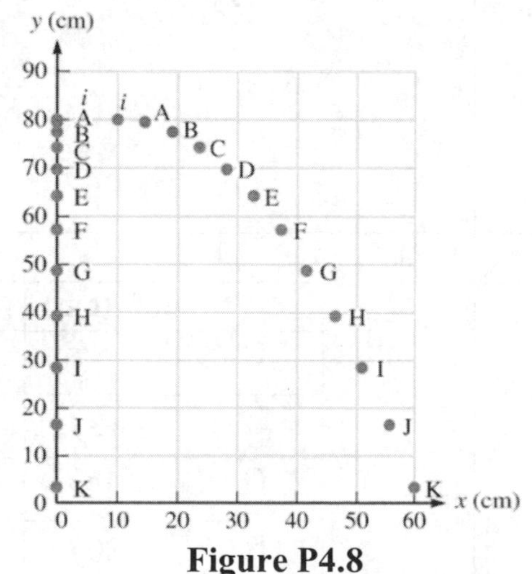

Figure P4.8

a. (C) Use the given coordinate system to write the position of points *i*, C, E, G, and K in component form for each ball.

Reading the points off the graph, we produce the following chart for the positions of the two balls:

	left (cm)	**right (cm)**
i	(0, 80)	(10, 80)
C	(0, 73)	(25, 73)
E	(0, 62)	(33, 62)
G	(0, 45)	(42, 45)
K	(0, 0)	(60, 0)

b. (N) Find the displacement of each ball from *i* to points C, E, G, and K.

INTERPRET and ANTICIPATE

From part (a), we have a list of positions for the ball at each of these times. The displacement is simply the change in position between two points, expressed as a vector, which we can calculate knowing the initial and final positions.

SOLVE

For each of the points, C, E, G, and K, we find the *x* and *y* displacement (final minus initial position) and then write the result as a vector. For instance, the approach is demonstrated for the ball on the right between point *i* and point C.

$$x_i = 10 \text{ cm}, \ y_i = 80 \text{ cm}$$

$$x_C = 25 \text{ cm}, \ y_C = 73 \text{ cm}$$

$$\Delta x = (25 \text{ cm} - 10 \text{ cm}) = 15 \text{ cm}$$

$$\Delta y = (73 \text{ cm} - 80 \text{ cm}) = -7 \text{ cm}$$

$$\Delta \vec{r} = (15\,\hat{\imath} - 7\,\hat{\jmath}) \text{ cm}$$

We continue with this procedure to write the displacement for all of the points relative to the initial position.

	left (cm)	**right (cm)**
C	$-7\hat{\jmath}$	$15\hat{\imath} - 7\hat{\jmath}$
E	$-18\hat{\jmath}$	$23\hat{\imath} - 18\hat{\jmath}$
G	$-35\hat{\jmath}$	$32\hat{\imath} - 35\hat{\jmath}$
K	$-80\hat{\jmath}$	$60\hat{\imath} - 80\hat{\jmath}$

CHECK and THINK

Given positions for the ball at two different times, we can determine the change in position for each component and write the displacement as a vector. As expected, the ball on the left has no x component of the displacement since its motion is entirely vertical.

c. (C) Compare your answers for the two balls in part (b). What similarities do you notice?

The y components of the motion are identical in both cases. This is expected when one ball is dropped from rest and another given an initial horizontal push. (Since, in the vertical component, they both have no initial velocity and are acted upon by gravity, the vertical positions will match as the balls fall.)

15. A glider is initially moving at a constant height of 3.59 m. It is suddenly subject to a wind such that its velocity at a later time t can be described by the equation $\vec{v}(t) = 15.72\hat{\imath} - 7.88(1+t)\hat{\jmath} + 0.79t^3\hat{k}$, where $\vec{v}$ and its components are in meters per second, t is in seconds, and the z axis is perpendicular to the level ground.

a. (N) What was the initial velocity of the glider?

INTERPRET and ANTICIPATE In order to determine the initial velocity, we need to solve the equation given in the problem statement for when $t = 0$.	
SOLVE Substituting $t = 0$ into the equation yields the initial velocity of the glider.	$\vec{v}_i = \vec{v}(0) = 15.72\hat{\imath} - 7.88(1+0)\hat{\jmath} + 0.79(0)^3\hat{k}$ $\vec{v}_i = \boxed{15.72\hat{\imath} - 7.88\hat{\jmath}}$
CHECK and THINK Given that the glider was initially moving at a constant height, the x and y axes must be parallel to the ground and there is no velocity in the z direction.	

b. (N) Write an expression for the acceleration of the glider in component form, when $t =$ 2.15 s.

INTERPRET and ANTICIPATE The acceleration is the time derivative of velocity. Once an expression for the acceleration is found, the time $t = 2.15$ s is substituted. We expect that the acceleration will have no x component since the velocity's x component does not depend on time.

SOLVE To determine the acceleration, we take the time derivative of the velocity as in Equation 4.14.	$\vec{a} = \frac{d\vec{v}}{dt} = \frac{d}{dt}\left[15.72\hat{i} - 7.88(1+t)\hat{j} + 0.79t^3\hat{k}\right]$ $\vec{a} = \frac{d}{dt}(15.72)\hat{i} - \frac{d}{dt}\left[7.88(1+t)\right]\hat{j} + \frac{d}{dt}(0.79t^3)\hat{k}$ $\vec{a} = -7.88\hat{j} + 3(0.79t^2)\hat{k}$ $\vec{a} = -7.88\hat{j} + 2.4t^2\hat{k}$
Now, solve for the acceleration when $t = 2.15$ s.	$\vec{a} = -7.88\hat{j} + 2.4(2.15)^2\hat{k}$ $\vec{a} = \boxed{(-7.88\hat{j} + 11\hat{k})\ \text{m/s}^2}$
The acceleration has no x component, as expected. We also notice that in the algebraic expression for acceleration, the y component does not depend on time, but the z component does. As time goes on, the acceleration will remain constant in the y direction, and will increase in the z direction.	

17. (A) If the vector components of a particle's position moving in the xy plane as a function of time are $\vec{x} = bt^2\hat{i}$ and $\vec{y} = ct^3\hat{j}$, where b and c are positive constants with the appropriate dimensions such that the components will be in meters, at what time t is the angle between the particle's velocity and the x axis equal to 45°?

INTERPRET and ANTICIPATE The x and y components of the velocity can be found by taking the derivative of the components of x and y position with respect to time. These x and y components can be related to the angle by the tangent function.	
SOLVE Our goal is to find the time that corresponds to when the velocity vector is at an angle of 45 degrees with respect to the x axis.	$\theta = 45°$ $t = ?$

We sketch the relevant variables in the problem. Using trigonometry, we can relate the tangent of the angle to the x and y components of the velocity. Since the angle is 45°, the ratio of the scalar components of the velocity must be equal to one. This means the velocity components are equal when the particle is moving at an angle of 45°.	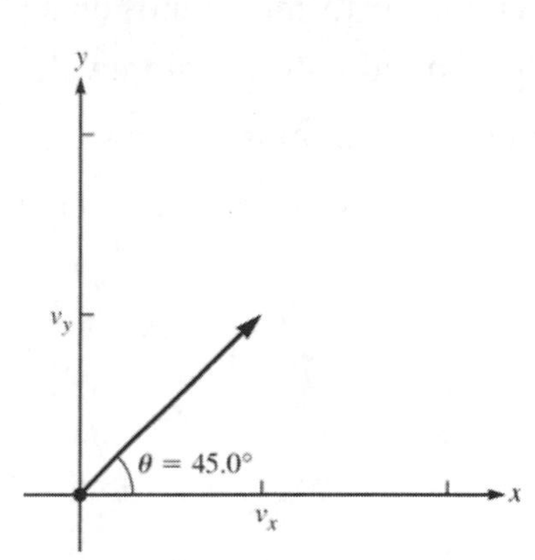 **Figure P4.17ANS** $\frac{v_y}{v_x} = \tan\theta = \tan(45°) = 1$ $v_y = v_x$
The components of the velocity are found by taking the derivative of the components of the position with respect to time (Equation 4.9).	$v_x = \frac{dx}{dt} = \frac{d}{dt}\left(bt^2\right) = 2bt$ $v_y = \frac{dy}{dt} = \frac{d}{dt}\left(ct^3\right) = 3ct^2$
Setting our expressions equal to one another, we solve for time. Mathematically, $t = 0$ is a possible solution. However, when $t = 0$, both components of the velocity are also zero, so it does not make sense to consider the velocity to have a direction at $t = 0$. Our final answer is $t = 2b/3c$.	$v_y = v_x$ $3ct^2 = 2bt$ $t = 0$ or $\boxed{t = \frac{2b}{3c}}$

CHECK and THINK

We determined the x and y components of the velocity and found when they are equal, the condition for the velocity vector to be inclined by 45 degrees with respect to the horizontal. We can double check that this time satisfies this condition by plugging the time back in to the expression for the angle:

$$\tan\theta = \frac{v_y}{v_x} = \frac{3ct^2}{2bt} = \frac{3c}{2b}t = \frac{3c}{2b}\frac{2b}{3c} = 1$$

20. (N) A circus performer stands on a platform and throws an apple from a height of 45 m above the ground with an initial velocity $\vec{v}_0$ as shown in Figure P4.20. A second, blindfolded performer must catch the apple. If $v_0 = 26$ m/s, how far from the end of the platform should the second performer stand?

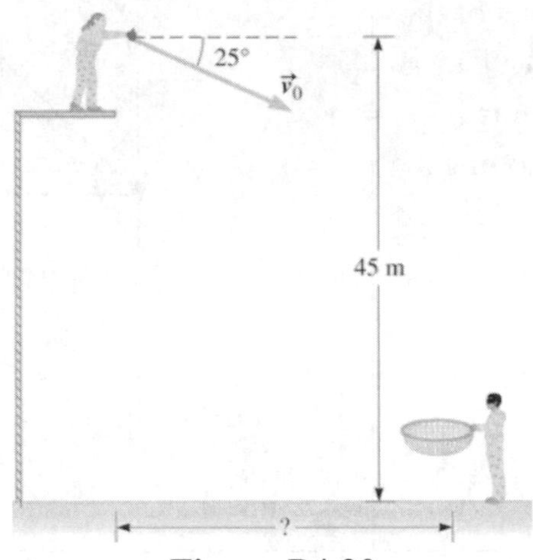

Figure P4.20

INTERPRET and ANTICIPATE	
This is a projectile motion problem. We can treat both the x and y components separately, with gravity acting in the vertical direction and zero acceleration in the horizontal direction. Using the vertical direction, we can determine how long it will take for the apple to reach the ground. With this time and the horizontal velocity, we can then determine how far from the platform the apple travels.	
SOLVE This is a 2D kinematics problem and we start by drawing a sketch. We can either consider the vector motion using Equation 4.25 or consider the x and y components separately and use our familiar 1D kinematic equations (for instance, Equation 2.11). We will treat the two components separately here.	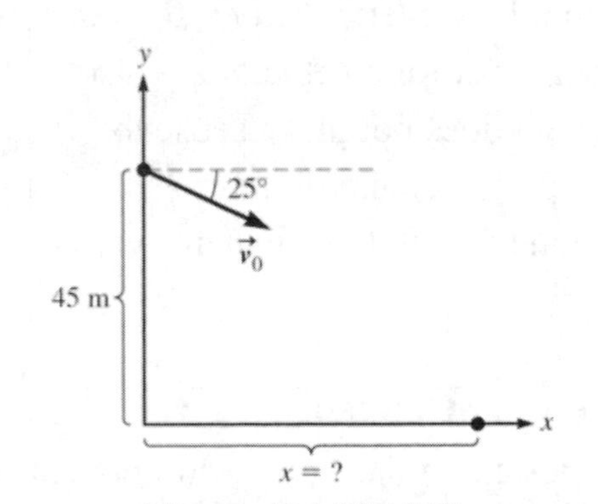 **Figure P4.20ANS**
The y component of this projectile motion determines how long the apple is in the air. We first keep track of our known variables. The initial position is 45 m	$y_i = 45$ m $y_f = 0$ m $g = -9.81$ m/s^2

above the final position (the ground, which we'll consider to be 0 m, with the positive direction pointing upwards). The acceleration is due to gravity. The initial y velocity can be found using trigonometry since we know both the total velocity and the angle with respect to the horizontal.	$v_{y,0} = -(25 \text{ m/s}) \sin 25° = -11 \text{ m/s}$
We can now use our kinematic equations to write an equation that can be solved for the time of flight.	$y_f = y_0 + v_{y,0}t + \frac{1}{2}a_y t^2$ $0 = 45 - 11t - \frac{9.81}{2}t^2$
This is a quadratic equation, which we can solve. There are two solutions, but only one has a physically meaningful value (a positive time), 2.1 s.	$t = \frac{11 \pm \sqrt{(11)^2 - 4\left(\frac{-9.81}{2}\right)45}}{2\left(\frac{-9.81}{2}\right)}$ $t = -4.4 \text{ s}$ or 2.1 s
The x velocity has a constant value, since there is no acceleration in the x direction, and we can find the initial x velocity using trigonometry. Since the apple travels at this constant x velocity throughout the time that it's falling, we can determine the final x position.	$v_x = -(25 \text{ m/s}) \cos 25° = 23 \text{ m/s}$ $x = v_x t = (23)(2.1) = \boxed{48 \text{ m}}$

CHECK and THINK

While we don't know precisely what to expect, it seems reasonable that the ball might be in the air for two seconds and travel 48 meters.

23. (N) During the battle of Bunker Hill, Colonel William Prescott ordered the American Army to bombard the British Army camped near Boston. The projectiles had an initial velocity of 45 m/s at 35° above the horizon and an initial position that was 35 m higher than where they hit the ground. How far did the projectiles move horizontally before they hit the ground? Ignore air resistance.

INTERPRET and ANTICIPATE

To find the horizontal distance traveled by the projectiles, we first find the time-of-flight using an equation of motion for the vertical components of the velocity and position. We then substitute the time of flight into an equation of motion for the horizontal component

<table>
<tr><td colspan="2">to determine the range.</td></tr>
<tr><td>SOLVE
First, let's sketch the situation and organize the information that we have. Since there is an acceleration due to gravity vertically and no acceleration horizontally, we consider the motion separately for these components. The components of the initial velocity can be found using trigonometry.</td><td>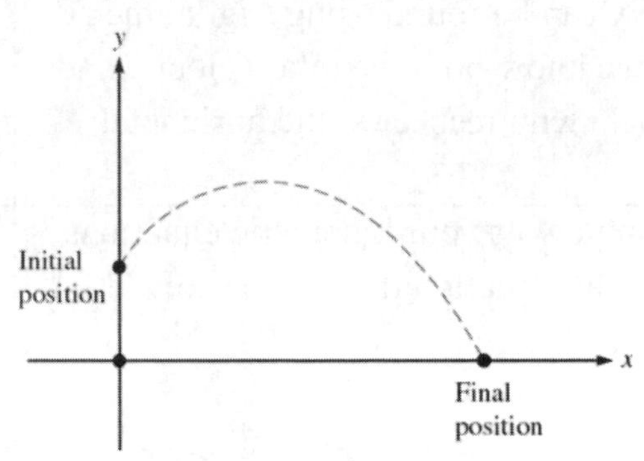

Figure P4.23ANS
$x_i = 0$ $y_i = h = 35$ m
$x_f =$ unknown $y_f = 0$
$v_{ix} = v_i \cos\theta$ $v_{iy} = v_i \sin\theta$
$a_x = 0$ $a_y = -9.81$ m/s^2</td></tr>
<tr><td>We can now use the y component to determine the time of flight for the projectile until it hits the ground (Eq. 2.11).</td><td>$y_f = y_i + v_{iy}t + \frac{1}{2}a_y t^2$
$0 = 35 + (45\sin 35^\circ)t + \frac{1}{2}(-9.81)t^2$</td></tr>
<tr><td>Solving this quadratic equation for time, we get two answers. The positive time (i.e. after the projectile is launched) is the physically meaningful answer.</td><td>$t = \dfrac{-45\sin 35^\circ \pm \sqrt{(45\sin 35^\circ)^2 - 4(35)\left(\frac{-9.81}{2}\right)}}{2\left(\frac{-9.81}{2}\right)}$
$t = 6.38$ s   or -1.11s</td></tr>
<tr><td>Since there is no horizontal acceleration, the x velocity is constant and we can calculate the total range travelled in this time.</td><td>$x = v_x t = (v_0 \cos 35^\circ)(t)$
$x = \left(45 \frac{\text{m}}{\text{s}} \cos 35^\circ\right)(6.38 \text{ s}) =$
$x = 240 \text{ m} = 2.4 \times 10^2 \text{ m}$</td></tr>
<tr><td colspan="2">CHECK and THINK
We've calculated the time of flight and range of the projectile. The answer of 240 meters sounds like a plausible distance for a large military projectile.</td></tr>
</table>

27. A circus performer throws an apple toward a hoop held by a performer on a platform (Fig. P4.27). The thrower aims for the hoop and throws with a speed of 24 m/s. At the exact moment the thrower releases the apple, the other performer drops the hoop. The hoop falls straight down.

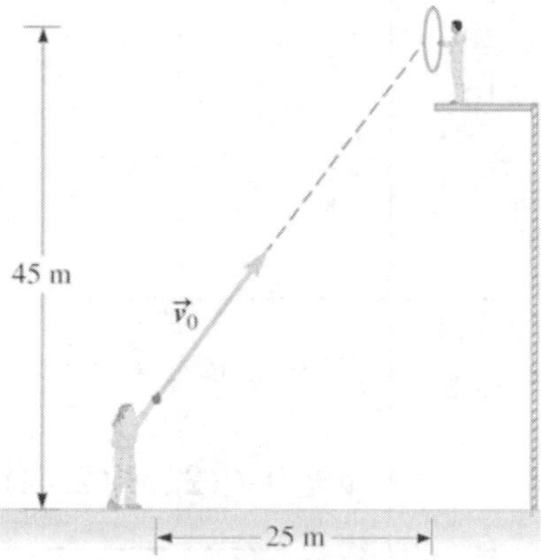

Figure P4.27

a. (N) At what height above the ground does the apple go through the hoop?

INTERPRET and ANTICIPATE The apple undergoes projectile motion. The *x* component determines how long the apple takes to reach the hoop and the *y* component determines vertical position of the apple at this time.	
SOLVE Using the figure, we can determine the angle at with the initial velocity $\vec{v}_0$ is elevated. The *x* component of the velocity can now be found using trigonometry.	$\theta = \tan^{-1}\left(\frac{45}{25}\right) = 61°$ $v_x = v_0 \cos\theta = (24 \text{ m/s})\cos 61° = 11.7 \text{ m/s}$
This *x* velocity is constant, since there is no acceleration in the *x* direction. Therefore, we can determine the time that it takes for the apple to travel the 25 m needed to reach the hoop. Assuming that the initial time was $t = 0$, we can also say $t = 2.14$ s.	$v_x = \frac{\Delta x}{\Delta t}$ $\Delta t = \frac{\Delta x}{v_x} = \frac{25 \text{ m}}{11.7 \text{ m/s}} = 2.14 \text{ s}$

The initial y velocity can similarly be found using trigonometry. The y position at the time the apple reaches the hoop can now be determined using the fact that the initial height is zero and the acceleration is due to gravity. We assume the positive direction is upwards.	$v_{y,0} = v_0 \sin\theta = (24 \text{ m/s})\sin 61° = 21.0 \text{ m/s}$ $g = -9.81 \text{ m/s}^2$ $y_i = 0 \text{ m}$ $y_f = ?$
We use a kinematic equation (2.11) to determine the final height. Since the final height is positive, it is above the initial position.	$y = y_0 + v_{y,0}t + \frac{1}{2}at^2$ $y = 0 + (21.0)(2.14) - \frac{9.81}{2}(2.14)^2 = 22.47 \text{ m}$ $y = \boxed{22 \text{ m above the ground}}$

CHECK and THINK

The resulting height is positive (above ground) at a distance that seems plausible for a circus act. This seems encouraging and suggests that we carried out the calculation correctly!

b. (C) If the performer on the platform did not drop the hoop, would the apple pass through it?

The apple would not pass through the hoop if the hoop was not dropped. The apple will cross the plane of the hoop at a height of 23 meters, which is 22 meters below the initial height of the loop. If the performers don't launch the apple and drop the hoop at the same time, the act will not be very impressive!

30. (A) A projectile is launched up and to the right over flat, level ground. If air resistance is ignored, its maximum range occurs when the angle between its initial velocity and the ground is 45°. Which angles would result in the range being equal to half the maximum?

INTERPRET and ANTICIPATE

As the angle of the projectile is changed from 45 degrees (either decreased or increased), the range will decrease from its maximum value. Therefore, there are two angles for which the range will be equal to half its maximum. One angle will be greater than 45° and the other will be less.

SOLVE We want to find launch angles that correspond to to a range that is half the maximum.	$R = \frac{R_{max}}{2}$ $\theta = ?$
Equation 4.28 gives the range, as a function of initial speed, angle and magnitude of acceleration due to gravity.	$R = \frac{v_0^2}{g}\sin 2\theta$
We can calculate the maximum range, which occurs when the angle is 45°.	$R_{max} = \frac{v_0^2}{g}\sin 2(45°) = \frac{v_0^2}{g}$
We now seek the angles for which the range is half this maximum value.	$R = \frac{R_{max}}{2}$ $\frac{v_0^2}{g}\sin 2\theta = \frac{v_0^2}{2g}$ $\sin 2\theta = \frac{1}{2}$ $2\theta = \sin^{-1}\left(\frac{1}{2}\right) = 30° \text{ or } 150°$ $\theta = \boxed{15° \text{ or } 75°}$
CHECK and THINK As we predicted, there are two angles for which the range is half the maximum value.	

35. (N) The bola is a traditional weapon used for tripping up or grounding an animal (Fig. P4.35). Once it is set into motion, each ball at the end of the bola can be thought of as a single object that is in uniform circular motion. Suppose it takes the bola 0.3250 s to traverse a circular path with a radius of 0.8661 m. What is the magnitude of the centripetal acceleration experienced by either ball at the end of the bola?

Figure P4.35

INTERPRET and ANTICIPATE

In order to determine the centripetal acceleration, we must determine the speed of the end of the bola. The time for an object to traverse a circular path is the period, *T*. Knowing this information and the radius of the circular path, we can determine the speed of the bola.

SOLVE	
The speed of the bola will depend on the circumference of the circular path and the period according to Equation 4.30. The time given in the problem statement is the period of the bola and the circumference can be determined from the radius *r*.	$v = \frac{2\pi r}{T} = \frac{2\pi(0.8661\text{ m})}{0.3250\text{ s}}$ $v = 16.74\text{ m/s}$
The magnitude of the centripetal acceleration is then found using Equation 4.39.	$a_c = \frac{v^2}{r} = \frac{(16.74\text{ m/s})^2}{0.8661\text{ m}}$ $a_c = \boxed{323.7\text{ m/s}^2}$

CHECK and THINK

The bola experiences an acceleration greater than 32 g's (or 32 times the gravitational acceleration on Earth)! Note that a longer bola, or longer radius, would result in the bola moving at a greater speed. This would aid the hunter when he or she lets go of the weapon and throws it at the target.

39. (A) Two particles A and B move at a constant speed in circular paths at the same angular speed ω. Particle A's circle has a radius that is twice the length of particle B's circle. What is the ratio v_A/v_B of their translational speeds?

Before we get started calculating, let's think about what we expect. If the particles move at the same *angular* speed, it means that they cover the same number of degrees per second and that they will travel around their circular paths in the same amount of time. The particle with the larger radius (A) will have to cover a larger distance in this time, so it must be going faster. Let's see:

The translational speed depends on the angular velocity and radius according to Equation 4.37, $v = \omega r$. We know that they have the same angular frequency ω but that $r_A = 2\ r_B$. Using this information, we can find the ratio of the translational velocity of A versus that

of B:
$$\frac{v_A}{v_B} = \frac{\omega r_A}{\omega r_B} = \boxed{2}.$$
Particle A is indeed going faster than B as we predicted.

43. The Moon's orbit around the Earth is nearly circular and has a period of approximately 28 days. Assume the Moon is moving in uniform circular motion.
a. (N) Find the angular speed of the Moon.

INTERPRET and ANTICIPATE We are given the time that it takes for the Moon to circle the Earth. That is, the time for the Moon to travel the entire 2π radians around its approximately circular trajectory. From this, we can deterimine its angular speed.	
SOLVE The period of the moon is approximately 28 days, but we convert this to the standard metric unit of seconds.	$T \approx 28 \text{ days}\left(\frac{24 \text{ h}}{1 \text{ day}}\right)\left(\frac{3600 \text{ s}}{1 \text{ h}}\right) \approx 2.4 \times 10^6 \text{s}$
It travels 360 degrees or 2π radians in this time, which allows us to calculate its angular velocity.	$\omega = \frac{2\pi}{T} \approx \frac{6.28}{2.4 \times 10^6 s} \approx \boxed{2.6 \times 10^{-6} \text{ rad/s}}$
CHECK and THINK The Moon makes one period (6.28 radians) in a month (around 10^6 seconds), so we should expect that the angular velocity will be quite small.	

b. (N) What is its centripetal acceleration?

INTERPRET and ANTICIPATE The centripetal acceleration can be calculated for an object in circular motion if we know how fast the object is moving and the radius of the orbit.	
SOLVE We look up the radius of the Moon's orbit and find $r = 3.84 \times 10^8 \text{m}$. The centripetal acceleration can now be	

calculated using Equation 4.38.	$a_c = \frac{v^2}{r} = \omega^2 r = (2.6\times10^{-6})^2(3.84\times10^8)$ $a_c = \boxed{2.6\times10^{-3}\ \text{m/s}^2}$

CHECK and THINK

This is quite a small acceleration, which may initially seem surprising, but we've carried out the calculation, which resulted in a value with units we expect for acceleration. (In fact, since it is the Earth's gravitational field keeping the Moon in orbit, which decreases in magnitude at points further from the Earth, the acceleration actually *must* be much smaller than the 9.81 m/s^2 we experience on Earth. It's difficult to have intuition about massive objects with very large orbits though!)

45. Pete and Sue, two reckless teenage drivers, are racing eastward along a straight stretch of highway. Pete is traveling at 98.0 km/h, and Sue is chasing him at 125 km/h. **a. (N)** What is Pete's velocity with respect to Sue?

INTERPRET and ANTICIPATE

If we know the velocities for both Pete and Sue with respect to the ground, we can find Pete's velocity relative to Sue.

SOLVE Both Pete and Sue move relative to the ground (G). Treating east as the positive direction, we express their velocities (in km/h) in the frame of the ground.	$(v_P)_G = 98.0$ (east) $(v_S)_G = 125$ (east)
The speed of the ground in Sue's frame of reference is the same as her speed relative to the ground, but in the opposite direction. The negative sign indicates that the direction is toward the west.	$(v_G)_S = -(v_S)_G = -125$ (or 125 to the west)
We can now write Pete's velocity with respect to Sue using Equation 4.42 and fill in the known values.	$(\vec{v}_P)_S = (\vec{v}_P)_G + (\vec{v}_G)_S$ $(v_P)_S = (v_P)_G + (v_G)_S$ $(v_P)_S = 98.0 - 125 = -27.0$ $(v_P)_S = \boxed{27.0 \text{ km/h to the west}}$

CHECK and THINK
Since Pete is traveling slower than Sue in the same direction, Pete appears to be going "backwards" in Sue's frame of reference. In other words, if you imagine Sue's perspective as she travels east, Pete is "heading backwards towards Sue" as she catches up to him.

b. (N) What is Sue's velocity with respect to Pete?

INTERPRET and ANTICIPATE	
Given Pete's velocity with respect to Sue from part (a), Sue's velocity with respect to Pete is the same magnitude with the opposite sign.	
SOLVE After part (a), we can easily write the velocity of Sue with respect to Pete.	$(v_S)_P = -(v_P)_S = +27.0$ $(v_S)_P = \boxed{27.0 \text{ km/h to the east}}$
CHECK and THINK From Pete's perspective, Sue is moving towards him (in the positive direction) as she catches up to him.	

c. (N) If Sue is initially 325 m behind Pete, how long will it take her to catch up to him?

INTERPRET and ANTICIPATE	
With the relative velocities and initial separation, it is straightforward to calculate the time needed for Sue to close the gap.	
SOLVE Sue is approaching Pete at 27.0 km/h and must travel the 325 m relative distance between them in order to catch up. (Note that they are *both* still moving to the right, so 325 m is just their separation and not the total distance that Sue actually travels in this time.)	$d = v\Delta t$ $\Delta t = \dfrac{d}{v} = \dfrac{.325 \text{ km}}{27.0 \text{ km/h}} = 0.0120 \text{ h}\left(\dfrac{3600 \text{ s}}{1 \text{ h}}\right)$ $\Delta t = \boxed{43.3 \text{ s}}$
CHECK and THINK Given the relative velocity and relative separation, we can calculate the time needed for Sue to reach the same location as Pete.	

49. (N) A man paddles a canoe in a long, straight section of a river. The canoe moves downstream with constant speed 3 m/s relative to the water. The river has a steady current of 1 m/s relative to the bank. The man's hat falls into the river. Five minutes later, he notices that his hat is missing and immediately turns the canoe around, paddling upriver with the same constant speed of 3 m/s relative to the water. How long does it take the man to reclaim his hat?

INTERPRET and ANTICIPATE

This problem is most easily done by considering the motion of the canoe and the hat in the frame of the river. In this frame, the hat is motionless, and the canoe moves with speed 3 m/s.

SOLVE

Both the hat and the canoe are carried by the water. We could determine the speeds of the canoe and the hat relative to the land, but that's not necessary. *Relative to the water*, the hat is not moving and the canoe is moving at 3 m/s. In fact, we could determine how far the hat has moved away from the canoe for the five minutes until he notices that it's missing, but this is also unnecessary. In the frame of the reference of the water, the man canoed for five minutes away from the stationary hat and therefore has to canoe for five minutes back to return to the hat (even though both are being carried downstream overall by the river).

CHECK and THINK

This problem initially appears to require a bit of calculation (and in fact, you might certainly find yourself calculating some of these quantities), but it's also possible to avoid calculating by considering the motion of the hat and the canoe relative to the water, without worrying about how the water is moving. Choosing the right reference frame can sometimes make a difficult problem into a simple one (or vice versa)!

52. An ant and a spider each move with constant velocity on a horizontal table. The velocity vectors and positions of the ant and the spider (with respect to the table) at time $t = 0$ s are shown in Figure P4.52.

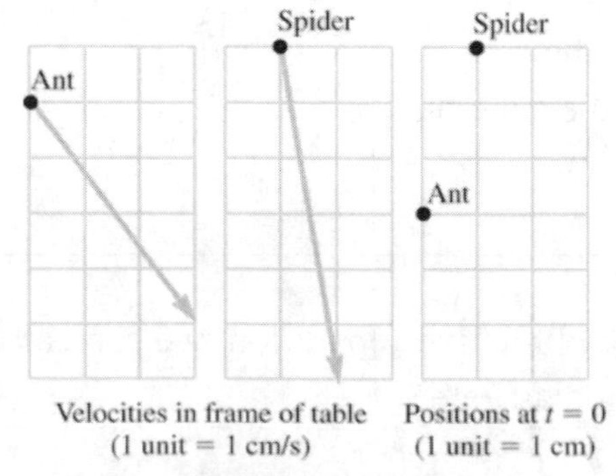

Figure P4.52

a. (G) Draw the velocity vector of the ant in the frame of the spider.

The equation relating the speed of the ant relative to the spider $(\vec{v}_A)_S$ in terms of the speed of the ant relative to the table $(\vec{v}_A)_T$ and the speed of the spider relative to the table $(\vec{v}_S)_T$ is:

$$(\vec{v}_A)_S = (\vec{v}_A)_T + (\vec{v}_T)_S = (\vec{v}_A)_T - (\vec{v}_S)_T$$

Using a graphical method, the relative velocity of the ant in the frame of the spider is found to point towards the right side and top of the table, as viewed from above in Figure P4.52aANS.

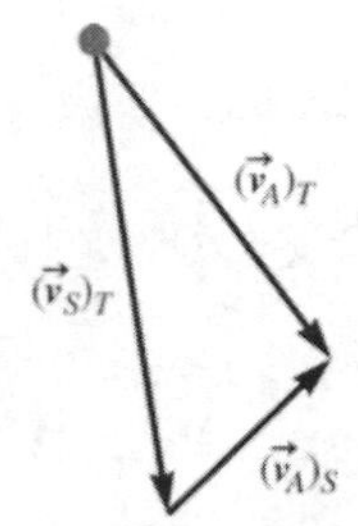

Figure P4.52aANS

b. (C) Is there a time at which the ant and the spider will have the same position? Explain.

We draw the ant's relative velocity vector in the frame of reference of the spider on the grid that shows the positions of the two bugs. The spider is at rest in this reference frame, so for the ant to reach the position of the spider, its trajectory must cross the position of the spider. Instead, the trajectory of the ant passes to the right of the spider. Therefore, the ant and the spider will never meet up.

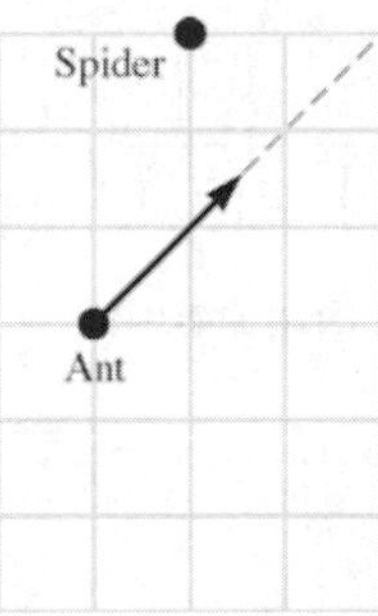

Figure P4.52bANS

53. (N) Suppose at one point along the Nile River a ferryboat must travel straight across a 10.3-mile stretch from west to east. At this location, the river flows from south to north with a speed of 2.41 m/s. The ferryboat has a motor that can move the boat forward at a constant speed of 20.0 mph in still water. In what direction should the ferry captain direct the boat so as to travel directly across the river?

INTERPRET and ANTICIPATE

Given the velocity of the boat relative to the water and the velocity of the water relative to the land, we can determine the speed of the boat relative to the land. We want the boat to have a velocity relative to the land that carries it straight across the river (with no component upstream or downstream). Here, we label the angle θ, our desired quantity, as the direction the boat should travel relative to a path directly across the river.

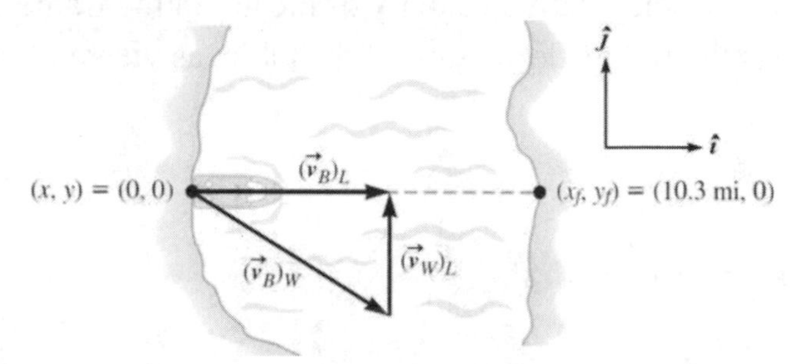

Figure P4.53ANS

SOLVE We first convert the boat's speed relative to the water into SI units and write the velocity of the boat with respect to the water using trigonometry and the unknown angle θ. We also note the speed of the current (i.e. the water relative to the land).	$(v_B)_W = (20.0 \text{ mi/h})\times\left(\frac{1609 \text{ m}}{1 \text{ mi}}\right)\times\left(\frac{1 \text{ h}}{3600 \text{ s}}\right) = 8.94 \text{ m/s}$ $(\vec{v}_B)_W = (8.94\cos\theta\hat{\imath} + 8.94\sin\theta\hat{\jmath}) \text{ m/s}$ $(\vec{v}_W)_L = 2.41\hat{\jmath} \text{ m/s}$
The y component of the boat velocity with respect to the land must be 0 in order for the boat to travel straight across the river.	$(\vec{v}_B)_L = (\vec{v}_B)_W + (\vec{v}_W)_L$ y components: $0 = v_{y,\text{boat relative to water}} + 2.41 \text{ m/s}$ $v_{y,\text{boat relative to water}} = -2.41 \text{ m/s}$
Knowing the y component of the boat's velocity and its magnitude, we can now determine the angle using the	$\sin\theta = \frac{v_{y,boat}}{v_{boat}} = \frac{-2.41}{8.94} = -0.270$ $\theta = -15.6°$ or $\boxed{15.6° \text{ south of east}}$

sine function.	

CHECK and THINK

Consistently pointing the boat upstream in this direction counters the effect of the current pushing the boat downstream, allowing it to travel straight across to the east bank of the river.

58. Two bicyclists in a sprint race begin from rest and accelerate away from the origin of an x–y coordinate system. Miguel's acceleration is given by $\left(-0.700\hat{\boldsymbol{\imath}}+1.00\hat{\boldsymbol{\jmath}}\right)$ m/s^2, and Lance's acceleration is given by $\left(1.20\hat{\boldsymbol{\imath}}+0.300\hat{\boldsymbol{\jmath}}\right)$ m/s^2.

a. (N) What is Miguel's acceleration with respect to Lance?

INTERPRET and ANTICIPATE

Given the accelerations for both Miguel and Lance, we can calculate relative accelerations in the same way that we would calculate relative velocities or positions.

SOLVE	
In general, Miguel's acceleration with respect to Lance can be expressed similar to the expression for relative velocities (Equation 4.45). Note that they are not moving at constant velocity, therefore the accelerations are not equal and Equation 4.46 does not hold.	$\left(\vec{a}_M\right)_L=\left(\vec{a}_M\right)_G+\left(\vec{a}_G\right)_L=\left(\vec{a}_M\right)_G-\left(\vec{a}_L\right)_G$ $\left(\vec{a}_M\right)_L=\left(-0.700\hat{\boldsymbol{\imath}}+1.00\hat{\boldsymbol{\jmath}}\right)\text{ m/s}^2-\left(1.20\hat{\boldsymbol{\imath}}+0.300\hat{\boldsymbol{\jmath}}\right)\text{ m/s}^2$ $\left(\vec{a}_M\right)_L=\boxed{\left(-1.90\hat{\boldsymbol{\imath}}+0.700\hat{\boldsymbol{\jmath}}\right)\text{ m/s}^2}$

CHECK and THINK

For two frames of reference (Miguel and Lance) accelerating relative to the ground, we are able to express their relative acceleration.

b. (N) What is Miguel's speed with respect to Lance after 4.50 s have elapsed?

INTERPRET and ANTICIPATE

With the accelerations provided, we can determine their velocities, which can be found from the integral of the acceleration. Once we have each of their velocities, we can find an expression for their relative velocity and determine its value at 4.5 seconds.

SOLVE Integrating the accelerations given, knowing that they both start from rest, we obtain their velocities (specifically, their velocities relative to the ground).	$\vec{v}_M = \left(-0.700t\,\hat{i} + 1.00t\,\hat{j}\right)$ m/s $\vec{v}_L = \left(-1.20t\,\hat{i} + 0.300t\,\hat{j}\right)$ m/s
We can now express their relative velocity using Equation 4.45.	$\left(\vec{v}_M\right)_L = \left(\vec{v}_M\right)_G + \left(\vec{v}_G\right)_L = \left(\vec{v}_M\right)_G - \left(\vec{v}_L\right)_G$ $\left(\vec{v}_M\right)_L = \left(-0.700t\,\hat{i} + 1.00t\,\hat{j}\right)$ m/s $- \left(1.20t\,\hat{i} + 0.300t\,\hat{j}\right)$ m/s $\left(\vec{v}_M\right)_L = \left(-1.90t\,\hat{i} + 0.700t\,\hat{j}\right)$ m/s
Substituting the desired time, we find their relative velocity at 4.5 seconds.	$\left(\vec{v}_M\right)_L = \left(-1.90(4.50\text{ s})\hat{i} + 0.700(4.50\text{ s})\hat{j}\right)$ m/s $\left(\vec{v}_M\right)_L = \left(-8.55\hat{i} + 3.15\hat{j}\right)$ m/s
Finally, Miguel's speed with respect to Lance at this time is simply the magnitude of this velocity.	$\left(v_M\right)_L = \sqrt{(-8.55\text{ m/s})^2 + (3.15\text{ m/s})^2} = \boxed{9.11\text{ m/s}}$

CHECK and THINK
In this case, we needed to find expressions for the velocities using the accelerations given, but with those, finding the relative velocity and speed (magnitude of the velocity) is fairly straightforward.

c. (N) What is the distance separating Miguel and Lance after 4.50 s have elapsed?

INTERPRET and ANTICIPATE Similar to part (b), where we needed to derive an expression for velocity using the expression for acceleration that was given, their positions are found by taking the integral of their velocities.	
SOLVE Integrating the expressions for velocity in part (b), and knowing that they start from the origin, we find	$\vec{r}_M = \left(-0.350t^2\,\hat{i} + 0.500t^2\,\hat{j}\right)$ m $\vec{r}_L = \left(0.600t^2\,\hat{i} + 0.150t^2\,\hat{j}\right)$ m

their positions.	
We can now follow a similar procedure as in parts (a) and (b) to find Miguel's position with respect to Lance using Equation 4.44. In addition, we are also seeking Miguel's position relative to Lance, which we might normally call Miguel's displacement from Lance.	$(\vec{r}_M)_L = (\vec{r}_M)_G + (\vec{r}_G)_L = (\vec{r}_M)_G - (\vec{r}_L)_G$ or $\vec{r}_{ML} = \vec{r}_M - \vec{r}_L$ $\vec{r}_{ML} = \left(-0.350t^2\,\hat{i} + 0.500t^2\,\hat{j}\right)\text{ m} - \left(0.600t^2\,\hat{i} + 0.150t^2\,\hat{j}\right)\text{ m}$ $\vec{r}_{ML} = \left(-0.950t^2\,\hat{i} + 0.350t^2\,\hat{j}\right)\text{ m}$
With this general expression (valid for any time after $t = 0$), we can determine their separation at 4.5 seconds.	$\vec{r}_{ML} = \left(-0.950(4.50\text{ s})^2\,\hat{i} + 0.350(4.50\text{ s})^2\,\hat{j}\right)\text{ m}$ $= \left(-19.2\,\hat{i} + 7.09\,\hat{j}\right)\text{ m}$
Finally, with the displacement calculated, we can now find the distance between them, which is the magnitude of the displacement.	$d = \sqrt{(-19.2\text{ m})^2 + (7.09\text{ m})^2} = \boxed{20.5\text{ m}}$

CHECK and THINK

For each of the kinematic quantities, the relative motion is determined in the same way. Other than needing to integrate to get from acceleration to velocity to displacement, the expressions for the relative motion all look similar.

61. (A) You are watching a friend practice archery when he misses the target completely, and the arrow sticks into the ground. Discouraged, your friend asks whether you can help by estimating the speed with which the arrow left the bow. Remembering your physics class, you realize that you can do so by measuring its height above the ground when it was launched, if you assume the arrow was launched horizontally and that the ground is level. For your analysis, you let h represent the arrow's height above the ground when it was launched, v_0 represent the launch speed, and θ represent the angle the arrow makes with the horizontal when it is stuck into the ground. Find an expression for v_0 in terms of h and θ.

<table>
<tr><td colspan="2">INTERPRET and ANTICIPATE
The arrow is a projectile during its flight. We will let t_0 be the time just after the arrow leaves the bow, and t_f be just before the arrow hits the ground. The angle at which the arrow sticks in the ground gives the direction of $\vec{v}_f$, the velocity vector of the arrow at the final time. Our goal is find an expression for v_0 in terms of h and θ.</td></tr>
<tr><td>SOLVE
First we draw a picture and establish a coordinate system. The origin is chosen to be at the arrow's initial location.</td><td>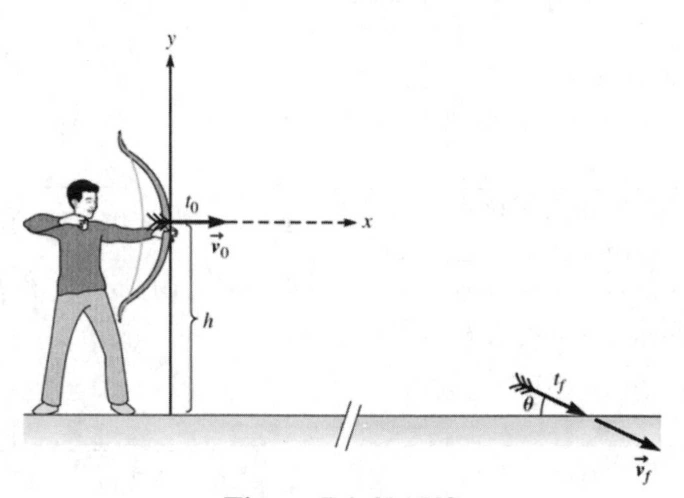

Figure P4.61ANS</td></tr>
<tr><td>In this projectile motion, the arrow falls a total height h with an acceleration due to gravity $a = -g$ and there is no acceleration in the horizontal direction. The initial y velocity is zero, since the arrow is shot horizontally. The x velocity remains unchanged, since there is no acceleration.</td><td>$a_x = 0$
$a_y = -g$
$v_{x,f} = v_{x,i} = v_0$
$v_{y,i} = 0$
$y_f - y_i = h$</td></tr>
<tr><td>We can draw the velocity vector of the arrow at t_f, just before it hits the ground. The measured angle θ can be used to relate the two components of this final velocity vector.</td><td>$\tan\theta = \frac{v_{y,f}}{v_{x,f}}$</td></tr>
</table>

Using kinematic equations (Equation 4.11) for the y component, we can relate some of the relevant variables.	$v_{y,f}^2 = v_{y,i}^2 + 2a\Delta y = 2gh$ $v_{y,f} = \sqrt{2gh}$
Now, putting these equations together, we can express the initial velocity in terms of the known quantities.	$\tan\theta = \dfrac{\sqrt{2gh}}{v_0}$ $v_0 = \boxed{\dfrac{\sqrt{2gh}}{\tan\theta}}$
CHECK and THINK We've derived the desired expression. You might convince yourself that if you shot an arrow with a larger speed, it would strike the ground at a lower angle relative to the horizontal (a really low speed would dive into the ground faster). Or, that to achieve the same landing angle from a larger height, you would need to shoot the arrow with a higher velocity (since otherwise it would dive down to a steeper angle from a larger height). These relationships are embedded in this algebraic equation.	

64. David Beckham has lined up for one of his famous free kicks from a point 25.0 m from the goal. He kicks the soccer ball with a speed of 22.0 m/s at 28.0° to the horizontal. The height of the goal is 2.44 m.

a. (N) What is the distance from the crossbar with which the ball will go into the goal or sail over?

INTERPRET and ANTICIPATE This is a projectile motion problem in which the ball accelerations at constant rate downward due to gravity and there is no acceleration in the horizontal direction. We need to determine when the ball reaches the goal (horizontal component) and then determine how high the ball is at this point in time (vertical component).	
SOLVE Since the ball is moving with constant velocity in the x direction, we can find the time needed for the ball to travel 25 m to the goal. The x component of the velocity is found using	$x = x_i + v_i \cos\theta_i t$ $t = \dfrac{x - x_i}{v_i \cos\theta_i} = \dfrac{25.0\text{ m}}{(22.0\text{ m/s})\cos 28.0^\circ} = 1.29\text{ s}$

trigonometry.	
Using this time, we can now determine the height of the ball when it crosses the goal line using kinematic equations (Eq. 2.11). The initial vertical velocity is found using kinematics, the initial position is 0 m (height of the ground), and the acceleration is –g (downward acceleration due to gravity).	$y = y_0 + v_{yi}t + \frac{1}{2}at^2$ $y = y_0 + v_i \sin\theta_i t - \frac{1}{2}gt^2$ $y = 0 + (22.0\text{ m/s})\sin 28.0^\circ(1.29\text{ s}) - \frac{1}{2}(9.81\text{ m/s}^2)(1.29\text{ s})^2$ $y = 5.17\text{ m}$
This height is above the 2.44 m height of the crossbar, so the ball sails over the crossbar by 2.73 m.	distance over crossbar = 5.17 – 2.44 m = $\boxed{2.73\text{ m}}$

CHECK and THINK
Well, David Beckham doesn't *always* score. This height seems believable for a distance over the crossbar though.

b. (N) Does the soccer ball reach the goal on its way up or on its way down?

INTERPRET and ANTICIPATE
There are a few ways to get this answer. For instance, we could figure out when the ball reaches the highest point. Easier though is just to see whether the vertical component of the velocity as it reaches the goal is positive (it's still going up) or negative (it's on its way down).

SOLVE	
We compute the *y* component of the ball's velocity at the crossbar. The *y* component of the velocity is negative, therefore the ball is on its way down.	$v_{yf} = v_{yi} - gt = 22.0\sin 28.0^\circ - (9.81\text{ m/s}^2)(1.29\text{ s})$ $v_{yf} = -2.33\text{ m/s}$

CHECK and THINK
By using the vertical velocity at the goal, we determined that the ball was on its way back down towards the ground.

76. A riverboat with a speed in still water of 20 knots (10.3 m/s) travels on a river that has a constant speed of 0.650 m/s.

a. (N) What is the time interval required for the riverboat to travel a distance of 4.00 km upstream and return to its starting point?

INTERPRET and ANTICIPATE	
The riverboat is moving faster relative to the land when it is going downstream and slower when it is moving upstream. If we can determine these speeds, we can calculate the time to travel the desired distance in each case and add them together to find the time for the entire trip.	
SOLVE The riverboat's speed in still water means "riverboat speed relative to the water." The river speed means "water speeds relative to land." With these, we can then determine the "riverboat speed relative to land" when its traveling downstream or upstream.	$(\vec{v}_R)_L = (\vec{v}_R)_W + (\vec{v}_W)_L$ $(\vec{v}_R)_L = 10.3\text{ m/s} + 0.650\text{ m/s}$ (downstream) $(\vec{v}_R)_L = 10.3\text{ m/s} - 0.650\text{ m/s}$ (upstream)
Knowing the speeds, we can calculate the time needed to travel a distance of 4 km.	$v = \frac{d}{t} \rightarrow t = \frac{d}{v}$ $t_{down} = \frac{4000\text{ m}}{(10.3+0.65)\text{ m/s}} = 365.3\text{ s}$ $t_{up} = \frac{4000\text{ m}}{(10.3-0.65)\text{ m/s}} = 414.5\text{ s}$
The time for the entire trip is the sum of the times for both segments of the trip.	$t_{total} = t_{up} + t_{down} = 414.5\text{ s} + 365.3\text{ s} = \boxed{780\text{ s}}$
CHECK and THINK The time to travel upstream against the current is larger than the time to travel downstream with the current, which sounds right.	

b. (N) What would be the time interval required for the same trip in still water?

<table>
<tr><td colspan="2">INTERPRET and ANTICIPATE
In this case, the riverboat needs to traverse the entire 8 km (there and back) at its still-water speed.</td></tr>
<tr><td>SOLVE
The total time in still water is found as in part (a), but the entire 8 km trip is made at the speed of 10.3 m/s.</td><td>$t_{total} = \frac{d}{v} = \frac{8000 \text{ m}}{10.3 \text{ m/s}} = \boxed{777 \text{ s}}$</td></tr>
<tr><td colspan="2">CHECK and THINK
The time in this case is very similar, though slightly longer with the current.</td></tr>
</table>

c. (C) Why does the boat trip take longer when there is a river current?

Driving the boat with the current does not compensate for the tine lost in driving against the current. In an extreme example, imagine if the river speed was the same as the maximum boat speed, 10.3 m/s. The downstream time would be cut in half, but the riverboat would move at a net speed of 0 m/s while heading upstream, so it would take an infinite amount of time to complete the trip!

79. (N) A circus cat has been trained to leap off a 12-m-high platform and land on a pillow. The cat leaps off at $v_0 = 3.5$ m/s and an angle $\theta = 25°$ (Fig P4.79).

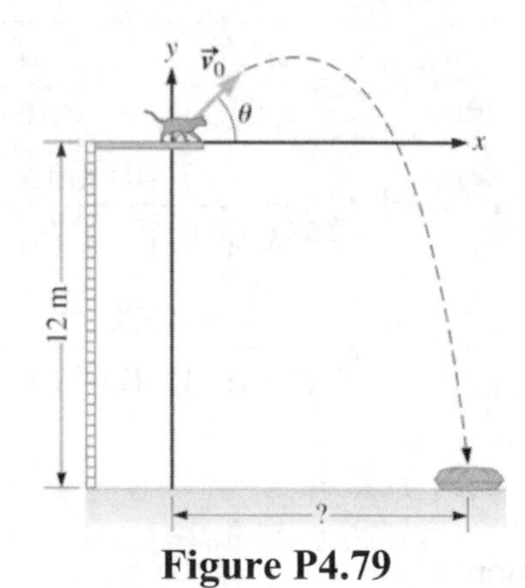

Figure P4.79

a. (N) Where should the trainer place the pillow so that the cat lands safely?

<table>
<tr><td colspan="2">INTERPRET and ANTICIPATE
This is a 2D projectile motion problem, which we can solve using kinematic equations. With the initial velocity and position specified, we can calculate where the cat lands. The vertical component will allow us to determine how long the cat is in the air (that is, the time of flight before the cat falls to the ground). The horizontal velocity remains constant and can then be used to determine how far the cat moves horizontally in this time.</td></tr>
<tr><td>SOLVE
We use the y component to determine the time of flight. The cat falls 12 m, accelerating due to gravity, with an initial y velocity that we can calculate using trigonometry. We choose the cat's initial position to be 0 m with the positive direction pointing upwards.</td><td>$v_{y,0} = v_0 \sin\theta = (3.5 \text{ m/s})(\sin 25°) = 1.48 \text{ m/s}$
$g = -9.81 \text{ m/s}^2$
$y_i = 0 \text{ m}$
$y_f = -12 \text{ m}$</td></tr>
<tr><td>We can now use kinematic equations (2.11) to calculate the time the cat is in the air. A quadratic equation is found, which we solve for time. There are two solutions mathematically, but only the positive time (1.72 s) makes sense physically since the cat must land after it takes off.</td><td>$y = v_{y,0}t - \frac{1}{2}gt^2$
$t = \dfrac{-v_0 \sin\theta \pm \sqrt{(v_0 \sin\theta)^2 - 2gy}}{-g}$
$t = \dfrac{-1.48 \text{ m/s} \pm \sqrt{(1.48 \text{ m/s})^2 - 2(9.81 \text{ m/s}^2)(-12 \text{ m})}}{-9.81 \text{ m/s}^2}$
$t = -1.42 \text{ or } 1.72 \text{ s}$</td></tr>
<tr><td>We then use this time with the constant x velocity to determine the total horizontal distance traveled by the cat.</td><td>$x = v_0 t \cos\theta = v_x t$
$x = (3.17 \text{ m/s})(1.72 \text{ s})$
$x = \boxed{5.5 \text{ m}}$</td></tr>
<tr><td colspan="2">CHECK and THINK
The result sound plausible for the distance that a cat could jump from a pretty high platform.</td></tr>
</table>

b. (N) What is the cat's velocity as she lands in the pillow?

INTERPRET and ANTICIPATE We now know a lot of the kinematic variables, such as the initial position and velocity, the time of flight, the acceleration due to gravity, etc. Since there is only an acceleration in the vertical direction, we know the horizontal velocity remains constant and we can calculate the final vertical velocity. Once we have both components, we can write the total velocity, which should be downward and to the right.	
SOLVE We know the constant x velocity v_x from part (a).	$v_x = 3.2$ m/s
We use the time from part (a) to determine the y velocity when the cat reaches the pillow.	$\vec{v} = v_0 \sin\theta - gt$ $\vec{v} = 1.48 \text{ m/s} - (9.81 \text{ m/s}^2)(1.72 \text{ s})$ $\vec{v} = -1.5 \times 10^1 \text{ m/s} = -15 \text{ m/s}$
With the two components of the velocity, we can easily write the velocity as a vector.	$\vec{v} = \boxed{(3.2\,\hat{\imath} - 15\hat{\jmath}) \text{ m/s}}$
CHECK and THINK The velocity is indeed to the right (positive x direction) and downward (negative y direction) as we expected.	

5

Newton's Laws of Motion

9. (N) Two forces act on an object with $\vec{F}_1 = (7.263\hat{\imath} + 8.889\hat{\jmath})$N and $\vec{F}_2 = (-13.452\hat{\imath} + 7.991\hat{\jmath})$N . What is the magnitude of the net force experienced by the object?

INTERPRET and ANTICIPATE The net force can be found by determining the vector sum of the two forces. We can then find the magnitude from the components of the net force. We expect a positive answer since magnitudes are positive by definition.	
SOLVE Equation 3.17 is used for the vector sum of the two forces.	$\vec{F}_{\text{tot}} = \vec{F}_1 + \vec{F}_2$ $\vec{F}_{\text{tot}} = (7.263\hat{\imath} + 8.889\hat{\jmath})\text{N} + (-13.452\hat{\imath} + 7.991\hat{\jmath})\text{N}$ $\vec{F}_{\text{tot}} = (-6.189\hat{\imath} + 16.880\hat{\jmath})\text{N}$
Now, use Equation 3.12 to find the magnitude of the net force.	$F_{\text{tot}} = \sqrt{(-6.189\text{ N})^2 + (16.880\text{ N})^2}$ $F_{\text{tot}} = \boxed{17.98\text{ N}}$
CHECK and THINK We have found the magnitude of the net force, which is the total force acting on the object. By looking at the components of the net force vector, we can also determine the direction. In this case, the net force is directed into the second quadrant, since the x component is negative and the y component is positive.	

11. (N) Three forces act on an object with $\vec{F}_1 = (6.03\hat{\imath} - 10.64\hat{\jmath})$N and $\vec{F}_2 = (-3.71\hat{\imath} - 12.93\hat{\jmath})$N . If the net force on the object is zero, what is the unknown force, $\vec{F}_3$?

INTERPRET and ANTICIPATE

The third vector can be found by recognizing that the vector sum must equal zero. Thus, the sum of both the x and y components must also be zero. We can sketch the problem graphically to get an idea of what the third force must be. To add vectors graphically, draw the first vector and then draw the next vector with its tail starting at the head of the first vector. We sketch $\vec{F}_1$, which points down and to the right. We then draw vector $\vec{F}_2$, which extends down and left from here. We now want the third vector, when added, to bring the resultant back to the origin, which is equivalent to saying that all three vectors add up to zero. The vector $\vec{F}_3$ indicated does this, so we expect that the resultant vector will point upwards and a little bit left. That is, it will have a negative x component and a larger, positive y component. Let's actually calculate what this vector is though.

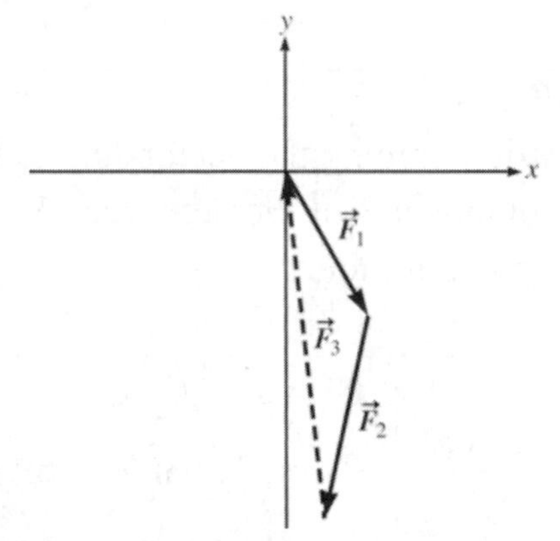

Figure P5.11ANS

SOLVE This is a vector addition problem that should be familiar from Chapter 3 (e.g. Equation 3.17), where the resultant vector is equal to 0.	$\vec{F}_1+\vec{F}_2+\vec{F}_3=0$ $(6.03\hat{i}-10.64\hat{j})\text{N}+(-3.71\hat{i}-12.93\hat{j})\text{N}+\vec{F}_3=0$
We can now rearrange the equation to solve for the unknown force.	$(6.03\hat{i}-10.64\hat{j})\text{N}+(-3.71\hat{i}-12.93\hat{j})\text{N}+\vec{F}_3=0$ $\vec{F}_3=-(6.03\hat{i}-10.64\hat{j})\text{N}-(-3.71\hat{i}-12.93\hat{j})\text{N}$ $\vec{F}_3=\left[(-6.03+3.71)\hat{i}+(10.64+12.93)\hat{j}\right]\text{N}$ $\vec{F}_3=\boxed{(-2.32\hat{i}+23.57\hat{j})\text{N}}$

CHECK and THINK

As expected, the third vector has a negative x component and a larger, positive y component. When the object experiences a net force of 0, we say that the object is in *equilibrium*. Each of these forces points in a different direction with a different

magnitude, but the net result is as if there was no force at all. When several forces are exerted on an object, what we observe is the net force, or the net effect of those forces.

22. A particle with mass $m = 4.00$ kg accelerates according to
$\vec{a} = \left(-3.00\hat{\imath} + 2.00\hat{\jmath}\right)$ m/s^2.

a. (N) What is the net force acting on the particle?

INTERPRET and ANTICIPATE The net force on the particle can be found using Newton's second law ($F = ma$) given the mass and acceleration of the object. We expect the net force to point in the same direction as the acceleration.	
SOLVE Newton's second law (Equation 5.1) relates the total, net force on an object to the acceleration of the object. Given the mass and acceleration, the total force is found directly using $F = ma$.	$\sum \vec{F} = \vec{F}_{\text{net}} = m\vec{a}$ $\vec{F}_{\text{net}} = (3.50 \text{ kg})\left[\left(-3.00\hat{\imath} + 2.00\hat{\jmath}\right) \text{ m/s}^2\right]$ $\vec{F}_{\text{net}} = \boxed{\left(-10.5\,\hat{\imath} + 7.00\,\hat{\jmath}\right) \text{ N}}$
CHECK and THINK As expected, the net force is up and to the left (positive y component and negative x component), the same as the acceleration.	

b. (N) What is the magnitude of this force?

INTERPRET and ANTICIPATE The magnitude of any vector must be a positive number that is at least as large as the magnitude of each component.	
SOLVE The magnitude of a vector can be calculated from the Pythagorean theorem, as the square root of the sum of the components squared.	$F_{\text{net}} = \sqrt{(-10.5)^2 + (7.00)^2}$ $F_{\text{net}} = \boxed{12.6 \text{ N}}$
CHECK and THINK The magnitude of this force is positive and has units of Newtons.	

23. (N) The x and y coordinates of a 4.00-kg particle moving in the xy plane under the influence of a net force F are given by $x = t^4 - 6t$ and $y = 4t^2 + 1$, with x and y in meters and t in seconds. What is the magnitude of the force F at $t = 4.00$ s?

INTERPRET and ANTICIPATE

We are given the x and y positions of the particle and are asked for the force. A net force on a particle is proportional to the acceleration of the particle according to Newton's second law ($F = ma$). It looks like we need to determine the acceleration from the position information and then find the net force on the particle.

SOLVE We are given equations that describe the x and y positions of the particle as functions of time. From Chapter 2, the acceleration of any object can be found as the time derivative of the object's velocity, and the velocity can be found as the time derivative of the position (Equations 2.4 and 2.7). We use these relationships to first find the velocity and then the acceleration for both the x and y components.	$v_x = \frac{dx}{dt} = \frac{d}{dt}\left(t^4 - 6t\right) = 4t^3 - 6$ $a_x = \frac{dv_x}{dt} = \frac{d}{dt}\left(4t^3 - 6\right) = 12t^2$ $v_y = \frac{dy}{dt} = \frac{d}{dt}\left(4t^2 + 1\right) = 8t$ $a_y = \frac{dv_y}{dt} = \frac{d}{dt}(8t) = 8\ \text{m/s}^2$
All of our variables are in metric base units. Given the mass in kg and the time in s we can use Newton's second law to determine the force in N. The components of the force on the particle can now be written assuming t is in s. The y component is constant while the x component increases in time.	$F_x = (4.00\ \text{kg})\left(12.0t^2\right) = 48.0t^2\ \text{N}$ $F_y = (4.00\ \text{kg})\left(8\ \text{m/s}^2\right) = 32.0\ \text{N}$
Now, the magnitude of the force can be found. We substitute $t = 4.00$ s to find the magnitude at this particular time.	$F_{\text{net}} = \sqrt{\left[(48.0)(4.00)^2\right]^2 + (32.0)^2} = \boxed{769\ \text{N}}$

CHECK and THINK

We have calculated a force in newtons. Conceptually, we used the position equations to find the velocity and then the acceleration and then $F = ma$ to determine the force on the particle.

27. A particle of mass m_1 accelerates at 4.25 m/s^2 when a force F is applied. A second particle of mass m_2 experiences an acceleration of only 1.25 m/s^2 under the influence of this same force F.

a. (N) What is the ratio of m_1 to m_2?

INTERPRET and ANTICIPATE

Given the same net force, a larger mass will experience a smaller acceleration. Since the second particle has a smaller acceleration, it must have a larger mass.

SOLVE

For the same force F acting on two different masses, using Newton's second law (Equation 5.1), it must be true that both $F = m_1 a_1$ and $F = m_2 a_2$. The ratio then can be found by equating the two expressions and solving.

$$F = m_1 a_1 = m_2 a_2$$

$$\frac{m_1}{m_2} = \frac{a_2}{a_1} = \frac{1.25 \text{ m/s}^2}{4.25 \text{ m/s}^2} = \boxed{0.294}$$

CHECK and THINK

Since $m_1/m_2 < 1$, the second particle has a mass that's larger than the first, which is what we predicted. In fact, since the variables in Newton's second law are linear, the mass and acceleration are inversely proportional to each other given the same force. Since m_1 has an acceleration about three times more than m_2, the mass must have been about one third that of m_2.

b. (N) If the two particles are combined into one particle with mass $m_1 + m_2$, what is the acceleration of this particle under the influence of the force F?

INTERPRET and ANTICIPATE

We can use the same approach as in part (a) for a total mass given by the sum of the two masses. The combined mass must be larger than each of the individual mass, so the acceleration must be lower than the acceleration of either of the individual masses if the same force is exerted.

SOLVE We use Newton's second law to relate the force to the acceleration in the case that the mass is the total mass of particles 1 and 2.	$F=(m_1+m_2)a$
From part (a), we can relate the masses of the two particles.	$\frac{m_1}{m_2}=0.294 \quad \rightarrow \quad m_1=0.294m_2$
Equating the force for the combined particle to either of the other expressions for the force and using the ratio of the masses allows for a solution for *a*.	$F=(m_1+m_2)a=(0.294m_2+m_2)a=1.294m_2a$ $1.294m_2a=m_2a_2$ $1.294a=a_2$ $a=\frac{a_2}{1.294}=\frac{1.25\ \text{m/s}^2}{1.294}=\boxed{0.966\ \text{m/s}^2}$

CHECK and THINK
As we predicted, the acceleration for the combined mass due to the applied force is lower than the acceleration if each of the smaller masses were accelerated using the same net force.

28. (N) Jim Bob needs to tow his truck home using his neighbor's SUV. The SUV pulls the truck by exerting a horizontal force of 3133 N. If this force alone causes the truck to accelerate at a rate of 2.91 m/s^2, what is the mass of Jim Bob's truck?

INTERPRET and ANTICIPATE
Being told both a force and the acceleration due to that force, we can use Newton's second law to find the mass of the truck. The acceleration will point in the same direction as the force, so we may simply compare the magnitudes. A pickup truck weighs about 2 or 3 tons. So, we expect the mass to be maybe a couple thousand kilograms.

SOLVE Newton's second law allows us to compare the force and accelerations to solve for the mass.	$F_{\text{tot}}=ma$ $m=\frac{F_{\text{tot}}}{a}=\frac{3133\ \text{N}}{2.91\ \text{m/s}^2}$ $m=\boxed{1.08\times10^3\ \text{kg}}$

CHECK and THINK
The SUV's mass is the right order of magnitude. If the force increased, the acceleration would increase. The mass is constant, as it represents the amount of matter making up the truck.

30. (N) Three forces $\vec{F}_1 = (62.98\hat{i} - 15.80\hat{j})\text{N}$, $\vec{F}_2 = (23.66\hat{i} - 78.05\hat{j})\text{N}$, and $\vec{F}_3 = (-86.64\hat{i} + 233.4\hat{j})\text{N}$ are exerted on a particle. The particle's mass is 14.23 kg. Find the particle's acceleration.

INTERPRET and ANTICIPATE

Once we find an expression for the total force, we use Newton's second law to find the particle's acceleration. Our result is a vector with dimensions of acceleration (length per time squared). It is easiest to give our answer in component form, and so that is what we do here.

SOLVE Find the total force by using vector addition.	$\vec{F}_{tot} = \vec{F}_1 + \vec{F}_2 + \vec{F}_3$ $\vec{F}_{tot} = (62.98\hat{i} - 15.80\hat{j})\text{ N} + (23.66\hat{i} - 78.05\hat{j})\text{ N} + (-86.64\hat{i} + 233.4\hat{j})\text{ N}$ $\vec{F}_{tot} = (62.98 + 23.66 - 86.64)\hat{i}\text{ N} + (-15.80 - 78.05 + 233.4)\hat{j}\text{ N}$ $\vec{F}_{tot} = (0\hat{i} + 139.55\hat{j})\text{ N}$ $\vec{F}_{tot} = 139.55\hat{j}\text{ N}$
Now, solve Newton's second law for acceleration.	$\vec{F}_{tot} = m\vec{a}$ $\vec{a} = \frac{1}{m}\vec{F}_{tot}$ $\vec{a} = \frac{1}{14.23\text{ kg}}(139.55\hat{j})\text{ N}$ $\vec{a} = \boxed{9.807\hat{j}\text{ m/s}^2}$

CHECK and THINK

Since the total force in the x direction is zero, there is no acceleration in that direction. If you did the previous problem, it is helpful to compare the results. In the previous problem, the total force in the y direction was smaller and in the negative direction. The particle's mass is the same in both problems. So in this problem, it makes sense that acceleration in the y direction here is greater and in the positive direction.

32. (N) If the vector components of the position of a particle moving in the xy plane as a function of time are $\vec{x}(t) = (2.5\text{ m/s}^2)t^2\hat{i}$ and $\vec{y}(t) = (5.0\text{ m/s}^3)t^3\hat{j}$, when is the angle between the net force on the particle and the x axis equal to 45°?

INTERPRET and ANTICIPATE The scalar components of the acceleration can be found by taking the second derivative of the scalar components of position with respect to time. The components of the net force can be found by multiplying the components of the acceleration by the object's mass, which isn't given. These components can be related to the angle by the tangent function.	
SOLVE The x component of the acceleration is found by taking the second derivative of the x component of the position with respect to time. Equivalently, you can take the derivative of the position with respect to time to find an expression for the velocity and then take the derivative of the velocity with respect to time to find the acceleration. (Here, we are using the definitions of instantaneous velocity and acceleration that were introduced in Chapter 2 – Equations 2.4 and 2.7.)	$a_x = \frac{d^2x}{dt^2} = \frac{d^2}{d^2t}\left[\left(2.5\text{ m/s}^2\right)t^2\right] = 5.0\text{ m/s}^2$ or $v_x = \frac{dx}{dt} = \frac{d}{dt}\left[\left(2.5\text{ m/s}^2\right)t^2\right] = 5.0\,t$ $a_x = \frac{dv}{dt} = \frac{d}{dt}\left[5.0\,t\right] = 5.0\text{ m/s}^2$
Similarly, the y component of the acceleration is found by taking the second derivative of the y component of the position with respect to time.	$a_y = \frac{d^2y}{dt^2} = \frac{d^2}{dt^2}\left[\left(5.0\text{ m/s}^3\right)t^3\right] = \left(30\text{ m/s}^3\right)t$
Using Newton's second law, the ratio of the components of the sum of the forces is equal to the ratio of the components of the acceleration because the mass cancels.	$\frac{\Sigma F_y}{\Sigma F_x} = \frac{ma_y}{ma_x} = \frac{a_y}{a_x}$
Since the angle required is 45°, the ratio of the components of the acceleration must be equal to one. (That is, if the x component of a vector is the same as the y	$\frac{a_y}{a_x} = \tan\theta = \tan(45°) = 1$ $a_y = a_x$ $\left(30\text{ m/s}^3\right)t = 5.0\text{ m/s}^2$

component, the vector is pointing at an angle of 45°.) Therefore, the components of the acceleration must be equal.	
We now solve this expression for time.	$t = \boxed{0.17\text{ s}}$

CHECK and THINK
The mass of the object wasn't given, but the answer can still be found because the mass cancels when the components of the net force are divided. Note that the angle is 45° only at this particular time. At other times, the angle is different. For example, you might be able to convince yourself that at $t = 0$ s the angle is 0° and that the angle of the acceleration vector is increasing in time.

39. (N) A student takes the elevator up to the fourth floor to see her favorite physics instructor. She stands on the floor of the elevator, which is horizontal. Both the student and the elevator are solid objects, and they both accelerate upward at 5.19 m/s^2. This acceleration only occurs briefly at the beginning of the ride up. Her mass is 80.0 kg. What is the normal force exerted by the floor of the elevator on the student during her brief acceleration?

INTERPRET and ANTICIPATE
We expect the magnitude of the normal force to be greater than the student's weight because the floor of the elevator must counteract the gravity of the Earth *and* accelerate the student upward.

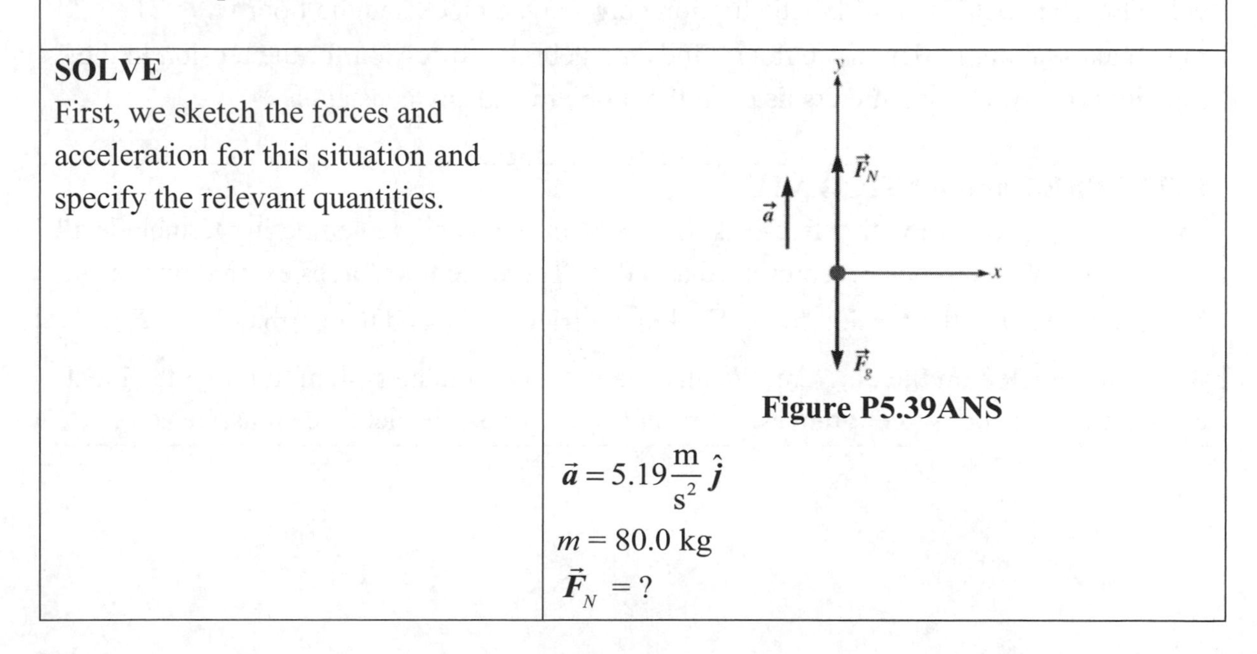

SOLVE First, we sketch the forces and acceleration for this situation and specify the relevant quantities.	**Figure P5.39ANS** $\vec{a} = 5.19\dfrac{\text{m}}{\text{s}^2}\hat{j}$ $m = 80.0$ kg $\vec{F}_N = ?$

Starting with Newton's second law, which relates the *net* force with the acceleration of an object, there are two forces acting on the student: the force of gravity and the normal force.	$\Sigma\vec{F} = m\vec{a}$ $\vec{F}_g + \vec{F}_N = m\vec{a}$
We can now solve for the normal force. Weight is equal to mass multiplied by the acceleration due to gravity (Equation 5.7).	$\vec{F}_N = m\vec{a} - \vec{F}_g$ $\vec{F}_N = m\vec{a} - m\vec{g} = m(\vec{a} - \vec{g})$
We now insert numerical values.	$\vec{F}_N = (80\text{ kg})\left(5.19\hat{j}\ \frac{\text{m}}{\text{s}^2} - \left(-9.81\hat{j}\ \frac{\text{m}}{\text{s}^2}\right)\right)$ $\vec{F}_N = \boxed{1.20 \times 10^3\,\hat{j}\text{ N}}$

CHECK and THINK

The magnitude of the normal force is greater than the weight of the student, which is 785 N, so the answer makes sense. If we had simply multiplied the student's mass by the acceleration to find the normal force, then the answer would've been far too small. This demonstrates the importance of including *all* of the forces acting on an object when using Newton's second law, which relates the *net* force to the acceleration.

43. (A) A woman uses a rope to pull a block of mass m across a level floor at a constant velocity. The coefficient of kinetic friction between the block and the floor is μ_k. The rope makes an angle θ with the floor. Find an algebraic expression for the tension in the rope in terms of the parameters listed in the problem and any constants.

INTERPRET and ANTICIPATE

Visualize the problem with a free-body diagram for the block, remembering to include all four elements. The block is represented as a dot. There are four forces exerted on the block: gravity $\vec{F}_g$, the tension force $\vec{F}_T$, kinetic friction $\vec{F}_k$ and the normal force $\vec{F}_N$, which are labeled on the diagram. We also choose a coordinate system. Finally, the block does not accelerate, since it moves at constant velocity, so the net force must be zero.

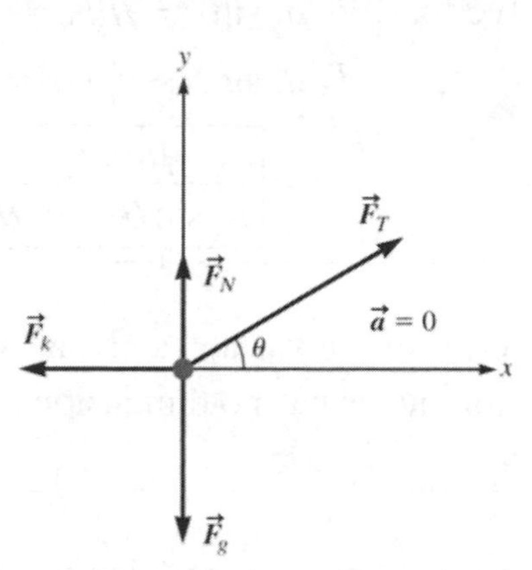

Figure P5.43ANS

SOLVE First, we apply Newton's second law in component form, Equation 5.2. There is no acceleration, since the block moves at constant velocity.	$\sum F_x = F_T \cos\theta - F_k = ma_x = 0$ (1) $\sum F_y = F_T \sin\theta + F_N - F_g = ma_y = 0$ (2)
We now write other relevant force equations. The magnitude of the gravitational force on the block equals its weight (Equation 5.7). The kinetic friction is proportional to the normal force (Equation 5.9).	$F_g = mg$ (3) $F_k = \mu_k F_N$ (4)
Now, we have some algebra to do. Eliminate kinetic friction from equation (1) with equation (4).	$F_T \cos\theta - F_k = 0$ $F_T \cos\theta - \mu_k F_N = 0$ $F_T \cos\theta = \mu_k F_N$ $F_N = \dfrac{F_T \cos\theta}{\mu_k}$ (5)
Eliminate weight (3) and normal force (5) from equation (2).	$F_T \sin\theta = F_g - F_N$ $F_T \sin\theta = mg - \dfrac{F_T \cos\theta}{\mu_k}$

Combine similar terms and solve for the tension.	$F_T\mu_k \sin\theta = \mu_k mg - F_T \cos\theta$ $F_T\mu_k \sin\theta + F_T \cos\theta = \mu_k mg$ $F_T = \boxed{\dfrac{\mu_k mg}{\mu_k \sin\theta + \cos\theta}}$

CHECK and THINK

The result is an expression in terms of the variables given. As a check, the dimension of the numerator is force. The denominator has no dimensions. So the dimension on both sides of the equation is force.

44. (N) A student working on a school project modeled a trampoline as a spring obeying Hooke's law and measured the spring constant of a certain trampoline as 4617 N/m. If a child of mass 27.0 kg compresses the trampoline vertically by a maximum of 0.25 m, while bouncing up and down, what is the child's acceleration at the moment of maximum compression?

INTERPRET and ANTICIPATE

Visualize the problem with a free-body diagram for the block. The child is represented as a dot. There are two forces exerted on the child: gravity $\vec{F}_g$ and the spring force $\vec{F}_H$ and the child accelerates upward. We need to find the net force in order to determine the acceleration with Newton's second law.

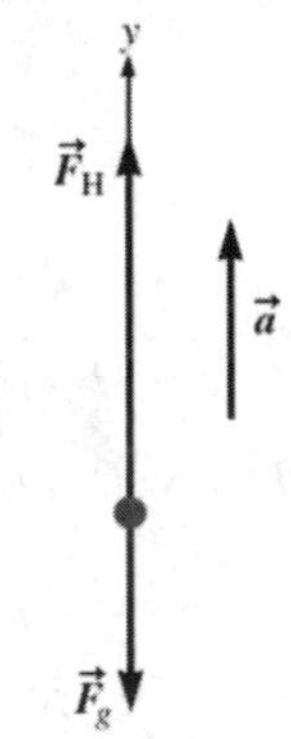

Figure P5.44ANS

SOLVE Using the free-body diagram, we can write an expression for the net force upwards and relate that to the acceleration upwards using Newton's second law.	$\sum F_y = F_H - F_g = ma_y = ma$

The magnitude of the gravitational force is the weight of the child (Eq. 5.7). According to Hooke's law the force exerted by a trampoline modeled as a spring is proportional to the amount it is compressed Δy (Eq. 5.8).	$F_g = mg$ $F_H = k\Delta y$
We can now substitute these into the expression for the net force and solve for acceleration.	$F_H - F_g = ma$ $k\Delta y - mg = ma$ $a = \dfrac{k\Delta y - mg}{m}$
Finally, we substitute values and write the final result as a vector.	$a = \dfrac{(4617\text{ N/m})(0.25\text{ m}) - (27.0\text{ kg})(9.81\text{ m/s}^2)}{(27.0\text{ kg})}$ $a = 32.9\text{ m/s}^2$ $\vec{a} = \boxed{32.9\hat{j}\text{ m/s}^2}$

CHECK and THINK

The value calculated has the correct units for acceleration, m/s^2 and the acceleration is about $3g$. However, while the acceleration might seem big, remember that it would only be this large for a brief moment when the spring is at its maximum compression. The acceleration will immediately begin to decrease as the spring relaxes.

46. A heavy crate of mass 50.0 kg is pulled at constant speed by a dockworker who pulls with a 345 N force at an angle θ with the horizontal (Fig. P5.46). The magnitude of the friction force between the crate and the pavement is 212 N.

Figure P5.46

a. (G) Draw a free-body diagram of the forces acting on the crate.

The free-body diagram includes all forces acting on the crate as shown. The dockworker exerts a pulling force with the frictional force opposing the motion of the crate. In addition, there is a gravitational force (weight of the crate) and the normal force of the ground pushing up on the crate.

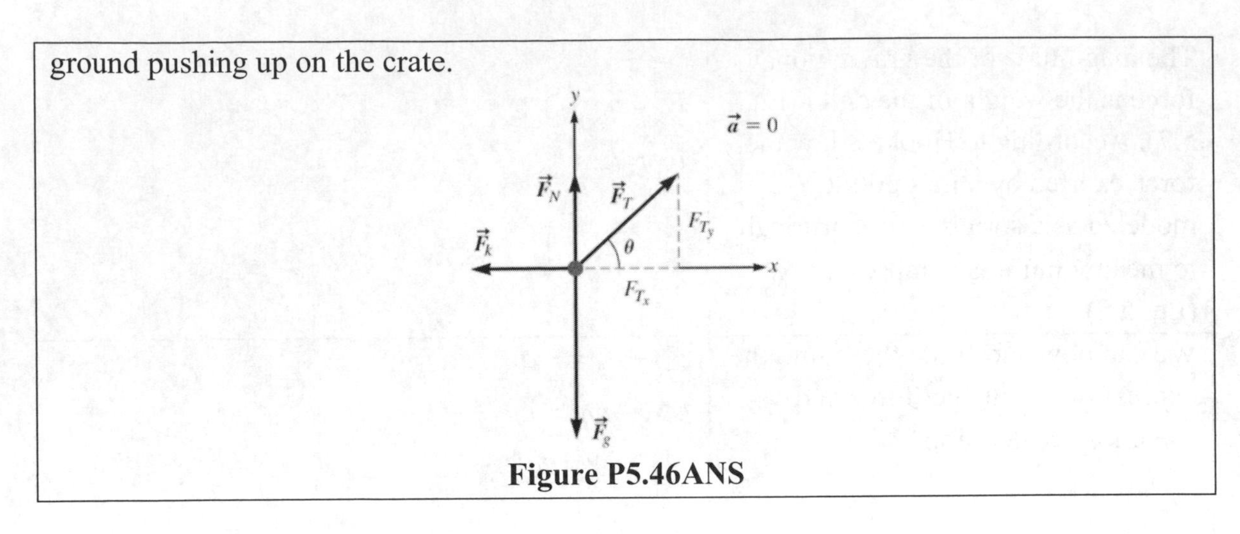

Figure P5.46ANS

b. (N) What is the angle θ of the rope with the horizontal?

INTERPRET and ANTICIPATE

Since the crate is moving at constant speed, it is not accelerating. If it is not accelerating, then according to Newton's second law there is no net (total) force acting on it. We can use this fact that the total force on the crate is zero to relate the forces we drew in the free-body diagram.

SOLVE We know that the crate is not accelerating, so the total force on the crate is zero. In order for the total force to be zero, the x and y components must also add up to zero, so we write out the sum of the forces for each component.	$\sum F_x = F_T \cos\theta - F_k = 0$ $\sum F_y = F_N + F_T \sin\theta - mg = 0$
To find the angle of the rope, it looks like we might be able to use either component. We aren't given the normal force, but we know the friction force and the pulling force, so we can use the x component to solve for the angle θ.	$F_T \cos\theta = F_k$ $\cos\theta = \frac{F_k}{F_T}$ $\theta = \cos^{-1}\frac{F_k}{F_T} = \cos^{-1}\frac{212\text{ N}}{345\text{ N}} = \boxed{52.1°}$

CHECK and THINK

We were able to use trigonometry and the fact that the total force is zero to determine the unknown angle. An angle of 52 degrees seems possible for this situation.

c. (N) What is the magnitude of the normal force exerted by the pavement on the crate?

INTERPRET and ANTICIPATE

From part (b), we have an expression for the y component of the force, which includes the normal force. We can use this expression to solve for this force.

SOLVE	
We start with the equation for the y component of the force from part (b). We can solve this for the normal force and insert values to calculate the force in Newtons.	$F_N + F_T \sin\theta - mg = 0$ $F_N = mg - F_T \sin\theta$ $F_N = (50.0\ \text{kg})(9.81\ \text{m/s}^2) - (345\ \text{N})\sin(52.1°)$ $F_N = \boxed{218\ \text{N}}$

CHECK and THINK

As in part (b), knowing that the net force is equal to zero allows us to relate the components and solve for the unknown normal force.

50. A block with mass m_1 hangs from a rope that is extended over an ideal pulley and attached to a second block with mass m_2 that sits on a ledge that is slanted at an angle of 20° (Fig. P5.49). Suppose the system of blocks is initially held motionless and, when released, begins to accelerate.

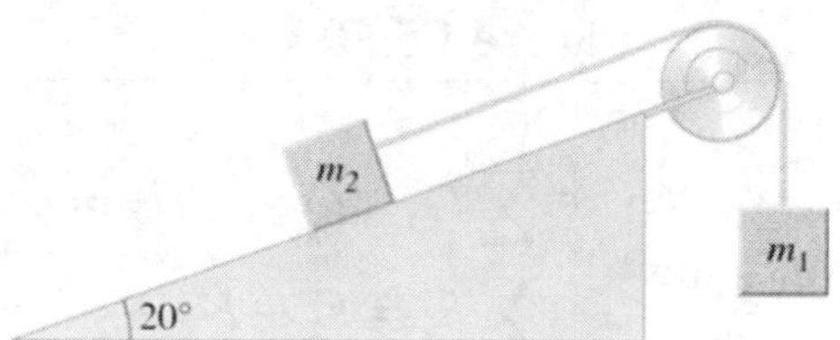

Figure P5.49

a. (N) If $m_1 = 7.00$ kg, $m_2 = 2.00$ kg, and the magnitude of the acceleration of the blocks is 0.134 m/s^2, find the magnitude of the kinetic friction force between the second block and the ledge.

INTERPRET and ANTICIPATE

We begin by drawing a free-body diagram for each mass in the picture. The first mass is expected to accelerate downward, so we point the positive y axis downward in its free-body diagram (either direction is fine, as long as we are consistent with signs throughout).

We choose to draw the coordinate system for m_2 in such a way that the x axis is parallel to the surface on which it rests. Because of the slanted ledge, the angle between the gravitational force and the y axis will be 20°. The forces acting on this mass are the tension of the rope, the force of gravity, and friction (pointing in the opposite direction of the motion, which we expect to be up the ramp).

We can find the net force on each block and relate it to its acceleration. We expect the kinetic friction force to be less than the tension, since the block is expected to slide up the ledge, while the first mass will fall vertically.

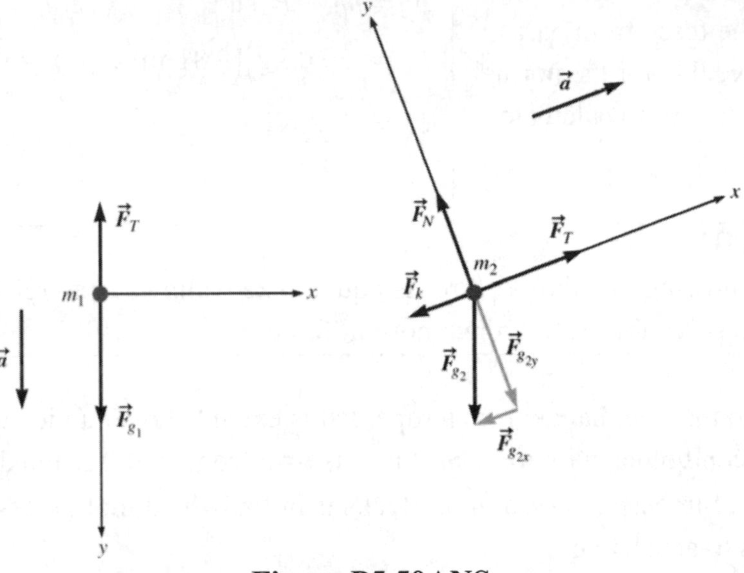

Figure P5.50ANS

SOLVE With the free-body diagram, we can write expressions for Newton's laws for each block along the expected direction of motion. The frictional force acts on m_2, so we might try to solve equation (2) for the frictional force, but we realize that we don't know the tension. We must first use Newton's second law on m_1 (equation 1), find the tension, and then use this information in the equation for m_2 (equation 2).	$\sum F_y = F_{g_1} - F_T = m_1 a \quad (1)$ $\sum F_x = F_T - F_{g_{2,x}} - F_k = m_2 a \quad (2)$
Since we will be using Newton's second law to relate the net force on each block to its acceleration, let's calculate the gravitational force (i.e. the weight) of each of the blocks.	$F_{g_1} = m_1 g = (7.00\text{ kg})(9.81\text{ m/s}^2) = 68.7\text{ N}$ $F_{g_2} = m_2 g = (2.00\text{ kg})(9.81\text{ m/s}^2) = 19.6\text{ N}$

Now, using equation (1), since we know the magnitude of the acceleration, we are able to find the tension.	$\sum F_y = F_{g_1} - F_T = m_1 a$ $68.7\text{ N} - F_T = (7.00\text{ kg})(0.134\text{ m/s}^2)$ $F_T = 68.7\text{ N} - (7.00\text{ kg})(0.134\text{ m/s}^2)$ $F_T = 68.7\text{ N} - 0.938\text{ N}$ $F_T = 68.7\text{ N} - 0.938\text{ N}$ $F_T = 67.8\text{ N}$
We also realize that to consider the x and y components for m_2, we need to find the components of the gravitational force on this mass.	$F_{g_{2,x}} = F_{g_2}\sin(20°) = (19.6\text{ N})\sin(20°) = 6.70\text{ N}$ $F_{g_{2,y}} = F_{g_2}\cos(20°) = (19.6\text{ N})\cos(20°) = 18.4\text{ N}$
We return our attention to the x component of the force for m_2 (equation 2). We can solve this for the magnitude of the kinetic friction force.	$\sum F_x = F_T - F_{g_{2,x}} - F_k = m_2 a$ $67.8\text{ N} - 6.70\text{ N} - F_k = (2.00\text{ kg})(0.134\text{ m/s}^2)$ $67.8\text{ N} - 6.70\text{ N} - F_k = 0.268\text{ N}$ $F_k = \boxed{60.8\text{ N}}$

CHECK and THINK

The kinetic friction force is less than the tension, as expected. If we make the incline angle smaller and smaller, the magnitude of the normal force will get bigger and bigger, and so will the magnitude of the kinetic friction force.

b. (N) What is the value of the coefficient of kinetic friction between the block and the ledge?

INTERPRET and ANTICIPATE

We know the magnitude of the kinetic friction force from part (a). If we knew the magnitude of the normal force on the object, we could use the formula for kinetic friction to get our answer. Coefficients of friction are positive, so we expect a positive answer. Because the kinetic friction force is much larger than the gravitational force on the object, we also expect the coefficient to likely be greater than 1.

SOLVE Apply Newton's second law to the y direction of the free-body diagram for m_2, using what we know from	

part (a). We can then find the magnitude of the kinetic friction force.	$\sum F_y = F_N - F_{g_{2,y}} = 0$ $F_N = F_{g_{2,y}}$ $F_N = 18.4\text{ N}$
The formula for the frictional force (Eq. 5.9) will allow us to determine the coefficient of kinetic friction, using our answer from part (a).	$F_k = \mu_k F_N$ $60.8\text{ N} = \mu_k (18.4\text{ N})$ $\mu_k = \dfrac{60.8\text{ N}}{18.4\text{ N}}$ $\mu_k = \boxed{3.30}$

CHECK and THINK

The coefficient of kinetic friction is positive and greater than 1, as expected. The fact that the coefficient is greater than one merely means that the friction force is greater in magnitude than the normal force. It indicates that the block will not slide easily on the ledge. Top fuel drag race car tires can achieve coefficients of kinetic friction near or greater than this value.

52. A runaway piano starts from rest and slides down a 20.0° frictionless incline 5.00 m in length.

a. (G) Draw a free-body diagram of the piano.

The two forces acting on the piano are the normal force, F_N, and the weight, *mg*. If the piano is considered to be a point mass and the *x* axis is chosen to be parallel to the plane, pointing down the incline, then the free-body diagram will be as shown in the figure. The angle θ is the angle of inclination of the plane.

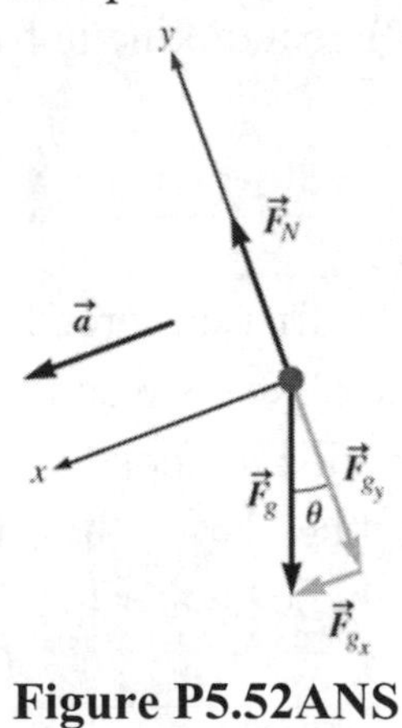

Figure P5.52ANS

b. (N) What is the acceleration of the piano?

INTERPRET and ANTICIPATE Using Newton's second law, we can determine the net force in the direction of motion and calculate the acceleration. Since the piano is sliding down the incline and is not in free fall, the acceleration should be less than 9.8 m/s^2.	
SOLVE Begin by applying Newton's second law to the piano using the free-body diagram. We choose a coordinate system such that the components of the weight are in the positive x direction and the negative y direction, as in our free-body diagram.	$\sum F_y = F_N - mg\cos\theta = 0$ $\sum F_x = mg\sin\theta = ma$
We now use the x component to solve for the acceleration.	$mg\sin\theta = ma$ $a = g\sin\theta$
Inserting numerical values, we determine the acceleration, which is positive, pointing down the plane.	$a = (9.81 \text{ m/s}^2)\sin 20° = \boxed{3.35 \text{ m/s}^2}$
CHECK and THINK As expected, the acceleration is positive (down the incline based on our chosen coordinate system) and less than if the object was in free fall.	

c. (N) What is the speed of the piano at the bottom of the incline?

INTERPRET and ANTICIPATE This is an example of linear motion with a constant acceleration and can be modeled using the kinematic equations to find the final speed if it started from rest.	
SOLVE The piano starts from rest and accelerates over a distance of 5 meters with the acceleration calculated in part (a).	$\Delta x = 5 \text{ m}$ $a = 3.35 \text{ m/s}^2$ $v_i = 0$ $v_f = ?$

We can use our kinematic equations from Chapter 2 to find a relationship among the variables above and solve for the final velocity (in this case, Equation 2.13 is one possibility).	$v_f^2 = v_i^2 + 2a\Delta x$ $v_f = \sqrt{2a\Delta x}$ $v_f = \sqrt{2(3.35\text{ m/s}^2)(5.00\text{ m})}$ $v_f = \boxed{5.79\text{ m/s}}$

CHECK and THINK

Using the acceleration in part (a) and finding an appropriate kinematic equation, we can solve for the final velocity. The speed (about 13 mph) sounds like a reasonable speed for the piano at the end of the ramp.

57. An astronaut is stationary while floating in space, a short distance from the safety of her spacecraft. The mass of the astronaut including all her gear is 106.4 kg. Her tether to the craft becomes disconnected, and she needs to get back before she runs out of air. She removes her backpack unit, which has a mass of 30.1 kg and, putting herself between the backpack and the craft, pushes the pack with a force of 212 N directly away from the craft.

a. (C) Explain how this action returns the astronaut to the craft.

Putting herself between the backpack and the spacecraft, when she pushes the backpack, the backpack exerts an equal and opposite force on her according to Newton's third law. This means that when she exerts a force that accelerates the backpack away from the craft, the backpack exerts an equal and opposite force on her that accelerates her towards the craft. This only happens while they are in contact and the astronaut is pushing on the pack. Once she releases the backpack and they are no longer in contact, the force is no longer exerted. According to Newton's first law, given no other significant forces acting on the astronaut, she will stop accelerating (not stop *moving*) and will continue moving at a constant velocity towards the craft.

b. (N) What is the magnitude of the acceleration experienced by the backpack while the force is applied?

INTERPRET and ANTICIPATE

Using the mass of the backpack and the force it experiences, we use Newton's second law to determine the acceleration.

SOLVE Newton's second law provides a relationship between the force and the acceleration. Writing this equation in a scalar form, we find our answer.	$F = ma = (30.1 \text{ kg})a$ $212 \text{ N} = (30.1 \text{ kg})a$ $a = \frac{212 \text{ N}}{30.1 \text{ kg}} = \boxed{7.04 \text{ m/s}^2}$
CHECK and THINK The backpack only experiences this acceleration for a short time, while the astronaut is pushing it. Once it is in motion though, we expect no other significant external forces to be acting on the pack, given the situation in the problem, and it will continue moving away from her and the craft.	

c. (N) What is the magnitude of the acceleration experienced by the astronaut while the force is applied?

INTERPRET and ANTICIPATE We can find the mass of the astronaut and her remaining gear, after she removes the pack. According to Newton's third law, the force due to the backpack on the astronaut must be equal in magnitude to the force of the astronaut on the backpack from part (b). Newton's second law allows us to determine the acceleration experienced by the astronaut. This should be less than the acceleration of the backpack because she is more massive and the forces are equal in magnitude.	
SOLVE First, we find the mass of the astronaut and her remaining gear after the backpack is removed.	$m_{\text{astro}} = 106.4 \text{ kg} - 30.1 \text{ kg}$ $m_{\text{astro}} = 76.3 \text{ kg}$
Using Newton's second law, we determine the magnitude of the acceleration.	$F = ma = (76.3 \text{ kg})a$ $212 \text{ N} = (76.3 \text{ kg})a$ $a = \frac{212 \text{ N}}{76.3 \text{ kg}} = \boxed{2.78 \text{ m/s}^2}$
CHECK and THINK The acceleration of the astronaut is indeed smaller than the acceleration of the less massive pack. The astronaut only experiences this force for a short time, but once she is in motion she will continue moving towards the craft. If the push takes half a second, she will be moving at over 1 m/s after losing contact with the backpack. Hopefully that is fast	

enough to get back before she completely runs out of air!

65. (N) A box with mass $m_1 = 6.00$ kg sliding on a rough table with coefficient of kinetic friction of 0.220 is connected by a massless cord strung over a massless, frictionless pulley to a second box of mass $m_2 = 12.0$ kg hanging from the side of the table (Fig. P5.51). What is the tension in the cord connecting the boxes?

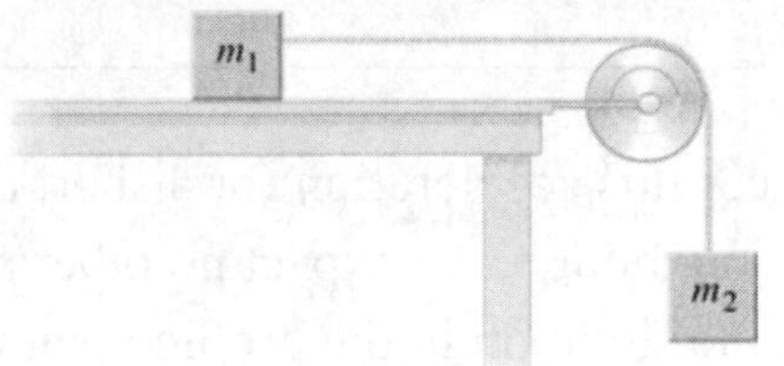

Figure P5.51

INTERPRET and ANTICIPATE

We begin by drawing a free-body diagram for both objects. The first mass is pulled by the cord (tension) with a frictional force opposing this motion. The normal force due to the table balances the weight (gravitational force) of the box. For the second mass, the tension exerts an upwards force while the weight of the box exerts a downward force. The net force on each box determines its acceleration based on Newton's second law. Since the boxes are attached by a cord and move together, the acceleration of each box is the same.

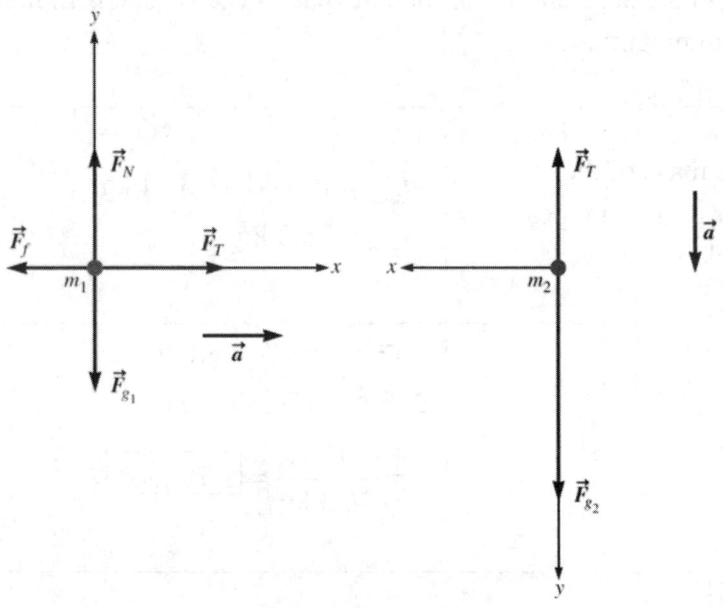

Figure P5.65ANS

SOLVE Begin by applying Newton's	$\sum F_x = F_T - F_f = m_1 a$ (1)

second law to each box for the component along which the acceleration occurs.	$\sum F_y = m_2 g - F_T = m_2 a \qquad (2)$
We can solve for the tension in the cord using equation (1). We need to substitute expressions for the friction force (Equation 5.9) and the acceleration, which we can get by solving equation (2).	$F_T = m_1 a + F_f \qquad (3)$ $a = \dfrac{m_2 g - F}{m_2} \qquad (4)$ $F_f = \mu_k m_1 g \qquad (5)$
Substituting equations (4) and (5) into (3), we have an expression for the tension force and we can insert numerical values.	$F_T = m_1\left(\dfrac{m_2 g - F_T}{m_2}\right) + \mu_k m_1 g$ $F_T = \dfrac{m_1 m_2 g(1+\mu_k)}{m_1 + m_2}$ $F_T = \dfrac{(6.00\text{ kg})(12.0\text{ kg})(9.81\text{ m/s}^2)(1.220)}{18.0\text{ kg}}$ $F_T = \boxed{47.9\text{ N}}$

CHECK and THINK

Finding the answer required a bit of algebra, but the basic steps were to draw a free-body diagram, write equations for the components of the force using Newton's second law, and solve for the desired quantity. The tension force is positive, as it must be.

68. A boulder of mass 80.0 kg rests directly on a spring with a spring constant of 6850 N/m.

a. (N) What is the compression of the spring?

INTERPRET and ANTICIPATE

In this case, the system is at rest, so the acceleration is zero and the net force must also be zero. The mass of the boulder presses down on the spring and its weight is balanced by the spring force of the compressed spring pushing upwards.

We can see this by drawing a free-body diagram for the boulder, which includes its weight and the force of the spring pushing upwards on it. For part (a), the acceleration is zero. For part (b), we are asked to consider the case when the system is in an elevator accelerating upwards.

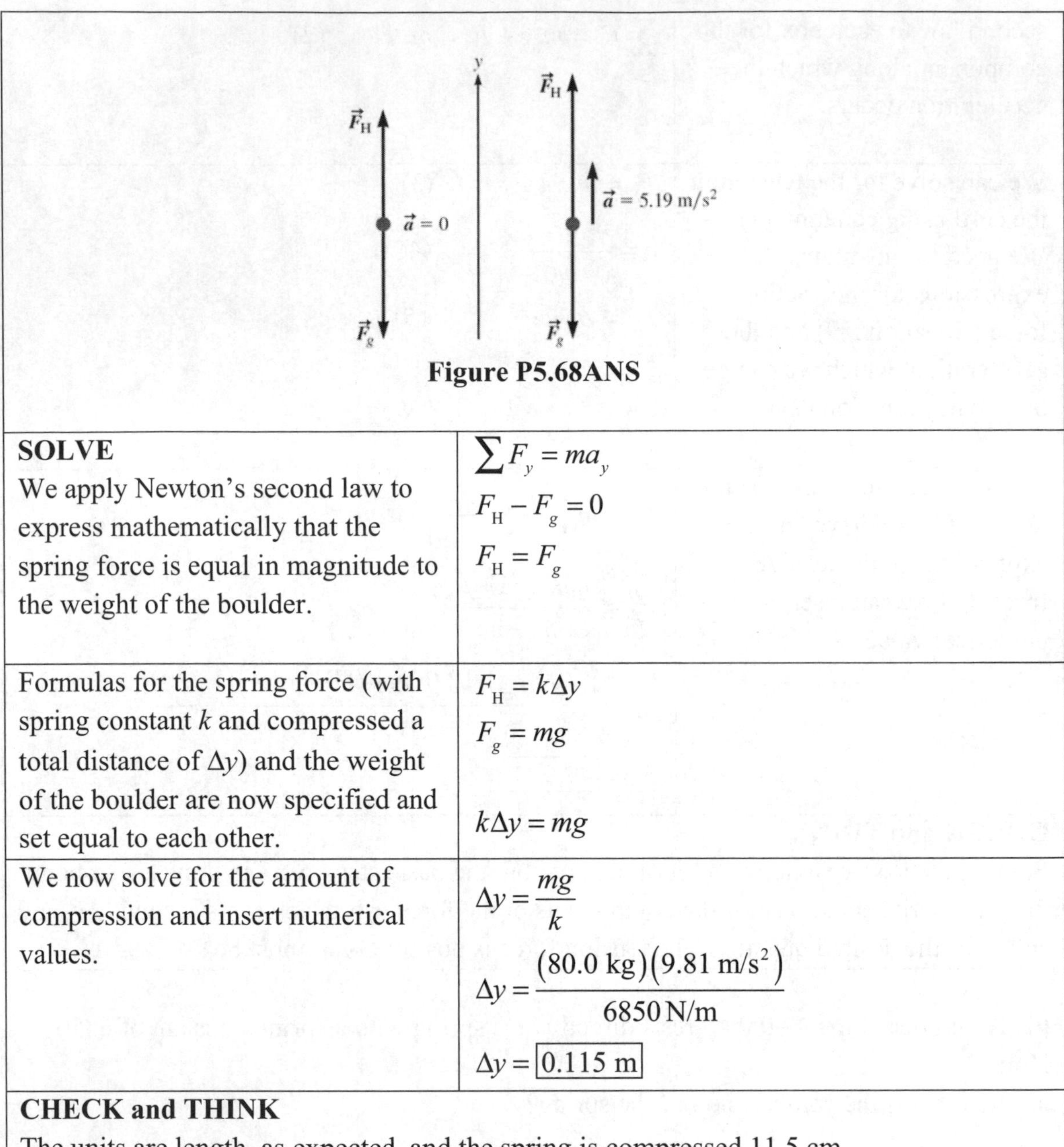

Figure P5.68ANS

SOLVE We apply Newton's second law to express mathematically that the spring force is equal in magnitude to the weight of the boulder.	$\sum F_y = ma_y$ $F_H - F_g = 0$ $F_H = F_g$
Formulas for the spring force (with spring constant k and compressed a total distance of Δy) and the weight of the boulder are now specified and set equal to each other.	$F_H = k\Delta y$ $F_g = mg$ $k\Delta y = mg$
We now solve for the amount of compression and insert numerical values.	$\Delta y = \dfrac{mg}{k}$ $\Delta y = \dfrac{(80.0\text{ kg})(9.81\text{ m/s}^2)}{6850\text{ N/m}}$ $\Delta y = \boxed{0.115\text{ m}}$

CHECK and THINK The units are length, as expected, and the spring is compressed 11.5 cm.

b. (N) Now, instead, the boulder and the spring are in an elevator accelerating upward at 5.19 m/s^2. What is the compression of the spring in this case?

INTERPRET and ANTICIPATE Looking at the free-body diagram, the weight of the boulder does not change, which means that the spring force must increase in order to produce the net force upwards needed to accelerate the boulder. In other words, as the elevator accelerates upwards, the spring will compress further. (If you were in the elevator, you might also feel "pushed down" towards the ground as the elevator accelerates upwards, the same as the boulder is

"pushed down" into the spring. In reality of course, the elevator is really exerting a larger upward push on you to accelerate you upwards.)	
SOLVE Applying Newton's second law, as in part (a), we now include a non-zero, upward acceleration *a*.	$\sum F_y = ma_y$ $F_H - F_g = ma$
Following the same steps as above, we can solve for the compression distance.	$k\Delta y - mg = ma$ $\Delta y = \dfrac{m(g+a)}{k}$
Inserting numerical values, we solve for the compression of the spring.	$\Delta y = \dfrac{80.0\text{ kg}\left(9.81\text{ m/s}^2 + 5.19\text{ m/s}^2\right)}{6850\text{ N/m}}$ $\Delta y = \boxed{0.175\text{ m}}$
CHECK and THINK The spring is indeed compressed more than when the system was stationary. The spring could also be the spring in a scale. If you were to stand on a scale in an elevator accelerating upwards, the spring in the scale would be compressed more and the scale would indicate a larger weight while accelerating.	

70. A 1.50-kg particle initially at rest and at the origin of an x–y coordinate system is subjected to a time-dependent force of $\vec{F}(t) = (3.00t\hat{\imath} - 6.00\hat{\jmath})$ N with t in seconds.

a. (N) At what time t will the particle's speed be 15.0 m/s?

INTERPRET and ANTICIPATE Given the force, we can calculate the acceleration using $F = ma$. The acceleration, exerted over time, determines the change in velocity of the particle. The force and acceleration are time-dependent though, which means that we can't use our constant acceleration kinematic equations and we will instead need to integrate the equation for acceleration.	
SOLVE We first write an expression for the acceleration of the particle, using Newton's second law $F = ma$.	$\vec{a} = \dfrac{\vec{F}}{m} = \dfrac{(3.00t\,\hat{\imath} - 6.00\hat{\jmath})\text{N}}{1.50\text{ kg}}$ $\vec{a} = (2.00t\,\hat{\imath} - 4.00\hat{\jmath})\text{m/s}^2$

To obtain an equation for the velocity of the particle as a function of time, we first write the equation for the instantaneous acceleration from Chapter 2 (Eq. 2.7). The change in velocity is equal to the integral of acceleration in time. The initial velocity is zero and the initial time is zero and we can calculate the final velocity $\vec{v}_f$ at time t.	$\vec{a} = \frac{d\vec{v}}{dt} = (2.00 \text{ m/s}^3)\ t\ \hat{\mathbf{i}} - (4.00 \text{ m/s}^2)\ \hat{\mathbf{j}}$ $d\vec{v} = (2.00 \text{ m/s}^3)\ t\ dt\ \hat{\mathbf{i}} - (4.00 \text{ m/s}^2) dt\ \hat{\mathbf{j}}$ $\int_0^{\vec{v}_f} d\vec{v} = \int_0^t \left(2.00 \text{ m/s}^3\right) t\ dt\ \hat{\boldsymbol{i}} - \int_0^t \left(4.00 \text{ m/s}^2\right) dt\ \hat{\boldsymbol{j}}$ $\vec{v}_f = \left(2.00 \text{ m/s}^3\right) \frac{t^2}{2}\ \hat{\boldsymbol{i}} - \left(4.00 \text{ m/s}^2\right) t\ \hat{\boldsymbol{j}}$
We now determine the magnitude of this vector and set the expression equal to the desired final velocity of 15 m/s. Since all quantities are in metric base units, units are omitted for clarity.	$\left\lvert \vec{v}_f \right\rvert = \sqrt{\left[(1.00)t^2\right]^2 + \left[(-4.00)t\right]^2} = 15.0$
We now square both sides.	$225 = 1.00\ t^4 + 16.0\ t^2$
We want to solve for time. One approach is to factor this equation. The expression is zero if either of the quantities in parentheses is zero. Since t^2 can't be negative, there are no times for which the second term is zero. The only real solutions are $t = \pm 3$ s, for which $(t^2 - 9) = 0$. Of the two possible solutions, we want to find a time *after* $t = 0$ s for which the velocity attains the desired speed, so the answer we are looking for is $t =$ 3 s.	$\left(t^2 - 9\right)\left(t^2 + 25\right) = 0$ $t = +3 \text{ s or } -3 \text{ s}$ $\rightarrow t = \boxed{3 \text{ s}}$

CHECK and THINK

While solving the problem involved a bit of algebra, the basic steps were to determine the acceleration, integrate the acceleration to find the velocity, and then find the time at which the velocity has the desired magnitude.

b. (N) How far from the origin will the particle be when its velocity is 15.0 m/s?

<table>
<tr><td colspan="2">INTERPRET and ANTICIPATE
In order to find the object's position at this time, we need to integrate the velocity equation to find the displacement of the object.</td></tr>
<tr><td>SOLVE
We use the definition of velocity (Eq. 2.4) and the expression for velocity determined in part (a) to write an equation for position, which we can integrate.</td><td>$\vec{v} = \frac{d\vec{r}}{dt} = \left(1.00\ \text{m/s}^3\right)t^2\hat{\imath} - \left(4.00\ \text{m/s}^2\right)t\,\hat{\jmath}$
$d\vec{r} = \left(1.00\ \text{m/s}^3\right)t^2 dt\,\hat{\imath} - \left(4.00\ \text{m/s}^2\right)t\,dt\,\hat{\jmath}$</td></tr>
<tr><td>We integrate from $t = 0$ s to $t = 3$ s (determined in part (a)) to find the change in position from 0 to $\vec{r}_f$. This is the final position after three seconds. The result of this integral is also the vector displacement from the origin, which is requested in part (c) of this problem.</td><td>$\int_0^{\vec{r}_f} d\vec{r} = \int_0^{3\text{s}} \left(1.00\ \text{m/s}^3\right)t^2\,dt\,\hat{\imath} - \int_0^{3\text{s}} \left(4.00\ \text{m/s}^2\right)t\,dt\,\hat{\jmath}$
$\vec{r}_f = \left(0.333\ \text{m/s}^3\right)t^3\Big|_0^{3\text{s}}\hat{\imath} - \left(2.00\ \text{m/s}^2\right)t^2\Big|_0^{3\text{s}}\hat{\jmath}$
$\vec{r}_f = 9.00\ \text{m}\,\hat{\imath} - 18.0\ \text{m}\,\hat{\jmath}$</td></tr>
<tr><td>We can now determine the magnitude of this displacement.</td><td>$r = \left|\vec{r}_f\right| = \sqrt{(9.00\ \text{m})^2 + (18.0\ \text{m})^2} = \boxed{20.1\ \text{m}}$</td></tr>
<tr><td colspan="2">CHECK and THINK
Similar to part (a), the definition of velocity was integrated to find the displacement.</td></tr>
</table>

c. (N) What is the particle's total displacement at this time?

<table>
<tr><td>The displacement was calculated in the process of determining part (b):
$\vec{r}_f = \boxed{9.00\ \text{m}\,\hat{\imath} - 18.0\ \text{m}\,\hat{\jmath}}$</td></tr>
</table>

73. (A) Ezra is pulling a sled, filled with snow, by pulling on a rope attached to the sled. The rope makes an angle θ with respect to the horizontal ground, and the sled is being pulled at a constant speed. If the sled and snow have a total mass of m, the acceleration due to gravity is g, the magnitude of the normal force is F_N, and the coefficient of kinetic friction between the sled and the ground is μ_k, what is the angle θ that the rope makes with the ground in terms of these five quantities?

INTERPRET and ANTICIPATE

Though we don't have numerical values, this sounds like a typical Newton's second law problem. The sled is moving at constant velocity, so that the acceleration is zero and the net force must be zero. We should start, as usual, by drawing a free-body diagram that includes all of the quantities mentioned in the problem.

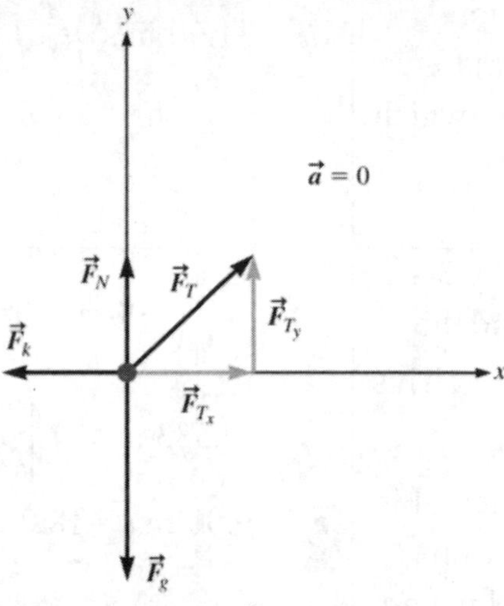

Figure P5.73ANS

SOLVE	
After indicating the necessary forces on the object, note that the sled is being pulled at a constant speed. This means it is not accelerating and the net force on the sled is zero (the sled is in equilibrium). By writing out the equilibrium conditions for both the x and y components (Equation 5.2), it is possible to find a function for θ in terms of the other quantities provided.	$\sum F_x = F_T \cos\theta - F_k = ma = 0$ $F_T \cos\theta - F_k = 0$ (1) $\sum F_y = F_T \sin\theta + F_N - F_g = ma = 0$ $F_T \sin\theta + F_N - F_g = 0$ (2)
We also need definitions for the gravitational (Eq. 5.7) and kinetic friction forces (Eq. 5.9).	$F_g = mg$ $F_k = \mu_k F_N$

From here, it is an algebra problem where we need to isolate the angle in terms of the quantities mentioned in the problem. Insert the definitions for the gravitational and kinetic friction forces into equations (1) and (2).	$F_T \cos\theta - \mu_k F_N = 0$ (3) $F_T \sin\theta + F_N - mg = 0$ (4)
Solve equation (3) for the tension.	$F_T \cos\theta = \mu_k F_N$ $F_T = \dfrac{\mu_k F_N}{\cos\theta}$
Now, substitute this expression into equation (4) and solve for the angle.	$F_T \sin\theta + F_N - mg = 0$ $\dfrac{\mu_k F_N}{\cos\theta}\sin\theta + F_N - mg = 0$ $\mu_k F_N \tan\theta = mg - F_N$ $\tan\theta = \dfrac{mg - F_N}{\mu_k F_N}$ $\theta = \boxed{\tan^{-1}\left[\dfrac{mg - F_N}{\mu_k F_N}\right]}$

CHECK and THINK

The quantity inside the brackets is just a number (i.e. all of the units cancel), as it should be in order to take the arctangent. We can look at a couple extreme examples to convince ourselves that this works as it should. If Ezra pulls the rope horizontally, $\theta = 0$, and the numerator must be zero, or $F_N = mg$. In other words, if Ezra pulls horizontally (no vertical component), the normal force must be exactly equal to the weight of the sled. That makes sense. As Ezra increases the angle, $F_N < mg$. In other words, the normal force is less than the weight because Ezra is supporting some of the weight by pulling up on the sled. This isn't proof that we have the right equation, but it seems to behave in a way that makes sense at least!

76. (N) Jamal and Dayo are lifting a large chest, weighing 207 lb, by using the two rope handles attached to either side. As they lift and hold it up so that it is motionless, each rope handle makes a different angle with respect to the vertical side of the chest (Fig. P5.76). If the angle between Jamal's handle and the vertical side is 25.0° and the angle between Dayo's handle and the vertical side of the chest is 30.0°, what are the tensions in each handle?

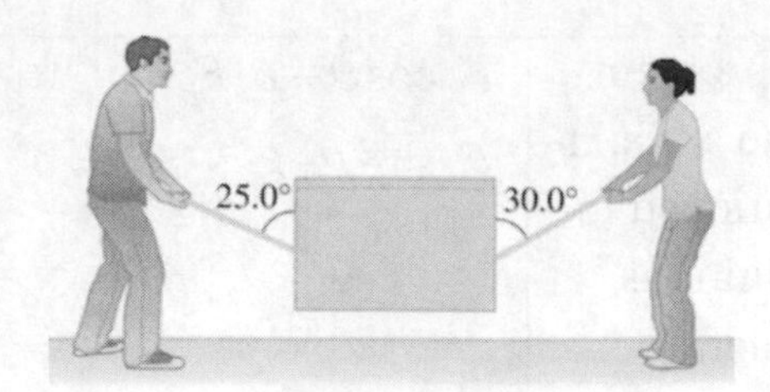

Figure P5.76

INTERPRET and ANTICIPATE Sketching a free-body diagram, we illustrate all of the forces we are to consider acting on the chest, as it is held motionless. There are two tension forces and the gravitational force on the chest pointing downward (its weight). The chest is in equilibrium ($a = 0 \rightarrow$ net force = 0), so we need only write out the equilibrium conditions for the chest and solve for the two tensions. We expect that the tensions will not be equal and that the tension in Jamal's handle will be greater because his handle is more vertical than Dayo's, and they are both working against the gravitational force. 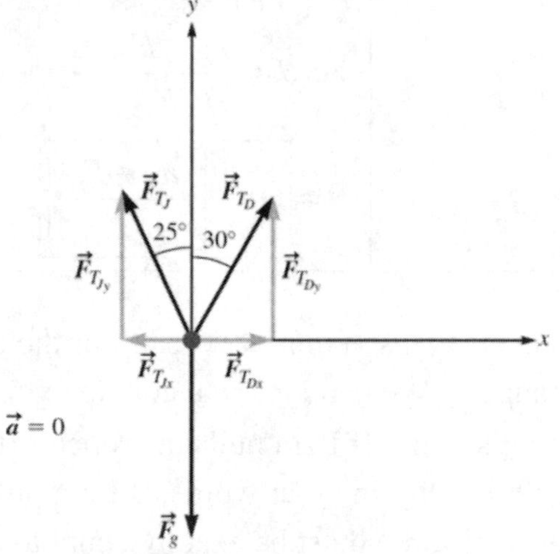 **Figure P5.76ANS**	
SOLVE Before continuing, we should convert the weight of the chest into the SI unit of Newtons.	$F_g = (207\text{ lbs}) \times \left(\frac{4.448\text{ N}}{1\text{ lbs}}\right) = 921\text{ N}$
Using the angles in the picture, we apply Newton's second law to the x direction.	$\sum F_x = F_{T_D} \sin(30°) - F_{T_J} \sin(25°) = 0$ $F_{T_D} \sin(30°) - F_{T_J} \sin(25°) = 0 \quad (1)$

We now write a similar equation for forces in the y direction.	$\sum F_y = F_{T_D}\cos(30°) + F_{T_J}\cos(25°) - F_g = 0$ $F_{T_D}\cos(30°) + F_{T_J}\cos(25°) - 921\text{ N} = 0 \qquad (2)$
Now, solve equation (1) for the tension in Dayo's rope to substitute it into equation (2).	$F_{T_D}\sin(30°) - F_{T_J}\sin(25°) = 0$ $F_{T_D} = \dfrac{F_{T_J}\sin(25°)}{\sin(30°)} \qquad (3)$
Make the substitution (equation 3 into 2) and solve for the tension in Jamal's handle.	$\left(\dfrac{F_{T_J}\sin(25°)}{\sin(30°)}\right)\cos(30°) + F_{T_J}\cos(25°) - 921\text{ N} = 0$ $(0.732)F_{T_J} + (0.906)F_{T_J} - 921\text{ N} = 0$ $(1.638)F_{T_J} = 921\text{ N}$ $F_{T_J} = \dfrac{921\text{ N}}{1.638} = \boxed{562\text{ N}}$
We can now get the tension in Dayo's handle by plugging known values into equation (3).	$F_{T_D} = \dfrac{(562\text{ N})\sin(25°)}{\sin(30°)} = \boxed{475\text{ N}}$
CHECK and THINK As expected, Jamal's handle is at a greater tension because his force is directed more vertically than Dayo's. If the angles were equal, the tension in each handle would turn out to be the same.	

80. Two blocks of mass $m_1 = 1.50$ kg and $m_2 = 5.00$ kg are connected by a massless cord passing over a frictionless pulley as shown in Figure P5.80, with $\theta = 35.0°$. The coefficient of kinetic friction between block 1 and the horizontal surface is 0.400, and the coefficient of kinetic friction between block 2 and the inclined surface is 0.330.

Figure P5.80

a. (G) Draw a free-body diagram for each of the two blocks.

Draw a free-body diagram for each block choosing appropriate coordinate axes in each case. Block 1 is accelerating due to the tension force, with a frictional force opposing motion, while the upwards normal force balances the gravitational force. For block 2, as it slides down the incline, both the tension force and the frictional force point up the incline. The normal force, like all normal forces, is perpendicular to the surface while the gravitational force is downward.

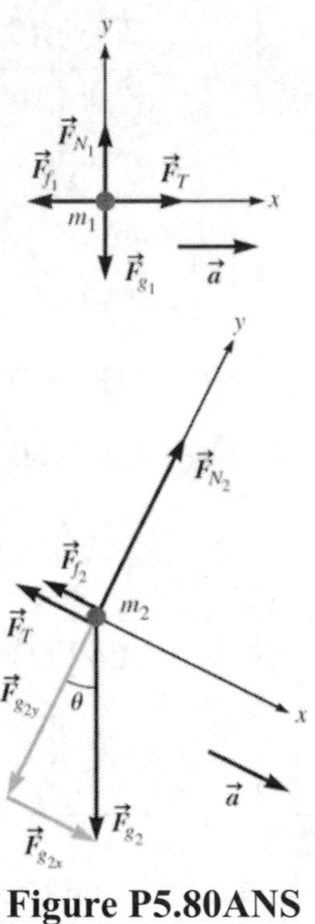

Figure P5.80ANS

b. (N) What is the acceleration of the system when it is released from rest?

INTERPRET and ANTICIPATE

With the free-body diagram in part (a), we can write formulas for the net force on each block and relate it to the acceleration of the block using Newton's second law. We expect the acceleration to be lower than 9.8 m/s^2 since the blocks are being accelerated by gravity, but are not in free fall.

SOLVE Apply Newton's second law to the first block to determine equations for both the x and y components.	$\sum F_y = F_{N_1} - m_1 g = 0$ $\sum F_x = F_T - F_{f_1} = m_1 a$

Using these equations, we can calculate the normal force on block and write an expression for the tension force. We've substituted the formula for kinetic friction in the process, $F_{f_1} = \mu_{k_1} F_{N_1} = \mu_{k_1} m_1 g$.	$F_{N_1} = m_1 g = (1.50 \text{ kg})(9.81 \text{ m/s}^2) = 14.7 \text{ N}$ $F_T = m_1 a + \mu_{k_1} m_1 g \quad (1)$
Applying Newton's second law to the second block, we derive similar equations.	$\sum F_y = F_{N_2} - m_2 g \cos\theta = 0$ $\sum F_x = m_2 g \sin\theta - F_T - F_{f_2} = m_2 a$
Using the equations for block 2, we can calculate the normal force on block 2 and derive another equation for the tension in the cord. We've again used the formula for kinetic friction for block 2, $F_{f_2} = \mu_{k_2} F_{N_2} = \mu_{k_2} m_2 g$.	$F_{N_2} = m_2 g \cos\theta = (5.00 \text{ kg})(9.81 \text{ m/s}^2)\cos 35.0°$ $F_{N_2} = 40.2 \text{ N}$ $F_T = m_2 g \sin\theta - \mu_{k_2} m_2 g \cos\theta - m_2 a \quad (2)$
Now, equate equation (1) and equation (2), since the tensions are both the same (the tension in the cord).	$m_1 a + \mu_{k_1} m_1 g = m_2 g \sin\theta - \mu_{k_2} m_2 g \cos\theta - m_2 a$ $a(m_1 + m_2) = m_2 g \sin\theta - \mu_{k_2} m_2 g \cos\theta - \mu_{k_1} m_1 g$ $a = \frac{g(m_2 \sin\theta - \mu_{k_2} m_2 \cos\theta - \mu_{k_1} m_1)}{(m_1 + m_2)}$

We can substitute numerical values to determine the acceleration.

$$a = \frac{(9.81 \text{ m/s}^2)[(5.00 \text{ kg})\sin 35.0° - (0.330)(5.00 \text{ kg})\cos 35.0° - (0.400)(1.50 \text{ kg})]}{(1.50 \text{ kg} + 5.00 \text{ kg})}$$

$$a = \boxed{0.141 \text{ m/s}^2}$$

CHECK and THINK

We were able to calculate the acceleration of the blocks. The acceleration is actually significantly lower than that of an object in free fall.

c. (N) What is the tension in the cord connecting the blocks?

<table>
<tr><td colspan="2">INTERPRET and ANTICIPATE
We've done all of the hard work at this point! We simply need to take one of our expressions for tension from part (b) and substitute known values.</td></tr>
<tr><td>SOLVE
From part (b), we can take either equation (1) or (2) and substitute numerical values.</td><td>$F_T = m_1 a + \mu_{k_1} m_1 g$
$F_T = (1.50\text{ kg})(0.141\text{ m/s}^2) + (0.400)(1.50\text{ kg})(9.81\text{ m/s}^2)$
$F_T = \boxed{6.10\text{ N}}$</td></tr>
<tr><td colspan="2">CHECK and THINK
Given the weights of the two blocks (around 15 N and 50 N), the tension force of 6 N sounds plausible. (You might not have much intuition about this, but it's the same order of magnitude, which sounds right.)</td></tr>
</table>

6

Applications of Newton's Laws of Motion

9. A man exerts a force of 16.7 N horizontally on a box so that it is at rest in contact with a wall as in Figure 6.3. The box weighs 6.52 N.
a. (N) Find the static friction exerted on the box, given the forces being applied.

INTERPRET and ANTICIPATE
Since the box is at rest, the total acceleration and net force are zero. We first draw a free-body diagram, similar to Figure 6.3(A). In order to have zero net force, the normal force must equal the pushing force and the friction force must equal the weight of the box.

SOLVE
We draw a free body diagram corresponding to the situation shown in Figure 6.3. The box is not accelerating, so the net force is zero.

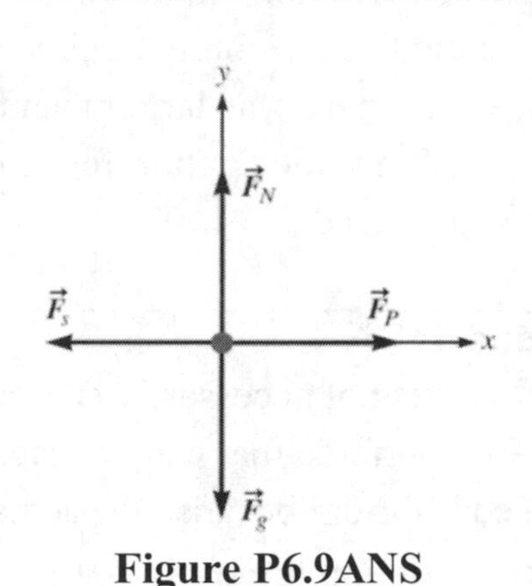

Figure P6.9ANS

Since the net force in the y direction must be zero, the friction force must equal the weight of the box, which is given.

$$F_s = F_g = \boxed{6.52\ \text{N}}$$

CHECK and THINK
As is often the case, once we have a free-body diagram for an object that is in static equilibrium, we can balance forces and find unknown values.

b. (N) If the coefficient of static friction between the wall and the box is 0.50, find the maximum static friction that may be exerted on the box. Comment on your results.

INTERPRET and ANTICIPATE	
The maximum static friction force depends on the coefficient of friction and the normal force. We can calculate this and compare it to the frictional force required to keep the box stationary found in part (a).	
SOLVE The normal force is equal to the pushing force, since the box is not accelerating (and, in fact, not moving at all) in the *x* direction.	$F_N = F_P = 16.7\text{ N}$
The maximum static friction force can be calculated as the coefficient of static friction times the normal force (Equation 6.1). The actual static friction force from part (a) is less than the maximum we've calculated, therefore the box will not slip down the wall. If the static friction force required to keep the box stationary was larger than the maximum possible static friction force, the box would slip instead.	$F_{s,\max} = \mu N = (0.5)(16.7\text{ N}) = \boxed{8.4\text{ N}}$
CHECK and THINK Equation 6.1 does not necessarily determine the static friction force. It determines the *maximum* friction force that can be reached before the object slides. In this case, the box remains in equilibrium because the actual friction force is less than the maximum possible static friction force $F_s < F_{s,\max} = \mu N$.	

11. A makeshift sign hangs by a wire that is extended over an ideal pulley and is wrapped around a large potted plant on the roof as shown in Figure P6.10. The mass of the sign is 25.4 kg, and the mass of the potted plant is 66.7 kg.

Figure P6.10

a. (N) Assuming the objects are in equilibrium, determine the magnitude of the static friction force experienced by the potted plant.

<table>
<tr><td colspan="2">INTERPRET and ANTICIPATE
Since the objects are in equilibrium, the total (net) force on each must be zero. We can draw a free-body diagram for each object and write equations such that each component of the net force is zero for each object.</td></tr>
<tr><td>SOLVE
We begin analyzing the forces on the pot and the sign by drawing free-body diagrams for each object.</td><td>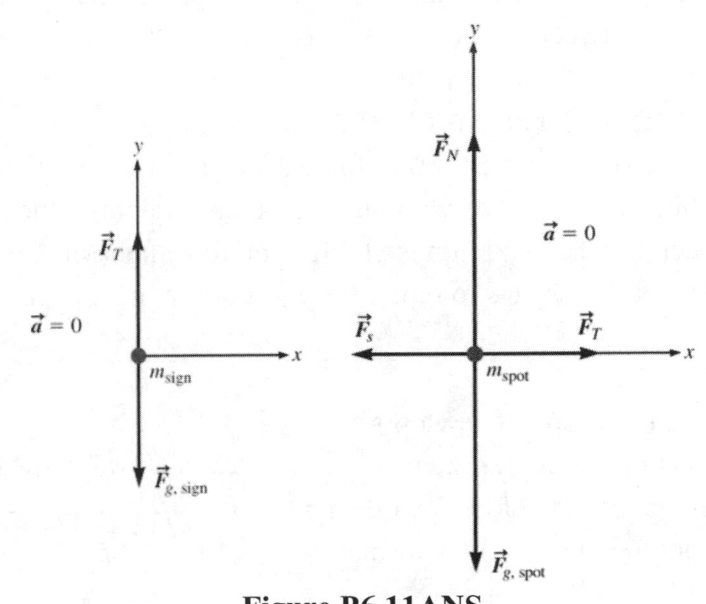

Figure P6.11ANS</td></tr>
<tr><td>Knowing that both objects are in equilibrium, we require that the net force is zero for both the x and y components. Therefore, the gravitational force on the sign must be equal in magnitude to the tension by Newton's second law.</td><td>$\sum F_y = F_T - F_{g,\text{sign}} = 0$
$F_T = F_{g,\text{sign}} = m_{\text{sign}} g = (25.4\text{ kg})(9.81\text{ m/s}^2) = 249\text{ N}$</td></tr>
<tr><td>Referring to the free-body diagram for the pot, the static friction force must be equal to the tension in the wire.</td><td>$\sum F_x = F_T - F_s = 0$
$F_s = F_T = \boxed{249\text{ N}}$</td></tr>
</table>

CHECK and THINK
For objects in equilibrium, with a free-body diagram we are easily able to relate components of the forces since the net force for each component must be zero. Ultimately, it's the static friction force that is actually holding up the sign, so it makes sense that the friction force is equal to the weight of the sign. If the surface was suddenly made very slippery such that it was nearly frictionless, then the pot would easily slide and the sign would fall.

b. (N) What is the maximum value of the static friction force if the coefficient of static friction between the pot and the roof is 0.572?

INTERPRET and ANTICIPATE	
The maximum static friction force depends on the coefficient of friction, which is given, and the normal force, which is equal to the weight of the pot in this case. We would expect that the maximum static friction force must be larger than the value of 249 N in order for the objects to remain in equilibrium in part (a).	
SOLVE We can now apply Newton's second law again for the y component of the forces on the pot the pot to determine the normal force.	$\sum F_y = F_N - F_{g,\text{pot}} = 0$ $F_N = F_{g,\text{pot}} = m_{\text{pot}} g = (66.7 \text{ kg})(9.81 \text{ m/s}^2) = 654 \text{ N}$
The maximum static friction force depends on the coefficient of static friction and the normal force (Equation 6.1).	$F_{s,\max} = \mu_s F_N = (0.572)(654 \text{ N}) = \boxed{374 \text{ N}}$
CHECK and THINK The maximum static friction force is larger than the value found in part (a). This makes sense, since the static friction force must be less than or equal to the maximum in order for the pot not to slide.	

14. A small steel I-beam (Fig. P6.14) is at rest with respect to the steel surface of a truck. The truck is accelerating with respect to the road. The mass of the I-beam is 5.8×10^3 kg.

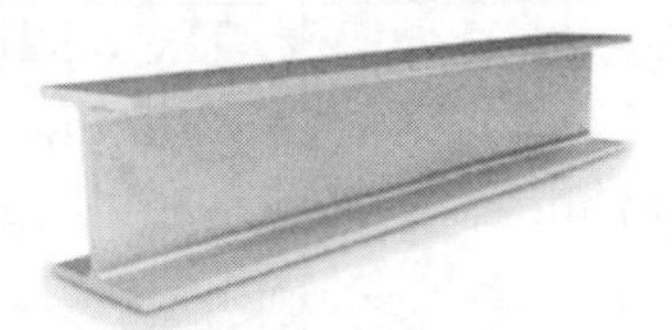

Figure P6.14

a. (G) Draw a free-body diagram for the I-beam.

b. (C) What force or forces accelerate the I-beam with respect to the ground?

The I-beam has a force of gravity (its weight) pointing downward. Since it is resting on a surface (the truck bed), there is a normal force pointing upwards.

There also must be a frictional force, but in which direction? If the truck is accelerating to the right and the I-beam is sitting in the truck, the I-beam must also be accelerating to the right. But what force is accelerating the I-beam with respect to the ground? It's the static friction force that actually allows the truck bed to "pull" the I-beam along, so the friction force is to the right. Another way to think about it is if there were *no* friction, the I-beam would just stay in place as the truck drove out from under it! Instead, the friction due to the truck bed exerts a force along the direction the truck is moving.

We should also take a moment and remember that forces don't *always* add up to zero. At this point, we're pretty used to systems in equilibrium where the total force *is* zero, but the I-beam is actually accelerating, so the total force on it *can't* be zero and must point in the direction it's accelerating.

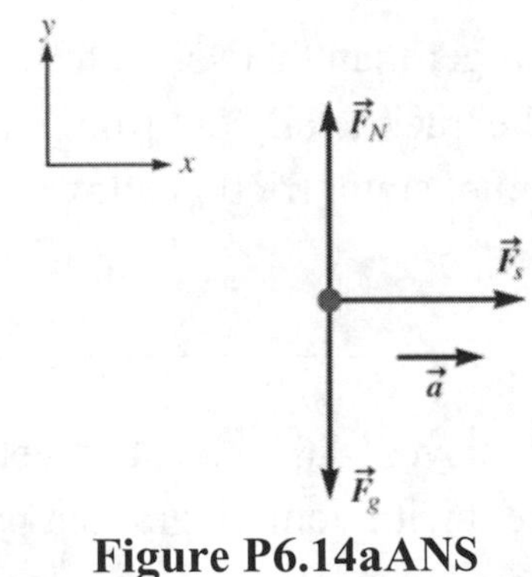

Figure P6.14aANS

c. (N) The I-beam must remain at rest with respect to the truck. What is the maximum acceleration of the truck? Evaluate your answer.

INTERPRET and ANTICIPATE

From the explanation above, it is actually the static friction force that accelerates the beam. If the truck accelerates a little, a small friction force is sufficient to keep the I-beam moving with the truck. As the acceleration increases, a larger static friction force is required to pull the I-beam along with the truck. The maximum acceleration occurs when the maximum static friction force is reached. Beyond this, if the truck accelerates too quickly, the I-beam would start to slide.

SOLVE To calculate the static friction force, we need the coefficient of friction and the normal force (Equation 6.1). Using the y component in the free-body diagram, the normal force is equal to the weight of the I-beam.	$\sum F_y = F_N - F_g = 0$ $F_N = F_g = mg$
We now want to find the acceleration at the maximum static friction force.	$\sum F_x = F_{s,\max} = ma_{\max}$ $a_{\max} = \frac{F_{s,\max}}{m} = \frac{\mu_s F_N}{m} = \frac{\mu_s mg}{m} = \mu_s g$
We are not given a static friction coefficient, but if we assume that this is a steel I-beam on a steel truck bed, we can use Table 6.1 to get an approximate coefficient of friction.	$\mu_s \approx 0.72$ $a_{\max} = \mu_s g \approx 0.72g \approx \boxed{7 \text{ m/s}^2}$

CHECK and THINK

If the truck accelerates at a rate larger than this value, for instance if you gun the engine at a green light, it would cause the truck to slip out from underneath the I-beam and leave it on the road. A larger coefficient of static friction allows a larger acceleration of the truck before slipping occurs.

d. (G) On the highway, the truck moves with a constant velocity. Draw a free-body diagram for the I-beam. Compare it with your diagram in part (a).

In this case, the truck and the I-beam are no longer accelerating, so the total (net) force on the I-beam must be zero. Since the static friction force is the only horizontal force, it must be zero! This might seems surprising, but (ignoring things like air resistance) if the truck and I-beam are both coasting along at the same velocity, it wouldn't matter if the surface was really slippery — they would both continue moving in the same direction and stay in contact. It's only with changes in motion (or accelerations) that the truck bed needs to exert a horizontal force on the I-beam.

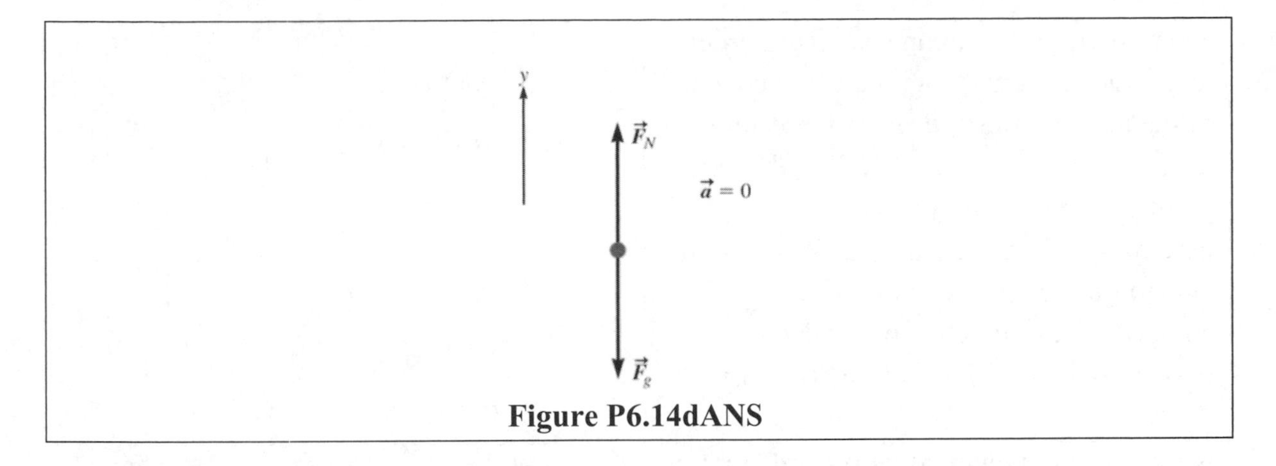

Figure P6.14dANS

16. (A) A filled treasure chest of mass m with a long rope tied around its center lies in the middle of a room. Dirk wishes to drag the chest, but there is friction between the chest and the floor with a coefficient of static friction, μ_s. If the angle between the rope and the floor is θ, what is the magnitude of the tension required to just get the chest moving? Express your answer in terms of m, μ_s, θ, and g.

INTERPRET and ANTICIPATE

There are a lot of variables given, but we approach this in the same way as previous problems: First, draw a free-body diagram. Then, we want to write Newton's second law force equations. The requirement is that Dirk is pulling at the maximum angle for which he is just able to move the box. That is, if the angle was any higher, he couldn't exert enough of a horizontal force to overcome the maximum static friction. It sounds like we need to apply Newton's second law to this special case where he pulls at this angle and the static friction force is at its maximum possible value. We can then determine the normal force, and eventually the tension force as well.

SOLVE

We begin by drawing a free-body diagram, labeling all forces acting on the chest. This includes the weight of the chest (gravitational force), normal force due to the ground, tension force from Dirk's rope, and the static friction force resisting this pull.

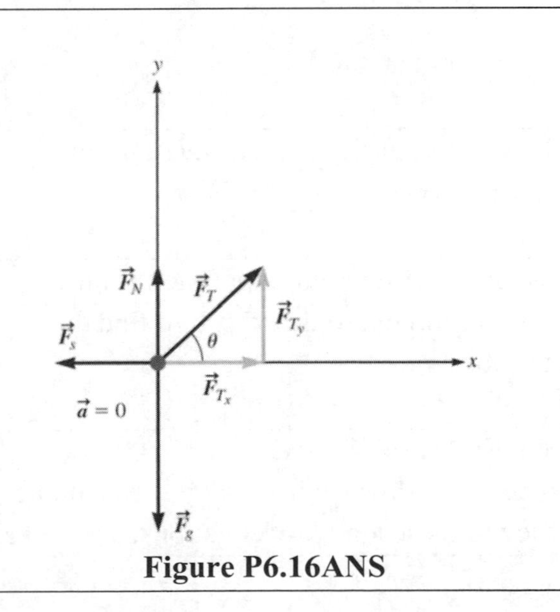

Figure P6.16ANS

From the figure, we can see that in order to get the chest moving, the *x* component of the tension must be slightly greater than the static friction force. We'll look at the point where they are just equal to determine this critical angle. We can write Newton's second law with the *x* component of the tension and the maximum static friction force (Eq. 6.1).	$\sum F_x = F_{T_x} - F_s = 0$ $F_T \cos\theta - \mu_s F_N = 0 \qquad (1)$
Now, let's write Newton's second law for the *y* components, including the *y* component of the tension.	$\sum F_y = F_N + F_{T_y} - F_g = 0$ $F_N + F_T \sin\theta - mg = 0 \qquad (2)$
Looking at equations (1) and (2), note that there are two unknown quantities, F_N and θ. In principle, we can solve two equations and two unknowns, though the sine and cosine make this a little less straightforward. Dividing one by the other to find the $\tan\theta$ will allow us to find an equation for the normal force.	$\cos\theta = \frac{\mu_s F_N}{F_T}$ $\sin\theta = \frac{mg - F_N}{F_T}$ $\tan\theta = \frac{\sin\theta}{\cos\theta} = \frac{(mg - F_N)/F_T}{\mu_s F_N / F_T} = \frac{mg - F_N}{\mu_s F_N}$
Now, solve for the normal force.	$\mu_s F_N \tan\theta = mg - F_N$ $F_N = \frac{mg}{1 + \mu_s \tan\theta}$
Now, use our Eq. (1) and solve for the tension force.	$F_T = \frac{\mu_s F_N}{\cos\theta}$
Then, substitute our expression for the normal force into this result to find the tension force.	$F_T = \frac{\mu_s}{\cos\theta} \frac{mg}{1 + \mu_s \tan\theta} = \boxed{\frac{mg\mu_s}{\cos\theta + \mu_s \sin\theta}}$

CHECK and THINK

The method was what we've seen a number of times: draw a free body diagram, write Newton's second law equations, and solve for the desired force.

23. (N) A block with a mass $M = 2.00$ kg is placed on an inclined plane. The plane makes an angle of 30° with the horizontal, and the coefficients of kinetic and static friction between the block and the plane are $\mu_k = 0.400$ and $\mu_s = 0.600$, respectively. Will the block slide down the plane, or will it remain motionless? Justify your answer.

INTERPRET and ANTICIPATE	
As with any inclined plane problem, a component of the gravitational force on the block will be directed down the plane and will exert a force that might cause the block to slip down the plane. If this force is less than the static friction force though, the object will not slip. We should draw a free-body diagram, write Newton's second law for force components along the incline, and see whether the component of the gravitational force is smaller or larger than the maximum friction force.	
SOLVE We first draw a free body diagram, which includes the gravitational force on the block, the normal force, and the static friction force.	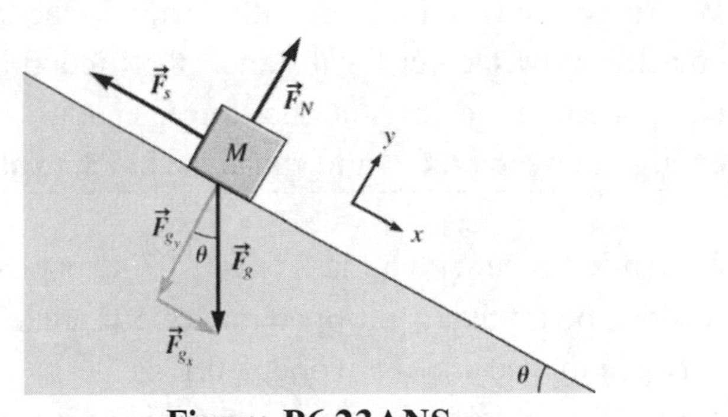 **Figure P6.23ANS**
We will assume the block remains in place and then see if the maximum static friction force is large enough to hold it in place. So, we need not consider the kinetic friction force because this acts only if the block slips. Consider the components of the force along the incline. To move, the x component of the gravitational force must exceed the maximum static friction force, so we can first determine the x component of the gravitational force.	$F_{g_x} = mg\sin\theta = (2.00\text{ kg})(9.81\text{ m/s}^2)\sin 30° = 9.81\text{ N}$

Now, we need to determine the maximum static friction force. The normal force can be found by writing Newton's second law for the y component.	$F_s = \mu_s N$ $\sum F_y = F_N - mg\cos\theta = 0$ $F_N = mg\cos\theta = (2.00\text{ kg})(9.81\text{ m/s}^2)\cos 30° = 17.0\text{ N}$
The static friction force has a maximum value larger than the force with which gravity pulls the block down the plane. Therefore, the block will remain in place and not slide down the incline.	$F_s = \mu_s N = (0.600)(17.0\text{ N}) = 10.0\text{ N}$

CHECK and THINK

We knew that this might go either way. In the last step, it's clear that if the friction coefficient was lower for instance, the friction force might become lower than the x component of the gravitational force. That is, if friction was less, gravity might be strong enough to overcome it and cause the block to slide.

25. An ice cube with a mass of 0.0507 kg is placed at the midpoint of a 1.00-m-long wooden board that is propped up at a 50° angle. The coefficient of kinetic friction between the ice and the wood is 0.133.

a. (N) How much time does it take for the ice cube to slide to the lower end of the board?

INTERPRET and ANTICIPATE

The ice cube must be sliding down the board. If we draw a free-body diagram and find the net force, we can determine the acceleration. With the acceleration, we should be able to determine the time for the ice cube to slide a particular distance using kinematic equations.

SOLVE

As it is usually helpful, we begin by drawing a free-body diagram for the ice cube. This includes the gravitational force (weight) of the ice, normal force due to the board, and friction (in this case, kinetic, since the ice cube is sliding). Using trigonometry, we can write the gravitational force in components in the direction of	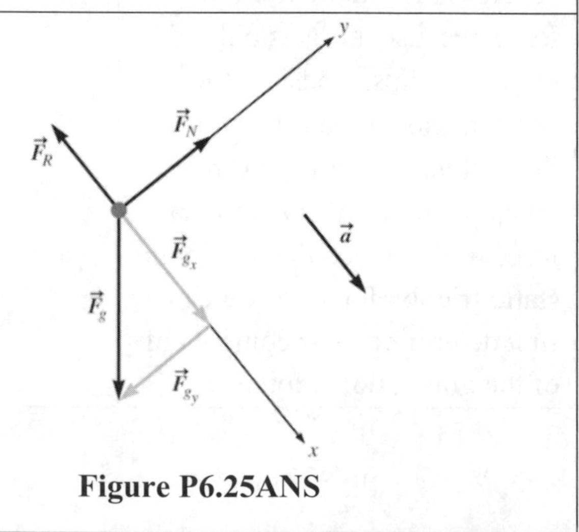 **Figure P6.25ANS**

motion along the board and perpendicular to it.	
Using a coordinate system with the x direction parallel to the board, and the y direction perpendicular to the board, it is the kinetic friction force and the x component of the gravitational force in competition along the board. Using Newton's second law and the expression for kinetic friction (Equation 6.2), we can write an equation for the x component of the forces that we can use to determine the acceleration along the board.	$\sum F_x = F_{g_x} - F_k = ma$ $mg\sin\theta - \mu_k F_N = ma \qquad (1)$
Now, write a similar equation for the y component of the forces. The y acceleration is zero.	$\sum F_y = F_N - F_{g_y} = 0$ $F_N - mg\cos\theta = 0 \qquad (2)$
Solving equation (2) for F_N and substituting into equation (1) allows us to solve for the acceleration.	$F_N = mg\cos\theta$ $mg\sin\theta - \mu_k mg\cos\theta = ma$ $a = g\sin\theta - \mu_k g\cos\theta \qquad (3)$
We can then use kinematics to solve for the time required to travel half the length of the board, $d = 0.500$ m. Using equations 3 and the kinematic equation $d = \frac{1}{2}at^2$ allows us to solve for the time.	$d = \frac{1}{2}(g\sin\theta - \mu_k g\cos\theta)t^2$ $t = \sqrt{\frac{2d}{(g\sin\theta - \mu_k g\cos\theta)}}$ $t = \sqrt{\frac{2(1.00\text{ m}/2)}{((9.81\text{ m/s}^2)\sin 50° - (0.133)(9.81\text{ m/s}^2)\cos 50°)}}$ $t = \boxed{0.387\text{ s}}$

CHECK and THINK

The time of 0.387 seconds sounds like it's in the right ballpark for an ice cube to slide half a meter. Finding a plausible answer encourages us that we didn't make any serious errors!

b. (N) If the ice cube is replaced with a 0.0507-kg wooden block, where the coefficient of kinetic friction between the block and the board is 0.275, at what angle should the board be placed so that the block takes the same amount of time to slide to the lower end as the ice cube does? You may find a spreadsheet program helpful in answering this question.

INTERPRET and ANTICIPATE

The approach will be similar: free-body diagram and finding force equations. In fact, we did the hard work in part (a) and we can use the expressions we found there. We want the block to take the same amount of time to travel the same distance. We can use the equations above, change the friction coefficient and find how that affects the angle needed, assuming all other quantities stay the same. Since friction is larger, we expect to have to tilt the board higher to get it to slide as fast.

SOLVE

The analysis from the part (a) applies here as well, except the value of the coefficient of kinetic friction is different. We can use the final relation for distance from part (a) and attempt to solve for θ.

$$d = (1/2)(g\sin\theta - \mu_k g\cos\theta)t^2$$

$$\sin\theta - \mu_k \cos\theta = \frac{2d}{gt^2}$$

$$\sin\theta - (0.275)\cos\theta = \frac{2(1.00 \text{ m}/2)}{(9.81 \text{ m/s}^2)(0.387 \text{ s})^2}$$

$$\sin\theta - (0.275)\cos\theta = 0.68063$$

This expression does not have an easy solution! We must rely on a numerical method, since a closed-form solution for the angle does not appear possible. A spreadsheet program may be a useful tool here. In one column, we consider possible angles for θ, and in the adjacent column we compute the left-hand side of this equation, looking for the angle that results in a value closest to 0.68063. A portion of the table is shown here. Using a precision to the tenth of a degree, the best possible answer is $\boxed{56.4°}$, given that the left-hand side of the equation above must be equal to at least 0.68063.

θ	**Equation**
56.1	0.676632
56.2	0.678003
56.3	0.679372
56.4	0.680739
56.5	0.682103
56.6	0.683466
56.7	0.684826

CHECK and THINK
Since the coefficient of friction is larger, we need to tilt the board to a higher angle to have the block slide down in the same amount of time (56.4° compared to 50°). This is exactly what we expected.

27. Curling is a game similar to lawn bowling except it is played on ice and instead of rolling balls on the lawn, stones are slid along ice. A curler slides a stone across a sheet of ice with an initial speed v_i in the positive x direction. The coefficient of kinetic friction between the stone and the curling lane is μ_k. Express your answers in terms of v_i, μ_k. and g only.
a. (A) What is the acceleration of the stone as it slides down the lane?

INTERPRET and ANTICIPATE	
To determine the acceleration of the stone, we need to determine the net force. We expect that friction will slow the stone and that the acceleration will be in the opposite direction of the velocity.	
SOLVE As usual, it is helpful to draw a free-body diagram for the stone, which includes the weight of the stone, the normal force, and the force due to kinetic friction.	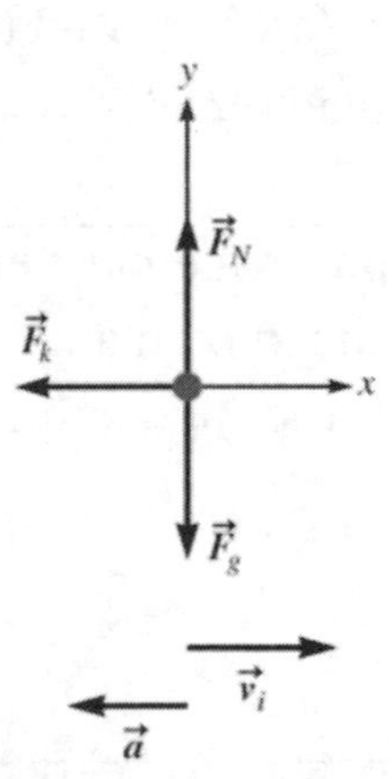 **Figure P6.27ANS**
There is only an acceleration in the x direction. We can use $f = ma$ for the y component, for which there is no acceleration, to equate the normal force and the weight of the stone.	$\sum F_y = F_N - mg = 0$ $F_N = mg$
Once the stone loses contact with the curling stick, the only horizontal force is a friction force in the negative x direction (to oppose the motion of the stone). The friction force is equal to the coefficient of kinetic friction times the normal force (Equation 6.2). Using this, we can determine the acceleration of the stone.	$a_x = \frac{\sum F_x}{m} = \frac{-F_f}{m}$ $a_x = \frac{-\mu_k mg}{m}$ $a_x = \boxed{-\mu_k g}$

CHECK and THINK
The acceleration is in the negative direction, as we expect. Since this is a deceleration of the stone, it also makes sense that the acceleration has a larger magnitude if the coefficient of kinetic friction is larger. The more friction, the more quickly the stone slows down.

b. (A) What distance does the curling stone travel?

INTERPRET and ANTICIPATE	
The frictional force and the acceleration are both constant as the stone slows. With constant acceleration, we can use the kinematic equations covered in Chapter 2.	
SOLVE We might first just keep track of what we know. The stone starts at a given speed v_i, comes to rest (final speed = 0), and the acceleration found in part (a). We want to solve for the distance traveled Δx.	v_i (initial speed) $v_f = 0$ $a_x = -\mu_k g$ $\Delta x =$ unknown
Now, we find a kinematic equation that relates the known and the desired quantities and solve for the distance.	$v_f^2 = v_i^2 + 2a_x \Delta x$ $0 = v_i^2 + 2(-\mu_k g)\Delta x$ $\Delta x = \boxed{\dfrac{v_i^2}{2\mu_k g}}$
CHECK and THINK While this might first look like a complicated expression, there are reasons to believe that it might be correct. If the initial speed is increased, the distance traveled will be larger. If the kinetic friction coefficient is increased, the distance traveled would be smaller. This sounds right at least!	

30. (N) A sled and rider have a total mass 56.8 kg. They are on a snowy hill accelerating at $0.7g$. The coefficient of kinetic friction between the sled and the snow is 0.18. What is the angle of the hill's slope measured upward from the horizontal? You may find a spreadsheet program helpful in answering this question.

INTERPRET and ANTICIPATE

The total force on the sled, which points downhill, leads to the acceleration. We should be able to draw a free-body diagram, find the net force, and calculate the acceleration. This will allow us to determine which angle is needed to result in the acceleration stated.

SOLVE As usual, we start by drawing a free body diagram for the forces on the sled/rider, which includes the weight of the sled/rider, the normal force due to the ground, and the force of kinetic friction. We choose the coordinate system so that there is an axis pointing downhill (along which the acceleration occurs) and an axis perpendicular to the hill (since there is no acceleration in that direction and therefore the net force must be zero).	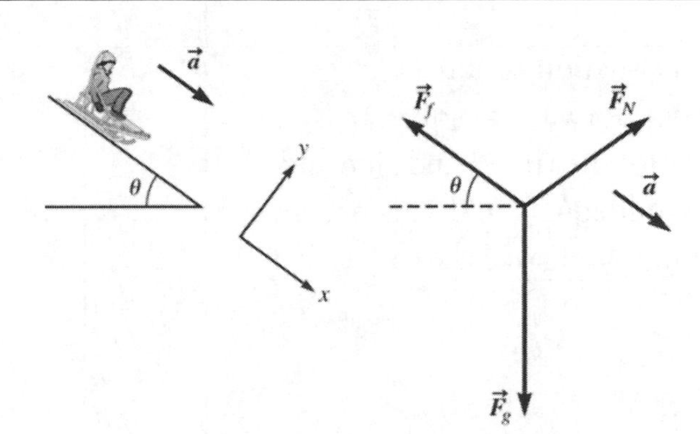 **Figure P6.30aANS**
Using the coordinate system indicated, we can express the acceleration of an object on the incline in terms of the coefficient of kinetic friction and the acceleration due to gravity. The net x component of the force leads to an acceleration while the net y component of the force must be zero.	$\sum F_y = F_N - F_g \cos\theta = 0$ $\sum F_x = F_g \sin\theta - F_k = ma$
The gravitational force is the weight of the object, the kinetic friction force depends on the coefficient of friction and the normal force (Eq. 6.2), and the acceleration is given.	$F_g = mg$ $F_k = \mu F_N = \mu F_g \cos\theta = \mu mg \cos\theta$ $a = 0.7g$

Using these relationships, we can write an expression in terms of the angle of the slope.	$a = \dfrac{F_g \sin\theta - F_k}{m}$ $0.7g = g\sin\theta - \mu g\cos\theta$ $0.7 = \sin\theta - 0.18\cos\theta$
This equation is not straightforward to solve, but we can plot the right hand side and see where it equals 0.7. This occurs at around 54 degrees.	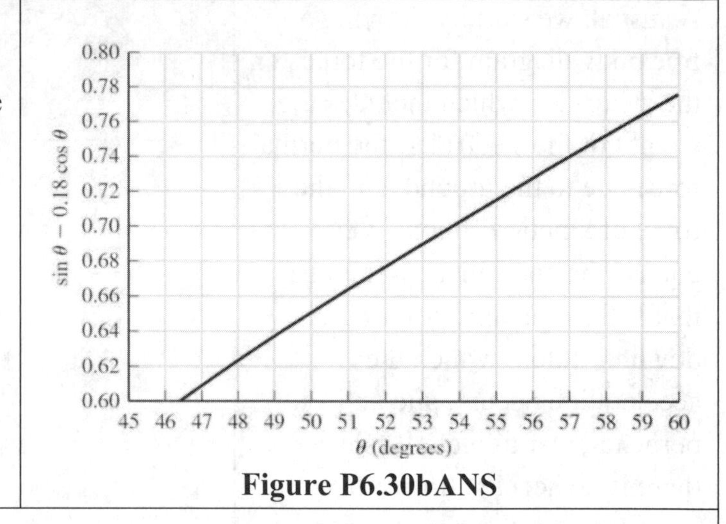 **Figure P6.30bANS**

CHECK and THINK

The answer sounds believable. The angle is pretty steep for a hill, but the acceleration is also pretty high compared to the acceleration due to gravity. Let's hope the hill is not *too* long, or the rider will be reaching a pretty high speed at the bottom of the hill!

35. A small sphere of mass $m = 0.500$ kg is dropped from rest into a viscous liquid in which the resistive force on the sphere can be expressed as $\vec{F}_D = -b\vec{v}$, and reaches one-fourth its terminal speed in 3.45 s.

a. (N) What is the terminal speed of the sphere?

INTERPRET and ANTICIPATE

This is a case of terminal velocity. As the sphere moves faster, the drag force increases. At some point, the drag force is equal to the force causing the sphere to fall through the liquid (weight of the sphere). At this point, the net force and acceleration are zero and the sphere continues to fall at a constant terminal velocity.

SOLVE

The forces on the sphere are the weight of the sphere and the drag force (we are ignoring the Buoyancy force in this problem). Terminal velocity occurs when those forces are equal, the acceleration is zero,	$bv_T = mg$ (at terminal velocity) $v_T = \dfrac{mg}{b}$

and the velocity is constant. We can equate these forces to determine this velocity.	
We aren't given a value for b, but we are told how long it takes (3.45 s) for the speed to reduce to ¼ of its original value. One approach is to try to find an expression for the velocity of the sphere over time. This is not a simple calculation, because the acceleration is not constant. We can write the net force and integrate to find an expression for the velocity. Once we have an expression for net force, we can separate our integration variables and integrate each side. The difficulty is that the speed depends on the time, so we need the *v* on the side with the *dv* in order to integrate it. (If we keep it on the side with the *dt*, we don't know *how* the speed depends on time, so we can't integrate this expression.)	$mg - bv = ma = m\frac{dv}{dt}$ $g\left(1 - \frac{bv}{mg}\right) = \frac{dv}{dt}$
Now, we integrate both sides. As the time goes from 0 to some time *t*, the speed goes from 0 to some speed *v* (or *v*(*t*)).	$-dt = \frac{dv}{\left(\frac{bv}{gm} - 1\right)}$ $-\int_0^t g\,dt = \int_o^v \frac{dv}{\left(\frac{bv}{gm} - 1\right)}$
The left side is straightforward, since the integral of *dt* is just *t*. The right side is not as easy, but the integral is a natural log. Finally, we can rearrange this expression to solve for the speed of the object at *any* time.	$-gt = \frac{gm}{b}\ln\left(\frac{bv}{gm} - 1\right)$ $-\frac{bt}{m} = \ln\left(\frac{bv}{gm} - 1\right)$ $\frac{bv}{gm} - 1 = \exp\left(\frac{-bt}{m}\right)$ (where exp(x) means e^x) $v = \frac{gm}{b}\left(1 + \exp\left(\frac{-bt}{m}\right)\right)$

As a quick reality check, notice that if you let the time approach infinite, we find that the speed equals the terminal velocity, as it should.	$v(t \to \infty) = \frac{gm}{b}$ (terminal velocity)
Ok, now that we have an expression for the velocity of the sphere in time, we can use the fact that the velocity reduces to ¼ of the original value to find out what the coefficient b must equal.	At $t = 3.45$ s: $\frac{1}{4}v_T = v_T\left[1 - \exp\left(\frac{-b(3.45 \text{ s})}{0.500 \text{ kg}}\right)\right]$ $\exp\left(\frac{-b(3.45 \text{ s})}{0.500 \text{ kg}}\right) = 0.750$ $\frac{-b(3.45 \text{ s})}{0.500 \text{ kg}} = \ln(0.750) = -0.288$ $b = \frac{(0.288)(0.500 \text{ kg})}{(3.45 \text{ s})} = 0.0417 \text{ kg/s}$
We can finally get a numerical answer for the terminal velocity.	$v_T = \left(\frac{mg}{b}\right) = \left(\frac{(0.500 \text{ kg})(9.80 \text{ m/s}^2)}{0.0417}\right) = \boxed{118 \text{ m/s}}$

CHECK and THINK

The hard part was finding an expression for the velocity of the sphere as it falls (in order to determine the coefficient b). Notice that finding an expression for the terminal velocity was actually pretty easy – the downward force equals the upward force and the velocity is constant.

b. (N) What is the distance traveled by the sphere in 3.45 s?

INTERPRET and ANTICIPATE

Again, the acceleration is not constant, so we unfortunately can't use our kinematic equations for constant acceleration. We do have an expression for the velocity though, so we could integrate that to get an expression for the position.

SOLVE To find the distance traveled, we begin with the equation for velocity	$v = \frac{dx}{dt} = \left(\frac{mg}{b}\right)\left[1 - \exp\left(\frac{-bt}{m}\right)\right]$

and integrate. Similar to above, as the time goes from 0 to t, the position changes from 0 to x (where we take the x axis to point downward).	$\int_0^x dx = \int_0^t \left(\frac{mg}{b}\right)\left[1-\exp\left(\frac{-bt}{m}\right)\right]dt$ $x = \frac{mgt}{b} + \left(\frac{m^2 g}{b^2}\right)\exp\left(\frac{-bt}{m}\right)\Big\|_0^t$ $x = \frac{mgt}{b} + \left(\frac{m^2 g}{b^2}\right)\left[\exp\left(\frac{-bt}{m}\right)-1\right]$

Ok, this is a complicated expression, but we have everything we need to substitute into it.

$$x = \frac{(0.500\text{ kg})(9.80\text{ m/s}^2)(3.45\text{ s})}{0.0417} + \left(\frac{(0.500\text{ kg})^2(9.80\text{ m/s}^2)}{(0.0417)^2}\right)\left[\exp\left(\frac{-(0.0417)(3.45\text{ s})}{(0.500\text{ kg})}\right)-1\right]$$

$$x = 405\text{ m} + 1409\text{ m}\left[\exp(-0.288)-1\right] = \boxed{52.7\text{ m}}$$

CHECK and THINK

Given the terminal velocity, the distance seems a reasonable distance.

43. Your sailboat has capsized! Fortunately, you are no longer aboard the boat. Instead, you are hanging onto the end of a long rope, the other end of which is attached to a Coast Guard helicopter. Model yourself as a particle of mass $M = 55.0$ kg with a diameter equal to 0.500 m. The density of the air is $\rho = 1.29\text{ kg/m}^3$. Assume the drag coefficient between you and the air is $C = 0.500$.

a. (N) First, ignore the drag force due to the air. If the helicopter is flying at a constant speed $v_0 = 35.0$ m/s, what angle will the rope make with the vertical?

INTERPRET and ANTICIPATE

If we ignore the drag force, the only forces on you are gravity and the tension in the rope. We start by drawing a free-body diagram and using the fact that you are moving at a constant speed to find the angle.

SOLVE

Let's first draw a free-body diagram that includes all possible forces, the force of gravity, the tension of the rope, and (only in the case of part **b.**) a drag force.

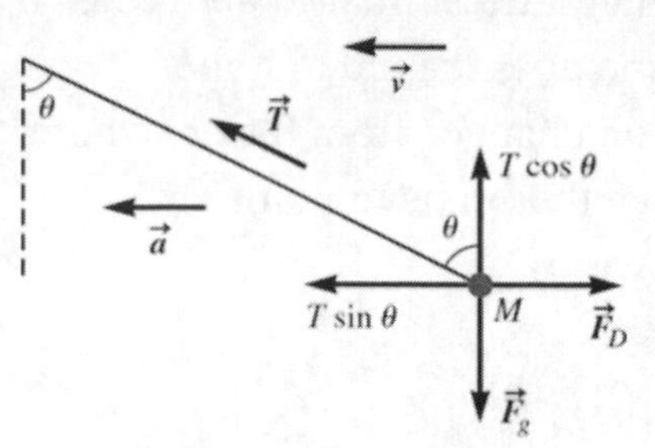

Figure P6.43ANS

We can also write force equations in general (again, for the moment, including the possibility of a drag force).

$$\sum F_y = T\cos\theta - Mg = 0$$
$$\sum F_x = T\sin\theta - F_D = 0$$
$$T\cos\theta = Mg \qquad (1)$$
$$T\sin\theta = F_D \qquad (2)$$

The helicopter has a constant horizontal velocity, so you must have the same constant horizontal velocity if you are to maintain your position with respect to the helicopter. Your acceleration is therefore zero. The net horizontal force is the difference between the horizontal component of the tension in the rope and the drag force. But the drag force is zero, so the horizontal component of the tension must be zero. The tension cannot be zero of course because the upward component of the tension must balance the gravitational force on you. Therefore, the angle must be zero.

Using equation (2):

$$T\sin\theta = 0$$
$$\boxed{\theta = 0}$$

CHECK and THINK

Although you might have a picture in your mind of the helicopter moving and the person lagging behind at the bottom of the rope, this is based on real life in which there is a drag force due to air resistance. That's the focus of part (b).

b. (N) Now, consider the drag force due to the air. What angle does the rope make with the vertical given the information in part (a)?

INTERPRET and ANTICIPATE We've written force equations in part (a) and now need to repeat that procedure without neglecting the drag force.	
SOLVE We must first calculate the drag force due to air resistance. This depends on the cross-sectional area, the speed, the density of air, and a coefficient related to how aerodynamic the shape is (Equation 6.5).	$F_D = \frac{1}{2} C\rho A v^2$
We assume the person is a spherical particle with a circular cross-section.	$A = \pi r^2 = \pi(0.250\text{m})^2$
We can now calculate the drag force.	$F_D = \frac{1}{2} C\rho A v^2 = \frac{1}{2}(0.500)(1.29)\pi(0.250)^2(35.0)^2$ N
Now, dividing Equation (2) by (1) above, we can determine the angle.	$\tan\theta = \frac{F_D}{Mg} = \frac{\frac{1}{2}(0.500)(1.29)\pi(0.250)^2(35.0)^2 \text{ N}}{(55.0 \text{ kg})(9.81 \text{ m/s}^2)}$ $\theta = \tan^{-1}\left[\frac{\frac{1}{2}(0.500)(1.29)\pi(0.250)^2(35.0)^2 \text{ N}}{(55.0 \text{ kg})(9.81 \text{ m/s}^2)}\right]$ $\theta = \boxed{8.18°}$
CHECK and THINK The angle sounds possible. As expected of the drag force is increased somehow, for instance because the helicopter moves faster, the angle will increase and the person will lag further behind the helicopter. This probably agrees with your intuition. The three significant figures is probably too precise for an estimate where we assume you are a sphere, but our calculation reflects the precision of the numbers given in the problem.	

49. (N) Artificial gravity is produced in a space station by rotating it, so it is a noninertial reference frame. The rotation means that there must be a centripetal force exerted on the occupants; this centripetal force is exerted by the walls of the station. The space station in Arthur C. Clarke's *2001: A Space Odyssey* is in the shape of a four-spoked wheel with a diameter of 155 m. If the space station rotates at a rate of 2.40 revolutions per minute, what is the magnitude of the artificial gravitational acceleration, provided to a space tourist walking on the inner wall of the station?

INTERPRET and ANTICIPATE

Let's first draw a sketch of the situation. The station is rotating, causing the person to accelerate towards the center of the circle. (The person must be accelerating since their velocity is changing.) If the station suddenly disappeared, the person would just fly off in a straight line. Instead, the station pushes up on the occupants to keep them moving in a circular trajectory. From the occupant's point of view, they would feel "pressed towards the center" by the wall of the space station, similar to gravity. It's the centripetal acceleration that tells us the acceleration of this "effective gravity."

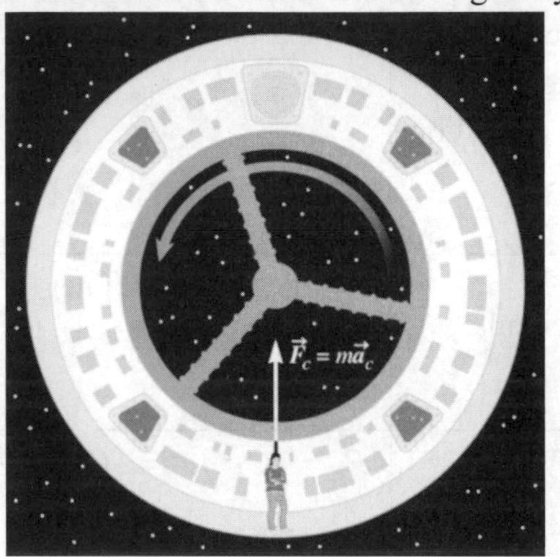

Figure P6.49ANS

SOLVE	
Centripetal acceleration depends on the speed of the object and the radius of the orbit (Eq. 4.38).	$a_c = \dfrac{v^2}{r}$
We are given the diameter of the station and the frequency in revolutions per minute. We first need to find the radius of the circular orbit (half the diameter) and convert the frequency into revolutions per second.	$r = \dfrac{155 \text{ m}}{2} = 77.5 \text{ m}$ $f = 2.40\dfrac{\text{rev}}{\text{min}}\left(\dfrac{1 \text{min}}{60 \text{ s}}\right) = 0.0400\dfrac{1}{\text{s}}$
The velocity of rotation can be calculated as the distance traveled over time (Eq. 4.30). The distance traveled is the circumference of the circular orbit ($2\pi r$) and the time is	$v = \dfrac{2\pi r}{T} = 2\pi r f$

the period of the orbit ($T = 1/f$).	
Now we can plug everything in to calculate an answer.	$a_c = \frac{(2\pi rf)^2}{r}$ $a_c = 4\pi^2 f^2 r$ $a_c = 4\pi^2 (0.0400\ \text{s}^{-1})^2 (77.5\ \text{m})$ $a_c = \boxed{4.90\ \text{m/s}^2}$
CHECK and THINK The acceleration is about ½ g. So, it is a reduced gravity environment, but a larger acceleration than found on the Moon, for instance, so it should work to provide a reasonable gravitational field.	

51. Harry Potter decides to take Pottery 101 as an elective to satisfy his arts requirement at Hogwarts. He sets some clay ($m = 3.25$ kg) on the edge of a pottery wheel ($r = 0.600$ m), which is initially motionless. He then begins to rotate the wheel with a uniform acceleration, reaching a final angular speed of 2.400 rev/s in 3.00 s.
a. (N) What is the speed of the clay when the initial 3.00 s has passed?

INTERPRET and ANTICIPATE We know the angular velocity at this moment in time, as well as the radius from the rotation axis, so we can calculate the linear speed directly.	
SOLVE In this case, we just need the relationship between the linear velocity and the angular velocity (Eq. 4.37). The angular velocity though is in *radians* per second, so we need to use the fact that there are 2π radians per revolution.	$v = r\omega$ $v = (0.600\ \text{m})\left(2.400\ \text{rev/s} \times \frac{2\pi\ \text{rad}}{1\ \text{rev}}\right)$ $v = \boxed{9.05\ \text{m/s}}$
CHECK and THINK We've calculated the speed using equations for rotational motion from Chapter 4.	

b. (N) What is the centripetal acceleration of the clay initially and when the initial 3.00 s has passed?

<table>
<tr><td colspan="2">INTERPRET and ANTICIPATE
Given the speed and orbital radius of a rotating object, we can calculate the centripetal acceleration. Since the clay is not moving initially, the initial centripetal acceleration must be $\boxed{0}$!</td></tr>
<tr><td>SOLVE
The centripetal acceleration is the velocity of the object squared divided by the radius of the orbit (Eq. 4.38). Use the speed at 3.00 s to find the centripetal acceleration at that time.</td><td>$$a_c = \frac{v^2}{r} = \frac{\left[(0.600\ \text{m})\left(2.400\ \text{rev/s} \times \frac{2\pi\ \text{rad}}{1\ \text{rev}}\right)\right]^2}{0.600\ \text{m}}$$
$$a_c = \boxed{136\ \text{m/s}^2}$$</td></tr>
<tr><td colspan="2">CHECK and THINK
In this case, we know all of the quantities needed to calculate the centripetal acceleration. This is pretty high at around 14 g's!</td></tr>
</table>

c. (N) What is the magnitude of the constant tangential acceleration responsible for starting the clay in circular motion?

<table>
<tr><td colspan="2">INTERPRET and ANTICIPATE
Similar to the connection between angular and linear velocity, there is a relationship between angular and linear acceleration.</td></tr>
<tr><td>SOLVE
The relationship between angular and linear acceleration looks like the relationship between angular and linear velocity.</td><td>$$a_t = r\alpha_c$$</td></tr>
<tr><td>We need to determine the angular acceleration, which is the change in angular velocity in time. This is the angular velocity at 3 seconds minus that at 0 seconds (when the velocity is zero), divided by the time interval (3 seconds). Again, we need to be sure to use radians for angular velocity and acceleration.</td><td>$$\alpha = \frac{\left[\left(2.400\ \text{rev/s} \times \frac{2\pi\ \text{rad}}{1\ \text{rev}}\right) - 0\ \frac{\text{rad}}{\text{s}}\right]}{3.00\ \text{s}}$$
$$\alpha = 5.03\ \text{rad/s}^2$$</td></tr>
</table>

Now, we can calculate the tangential acceleration.	$a_t = r\alpha = (0.600\text{ m})(5.03\text{ rad/s}^2) = \boxed{3.02\text{ m/s}^2}$

CHECK and THINK

The relationship for the tangential/linear accelerations is similar to the relationship for the tangential/linear velocities. We have to be careful to use radians for the angles, but otherwise, this is a straightforward calculation.

58. (N) A satellite of mass 16.7 kg in geosynchronous orbit at an altitude of 3.58×10^4 km above the Earth's surface remains above the same spot on the Earth. Assume its orbit is circular. Find the magnitude of the gravitational force exerted by the Earth on the satellite. *Hint*: The answer is not 163 N.

INTERPRET and ANTICIPATE

Since we don't have an expression for the gravitational force between these objects, it must be something we can figure out knowing that the satellite is in orbit. The key is that the satellite is in circular orbit. *Anything* in circular motion is accelerating and there *must* be a centripetal force keeping the object in orbit. In this case, it's the force of gravity that is keeping the satellite in orbit. That is, the gravitational force must be equal to the centripetal force, which we can calculate. We're told that the answer is not 163 N, which is what you would expect for the weight of the satellite if it were on the Earth's surface.

SOLVE We use the fact that the speed is related to the orbital radius and period (Eq. 4.30).	$v = \dfrac{2\pi r}{T} = 2\pi r f$
Now, we can write an expression for the centripetal force (Eq. 6.7).	$F_c = ma_c = m\dfrac{v^2}{r} = m\dfrac{4\pi^2 r}{T^2}$
The period for geosynchronous orbit is 24 hours, or 86,400 s and the radius includes the radius of the Earth and is converted to meters to calculate a force in newtons.	$F_c = (16.7)\dfrac{4\pi^2(4.22\times10^7)}{(8.64\times10^4)^2} = \boxed{3.7\text{ N}}$

CHECK and THINK

This is *much* less than the force of gravity for the object on the surface of the Earth (which is 163 N). The object is further from the Earth, so the gravitational force due to the Earth is smaller.

61. (N) A car with a mass of 1453 kg is rolling along a flat stretch of road and eventually comes to a stop due to rolling friction. If the car begins with a speed of 10.0 m/s and the car comes to a stop in 6.88 s, what is the coefficient of rolling friction between the tires and the road?

INTERPRET and ANTICIPATE

The rolling friction force is solely responsible for stopping the car. It looks like we might have enough information to determine the acceleration of the car, use this to determine the force of rolling friction, and then find the coefficient of friction.

SOLVE	
The free-body diagram includes the normal force, gravitational force, and force of friction, which points in the opposite direction the car is moving (assumed to be to the left in this figure), since the car is decelerating.	y, $\vec{F}_N$, $\vec{a}$, $\vec{F}_r$, x, $\vec{F}_g$ **Fig P6.61ANS**
Based on the y components of the force, the normal force is equal to the weight of the car.	$\sum F_y = F_N - F_g = F_N - mg = 0$ $F_N = mg$
Using the x components of the force, we can relate the acceleration to the rolling friction force, which depends on the coefficient of rolling friction and the normal force (Equation 6.3).	$\sum F_x = F_r = ma$ $\mu_r F_N = \mu_r mg = ma$
Using the definition of acceleration, we can write the average acceleration as a change in velocity over time and solve for the coefficient of rolling friction between the tires and the road. We equate the magnitude of the friction force to the magnitude of ma for simplicity. If we are	$\mu_r mg = m\left(\left\|\frac{\Delta v}{\Delta t}\right\|\right)$ $\mu_r = \left\|\frac{\Delta v}{g\Delta t}\right\| = \left\|\frac{(0-10.0\text{ m/s})}{(9.81\text{ m/s}^2)(6.88\text{ s})}\right\| = \boxed{0.148}$

careful and not simply taking the magnitudes, we have to remember that the initial speed is actually in the negative direction based on our coordinate system and $\Delta v = (0-(-10\text{ m/s}))$, so the sign is indeed positive anyway.	

CHECK and THINK

We know that a coefficient of friction around 1 would be a pretty strong sliding friction coefficient, so a coefficient of around 0.15 for this rolling friction sounds plausible.

65. (A) A box of mass m rests on a rough, horizontal surface with coefficient of static friction μ_s. If a force $\vec{F}_p$ is applied to the box at an angle θ as shown, what is the minimum value of θ for which the crate will not move regardless of the magnitude of $\vec{F}_p$?

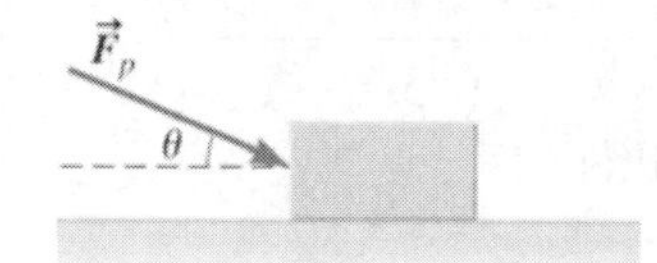

Figure P6.65

INTERPRET and ANTICIPATE

The box is in equilibrium, just before it starts to move. If the angle is very low, we will be able to push the box. If the angle is large, we will mostly push the box into the ground and the horizontal component will not be enough to cause the box to move. We should draw a free-body diagram and determine a condition for when the horizontal component of the pushing force cannot be larger than the static friction force. That is, we will look for the angle where this condition is just met.

SOLVE

Let's first draw a free-body diagram that includes all the forces, the gravitational, normal, friction, and pushing forces.	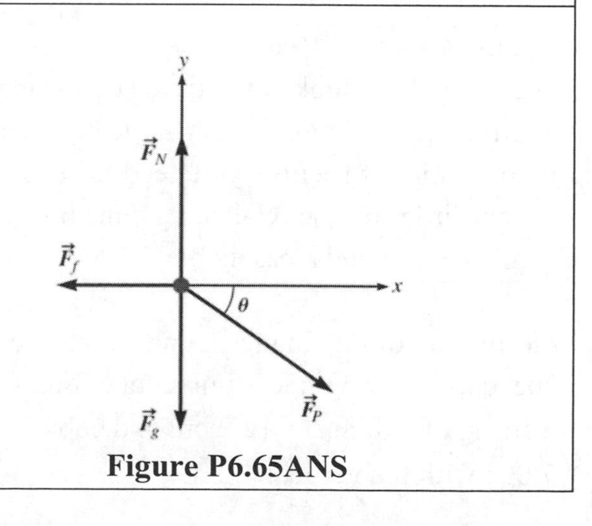 **Figure P6.65ANS**

The condition we want to satisfy such that the box does *not* move, is that the *x* component of the pushing force must be less than the force of static friction. The point where it is just able to overcome static friction is when these quantities are equal.	$F_p \cos\theta \leq \mu_s F_N$ $F_p \cos\theta = \mu_s F_N$
We don't know the normal force, but writing the *y* components of Newton's second law, we get an expression for the normal force.	$\sum F_y = F_N - mg - F_p \sin\theta = 0$ $F_N = mg + F_p \sin\theta$
We can substitute this into our condition for the minimum angle for which motion occurs.	$F_p \cos\theta = \mu_s \left(mg + F_p \sin\theta\right)$ $F_p (\cos\theta - \mu_s \sin\theta) = \mu_s mg$
We want to find an angle for which there is *no* force that can satisfy this relationship. If the expression in the parentheses is positive, then there is always *some* force F_p that will make the left side equal to the right. If the expression in parentheses is negative though, there is *no* force that will allow the left and right sides to be equal. This satisfies our requirement.	$(\cos\theta - \mu_s \sin\theta) < 0$ $\mu_s \tan\theta > 1$ $\tan\theta > \frac{1}{\mu_s}$ $\theta > \boxed{\tan^{-1}\left[1/\mu_s\right]}$

CHECK and THINK

We can at least look at the limits of this expression and see if they make sense. If the coefficient of friction is very large then the angle is very small and vice versa. That is, if there is a lot of friction, you need to keep a shallow angle to get the box moving. If there is very little friction, you can get the box moving by pushing down at a much higher angle. This sounds reasonable.

68. Instead of moving back and forth, a conical pendulum moves in a circle at constant speed as its string traces out a cone (Fig. P6.68). One such pendulum is constructed with a string of length $L = 12.0$ cm and bob of mass 210 g. The string makes an angle of $\theta = 7.00°$ with the vertical.

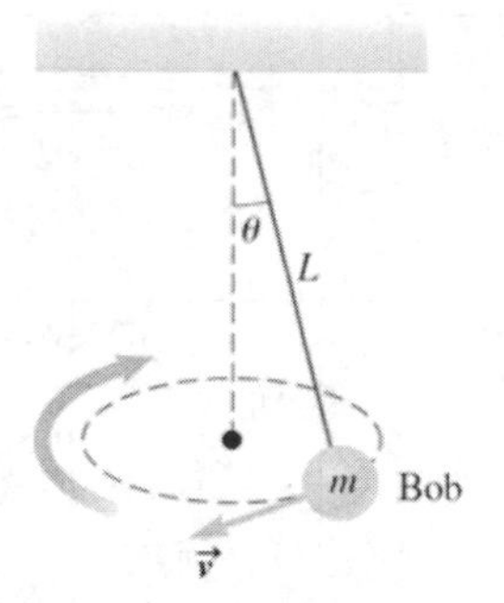

Figure P6.68

a. **(N)** What is the radial acceleration of the bob?

INTERPRET and ANTICIPATE The pendulum bob is tracing out a circular orbit. Therefore, the radial acceleration is actually a centripetal acceleration, which we can calculate.	
SOLVE The free-body diagram shows that the vertical component of the tension force balances gravity, and the horizontal component supplies the centripetal acceleration of the bob. The diagram is drawn for a moment in time when the centripetal acceleration is along the positive x axis and the y axis points straight up.	**Figure P6.68ANS**
In previous centripetal force problems, we were usually seeking a speed and radius of the orbit to calculate v^2/r. We aren't readily given these quantities, so we can also use $F = ma$ and determine the radial acceleration. We can write Newton's second law for the x component. This will allow us to calculate the centripetal acceleration, but we need an expression for the tension.	$\sum F_x = F_T \sin\theta = ma_c$
Now using the y components, we can relate the tension to other known quantities.	$\sum F_y = F_T \cos\theta - mg = 0$

	$F_T = \frac{mg}{\cos\theta} = \frac{(0.200\text{ kg})(9.80\text{ m/s}^2)}{\cos 7.00°} = 1.97\text{ N}$
Finally, using the x component equation, we can determine the acceleration.	$a_c = \frac{F_T \sin\theta}{m}$ $a_c = \frac{(1.97\text{ N})\sin 7.00°}{(0.200\text{ kg})}$ $a_c = \boxed{1.20\text{ m/s}^2\text{ inward}}$

CHECK and THINK

We don't have a good way to predict the answer, but we've calculated a centripetal acceleration using the inward component of the tension, which keeps the bob in circular motion.

b. (N) What are the horizontal and vertical components of the tension force exerted by the string on the bob?

INTERPRET and ANTICIPATE

From part (a), we know the magnitude of the force and, from the problem statement, we know the angle of the string. That is, we know the magnitude and direction and now simply need to write it in vector form.

SOLVE	
The tension has a magnitude of F_T = 1.97 N from part (a) and makes an angle of 7 degrees with respect to the vertical.	$\vec{F}_T = F_T \sin\theta\,\hat{\imath} + F_T \cos\theta\,\hat{\jmath}$ $\vec{F}_T = (1.97\text{ N})\sin 7.00°\hat{\imath} + (1.97\text{ N})\cos 7.00°\hat{\jmath}$ $\vec{F}_T = \boxed{\left(0.241\,\hat{\imath} + 1.96\,\hat{\jmath}\right)\text{N}}$

CHECK and THINK

Given the small angle of the string from vertical, the y component of the tension is much larger than the x component.

71. (A) A block of mass M is placed on a plane, where μ_s is the coefficient of static friction between the block and the plane. The plane is inclined at an angle θ. When the plane is stationary, the block slips down the plane. We can keep the block in place on the incline by revolving the plane around the vertical axis shown in Figure P6.67. In this case, what is the *maximum* period of revolution (that is, the slowest rotation) that will keep the block on the incline a distance d from the lower end?

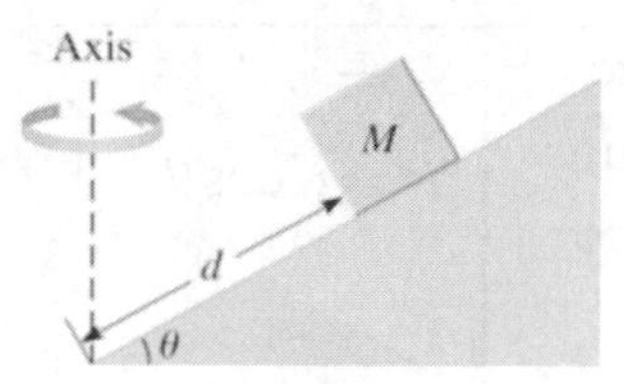

Figure P6.67

<table>
<tr><td colspan="2">INTERPRET and ANTICIPATE
In the case where the block is placed on a non-revolving incline, the static friction force points up the plane to counter the component of the gravitational force down the plane. When the plane revolves around the vertical axis, as in Problem 6.67, the normal force is increased because it now must contribute to a net force toward the center to maintain circular motion and the normal force is increased. In addition, the centripetal acceleration is larger, so a larger inward net force is needed just to keep the block moving at a constant distance from the rotation axis. These both help keep the block in place relative to the plane. We are now asked to calculate a maximum period. Why is there a maximum? If the period is long enough (and the revolution is slow enough), we approach the no-revolution case, and we know that the block will slip. So there must be a maximum period beyond which we cannot go without the block slipping.</td></tr>
<tr><td>SOLVE
The free-body diagram is drawn here, which includes the gravitational, normal, and static friction forces. There must be a total (net) force since there is a centripetal force to the left maintaining the circular motion of the block. Refer also to Problem 67 for a similar problem on a frictionless plane.</td><td>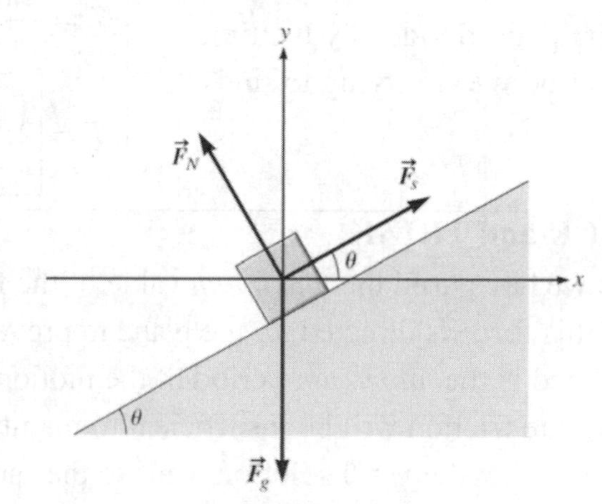

Figure P6.71ANS</td></tr>
<tr><td>First, write Newton's second law for the vertical and horizontal directions. There is no vertical acceleration, but there is a horizontal (centripetal) acceleration.</td><td>$\sum F_{vert} = F_N \cos\theta + F_s \sin\theta - Mg = 0$
$\sum F_{horiz} = F_N \sin\theta - F_s \cos\theta = Ma_c = M\frac{v^2}{R}$</td></tr>
</table>

Now use the formula for the static friction force and express the radius of the circular orbit in terms of the distance along the plane, using trigonometry.	$F_s = \mu_s F_N$ $R = d\cos\theta$
We can now use the vertical components to determine an expression for the normal force.	$F_N \cos\theta + \mu_s F_N \sin\theta - Mg = 0$ $F_N = \dfrac{Mg}{\cos\theta + \mu_s \sin\theta}$
This can now be substituted into the formula for the horizontal components of the force.	$F_N \sin\theta - \mu_s F_N \cos\theta = M\dfrac{v^2}{d\cos\theta}$ $v^2 = F_N \dfrac{d\cos\theta}{M}(\sin\theta - \mu_s \cos\theta)$ $v = \sqrt{gd\cos\theta\left(\dfrac{\sin\theta - \mu_s\cos\theta}{\cos\theta + \mu_s \sin\theta}\right)}$
Finally, we find the period by defining the speed as the distance traveled (circumference of the circular path) divided by the time (the period we are trying to find).	$v = \dfrac{2\pi R}{T} = \dfrac{2\pi d\cos\theta}{T}$ $T = \dfrac{2\pi d\cos\theta}{v}$ $T = \boxed{2\pi d\cos\theta\sqrt{\dfrac{\cos\theta + \mu_s\sin\theta}{gd\cos\theta(\sin\theta - \mu_s\cos\theta)}}}$

CHECK and THINK

Since we have used the *maximum* value of the force of static friction ($F_s = \mu_s F_N$) and since this force is directed *up* the plane to prevent slipping down the plane, what we have calculated is the *maximum* period of the motion. At any *slower* rotation, with a *longer* period, the friction will be insufficient to maintain circular motion at this same radius. Note also that the $v = 0$ solution leads to the same result as in Problem 70 for the none-rotating case. This is a good sign!

74. A car is driving around a banked, circularly curved road with a radius of 5.00×10^2 m. The mass of the car is 1500 kg, and the road is banked at an angle of 20°.

a. (N) Using the results of Example 6.8 (pages 204-205) and ignoring friction, what is the maximum speed the car can have without slipping?

INTERPRET and ANTICIPATE	
In this case, we are told to use the results of Example 6.8 to find an appropriate relationship and apply it in this situation.	
SOLVE Taking the result of Example 6.8 directly, inserting the values given. The maximum speed depends on the radius of the track and the angle that the track is inclined.	$v_{max} = \sqrt{gr\tan\theta}$ $v_{max} = \sqrt{(9.81\ \text{m/s}^2)(5.00\times10^2\ \text{m})(\tan 20^\circ)}$ $v_{max} = \boxed{42.3\ \text{m/s}}$
CHECK and THINK This is a straight substitution problem, but the speed sounds like a reasonable value for a car.	

b. (N) What is the magnitude of the centripetal force on the car if it were to travel at the maximum speed around the track?

INTERPRET and ANTICIPATE	
The centripetal force depends on the mass, speed, and radius of the circular motion. We have the necessary ingredients to calculate this.	
SOLVE The centripetal force depends on the mass, speed, and radius according to Equation 6.7. The speed was calculated in part (a) and the other quantities are given.	$F_c = m\dfrac{v^2}{r}$ $F_c = (1500\ \text{kg})\dfrac{(42.3\ \text{m/s})^2}{(5.00\times10^2\ \text{m})}$ $F_c = \boxed{5.4\times10^3\ \text{N}}$
CHECK and THINK With the speed from part (a) and the known mass and radius of the circular motion, we calculate a centripetal force in Newtons.	

c. (C) Would including friction in this problem increase the maximum speed allowed or decrease it? Explain your answer.

Adding friction would increase the maximum speed allowed since there would be an additional force contributing to the centripetal force, helping accelerate the car towards the center of motion and reducing the chance that the car will slide off of the track. If you imagine that the track is extremely slippery, then clearly the car will more easily slide off the track.

78. An aircraft carrier approaching home port cuts its engines and coasts to a full stop at the dock. Upon cutting engine power, the aircraft carrier's speed is seen to be decreasing exponentially, given by $v = v_i e^{-0.050t}$, where t is in seconds.

a. (N) If the aircraft carrier's initial speed is 22 knots (11.3 m/s), what is its speed at $t =$ 30.0 s after cutting its engines?

INTERPRET and ANTICIPATE	
We are provided an expression for the speed of the aircraft carrier versus time. With the initial speed given, we can substitute the desired time to determine the speed at that time.	
SOLVE Using the formula given, we substitute the initial velocity and the time of 30 seconds.	$v(t = 30.0\text{ s}) = (11.3\text{ m/s})e^{-0.050(30.0\text{ s})}$ $v(30.0\text{ s}) = \boxed{2.53\text{ m/s}}$
CHECK and THINK The speed is at least slower than the initial speed, which we expect for this ship that is slowing down in time.	

b. (A) What is the acceleration of the aircraft carrier as a function of time?

INTERPRET and ANTICIPATE	
The velocity depends on time and we are given an algebraic expression for this velocity as a function of time. The acceleration of an object is equal to the derivative of the velocity, so all we need to do is take the derivative.	
SOLVE We differentiate the expression for v with respect to time to determine the acceleration.	$a = \dfrac{dv}{dt}$ $a = -0.050 v_i e^{-0.050t}$ $a = \boxed{-0.050v}$
CHECK and THINK The acceleration is negative, which makes sense, since the ship is moving in the positive direction and slowing down.	

80. (N) A particle of dust lands 45.0 mm from the center of a compact disc (CD) that is 120 mm in diameter. The CD spins up from rest, and the dust particle is ejected when the CD is rotating at 90.0 revolutions per minute. What is the coefficient of static friction between the particle and the surface of the CD?

<table>
<tr><td colspan="2">INTERPRET and ANTICIPATE
As the CD spins up, the dust particle rotates in a circular trajectory and has an increasing centripetal acceleration (and centripetal force). The moment the dust particle loses contact and flies off in a straight line is the moment that this centripetal force is too large for the static friction force to keep the dust particle stuck on the CD.</td></tr>
<tr><td>SOLVE
We will need to know the linear speed of rotation of the dust particle when it is ejected to calculate centripetal acceleration.</td><td>$v = 2\pi r f$
$v = 2\pi(0.045\text{ m})(90.0\text{ rpm})\left(\frac{1\text{ min}}{60\text{ s}}\right)$
$v = 0.424\text{ m/s}$</td></tr>
<tr><td>Taking x in the radially inward direction and y in the vertical direction and applying Newton's second law in the x direction allows us to write the normal force in terms of the weight of the dust particle.</td><td>$\sum F_y = F_N - mg = 0$
$F_N = mg$</td></tr>
<tr><td>Applying Newton's second law in the y direction provides an expression for the coefficient of static friction.</td><td>$\sum F_x = F_f = ma_c$
$F_f = \frac{mv^2}{r}$
$\mu F_N = \frac{mv^2}{r}$
$\mu mg = \frac{mv^2}{r}$
$\mu = \frac{v^2}{gr}$</td></tr>
<tr><td>Finally, we can substitute numerical values to determine this coefficient.</td><td>$\mu = \frac{v^2}{gr} = \frac{(0.424\text{ m/s})^2}{(9.81\text{ m/s}^2)(0.045\text{ m})} = \boxed{0.407}$</td></tr>
<tr><td colspan="2">CHECK and THINK
The coefficient of static friction sounds like a reasonable number based on other coefficients we've encountered.</td></tr>
</table>

7

Gravity

5. (N) You are given a string of length 10 cm, two tacks, and a pencil. You are to use these to draw an ellipse as in Figure 7.5B. How far apart must you place the tacks to ensure that the semimajor axis of the ellipse is 5 cm? Explain your answer. If possible, try it yourself.

INTERPRET and ANTICIPATE

We can tie each end of the string to the tacks and then trace an ellipse with the pencil by tracing the path of the largest distance we can reach while keeping the string taught. We require the semimajor axis, half the distance across the longest axis of the ellipse, to be 5 cm. It will be easiest to sketch an ellipse and think about the conditions needed to ensure that this is 5 cm while the total length of the string is 10 cm.

SOLVE

Let's first sketch the position of the tacks and string when the pencil is at the furthest point along the semimajor axis, since this is the point that will define the semimajor axis when we trace the ellipse. We call the separation between the tacks d and the distance from the tack to the end of the major axis B.	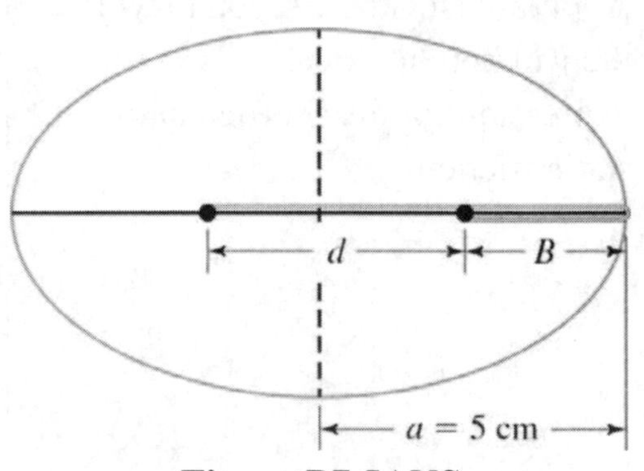 **Figure P7.5ANS**
Referring to the drawing, the semimajor axis distance (5 cm) is half the distance between the tacks ($d/2$) plus the distance from a tack to the end of the ellipse (B).	$B+\frac{d}{2}=5\text{ cm}$
We can also write an equation guaranteeing that the total length of the string is 10 cm: Inspecting the drawing, the string covers the distance between the two tacks (d) plus the distance to the edge of the ellipse (B) and back to the second tack (B), or in total $d+2B$. This is actually the	$2B+d=10\text{ cm}\rightarrow B+\frac{d}{2}=5\text{ cm}$

same equation we just found, which means that the length of the string does not put any new constraints on the parameters.	
Now we have one equation with two unknowns. There is a range of possibilities that will satisfy this equation. If $d = 0$ cm and $B = 5$ cm, we will trace out a circle with a radius of 5 cm. If $d = 10$ cm and $B = 0$ cm, we will get a straight line and half the length of this line is 5 cm. For any d between 0 and 10 cm, we will trace out an ellipse with a semimajor axis of 5 cm.	$\boxed{0\text{ cm} < d < 10\text{ cm}}$
CHECK and THINK In this case, there is not a single answer, but rather a range of ellipses with a semimajor axis of 5 cm that can be constructed in this way.	

6. (N) Io and Europa are two of Jupiter's many moons. The mean distance of Europa from Jupiter is about twice as far as that for Io and Jupiter. By what factor is the period of Europa's orbit bigger than that of Io's?

INTERPRET and ANTICIPATE Kepler's third law relates the cube of the semimajor axis to the square of the orbital period of an orbiting body. Given the ratio of the distances, we can use Kepler's law to determine the ratio of the periods. The larger the orbit, the larger the period of the orbiting body.	
SOLVE When using astronomical units, with the semimajor axis a represented in AU (Astronomical Unit, approximately the mean Earth-Sun distance) and the period in years, we can write Kepler's third law as in Equation 7.2, $T^2 = a^3$. Both Europa and Io satisfy this relationship, so we write this equation for each of the moons.	$T_{\text{Eu}}^2 = a_{\text{Eu}}^3$ $T_{\text{Io}}^2 = a_{\text{Io}}^3$

Then, the ratio of the periods can be solved for in terms of the distances between each of the moons and Jupiter, using the fact that $a_{\text{Eu}} = 2a_{\text{Io}}$.	$\frac{T_{\text{Eu}}^2}{T_{\text{Io}}^2} = \frac{a_{\text{Eu}}^3}{a_{\text{Io}}^3} = \frac{(2a_{\text{Io}})^3}{a_{\text{Io}}^3} = 8$ $\frac{T_{\text{Eu}}}{T_{\text{Io}}} = \boxed{\sqrt{8}}$
CHECK and THINK Since Europa has a larger orbital distance, it has a larger period as expected.	

11. (N) One method for estimating the travel time from the Earth to the Moon is to use Kepler's third law to find half the period of an orbit whose perigee is at the Earth and whose apogee is at the Moon. What is the travel time to the Moon, a distance of 3.84×10^5 km from the Earth's center, for a spacecraft initially at an altitude of 200 km in the Earth's orbit, assuming it does not use any means of propulsion?

INTERPRET and ANTICIPATE

Kepler's third law gives the relation between the orbital period T and the semimajor axis of the orbit. We can put the spacecraft on an orbit around the Earth such that its perigee is at the Earth and apogee is at the Moon as shown in the figure. For the spacecraft to reach the moon, it must travel for half of the total period for this orbit.

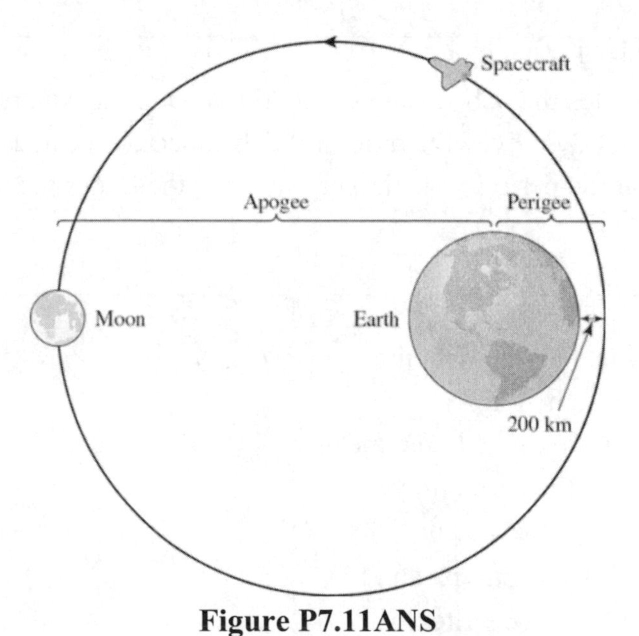

Figure P7.11ANS

SOLVE For an object in orbit about Earth, Kepler's third law gives the relation between the orbital period T and the	$T^2 = \left(\frac{4\pi^2}{GM_\oplus}\right)a^3$

semimajor axis of the orbit (Equation 7.6).	
The semimajor axis is half the major axis (apogee plus perigee distance, as in Equation 7.1). The apogee distance is from the Earth to the Moon $r_{\text{apogee}} = 3.84 \times 10^8$ m and the perigee distance is the radius of Earth plus the 200 km orbital height $r_{\text{perigee}} = 6.37 \times 10^3 + 200\text{ km} = 6.57 \times 10^6$ m).	$a = \dfrac{r_{\text{apogee}} + r_{\text{perigee}}}{2}$ $a = \dfrac{(3.84\times10^8\text{ m}) + (6.57\times10^6\text{ m})}{2}$ $a = 1.95\times10^8\text{ m}$
Then the period of a complete orbit, or a round trip to the moon can be found with Kepler's third law.	$T = 2\pi\sqrt{\dfrac{a^3}{GM_\oplus}}$ $T = 2\pi\sqrt{\dfrac{(1.95\times10^8\text{ m})^3}{\left(6.67\times10^{-11}\text{ N}\cdot\text{m}^2/\text{kg}^2\right)(5.98\times10^{24}\text{ kg})}}$ $T = 8.59\times10^5\text{ s}$
The time to reach the Moon on a one-way trip is half of this period.	$\Delta t = \dfrac{1}{2}T$ $\Delta t = \dfrac{8.59\times10^5\text{ s}}{2}$ $\Delta t = 4.295\times10^5\text{ s} = \boxed{4.97\text{ days}}$

CHECK and THINK

We've estimated the time to reach the Moon by putting a spacecraft in an orbit around the Earth that reaches the Moon. So how does our estimate of five days compare to actual missions to the moon? There is a wide range, but the Apollo missions that first landed men on the Moon took a little over three days to travel from Earth to orbit the Moon.

14. (N) Since 1995, hundreds of extrasolar planets have been discovered. There is the exciting possibility that there is life on one or more of these planets. To support life similar to that on the Earth, the planet must have liquid water. For an Earth-like planet orbiting a star like the Sun, this requirement means that the planet must be within a habitable zone of 0.9 AU to 1.4 AU from the star. The semimajor axis of an extrasolar planet is inferred from its period. What range in periods corresponds to the habitable zone for an Earth-like planet orbiting a Sun-like star?

Chapter 7 – Gravity

INTERPRET and ANTICIPATE

Kepler's third law relates the semimajor axis of an orbiting planet to its period. Given the range of orbital distances provided, we can determine the corresponding range of periods.

SOLVE	
When using astronomical units, with the semimajor axis *a* represented in AU (Astronomical Units) and the period in years, we can write Kepler's third law as in Equation 7.2, $T^2 = a^3$. We can calculate the period using each of the two limits required.	$T = a^{3/2} = (0.9)^{3/2} \approx \boxed{0.9 \text{ yr}}$ $T = a^{3/2} = (1.4)^{3/2} \approx \boxed{1.7 \text{ yr}}$

CHECK and THINK

With Kepler's third law and the desired range for the semimajor axis, we are able to calculate a range of orbital periods. Since the semimajor axes are around the mean radius of Earth's orbit (1 AU), the range of periods for habitable planets are also around the Earth's period (1 year). So, the Earth's period is in the range for habitable planets, as we'd expect!

17. The mass of the Earth is approximately 5.98×10^{24} kg, and the mass of the Moon is approximately 7.35×10^{22} kg. The Moon and the Earth are separated by about 3.84×10^{8} m.

a. (N) What is the magnitude of the gravitational force that the Moon exerts on the Earth?

INTERPRET and ANTICIPATE

The Earth and the Moon (or any two masses) exert a gravitational force on each other. Newton's Law of Universal Gravity can be used to calculate this gravitational force between any two masses.

SOLVE	
Newton's Law of Universal Gravity tells us that the gravitational force between any two objects is proportional to each mass and inversely proportional to their separation squared (Equation 7.4).	$F_G = G\frac{m_1 m_2}{r^2}$

We are given the masses and separation in metric base units and can plug in the gravitational constant to get a force in Newtons.	$F_G = \left(6.67\times10^{-11}\ \frac{\text{N}\cdot\text{m}^2}{\text{kg}^2}\right)\frac{(5.98\times10^{24}\ \text{kg})(7.35\times10^{22}\ \text{kg})}{(3.84\times10^{8}\ \text{m})^2}$ $F_G = \boxed{1.99\times10^{20}\ \text{N}}$
CHECK and THINK The force is quite large for these two massive objects. These are a Newton's third law pair, so this is both the gravitational force of the Earth on the Moon *and* the gravitational force of the Moon on the Earth.	

b. (N) If Serena is on the Moon and her mass is 25 kg, what is the magnitude of the gravitational force on Serena due to the Moon? The radius of the Moon is approximately 1.74×10^6 m.

INTERPRET and ANTICIPATE In this case, we are asked about the force on Serena due to the Moon. Therefore, the two masses are *Serena* and the *Moon* and the separation distance we need is the distance between Serena and the Moon (i.e. essentially the radius of the Moon).	
SOLVE We again start with Newton's Universal Law of Gravitation, which can be used to find the distance between any two masses.	$F_G = G\frac{m_1 m_2}{r^2}$
We now insert the mass of Serena and the Moon. In addition, the distance between Serena and the Moon is, more precisely, the distance between the centers of mass of Serena and the Moon. Since the Moon is much larger tham Serena, this is essentially the distance from the center of the Moon to its surface, or the radius of the Moon.	$F_G = \left(6.67\times10^{-11}\ \frac{\text{N}\cdot\text{m}^2}{\text{kg}^2}\right)\frac{(25\ \text{kg})(7.35\times10^{22}\ \text{kg})}{(1.74\times10^{6}\ \text{m})^2}$ $F_G = \boxed{40\ \text{N}}$
CHECK and THINK Even though the distance is much smaller than in part (a), Serena's mass is also much smaller than the Earth's and the resulting force is much smaller than we found in part (a).	

This is about the right order of magnitude for what we would expect though. If Serena was instead on Earth, we know that the gravitational force due to the Earth on her (which is also called her weight) would be $mg = (25\text{ kg})(9.81\text{ m/s}^2) \approx 250\text{ N}$. That is, on the moon, her weight would be about $1/6^{\text{th}}$ of what it is on Earth, which sounds reasonable, since we know that gravity is weaker on the moon.

24. (A) Suppose a planet with a mass *m* is orbiting a star with a mass *M* and the mean distance between the planet and the star is *a*. Using Newton's law of universal gravity, derive an algebraic expression for the speed of the planet when it is at the mean distance from the star.

INTERPRET and ANTICIPATE

We have so far used Newton's Universal Law of Gravitation, but now we must relate it to the orbital speed. But, how does the orbital speed relate to this force? After Chapter 6, *orbit*, *speed*, and *force* should trigger us to think about circular motion as a related concept. The centripetal force is related to the orbital properties (including speed) of a satellite as $F_c = \frac{m_{\text{satellite}} v_{\text{satellite}}^2}{r_{\text{orbit}}}$. In the case of a planet orbiting a star, it's the gravitational force that is the centripetal force keeping the planet in orbit.

SOLVE

The gravitational force on the planet due to the star is the force that keeps the planet in orbit. That is, the gravitational force is equal to the centripetal force. If we use the mean orbital distance and assume that the planet moves in circular motion, we can write these expressions in terms of the variables required.	$F_G = F_c$ $G\frac{mM}{a^2} = m\frac{v^2}{a}$
Now, solve for the speed *v*.	$v^2 = G\frac{M}{a}$ $\boxed{v = \sqrt{\frac{GM}{a}}}$

CHECK and THINK

This is also the result found in Example 7.8(B). The answer does not actually depend at all on the mass of the planet. The more massive the star, the faster the planet must move to maintain the same circular orbit (and not get sucked into the star) and the further away

the planet is, the slower it must move (in order to not fly out of orbit since the gravitational force is weaker further from the star).

27. Saturn's ring system forms a relatively thin, circular disk in the equatorial plane of the planet. The inner radius of the ring system is approximately 92,000 km from the center of the planet, and the outer edge is about 137,000 km from the center of the planet. The mass of Saturn itself is 5.68×10^{26} kg.
a. (N) What is the period of a particle in the outer edge compared with the period of a particle in the inner edge?

INTERPRET and ANTICIPATE We are given a central object of known mass and orbiting particles at known distances. We are asked for the periods. These three quantities are related by Kepler's third law. We need to write the third law and then solve for the ratio of the periods of particles at different radii.	
SOLVE For a circular orbit, Kepler's third law relates the radius of the orbit to the orbital period. The period squared is proportional to the radius of the orbit cubed (Equation 7.5).	$T^2 = \left(\frac{4\pi^2}{GM}\right)r^3$
The quantities in the parentheses are all constants. They have the same value whether we are making calculations for the inner particles or the outer particles. These constants will cancel if we find the ratio of the outer and inner periods.	$\frac{T_o^2}{T_i^2} = \frac{\left(\frac{4\pi^2}{GM}\right)r_o^3}{\left(\frac{4\pi^2}{GM}\right)r_i^3}$ $\left(\frac{T_o}{T_i}\right)^2 = \left(\frac{r_o}{r_i}\right)^3$
We then isolate the ratio of the outer to inner periods by taking the square root of both sides.	$\frac{T_o}{T_i} = \left(\frac{r_o}{r_i}\right)^{3/2}$
Inserting the outer and inner radii results in a value for the ratio of the outer to the inner periods.	$\frac{T_o}{T_i} = \left(\frac{r_o}{r_i}\right)^{3/2} = \left(\frac{137{,}000}{92{,}000}\right)^{3/2} = \boxed{1.8}$

CHECK and THINK
The outer radius is larger, therefore the period for the outer edge of the ring is larger (about 1.8 times larger than the period of the inner edge of the ring).

b. (N) How long does it take a particle in the inner edge to move once around Saturn?

INTERPRET and ANTICIPATE	
Using Kepler's third law for the inner edge of the ring, we can determine the period of the orbit if we know the semi-major axis, mass of Saturn, and the gravitational constant.	
SOLVE We again start with Newton's third law.	$T^2 = \left(\frac{4\pi^2}{GM}\right)r^3$
Solve this expression for the period.	$T = 2\pi\sqrt{\frac{r^3}{GM}}$
Insert values for the radius of the inner edge, the gravitational constant, and the mass of Saturn. Using metric base units, the answer will be in metric base units for time (seconds).	$T = 2\pi\sqrt{\frac{(9.2\times10^7\ \text{m})^3}{(6.67\times10^{-11}\ \frac{\text{N}\cdot\text{m}^2}{\text{kg}^2})(5.68\times10^{26}\ \text{kg})}}$ $T = 2.8\times10^4\ \text{s} = \boxed{7.9\ \text{h}}$
CHECK and THINK As a comparison, the orbital period of the Moon around the Earth is about 27 days.	

c. (N) While this inner-edge particle is completing one orbit abound Saturn, how far around Saturn does a particle on the outer edge move?

From our calculation in part (a), the outer edge particles have a period that is 1.8 times the period of the inner edge particles. So, when the inner particles have made one revolution, the outer particles have gone through only a portion of their orbit, specifically $\frac{1}{1.8}$ of the way around Saturn. This is 0.56 or 56 percent of the total path for one period, or about 200 degrees around the circular orbit. Since the outer edge particles have a longer period, they have not completed a full orbit during this time. If you multiply this fraction by the circumference for the 137,000 km radius you get a total distance traveled of $\boxed{4.74\times10^8\ \text{m}}$.

31. (N) When first detected near-Earth asteroid 2011 MD was at its closest approach of only 12,000 km above the Earth's surface. What was the asteroid's acceleration due to the Earth's gravity at this point in its trajectory?

INTERPRET and ANTICIPATE	
The acceleration due to Earth's gravity depends on the distance of the object from Earth, the mass of the Earth, and the gravitational constant. Since the asteroid is well above the surface of the Earth, we expect the acceleration to be smaller than 9.8 m/s^2.	
SOLVE The acceleration due to Earth's gravity decreases with distance from the Earth. It also depends on the mass of the Earth and *not* the mass of the object in the Earth's gravitational field. Equation 7.8 expresses this relationship.	$g = \frac{GM_E}{r^2}$
The total distance of the asteroid from the center of mass of the Earth is the radius of the Earth plus its distance above the Earth's surface.	$r = 6.37 \times 10^3$ km + 12,000 km $r = 1.84 \times 10^4$ km $r = 1.84 \times 10^7$ m
Plugging in values for the radius, mass of Earth, and the gravitational constant, we find an acceleration in m/s^2.	$g = \frac{\left(6.67\times10^{-11}\ \text{N}\cdot\text{m}^2/\text{kg}^2\right)(5.98\times10^{24}\ \text{kg})}{(1.84\times10^7\ \text{m})^2}$ $g = \boxed{1.18\ \text{m/s}^2\ \text{towards Earth}}$
CHECK and THINK As predicted, the acceleration is less than 9.8 m/s^2, the value for objects near the Earth's surface. As objects get further from the center of the Earth, the gravitational acceleration due to the Earth decreases.	

33. (A) Three particles, each with mass *m*, are located at coordinates (0, *L*), (*L*, 0), and (*L*, *L*) as shown in Figure P7.33. What are the magnitude and direction of the gravitational field due to the three particles at the origin?

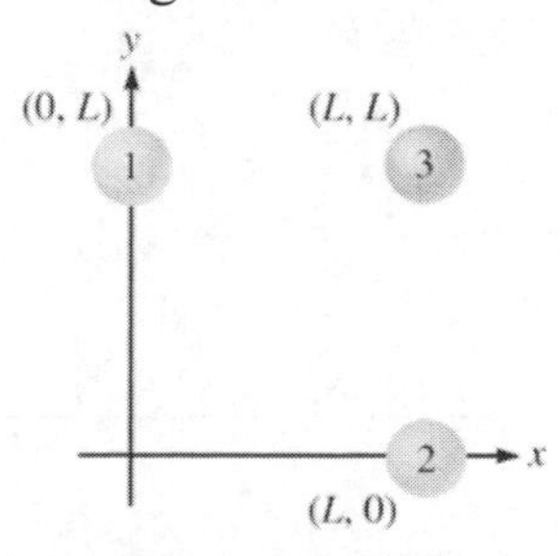

Figure P7.33

<table>
<tr><td colspan="2">INTERPRET and ANTICIPATE
The gravitational field is the vector sum due to each particle. We need to determine the gravitational field at the origin due to each particle, write it as a vector, and then find the resultant vector.</td></tr>
<tr><td>SOLVE
The magnitude gravitational field at a particular location due to a mass depends on the mass and the distance to the location according to Equation 7.8. The direction is always radially towards the mass, so we need to express each vector in x and y coordinates by referring to the figure and expressing the unit vector at the origin, towards the mass, in terms of $\hat{\imath}$ and $\hat{\jmath}$.</td><td>$\vec{g} = -\frac{Gm}{r^2}\hat{r}$ (towards the mass)</td></tr>
<tr><td>For the mass at (L, 0), the gravitational field at the origin is to the right and the distance from the mass to the origin is L.</td><td>$\vec{g}_{(L,0)} = \frac{Gm}{L^2}\hat{\imath}$</td></tr>
<tr><td>For the mass at (0, L), the gravitational field at the origin is upwards and the distance from the mass to the origin is L.</td><td>$\vec{g}_{(0,L)} = \frac{Gm}{L^2}\hat{\jmath}$</td></tr>
<tr><td>For the mass at (L, L), the gravitational field at the origin is up and to the right, at an angle of 45 degrees, and the distance from the mass to the origin is $L\sqrt{2}$ using the Pythagorean theorem.</td><td>$\vec{g}_{(L,L)} = \frac{Gm}{\left(L\sqrt{2}\right)^2}(\cos 45°\hat{\imath} + \sin 45°\hat{\jmath})$</td></tr>
</table>

To get the total gravitational field, we find the vector sum of these three contributions.	$\vec{g} = \frac{Gm}{L^2}\hat{i} + \frac{Gm}{L^2}\hat{j} + \frac{Gm}{\left(L\sqrt{2}\right)^2}(\cos 45°\hat{i} + \sin 45°\hat{j})$ $\vec{g} = \frac{Gm}{L^2}\left(1 + \frac{1}{2\sqrt{2}}\right)(\hat{i} + \hat{j})$
The magnitude can be found from the Pythagorean theorem. Specifically, the magnitude of $\hat{i} + \hat{j}$ is $\sqrt{2}$. The direction is the $\hat{i} + \hat{j}$ direction, which is at 45 degrees with respect to the x axis.	$\lvert\vec{g}\rvert = \frac{Gm}{L^2}\left(1 + \frac{1}{2\sqrt{2}}\right)\left(\sqrt{2}\right)$ $\boxed{\vec{g} = \frac{Gm}{L^2}\left(\sqrt{2} + \frac{1}{2}\right) \text{ at an angle of } 45° \text{ to the horizontal}}$
CHECK and THINK To find the total gravitational field of any number of masses at a particular position, find the vector gravitational field due to each mass at this position and then add the vectors to find the resultant. Since the masses are generally up and to the right, the direction of the resultant vector sounds right.	

47. The Moon rotates on its axis as it orbits the Earth. The same side of the Moon always faces the Earth.

a. (G) Sketch a motion diagram for the Moon. Be sure to include the Earth on your sketch.

We can sketch the position of the moon at a few points in its orbit, where the same side of the moon faces the Earth at each position.

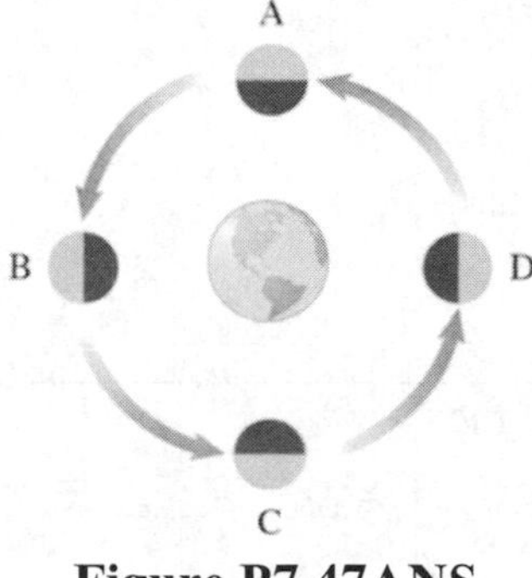

Figure P7.47ANS

b. (N) Find the Moon's rotational period.

Inspecting the diagram from part (a), notice that each time that the Moon orbits the Earth, it also must rotate once on its axis in order to keep the same side facing the Earth. Therefore, the Moon's rotational period must be the same as the time for it to orbit the Earth, $T \approx \boxed{28 \text{ days} \approx 2.4 \times 10^6 \text{ s}}$.

c. (N) Compare (by finding $w_{\text{pole}} - w_{\text{equator}}$) the apparent weight of a 1-kg object on the Moon's equator with its apparent weight at one of the Moon's poles. Assume the same elevation.

INTERPRET and ANTICIPATE The difference between the apparent weights is related to the rotation of the Moon, since this Moon is a non-inertial reference frame. We can follow Example 7.11, in which we answered the same question for effective gravity on Earth.	
SOLVE The apparent weight of an object depends on where it is on the Moon. Objects at the Equator have a larger centripetal acceleration around the Moon compared to objects at either of the poles. Because of this rotation, there must be a net force on the object to keep it in circular motion, so the normal force (which is one way we might experience the strength of the gravitational field) must be smaller than the gravitational force. We can use Equation 7.17 and Example 7.11 as a guide.	$g_{pole} - g_{equator} = \dfrac{v^2}{R} = \dfrac{4\pi^2 R}{T^2}$
Using the radius of the moon and the period from part (b), we calculate the answer.	$g_{pole} - g_{equator} = \dfrac{4\pi^2 \left(1.74 \times 10^6 \text{ m}\right)}{\left(2.4 \times 10^6 \text{ s}\right)^2}$ $g_{pole} - g_{equator} = 1.2 \times 10^{-5} \text{ m/s}^2$

For a 1 kg object, the difference in weight can be found.	$\Delta F_G = m\left(g_{pole} - g_{equator}\right)$ $\Delta F_G = (1\text{ kg})(1.2\times10^{-5}\text{ m/s}^2)$ $\Delta F_G = \boxed{1.2\times10^{-5}\text{ N}}$

CHECK and THINK

This is a very small difference. The weight of the object is around 1.6 N (since gravity on the moon is around 1/6th the magnitude on Earth). We are safe to continue neglecting the fact that the moon is a non-inertial reference frame.

d. (N) How does your answer compare with what you would expect to find for a 1-kg object's apparent weight measure on the Earth's equator and at one of its poles? Explain your results.

From Example 7.11, $\Delta g_{lat} = 3.4\times10^{-2}\text{ m/s}^2$ so the difference in weight would be $\boxed{3.4\times10^{-2}\text{ N}}$. Though still quite small, this is 2800 times larger than on the moon. This is in part because the moon has a smaller radius and, more importantly, rotates at a much slower rate than the Earth.

49. In 1851, Leon Foucault built small a pendulum in his basement that confirmed the theory that the Earth rotates. Today, you can see his much larger pendulum at the Pantheon in Paris, France (Fig. P7.49). To an inertial observer, the pendulum appears to swing back and forth along a single plane. To an observer on the Earth, this plane appears to rotate.

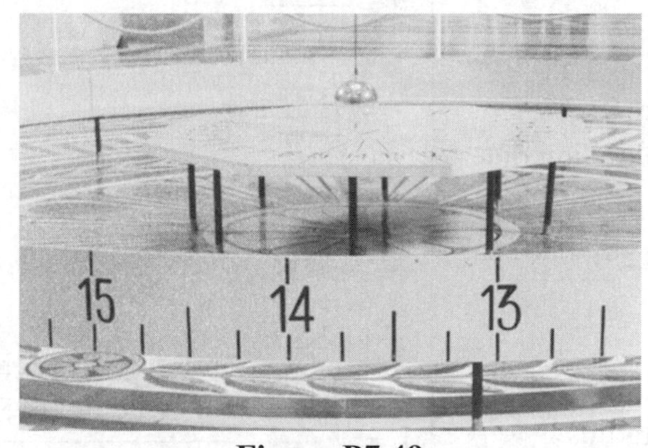

Figure P7.49

a. (N) Imagine that such a pendulum is at the North Pole. Through how many degrees would the plane appear to rotate in a single day?

Imagine that the pendulum is swinging at the North Pole. As the pendulum swings back and forth in the same plane, the Earth will simply rotate underneath the pendulum. In one day, the Earth will make a complete revolution as the pendulum continues to stay in the same plane. From the point of view of an observer on Earth, the pendulum would seem to rotate through $\boxed{360^\circ}$ each day as the Earth makes one complete rotation relative to it.

b. (N) Now imagine that such a pendulum is on the equator. Through how many degrees would the plane appear to rotate in a single day?

In this case, the pendulum would not appear to rotate. Imagine, for instance, that the pendulum rotates in a plane along the equator. As the pendulum is carried around the Earth, the plane of oscillation continues to be in the same plane as the equator. The motion of the Earth carries the plane of the pendulum motion in a large circle around the Earth, but the plane does not rotate relative to the Earth observer.

51. (N) The International Space Station (ISS) experiences an acceleration due to the Earth's gravity of 8.83 m/s^2. What is the orbital period of the ISS?

INTERPRET and ANTICIPATE

Both the acceleration due to gravity and the orbital period depend on the mass of the Earth and the radius of the orbit. Knowing the acceleration due to gravity, we can determine the orbital radius and use this to calculate the period.

SOLVE	
We can determine the orbital radius given the acceleration due to gravity by using Equation 7.8, which relates the acceleration due to gravity to the mass of the Earth, orbital radius, and the gravitational constant.	$g = \frac{GM_\oplus}{r^2}$ $r = \sqrt{\frac{GM_\oplus}{g}}$
Inserting known values, we determine the orbital radius.	$r = \sqrt{\frac{(6.67\times10^{-11}\ \mathrm{N\cdot m^2/kg^2})(5.98\times10^{24}\ \mathrm{kg})}{8.83\ \mathrm{m/s^2}}}$ $r = 6.72\times10^6\ \mathrm{m}$
Kepler's third law relates the orbital period to the orbital radius (Eq. 7.5, for a circular orbit). We can solve this expression to calculate the	$T^2 = \left(\frac{4\pi^2}{GM}\right)r^3$

period using the radius we've calculated.	$T = 2\pi\sqrt{\frac{r^3}{GM_\oplus}}$
Finally, plug in metric numerical values to calculate the period in seconds.	$T = 2\pi\sqrt{\frac{(6.72\times10^6\ \text{m})^3}{(6.67\times10^{-11}\ \text{N}\cdot\text{m}^2/\text{kg}^2)(5.98\times10^{24}\ \text{kg})}}$ $T = 5.48\times10^3\ \text{s}$ $T = (5.48\times10^3\ \text{s})\left(\frac{1\ \text{h}}{3600\ \text{s}}\right) = \boxed{1.52\ \text{h} = 91.4\ \text{min}}$

CHECK and THINK
The period is about an hour and a half. Astronauts in the space station get to see a sunrise and sunset every 91.4 minutes!

52. If you were to calculate the pull of the Sun on the Earth and the pull of the Moon on the Earth, you would undoubtedly find that the Sun's pull is much stronger than that of the Moon, yet the Moon's pull is the primary cause of tides on the Earth. Tides exist because of the *difference* in the gravitational pull of a body (Sun or Moon) on opposite sides of the Earth. Even though the Sun's pull is stronger, the difference between the pull on the near and far sides is greater for the Moon.
a. (A) Let $F(r)$ be the gravitational force exerted on one mass by a second mass a distance r away. Calculate $dF(r)/dr$ to show how F changes as r is changed.

INTERPRET and ANTICIPATE
With an expression for the gravitational force, we can take the derivative to find its rate of change.

SOLVE The gravitational force depends on the inverse square of the distance between two objects (Eq. 7.4).	$F_G = G\frac{m_1 m_2}{r^2}$
We now take the derivative with respect to the separation distance.	$\frac{dF_G}{dr} = \frac{d}{dr}\left(\frac{GMm}{r^2}\right)$ $\frac{dF_G}{dr} = \boxed{-2\frac{GMm}{r^3}}$

CHECK and THINK
We see that the rate of change of the gravitational field depends on the inverse cube of

the separation distance. The negative sign indicates that the magnitude of the force decreases as r increases.

b. (N) Evaluate this expression for $dF(r)/dr$ for the force of the Sun at the Earth's center and for the Moon at the Earth's center.

INTERPRET and ANTICIPATE We can use the result from part (a) to determine this quantity in both cases.	
SOLVE Let's first keep track of the relevant variables for each case.	Sun-Earth distance = 1.5×10^{11} meters Mass of Sun $= 2.0 \times 10^{30}$ kg Mass of Earth $= 6.0 \times 10^{24}$ kg Earth-Moon distance 3.8×10^{8} meters Mass of Moon $= 7.4 \times 10^{22}$ kg
Now, substitute values for each situation into the expression from part (a).	$\left(\frac{dF}{dr}\right)_{Sun} = -2\frac{\left(6.67\times10^{-11}\,\frac{\text{N}\cdot\text{m}^2}{\text{kg}^2}\right)\left(2.0\times10^{30}\text{ kg}\right)\left(6.0\times10^{24}\text{ kg}\right)}{\left(1.5\times10^{11}\text{ m}\right)^3}$ $\left(\frac{dF}{dr}\right)_{Sun} = \boxed{-4.7\times10^{11}\text{ N/m}}$ $\left(\frac{dF}{dr}\right)_{Moon} = -2\frac{\left(6.67\times10^{-11}\,\frac{\text{N}\cdot\text{m}^2}{\text{kg}^2}\right)\left(7.4\times10^{22}\text{ kg}\right)\left(6.0\times10^{24}\text{ kg}\right)}{\left(3.8\times10^{8}\text{ m}\right)^3}$ $\left(\frac{dF}{dr}\right)_{Moon} = \boxed{-10.8\times10^{11}\text{ N/m}}$
CHECK and THINK Though the Moon is less massive, the gradient (the rate of change with respect to distance) in the gravitational force due to the Moon is larger than that of the Sun. That makes sense, since we associate the tides with the position of the Moon. The tidal force due to the Sun is not insignificant, but it is definitely weaker than the tidal force due to the Moon.	

c. (N) Suppose the Earth–Moon distance remains the same, but the Earth is moved closer to the Sun. Is there any point where $dF(r)/dr$ for the two forces has the same value?

INTERPRET and ANTICIPATE

If the Earth–Moon distance remains the same, $\left(\frac{dF}{dr}\right)_{Moon}$ remains the same. Our goal then is to find a Sun-Earth distance that results in $\left(\frac{dF}{dr}\right)_{Sun}$ having the same value.

SOLVE

We want dF/dr due to the Sun to have the same value as that due to the Moon. We are keeping all of the Earth-Moon parameters the same and only changing the Sun-Earth distance.

$$\left(\frac{dF}{dr}\right)_{Sun} = \left(\frac{dF}{dr}\right)_{Moon}$$

$$\left(\frac{dF}{dr}\right)_{Sun} = -10.8\times10^{11}\,\text{N/m}$$

$$-2\frac{\left(6.67\times10^{-11}\,\frac{\text{N}\cdot\text{m}^2}{\text{kg}^2}\right)\left(2.0\times10^{30}\,\text{kg}\right)\left(6.0\times10^{24}\,\text{kg}\right)}{r^3} = -10.8\times10^{11}\,\text{N/m}$$

$$\boxed{r = 1.1\times10^{11}\,\text{m}}$$

CHECK and THINK

The Earth would have to move around 30% of the way closer to the sun for the gradient of the gravitational force due to the Sun to become equal to that of the Moon.

57. Consider the Earth and the Moon as a two-particle system.

a. (N) Find the gravitational field of this two-particle system at the point that is exactly halfway between the Earth and the Moon.

INTERPRET and ANTICIPATE

The Earth is more massive than the Moon, so we expect the net gravitational field halfway between them will be in the direction of the Earth.

SOLVE

First, let's sketch the situation to be clear about the coordinate system. We will use an axis with an origin at the Earth, which points towards the Moon.

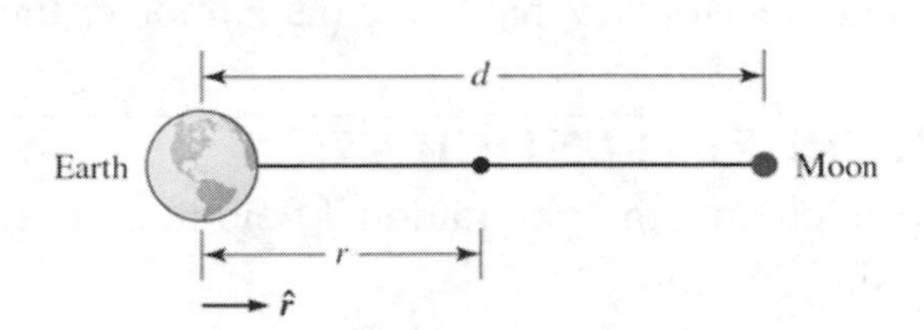

Figure P7.56aANS

We need to evaluate the gravitational field due to both the Earth and Moon at the midpoint (Eq. 7.13), write these as vectors, and then add them to find the resultant. We will write the vectors in terms of the radial vector pointing from the Earth towards the Moon.	$\vec{g}(r) \equiv -\frac{GM}{r^2}\hat{\boldsymbol{r}}$
The Earth and Moon are each a distance $d/2$ from the midpoint. The gravitational field due to the Earth is in the negative direction (towards the Earth), while the field due to the Moon points in the positive direction (towards the Moon).	$\vec{g} = \left[-G\frac{M_{\oplus}}{(d/2)^2} + G\frac{M_{moon}}{(d/2)^2}\right]\hat{\boldsymbol{r}}$ $\vec{g} = \frac{4G}{d^2}\left[M_{moon} - M_{\oplus}\right]\hat{\boldsymbol{r}}$
We are now able to insert numerical values to calculate the gravitational field. The answer can be expressed either in N/kg or m/s^2, since 1 N/kg = 1 m/s^2.	$\vec{g} = \frac{4\left(6.67\times10^{-11}\ \frac{\text{N}\cdot\text{m}^2}{\text{kg}^2}\right)}{\left(3.84\times10^{8}\ \text{m}\right)^2}\left[7.35\times10^{22}\ \text{kg} - 5.98\times10^{24}\ \text{kg}\right]\hat{\boldsymbol{r}}$ $\vec{g} = \boxed{-1.07\times10^{-2}\,\hat{\boldsymbol{r}}\ \text{N/kg}}$

CHECK and THINK

The gravitational field is pointing towards the Earth, as we predicted. It is also much smaller than 9.8 m/s^2, as we'd expect. This is primarily because this point in space is much further away than the surface of the Earth. (The Moon's gravitational field also acts in the opposite direction, but is still relatively weak compared to 9.8 m/s^2.)

b. (N) An asteroid of mass 6.69×10^{15} kg is at the point exactly halfway between the Earth and the Moon. What is the magnitude of the gravitational force on it?

INTERPRET and ANTICIPATE

We've calculated the gravitational field in part (a). Now, we just need to determine the force using $F = mg$.

SOLVE The magnitude of the force is equal to the mass times the magnitude of the gravitational acceleration. (On the Earth, we'd call this the weight, *mg*).	$F = m\left\|\vec{g}\right\| =$ $F = \left(6.69\times10^{15}\text{kg}\right)\left(1.07\times10^{-2}\text{ N/kg}\right)$ $F = \boxed{7.15\times10^{13}\text{ N}}$
CHECK and THINK This force is still quite large, but is much smaller than the weight of the asteroid if it were on the Earth's surface (which would be over 10^{16} N).	

63. Planetary orbits are often approximated as uniform circular motion. Figure P7.9 is a scaled representation of Mars's orbit. Mars's semimajor axis is 1.524 AU.

a. (G) Use the Figure P7.9 to find the ratio of the Sun's maximum gravitational field to its minimum gravitational field on the planet's orbit.

Based on Figure P7.9, we estimate the perihelion and aphelion distances in AU:

$r_P = 1.2$ AU

$r_A = 1.8$ AU

Using the equation for the gravitational field, $g = G\frac{M_\odot}{r^2}$ (Eq. 7.13), we can find the ratio at the minimum and maximum distance.

$$\frac{g_{\max}}{g_{\min}} = \frac{G\frac{M_\odot}{r_P^2}}{G\frac{M_\odot}{r_A^2}} = \frac{r_A^2}{r_P^2} = \left(\frac{1.8}{1.2}\right)^2 = \boxed{2.3}$$

b. (N) What is the ratio of Mars's maximum speed to its minimum speed?

INTERPRET and ANTICIPATE

The orbital speed depends on the mass of the Sun and the distance to Mars. From part (a), we can determine a ratio of the perihelion and aphelion distances and therefore can find a ratio for the orbital speeds. We expect that Mars travels faster when it is closer to the Sun.

SOLVE The orbital speed for Mars around the Sun can be determined by setting the gravitational force equal to the centripetal force. Here, we assume the orbit is circular and use the fact that the gravitational force is the force that keeps the satellite in circular motion. (See Problem 7.24.)	$G\frac{mM}{r^2} = m\frac{v^2}{r}$ $v = \sqrt{\frac{GM}{r}}$
Now, we take the ratio of the speeds for perihelion and aphelion distances.	$\frac{v_P}{v_A} = \frac{\sqrt{G\frac{M_\odot}{r_P}}}{\sqrt{G\frac{M_\odot}{r_A}}} = \sqrt{\frac{r_A}{r_P}} = \sqrt{\frac{1.8}{1.2}} = \boxed{1.2}$

CHECK and THINK
The speed changes by about 20% as Mars orbits. As expected, this ratio is larger than 1, meaning that the speed is larger at the perihelion distance, when Mars is closer to the sun.

c. (C) Comment on the validity of approximating Mars's orbit as uniform circular motion.

The circular motion model is reasonable when estimating quantities related to the orbit or completing physics problems, but would not be sufficient for accurate predictions, for instance if you were launching a rocket to Mars. The speed and acceleration vary by about 20% and 120% respectively, which would lead to significant errors if ignored.

66. (E) Some people insist on believing that the positions of the planets at the time of your birth can have an influence on your life, a subject sometimes referred to as *astrology*. Some astrologers even insist that there is a physical basis in astrology because the gravitational pull of the various planets affects a newborn's body. Estimate the magnitude of the gravitational force on a newborn from the planet Jupiter and compare it with the magnitude of the gravitational force due to the pediatrician who happens to be present during the birth. Which is more likely to have a gravitational "influence" on a newborn child? Is this method a valid scientific test of the gravitational basis of astrology?

We need to estimate the masses involved and insert them into the expression for Newton's Law of Universal Gravity. In the case of Jupiter, we can look up its mass $m_{Jupiter} = 1.90\times10^{27}$ kg and its semimajor axis, $r = a_{Jupiter} = 7.778\times10^{11}$ m.

We can assume that Earth's smaller orbit means half the time Earth is closer to Jupiter than the Sun and half the time it is farther away. So, using this distance as an estimated distance between Jupiter and the newborn on Earth is reasonable. For the mass of the newborn child, we estimate it to be $m_{newborn} = 3.5 \text{ kg}$. Putting this all together:

$$\left(F_G\right)_{Jupiter} = \left(6.67\times10^{-11}\ \tfrac{\text{N m}^2}{\text{kg}^2}\right)\frac{(3.5\text{ kg})\left(1.90\times10^{27}\text{ kg}\right)}{\left(7.778\times10^{11}\text{ m}\right)^2} = \boxed{7.3\times10^{-7}\text{ N}}$$

We repeat this calculation considering the force between the obstetrician and the newborn. We estimate the mass of the obstetrician to be $m_{Obs} = 80 \text{ kg}$ and the center of mass of the pediatrician is about 1 meter from the center of mass of the newborn during the birth:

$$\left(F_G\right)_{Obstetrician} = \left(6.67\times10^{-11}\ \tfrac{\text{N}\cdot\text{m}^2}{\text{kg}^2}\right)\frac{(3.5\text{ kg})(80\text{ kg})}{(1\text{ m})^2} = \boxed{2\times10^{-8}\text{ N}}$$

The gravitational force from Jupiter is about an order of magnitude larger than the gravitational force of the obstetrician on a newborn. Of course, this doesn't really test the idea that the gravitational force of Jupiter affects a newborn's life. To test that, we would have to change the gravitational force acting on newborns in a systematic manner. That said, it does suggest that more massive, nearby objects (such as mountains or buildings) may have a significant influence compared to distant planets in this 'gravitational model' of astrology. Astrologers focusing on the influence of planets seem to be selective about *which* gravitational forces to include.

8

Conservation of Energy

5. (N) A 0.430-kg soccer ball is kicked at an initial speed of 34.0 m/s at an angle of 35.0° to the horizontal. What is the kinetic energy of the soccer ball when it has reached the apex of its trajectory?

<table>
<tr><td colspan="2">INTERPRET and ANTICIPATE
From our earlier practice with kinematic equations, we know that the apex occurs when the y component of the velocity is zero and that the x component does not change. From this, we can determine the velocity of the ball and calculate the kinetic energy.</td></tr>
<tr><td>SOLVE
At the apex, the y velocity is zero and the x velocity is equal to the initial x velocity (since there is no x acceleration). We can use trigonometry to determine the initial x velocity, which is equal to the speed at the apex.</td><td>$\vec{v}_i = (34.0\text{ m/s})\cos 35.0°\hat{\imath} + (34.0\text{ m/s})\sin 35.0°\hat{\jmath}$
$\vec{v}_i = (27.9\text{ m/s})\hat{\imath} + (19.5\text{ m/s})\hat{\jmath}$
$v_{\text{apex}} = v_{i,x} = 27.9\text{ m/s}$</td></tr>
<tr><td>We can now calculate the kinetic energy, which depends on the mass and speed of the ball according to Equation 8.1.</td><td>$K = \frac{1}{2}mv^2$
$K = \frac{1}{2}(0.430\text{ kg})(27.9\text{ m/s})^2$
$K = \boxed{167\text{ J}}$</td></tr>
<tr><td colspan="2">CHECK and THINK
The kinetic energy is a scalar quantity that depends on the mass and square of the speed. In this case, we determine the speed at the apex, which we use to calculate the kinetic energy.</td></tr>
</table>

6. Both a car ($m_c = 1550$ kg) and a truck ($m_t = 8150$ kg) are initially at rest and are each sped up to a final speed of 30 mph.

a. (N) What is the final kinetic energy of the car?

INTERPRET and ANTICIPATE The kinetic energy of the car depends on its mass and speed. We are given both quantities, so we just need to make sure they are in metric units and apply the formula for kinetic energy.	
SOLVE The kinetic energy of an object depends on its mass and speed squared (Equation 8.1).	$K = \frac{1}{2}mv^2$
The mass of the car is given. The final speed is also given, but we must convert this to metric base units in order to calculate an energy in Joules.	$m = 1550 \text{ kg}$ $v = \left(30.0 \frac{\text{mi}}{\text{h}}\right)\left(\frac{1 \text{ h}}{3600 \text{ s}}\right)\left(\frac{1609 \text{ m}}{1 \text{ mi}}\right) = 13.4 \text{ m/s}$
Now, substitute known values into the formula for kinetic energy.	$K = \frac{1}{2}(1550 \text{ kg})(13.4 \text{ m/s})^2$ $K = \boxed{1.39\times10^5 \text{ J}}$
CHECK and THINK With the mass and speed, we've calculated the kinetic energy. We'll be able to compare the result to the truck in part (b).	

b. (N) What is the final kinetic energy of the truck?

INTERPRET and ANTICIPATE We follow the same procedure as in part (a). The truck has a larger mass and the same speed, so it should have a larger kinetic energy.	
SOLVE We can use the formula for kinetic energy with the mass provided and the same speed as in part (a).	$K = (1/2)(8150 \text{ kg})(13.4 \text{ m/s})^2$ $K = \boxed{7.31\times10^5 \text{ J}}$
CHECK and THINK As expected, the kinetic energy of the truck is larger than the car. Since they are traveling at the same speed, the kinetic energy is larger by about a factor of five, the same as the ratio of their masses.	

c. (C) Are we able to determine which vehicle was subject to a greater acceleration? If so, explain how. If not, what would we need to know to make that determination?

No. While both the truck and the car have definitely accelerated, because their speeds changed, we don't know the time frame over which these changes occurred. Did one take longer to reach the 30 mph-speed than the other? Acceleration is the *rate* of change of the velocity. If we knew the time over which the acceleration occurred in each case, we could calculate the change in velocity per time and determine which was subject to a greater acceleration. The final kinetic energy depends on the final speed – not how quickly that speed was reached.

11. (N) A ball ($m = 2.5$ kg) is given an initial velocity upwards of 15 m/s. What is the change in the gravitational potential energy of the ball from its initial height to when the ball is at the peak of its motion?

INTERPRET and ANTICIPATE Using kinematic equations, we can determine the height that the ball reaches. This, along with the mass, will allow us to determine the gravitational potential energy relative to the ground.	
SOLVE The gravitational potential energy depends on the mass and height of an object, as well as the acceleration due to gravity, as described by Equation 8.5.	$U_g(y) = mgy$
We know the mass and therefore must determine the *y* position for the initial and final times. We are free to set a reference configuration (since only *changes* in potential energy will be relevant for the motion of objects), so we set the initial height to zero.	$\Delta U_g = mgy_f - mgy_i$ let $y_i = 0$

We now have a kinematics problem. We know the initial velocity and acceleration due to gravity and the final velocity at the top of the trajectory is zero. Therefore, we can solve a kinematic equation for the distance traveled to get the final height.	$v_{y0} = 15\text{ m/s}$ $v_y = 0$ $a_y = -9.81\text{ m/s}^2$ $y?$ $v_y^2 = v_{y0}^2 + 2a_y y$ $0 = (15\text{ m/s})^2 + 2(-9.81\text{ m/s}^2)y$ $y = 11.5\text{ m}$
Now, we can calculate the change in potential energy.	$\Delta U_g = mgy_f - mgy_i = (2.5\text{ kg})(9.81\text{ m/s}^2)(11\text{ m}) - 0$ $\Delta U_g = \boxed{280\text{ J}}$

CHECK and THINK

Using kinematic equations, we can determine the change in height, and using this with the mass and acceleration due to gravity, we can determine the change gravitational potential energy. It is positive because the mass gains potential energy as it moves to a higher position.

We will see later in the chapter that the total energy is *conserved.* Therefore, the change in potential energy has the same magnitude as the change in kinetic energy, which we can calculate directly:

$$\Delta U = -\Delta K = -\left(\frac{1}{2}mv_f^2 - \frac{1}{2}mv_i^2\right) = \frac{2.5\text{ kg}}{2}\left((15\text{ m/s})^2 - 0\right) = \boxed{280\text{ J}}$$

In other words, we will soon have a way of avoiding the kinematic equations entirely in cases where the total energy of the system is conserved.

16. (A) Two blocks of masses m_1 and m_2 are connected by a cord that passes over a pulley as shown in Figure P8.16. They start at the same height, and the hanging block is lowered through a distance h, causing the block on the incline to rise. Calculate the change in gravitational potential energy for each of the blocks.

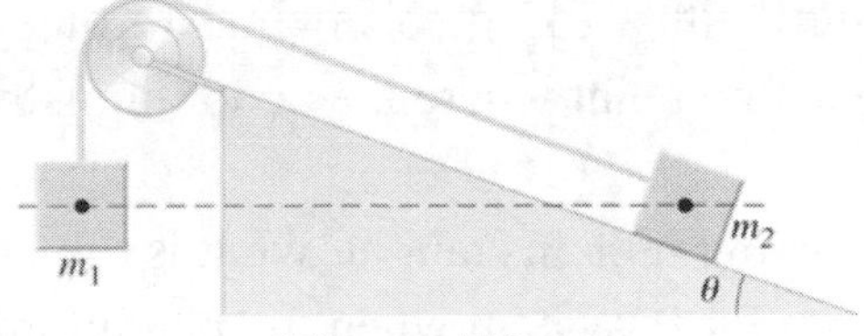

Figure P8.16

<table>
<tr><td colspan="2">INTERPRET and ANTICIPATE
The change in gravitational potential energy depends only on the mass and the vertical displacement, so we need to use trigonometry to find how the height of block 2 changes as it slides a distance h up the ramp. For each block the system consists of that block and the earth. The sliding mass slides up the ramp and will therefore gain potential energy, while the hanging mass falls to a lower position and will lose potential energy.</td></tr>
<tr><td>SOLVE
The gravitational potential energy depends on its mass, height, and the acceleration due to gravity (Equation 8.5). We choose the initial position of both blocks to be $y = 0$ (reference configuration), so their initial potential energies are both zero.</td><td>$U_g(y) = mgy$
$\Delta U = U_f - U_i = mgy_f - 0$</td></tr>
<tr><td>Mass m_1 has a final vertical position $y_{1f} = -h$.</td><td>$\Delta U_1 = \boxed{-m_1 gh}$</td></tr>
<tr><td>When m_1 descends, m_2 moves a distance h along the incline, which results in a vertical rise that is less than h. We find its final vertical position using trigonometry, and then calculate change in energy.</td><td>$y_{2f} = h\sin\theta$
$\Delta U_2 = \boxed{m_2 gh\sin\theta}$</td></tr>
<tr><td colspan="2">CHECK and THINK
Since the hanging mass descends, it loses gravitational potential energy, as expected. The sliding mass rises and gains gravitational potential energy. The magnitudes of these changes are different in general because their masses and vertical displacements are both different.</td></tr>
</table>

20. A 34.0-cm-long pendulum with a 245-g bob swings through an angle of 180°. Choose the reference configuration as the point where the pendulum string is vertical.

a. (N) Assuming pendulum string remains taught, what is the gravitational potential energy of the pendulum bob – Earth system when the pendulum string makes an angle of 90° with the vertical?

Interpret and Anticipate We need to determine the height of the bob above the reference configuration and use the formula for gravitational potential energy. It will be positive since the pendulum bob is above the reference position.	
SOLVE The gravitational potential energy of a mass near the surface of the Earth depends on its mass, height, and the acceleration due to gravity (Equation 8.5).	$U_g = mgy$
We sketch the pendulum, indicating the reference position (r) and the positions corresponding to parts (a) and (b) of this problem.	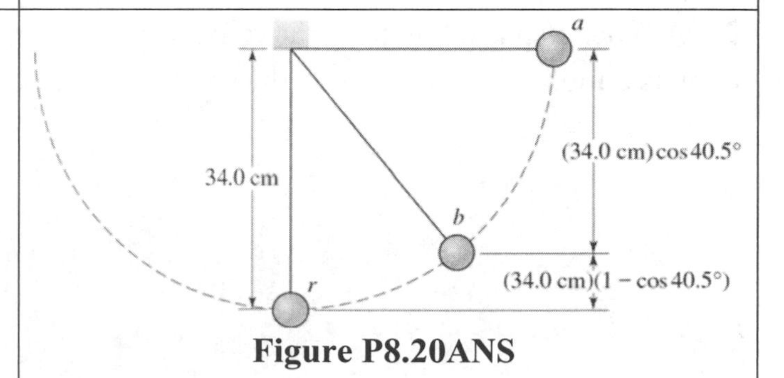 **Figure P8.20ANS**
When the pendulum bob is at 90º (position a), the bob has moved up a distance equal to the radius of the circle, 34.0 cm = 0.340 m. We can now calculate the potential energy for position a.	$U_g = mgy$ $U_g = (0.245\text{ kg})(9.81\text{ m/s}^2)(0.340\text{ m})$ $U_g = \boxed{0.817\text{ J}}$
CHECK and THINK The gravitational potential energy is positive as expected.	

b. (N) What is the gravitational potential energy of the pendulum bob–Earth system when the pendulum string makes an angle of 40.5° with the vertical?

INTERPRET and ANTICIPATE We repeat the steps for part (a). Since the pendulum bob is lower (though still above the reference position), the potential energy will be smaller than in part (a) (but still positive).	
SOLVE Referring to the figure, we can determine the height of the bob at position b, and therefore the gravitational potential energy.	$y = (0.340\text{ m})(1 - \cos 40.5°)$ $U_g = mgy$ $U_g = (0.245\text{ kg})(9.81\text{ m/s}^2)(0.340\text{ m})(1 - \cos 40.5°)$ $U_g = \boxed{0.196\text{ J}}$

CHECK and THINK
The potential energy is indeed positive and smaller in magnitude than in part (a).

c. (N) What is the gravitational potential energy of the pendulum bob–Earth system at the lowest point of the pendulum's motion?

This is the reference configuration chosen for which the height is zero, so $y = 0$ and $U_g = mgy = \boxed{0}$.

24. A block is placed on top of a vertical spring, and the spring compresses. Figure P8.24 depicts a moment in time when the spring is compressed by an amount *h*.

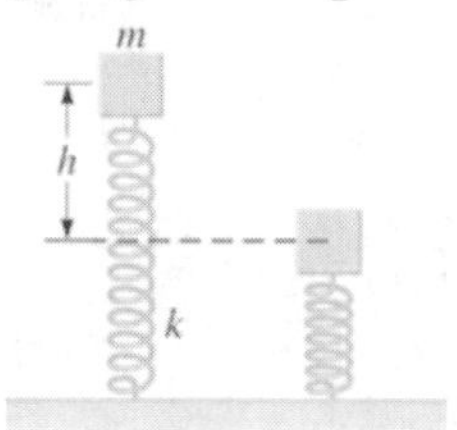

Figure P8.24

a. (C) To calculate the change in the gravitational and elastic potential energies, what must be included in the system?

We must include all sources of the forces that act on the block as it moves. So the block, spring, and Earth must be in the system to account for the elastic energy of the spring and the gravitational potential energy. You might wonder if we should consider the floor to be a part of the system, for instance, since without it the spring would not compress. We typically include only the objects that directly lead to the relevant forces when referring to the system, so we would not worry about what is holding the spring up in this spring-mass-Earth system.

b. (A) Find an expression for the change in the system's potential energy in terms of the parameters shown in Figure P8.24.

INTERPRET and ANTICIPATE
Let's set the reference to configuration (for gravity) to the relaxed position of the spring. Then the gravitational potential energy is negative when the spring is compressed. The energy stored in the spring (regardless of whether the spring is compressed or extended) is positive.

SOLVE First, express the system's change in gravitational energy based on its mass and vertical position (Equation 8.5).	$\Delta U_g = U_f - U_i$ $\Delta U_g = 0 - mgh$ $\Delta U_g = -mgh$
We can now account for the energy in the compressed spring, which depends on the spring constant and displacement from the relaxed position (Equation 8.9).	$\Delta U_e = \frac{1}{2}kx_f^2 - \frac{1}{2}kx_i^2$ $\Delta U_e = \frac{1}{2}kh^2 - 0$ $\Delta U_e = \frac{1}{2}kh^2$
The system's total potential energy is the sum of its gravitational and elastic potential energies.	$\Delta U = \Delta U_g + \Delta U_e$ $\Delta U = \boxed{-mgh + \frac{1}{2}kh^2}$

CHECK and THINK
Our expression makes sense. As the block moves downward, the gravitational potential energy decreases. And as the spring is compresses the elastic potential energy increases.

c. (N) If $m = 0.865$ kg and $k = 125$ N/m, find the change in the system's potential energy when the block's displacement is $h = 0.0650$ m, relative to its initial position.

INTERPRET and ANTICIPATE
Here we simply need to substitute values using the formula we derived in part (b). The change may be positive or negative depending on the relative importance of gravitational and elastic potential energy.

SOLVE We substitute our known values into the formula from part (b).	$\Delta U = -(0.865\text{ kg})(9.81\text{ m/s}^2)(0.650\text{m}) + \frac{1}{2}(125\text{ N/m})(0.650\text{ m})^2$ $\Delta U = \boxed{20.9\text{ J}}$

CHECK and THINK
The change in potential energy is positive, which means the elastic potential energy dominates. If the block had more mass and/or if the spring constant were lower (with all other factors remaining the same), it is possible for the change in potential energy to be negative. Try this out for yourself – see what happens if the block's mass is 1865 kg and if $k = 1.25$ N/m.

27. (A) A spring of spring constant k lies along an incline as shown in Figure P8.27. A block of mass m is attached to the spring. The spring compresses, and the block comes to rest as shown. Find an expression for the change in the Earth–block–spring system's potential energy in terms of the parameters given in the figure.

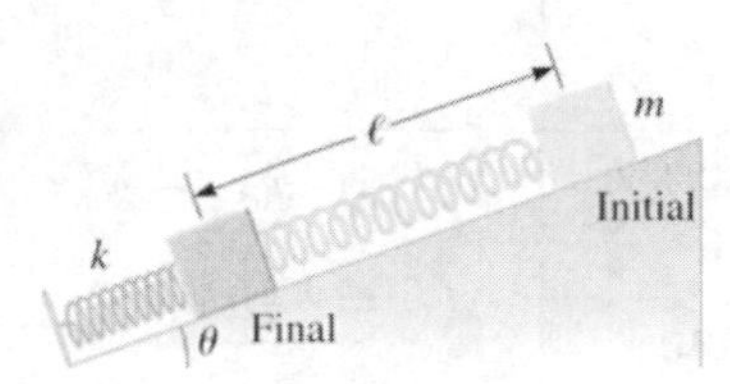

Figure P8.27

INTERPRET and ANTICIPATE

We need to determine both the gravitational potential energy and the energy stored in the spring. We set the reference configuration to the relaxed position of the spring, and then the gravitational potential energy is negative when the spring is compressed.

SOLVE

The change in the block's gravitational potential energy depends on its mass and vertical displacement (Eq. 8.5). Setting the initial height as $y_i = 0$, the initial potential energy is zero. We need the *vertical* displacement, which we find using trigonometry: $y_f = -\ell \sin\theta$.	$\Delta U_g = U_f - U_i$ $\Delta U_g = mgy_f - mgy_i$ $\Delta U_g = mg(-\ell \sin\theta) - 0$ $\Delta U_g = -mg\ell \sin\theta$
Find the change in the system's elastic potential energy. The spring is compressed along the plane and the initial state is that the spring is relaxed, $x_i = 0$.	$\Delta U_e = \frac{1}{2}kx_f^2 - \frac{1}{2}kx_i^2$ $\Delta U_e = \frac{1}{2}k\ell^2 - 0$ $\Delta U_e = \frac{1}{2}k\ell^2$
The system's total potential energy is the sum of its gravitational and elastic potential energies.	$\Delta U = \Delta U_g + \Delta U_e$ $\Delta U = \boxed{-mg\ell \sin\theta + \frac{1}{2}k\ell^2}$

CHECK and THINK

To check our answer imagine that the spring were vertical. In that case, $\theta = 90°$, and we'd find the same expression as in Problem 8.24 with $\ell \to h$.

31. (N) A newly established colony on the Moon launches a capsule vertically with an initial speed of 1.50 km/s. Ignoring the rotation of the Moon, what is the maximum height reached by the capsule?

<table>
<tr><td colspan="2">INTERPRET and ANTICIPATE
The initial speed is large and gravity on the Moon is weaker than Earth, so we expect that the capsule will travel quite high. Therefore, using $U_g = mgy$ (which is an approximation for masses near the surface of a planet or moon) may not be a good approximation. We will use the formula for universal gravitational potential energy (which is accurate for the potential energy for any two masses) to be safe. Using this, we can apply the conservation of energy to equate the change in kinetic energy with the change in potential energy.</td></tr>
<tr><td>SOLVE
Universal gravitational potential energy for two masses depends on the masses and inversely on their separation according to Equation 8.7.</td><td>$U = -\frac{GM_1M_2}{r}$</td></tr>
<tr><td>We can use conservation of energy. Initially, the capsule has kinetic energy (Eq. 8.1) and is on the surface of the Moon (at a distance equal to the radius of the Moon R_M from the Moon's center). At the top of its trajectory, it is momentarily at rest ($K = 0$) and it reaches a final position of $R_M + h$, where h is the final height above the surface of the Moon.</td><td>$K_i + U_i = K_f + U_f$

$\frac{1}{2}M_c v_i^2 - \frac{GM_cM_M}{R_M} = 0 - \frac{GM_cM_M}{R_M + h}$</td></tr>
<tr><td colspan="2">We know the quantities in the formula, other than the unknown height, so we can cancel out the mass of the capsule, substitute values, and solve for h.

$$\frac{1}{2}(1.50\times10^3 \text{ m/s})^2 - \left(6.67\times10^{-11} \tfrac{\text{N}\cdot\text{m}^2}{\text{kg}^2}\right)\left(\frac{7.35\times10^{22} \text{ kg}}{1.74\times10^6 \text{ m}}\right)$$
$$= -\left(6.67\times10^{-11} \tfrac{\text{N}\cdot\text{m}^2}{\text{kg}^2}\right)\left(\frac{7.35\times10^{22} \text{ kg}}{1.74\times10^6 \text{ m} + h}\right)$$</td></tr>
</table>

Continuing to simplify this expression, we calculate the final height of the capsule.	$1.13\times10^{6}-2.82\times10^{6}=-\dfrac{4.90\times10^{12}\text{ m}}{1.74\times10^{6}\text{ m}+h}$ $1.74\times10^{6}\text{ m}+h=\dfrac{4.90\times10^{12}\text{ m}}{1.69\times10^{6}}=2.90\times10^{6}\text{ m}$ $h=\boxed{1.16\times10^{6}\text{ m}}$
CHECK and THINK The height of around 1000 km is indeed substantial, on the order of the radius of the Moon! We are likely to have significant errors if we us the approximation $U_g = mgy$. (See Problem 8.74 for an example of this error.)	

37. In the Marvel comic series *X-Men*, Colossus would sometimes throw Wolverine toward an enemy in what was called a *fastball special.* Suppose Colossus throws Wolverine at an angle of 30.0° with respect to the ground (Fig. P8.37). Wolverine is 2.15 m above the ground when he is released, and he leaves Colossus's hands with a speed of 20.0 m/s.

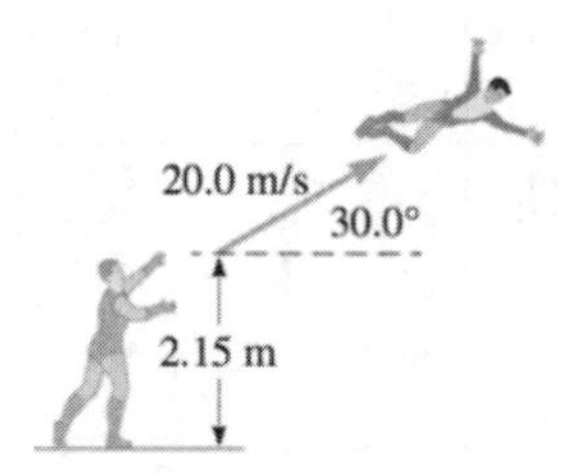

Figure P8.37

a. (N) Using conservation of energy and the components of the initial velocity, find the maximum height attained by Wolverine during the flight.

INTERPRET and ANTICIPATE Mechanical energy is conserved. We can write the total energy at the beginning and at the highest point of the trajectory. At this point, the vertical velocity is zero and the horizontal velocity is the same as the initial value (since there is no horizontal acceleration).	
SOLVE The components of the initial velocity can be found by using the initial speed and the angle at which Wolverine is thrown.	$v_{0x}=(20.0\text{ m/s})\cos30°$ $v_{0y}=(20.0\text{ m/s})\sin30°$

We'll choose $y = 0$ to be even with the ground such that the initial height of Wolverine is 2.15 m. When he is at the peak of the trajectory, the y velocity is zero while the x velocity is unchanged.	$y_i = 2.15$ m $v_y = 0$ $v_x = 17.3$ m/s
We now express the conservation of mechanical energy (Equation 8.14).	$\frac{1}{2}m(20.0\text{ m/s})^2 + mg(2.15\text{ m}) = \frac{1}{2}m((20.0\text{ m/s})\cos 30°)^2 + mgy_{peak}$
After canceling the mass from each term and using $g = 9.81$ m/s^2, the equation can be solved for the peak height above the ground.	$y_{peak} = \boxed{7.25\text{ m}}$

CHECK and THINK

We have successfully used conservation of energy. You may wonder if you can use just the vertical components, since we know from kinematics that it is the vertical component of the initial velocity that determines the height reached. In fact, it turns out that it does work, because the kinetic energy associated with the horizontal velocity remains the same and appears on each side of the equation. So, if we used only the vertical components of velocity:

$$\frac{1}{2}m(10.0\text{ m/s})^2 + mg(2.15\text{ m}) = \frac{1}{2}m(0)^2 + mgy_{peak}$$

$$y_{peak} = 7.25\text{ m}$$

However, we should be careful to note that kinetic energy is a *scalar*, not a *vector*, and does not have components. (There is no such thing as the "y component of the kinetic energy" and you should never use trigonometry to try to find a component of energy.) For this reason, it is safest to always consider the total energy initially and the total energy in the final state in order to solve problems using conservation of energy.

b. (N) Using conservation of energy, what is Wolverine's speed the instant before he hits the ground?

INTERPRET and ANTICIPATE The instant before Wolverine hits the ground, his height is 0. We can use the conservation of energy equation from part (a), using this instant instead of the moment that he reaches the peak height.	
SOLVE Using the conservation of energy equation from part (a) with the final height of 0, we can solve for the final speed.	$\frac{1}{2}m(20.0\text{ m/s})^2 + mg(2.15\text{ m}) = \frac{1}{2}mv_f^2 + mg(0)$ $v_f = \boxed{21.0\text{ m/s}}$
CHECK and THINK This result makes sense. When Wolverine returns back to his initial height (2.15 m), he will have the same speed as his initial speed (20 m/s). Since he falls a little further to reach the ground, his speed increases a little beyond this initial value.	

41. (N) If a spacecraft is launched off the Moon at the escape speed of the Earth, how fast will the spacecraft be going when it is very far away from the Moon, ignoring the effects of other celestial bodies?

INTERPRET and ANTICIPATE This is similar to Example 8.8 and so we make several similar choices: **(1)** the system consists of the Moon and the spacecraft, **(2)** the origin of the coordinate axes is at the center of the Moon and the reference configuration is infinity, and **(3)** the initial time is when the spacecraft is launched, the final time is when the spacecraft is very far from the Moon. Thus we can start our solution with step **(4)**.	
SOLVE We can express the initial kinetic energy of the spacecraft (Eq. 8.1). We denote the launch speed as $v_{\text{esc},\oplus}$.	$K_i = \frac{1}{2}mv_{\text{esc},\oplus}^2$
As in Example 8.8, the initial gravitational potential energy of the system is given by Equation 8.7, but now the masses involved are that of the Moon and the spacecraft. At launch, the spacecraft begins on the Moon's surface at a distance R_{Moon} from the origin.	$U_{gi} = -G\frac{M_{\text{moon}}m}{R_{\text{moon}}}$

The final kinetic energy is unknown; we need to find the final speed v_f. The final gravitational potential energy is zero, because infinite separation ("very far away") is the reference configuration.	$K_f = \frac{1}{2}mv_f^2$ $U_{gf} = 0$
We sketch a new bar chart, which shows that when the spacecraft is launched faster than the Moon's escape speed, it will still have some kinetic energy when far from the Moon.	No spring; zero; No spring K_i U_{gi} U_{ei} K_f U_{gf} U_{ef} **Figure P8.41ANS**
We now express conservation of mechanical energy (Eq. 8.14) and substitute expressions from above.	$K_i + U_{gi} = K_f$ $\frac{1}{2}mv_{\text{esc},\oplus}^2 - G\frac{M_{\text{moon}}m}{R_{\text{moon}}} = \frac{1}{2}mv_f^2$ $v_{\text{esc},\oplus}^2 - 2G\frac{M_{\text{moon}}}{R_{\text{moon}}} = v_f^2$
It is possible to use this expression directly, but we also note that the gravitational potential term is actually related to the escape velocity of the *Moon*.	$v_{\text{esc, moon}}^2 = 2G\frac{M_{\text{moon}}}{R_{\text{moon}}}$
We can now express the final velocity in a succinct manner.	$v_f^2 = v_{\text{esc},\oplus}^2 - v_{\text{esc, moon}}^2$ $v_f = \sqrt{v_{\text{esc},\oplus}^2 - v_{\text{esc, moon}}^2}$
Substituting numerical values, we find the final speed of the spacecraft.	$v_f = \sqrt{\left(1.12\times10^4 \text{ m/s}\right)^2 - \left(2.38\times10^3 \text{ m/s}\right)^2}$ $v_f = \boxed{1.09\times10^4 \text{ m/s}}$

CHECK and THINK

Because the spacecraft was launched at a speed so much greater than the escape speed of the Moon, the spacecraft still has a high speed when it is very far away.

44. (N) Starting at rest, Tina slides down a frictionless waterslide with a horizontal section at the bottom that is 4.00 ft above the surface of the swimming pool and strikes the water a distance of 15.0 ft away from the end of the slide. Using conservation of energy, what is Tina's initial height on the waterslide?

<table>
<tr><td colspan="2">INTERPRET and ANTICIPATE
Using conservation of energy, we can relate Tina's initial potential energy with her speed when she leaves the slide. Given her horizontal speed when she leaves the slide, we can relate it to the distance she travels using kinematics.</td></tr>
<tr><td>SOLVE
Let's first sketch the problem. We take $y = 0$ at the surface of the swimming pool. We denote the height of the slide as H and the height of the horizontal section above the swimming pool as h.</td><td>
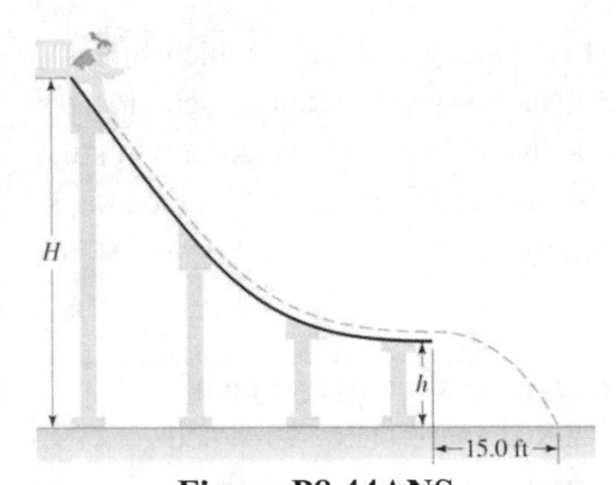

Figure P8.44ANS</td></tr>
<tr><td>Then, we can determine the horizontal speed v at the bottom of the slide, since the initial potential energy at the top of the slide is converted to kinetic energy at the bottom.</td><td>$E_{bottom} = E_{top}$
$\frac{1}{2}mv^2 + mgh = 0 + mgH$
$v^2 = 2g(H - h) \qquad (1)$</td></tr>
<tr><td>This is the horizontal speed with which Tina leaves the waterslide. She has no vertical velocity when she leaves the horizontal bottom of the slide. We use the y component of her motion after leaving the slide to determine the time it takes for her to strike the water and use this time and the horizontal speed to determine the horizontal distance she travels.</td><td>$\Delta y = v_{yi}t - \frac{1}{2}gt^2$
$-h = -\frac{1}{2}gt^2$
$t = \sqrt{\frac{2h}{g}}$</td></tr>
<tr><td>With an expression for the time of flight, we can now calculate her range from the x component of the motion.</td><td>$x = vt = v\sqrt{\frac{2h}{g}} \qquad (2)$</td></tr>
<tr><td>We want to relate the distance traveled off the end of the slide (x) to the height of the slide (H). One way to do this is</td><td>$v = x\sqrt{\frac{g}{2h}} = \sqrt{\frac{gx^2}{2h}}$</td></tr>
</table>

to express equation (2) in terms of velocity squared and then use equation (1) to write the velocity squared in terms of H.	$v^2 = \frac{gx^2}{2h}$ $2g(H-h) = \frac{gx^2}{2h}$
We now solve this equation for the height of the slide and substitute numerical values.	$H = h + \frac{x^2}{4h} =$ $H = 4.00\text{ ft} + \frac{(15.0\text{ ft})^2}{4(4.00\text{ ft})}$ $H = \boxed{18.1\text{ ft}}$

CHECK and THINK
18 feet sounds like a reasonable height for a water slide. At the very least, it is above the bottom of the slide (4 ft) and is of the right order of magnitude!

45. (N) Karen and Randy are playing with a toy car and track. They set up the track on the floor as shown in Figure P8.45, where the apparatus on the far left is used to launch the car forward by pressing down on the top portion. After the car is launched, it follows the track and continues upward, leaving the track. If the car has a mass of 130 g and it reaches a maximum vertical height of 1.2 m above the floor, what was the speed of the car while it was moving along the floor? Ignore the effects of friction and air resistance.

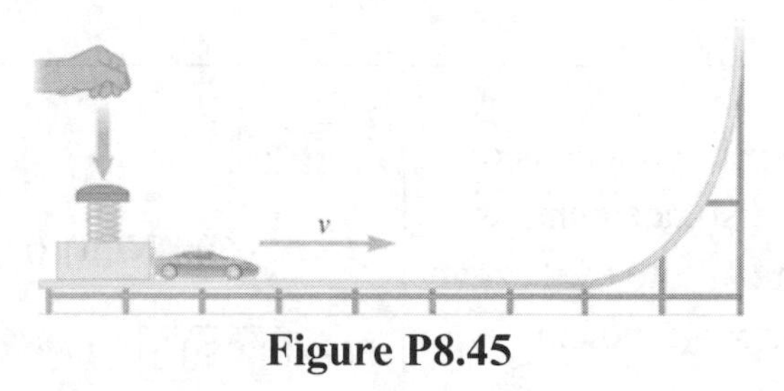

Figure P8.45

INTERPRET and ANTICIPATE
During the motion, the only force doing work on the car will be gravity. Choosing the floor to be the height where $y = 0$, and recognizing that the speed of the car is 0 at the peak height, we can write an expression for conservation of energy.

SOLVE Writing an expression for the conservation of energy (Eq. 8.14), the car starts with kinetic energy at zero potential and reaches its highest potential energy and stops	$(1/2)mv_i^2 + mg(0) = (1/2)m(0)^2 + mg(1.2\text{ m})$

instantaneously at the highest point it reaches.	
Canceling the mass from each term, we can solve for the initial speed.	$v_i = \boxed{4.9\text{ m/s}}$
CHECK and THINK In this case, we see that the energy approach can be quite useful if total energy is conserved, even when the direction of an object changes. We would need a lot more than this one-line derivation if we were to attempt to use kinematics!	

48. A block of mass $m = 1.50$ kg attached to a horizontal spring with force constant $k =$ 600 N/m that is secured to a wall is stretched a distance of 5.00 cm beyond the spring's relaxed position and released from rest.

a. (N) What is the elastic potential energy of the block–spring system just before the block is released?

INTERPRET and ANTICIPATE The spring contains elastic potential energy that depends on the spring constant of the spring and the displacement from the relaxed position.	
SOLVE The energy stored in a spring depends on the spring constant and displacement, as seen in Equation 8.9. Substitute the known values, $k = 600$ N/m and $x = 5.00$ cm = 0.0500 m.	$U = \frac{1}{2}kx^2$ $U = \frac{1}{2}(600\text{ N/m})(0.050\text{ m})^2 =$ $U = \boxed{0.750\text{ J}}$
CHECK and THINK The spring potential energy depends only on the spring constant and displacement from the relaxed length.	

b. (N) What is the elastic potential energy of the block–spring system when the block passes through the spring's relaxed position?

At the relaxed position, by definition, $x = 0$ and $U = \boxed{0}$.

c. (N) What is the speed of the block as it passes through the spring's relaxed position?

INTERPRET and ANTICIPATE	
Total energy, which includes the spring potential energy and kinetic energy, is conserved. When stretched, the spring contains potential energy, which is converted to kinetic energy when the block is released.	
SOLVE Taking into account both the spring potential energy and kinetic energy, we can write an expression for conservation of energy (Eq. 8.16). The initial kinetic energy is zero when the block is first released and the spring potential energy is zero as the block passes through the position corresponding to the relaxed length.	$U_1 + K_1 = U_2 + K_2$ $\frac{1}{2}kx_1^2 + 0 = 0 + \frac{1}{2}mv_2^2$
Solve this equation for speed and substitute numerical values.	$v_2 = \sqrt{\frac{k}{m}}x_1$ $v_2 = \sqrt{\frac{(600\text{ N/m})}{(1.50\text{ kg})}}(0.050\text{ m})$ $v_2 = \boxed{1.00\text{ m/s}}$
CHECK and THINK This is very similar to previous examples in which gravitational potential energy is converted to kinetic energy. In this case, for a horizontal system with a spring, there is no gravitational potential and only spring potential energy.	

d. (N) What is the speed of the block when it has compressed the spring 2.50 cm beyond its relaxed position?

INTERPRET and ANTICIPATE	
This is another example where we might have significant work ahead of us if we attempted to take a kinematics approach and integrate Newton's laws. In this case, we just set the initial potential energy equal to the total energy at this point in time and solve for velocity.	
SOLVE Using the conservation of energy	$U_1 + K_1 = U_3 + K_3$

equation from above, the potential energy when the block is at this position can be expressed to allow us to solve for the speed of the block at this moment.	$\frac{1}{2}kx_1^2 + 0 = \frac{1}{2}kx_3^2 + \frac{1}{2}mv_3^2$ $v_3 = \sqrt{\frac{k}{m}(x_1^2 - x_3^2)}$ $v_3 = \sqrt{\frac{(600\text{ N/m})}{(1.50\text{ kg})}\left[(0.050\text{ m})^2 - (0.025\text{ m})^2\right]}$ $v_3 = \boxed{0.866\text{ m/s}}$

CHECK and THINK

The block moves fastest as it passes through the relaxed position. The velocity at this position is comparable to, but smaller than, this maximum speed.

57. Figure P8.57A shows the potential energy curve for a two-particle system. Particle 1 remains at rest at the origin, and particle 2 ($m = 3.75$ kg) is initially at $\vec{x}_i = 24.0\hat{i}$ m with a velocity of $\vec{v}_i = 3.00\hat{i}$ m/s as shown in Figure P8.57B. Report numerical answers to three significant figures.

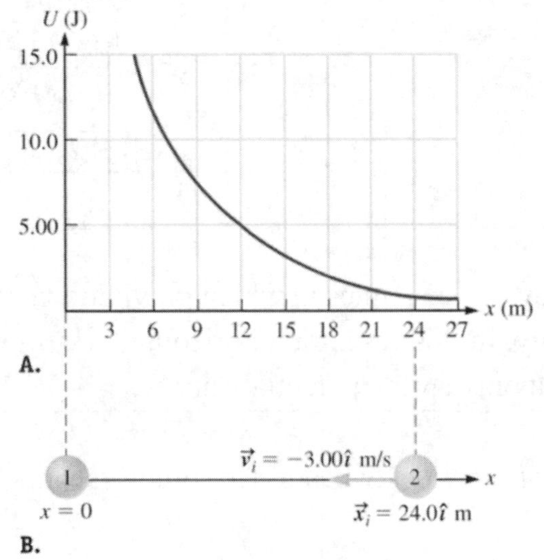

Figure P8.57

a. (N) What is the mechanical energy of this system?

INTERPRET and ANTICIPATE

The mechanical energy is constant, so can find the mechanical energy at *any* point in time, for instance the initial time, by adding the initial kinetic and potential energies.

SOLVE We use the initial speed to calculate the kinetic energy and read the potential energy off the plot.	$K_i = \frac{1}{2}mv_i^2 = \frac{1}{2}(2.75 \text{ kg})(3.00 \text{ m/s})^2 = 12.4 \text{ J}$ $U = 1.0 \text{ J}$ $E = K + U = 12.4 \text{ J} + 1.0 \text{ J} = \boxed{13.4 \text{ J}}$
CHECK and THINK The total energy, kinetic plus potential, is calculated for the initial time and is conserved in this system.	

b. (N) What is the kinetic energy of the system when particle 2 is at $\vec{x} = 9.00\hat{\imath}$ m?

INTERPRET and ANTICIPATE Since the total energy is conserved, we can read the potential energy at this location off the plot and then determine the kinetic energy required to maintain a constant total energy.	
SOLVE We read the potential energy at the position 9m from the plot.	$U = 7.5 \text{ J}$
We now require the total energy to be 18.4 J, as calculated in part (a).	$E = K + U$ $K = E - U = 13.4 \text{ J} - 7.5 \text{ J} = \boxed{5.9 \text{ J}}$
CHECK and THINK The total energy found in part (a) is conserved, so the kinetic energy equals the total energy minus potential.	

c. (C) Will the two particles ever touch? Explain why or why not.

> The particles will not touch because the system does not have enough energy. When the potential energy equals the mechanical energy (roughly at $x = 5$ m), particle 2 will momentarily stop, reverse direction, and then move in the positive x direction. That is, it does not have enough energy to overcome the potential energy needed to touch particle 1.

63. CASE STUDY: Comet Hale-Bopp's closest approach to the Sun is 0.914 AU, and its aphelion, or maximum distance from the Sun, is 371 AU in an elliptical orbit. (The average distance of the Earth from the Sun is 1 AU = 1.496×10^{11} km).

a. (N) What is the eccentricity of Comet Hale-Bopp's orbit?

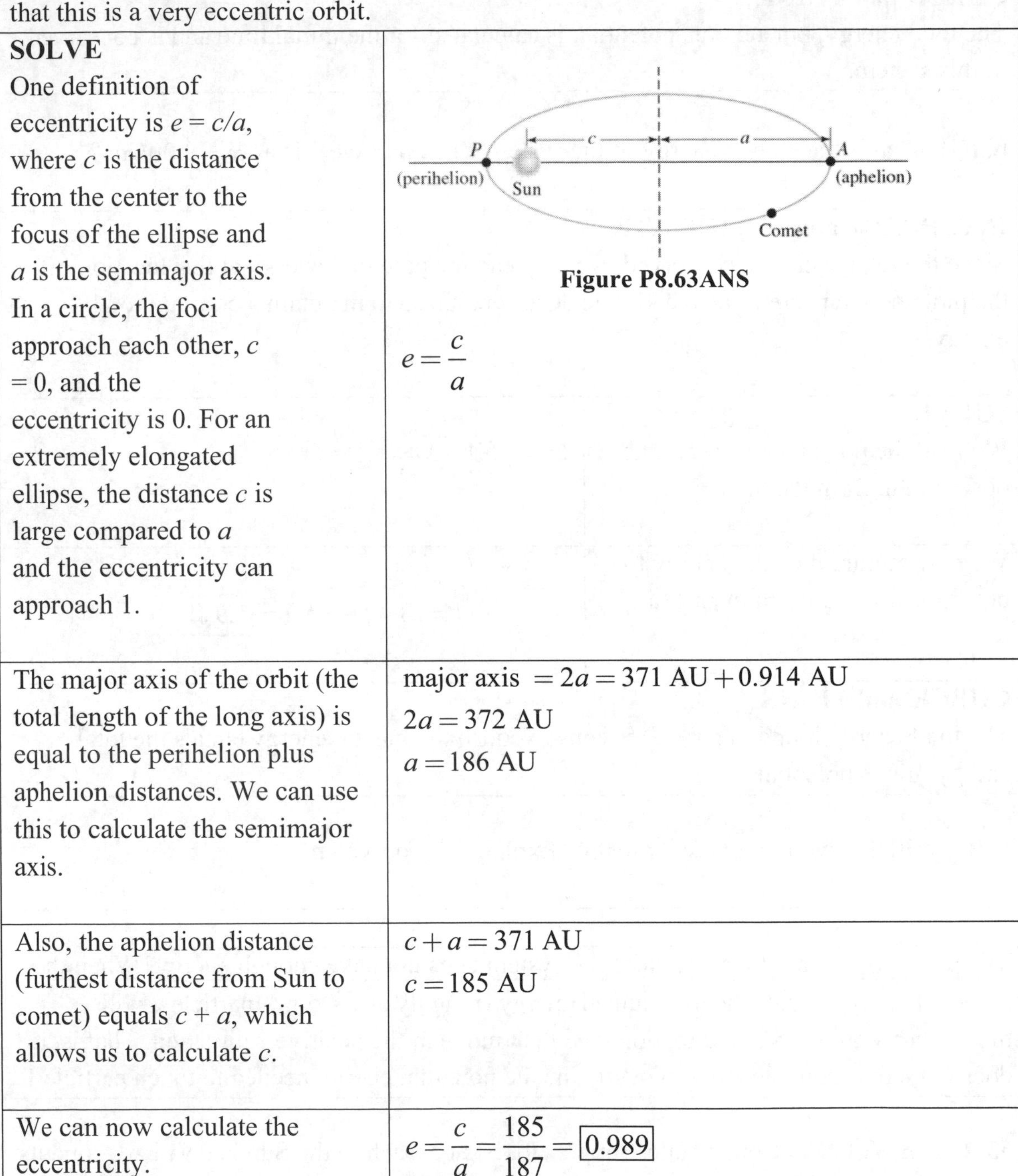

INTERPRET and ANTICIPATE

The eccentricity describes the degree to which the orbit is elliptical. An eccentricity of 0 corresponds to a circular orbit while a value close to 1 indicates an extremely stretched ellipse. The Sun is at one focus of the ellipse. Given the relatively small perihelion distance (which is also the distance from the focus to the end of the ellipse) compared to the aphelion distance (which spans the rest of the major axis of the ellipse), we expect that this is a very eccentric orbit.

SOLVE	
One definition of eccentricity is $e = c/a$, where c is the distance from the center to the focus of the ellipse and a is the semimajor axis. In a circle, the foci approach each other, $c = 0$, and the eccentricity is 0. For an extremely elongated ellipse, the distance c is large compared to a and the eccentricity can approach 1.	 **Figure P8.63ANS** $e = \frac{c}{a}$
The major axis of the orbit (the total length of the long axis) is equal to the perihelion plus aphelion distances. We can use this to calculate the semimajor axis.	major axis $= 2a = 371\ \text{AU} + 0.914\ \text{AU}$ $2a = 372\ \text{AU}$ $a = 186\ \text{AU}$
Also, the aphelion distance (furthest distance from Sun to comet) equals $c + a$, which allows us to calculate c.	$c + a = 371\ \text{AU}$ $c = 185\ \text{AU}$
We can now calculate the eccentricity.	$e = \frac{c}{a} = \frac{185}{187} = \boxed{0.989}$

CHECK and THINK
This orbit is extremely eccentric – that is, very far from a circular orbit.

b. (N) What is the period of this comet's orbit?

INTERPRET and ANTICIPATE	
Kepler's law relates the period of the orbit to the semimajor axis, which we know from part (a).	
SOLVE We use Kepler's Third Law in astronomical units, $T^2 = a^3$ with T in years and a in AU.	$T^2 = (186\ \text{AU})^3$ $T = \boxed{2540\ \text{yr}}$
CHECK and THINK Not only is this an eccentric orbit, the comet is making a very large excursion away from our Sun and returns only once every couple millennia!	

c. (N) The comet's mass is estimated to be 1.30×10^{16} kg. What is the potential energy of the Comet Hale-Bopp–Sun system when it is at its farthest distance from the Sun?

INTERPRET and ANTICIPATE	
We have all the information we need to calculate the gravitational potential energy using the Sun and comet separated by the maximum distance.	
SOLVE The potential energy of the Comet-Sun system depends on their masses and separation (Equation 8.7).	$U = -\dfrac{GmM_S}{r}$
We substitute numerical values and calculate.	$U = -\dfrac{(6.67\times10^{-11}\ \text{N}\cdot\text{m}^2/\text{kg}^2)(1.30\times10^{16}\ \text{kg})(1.991\times10^{30}\ \text{kg})}{371(1.496\times10^{11}\ \text{m})}$ $U = \boxed{-3.11\times10^{22}\ \text{J}}$
CHECK and THINK Though this is still quite a large number, the huge distance means that the potential	

energy is much smaller than potential energy calculations for other bodies in our solar system. See Problem 8.67, for example.

67. (N) The Earth's perihelion distance (closest approach to the Sun) is $r_P = 1.48 \times 10^{11}$ m, and its aphelion distance (farthest point) is $r_A = 1.52 \times 10^{11}$ m. What is the change in the Sun–Earth's gravitational potential energy as the Earth moves from aphelion to perihelion? What is the change in its gravitational potential energy from perihelion to aphelion?

INTERPRET and ANTICIPATE When the Earth moves towards the Sun, the gravitational potential energy must decrease just as when a hammer falls from the top of a building towards the Earth. We can calculate the gravitational potential energy at both points and determine the difference.	
SOLVE The gravitational potential between any two objects depends on their masses and separation according to Equation 8.7.	$U_g = -\frac{GMm}{r}$
Calculating the difference, we can determine the change in potential energy from aphelion to perihelion.	$\Delta U = \left(-\frac{GMm}{r_f}\right) - \left(-\frac{GMm}{r_i}\right)$ $\Delta U_{[\text{A to P}]} = GM_{\odot}M_{\oplus}\left(\frac{1}{r_A} - \frac{1}{r_P}\right)$
To simplify the calculation, we first calculate the terms in front of the parentheses.	$GM_{\odot}M_{\oplus} = \left(6.67 \times 10^{-11}\,\frac{\text{N}\cdot\text{m}^2}{\text{kg}^2}\right)\left(1.99 \times 10^{30}\,\text{kg}\right)\left(5.98 \times 10^{24}\,\text{kg}\right)$ $GM_{\odot}M_{\oplus} = 7.94 \times 10^{44}\ \text{N}\cdot\text{m}^2$
Inserting the distances, we calculate the change in potential energy.	$\Delta U_{[\text{A to P}]} = 7.94 \times 10^{44}\ \text{N}\cdot\text{m}^2\left(\frac{1}{1.52 \times 10^{11}\ \text{m}} - \frac{1}{1.48 \times 10^{11}\ \text{m}}\right)$ $\Delta U_{[\text{A to P}]} = \boxed{-1.41 \times 10^{32}\ \text{J}}$

Repeating the calculation for the perihelion to aphelion, we find the change in energy is the same magnitude, but opposite sign.	$\Delta U_{[\text{P to A}]} = -GM_{\odot}M_{\oplus}\left(\frac{1}{r_A} - \frac{1}{r_P}\right)$ $\Delta U_{[\text{P to A}]} = -\Delta U_{[\text{A to P}]}$ $\Delta U_{[\text{P to A}]} = \boxed{1.41\times10^{32}\ \text{J}}$
CHECK and THINK We found just what we expected. When the Earth moves toward the Sun ("falls into the sun") the change in the gravitational potential energy is negative $\Delta U_{[\text{A to P}]} < 0$, and when the Earth moves away from the Sun it is positive $\Delta U_{[\text{P to A}]} > 0$. We have also found that they have the same magnitude, so in one complete orbit there is no net change in system's gravitation potential energy.	

72. At the start of a basketball game, a referee tosses a basketball straight into the air by giving it some initial speed. After being given that speed, the ball reaches a maximum height of y_{max} above where it started.

a. (A) Using conservation of energy, find an expression for the ball's initial speed in terms of the gravitational acceleration g and the maximum height, y_{max}.

INTERPRET and ANTICIPATE During the motion, the only force doing work on the ball will be gravity. We can equate the mechanical energy of the ball at the beginning to the mechanical energy at the peak of the motion.	
SOLVE Choosing the release point to be the height where $y = 0$, and recognizing that the speed of the ball is 0 at the peak of the motion, we can approach this problem using conservation of energy (Eq. 8.14).	$(1/2)mv_i^2 + mg(0) = (1/2)m(0)^2 + mgy_{max}$
Cancel the mass from each term and solve.	$(1/2)v_i^2 = gy_{max}$ $v_i = \boxed{\sqrt{2gy_{max}}}$
CHECK and THINK The larger the initial speed, the higher the ball will travel.	

b. (A) Using conservation of energy, find an expression for the height of the ball when it has a speed of v in terms of its current height y, the gravitational acceleration g, and the maximum height, y_{max}.

INTERPRET and ANTICIPATE We can repeat the approach of part (a), this time considering a general height y (between 0 and y_{max}).	
SOLVE At a height y between 0 and y_{max}, the ball will have a potential energy mgy and a speed v. The total energy must equal the total energy when the ball reaches the highest point.	$\frac{1}{2}mv^2 + mgy = \frac{1}{2}m(0)^2 + mgy_{max}$
Cancel the mass from each term and solve.	$(1/2)v^2 + gy = gy_{max}$ $gy = gy_{max} - (1/2)v^2$ $y = \boxed{y_{max} - v^2/(2g)}$
CHECK and THINK This formula is valid for $0 \le v \le \sqrt{2gy_{max}}$. As v decreases from it's initial value of $\sqrt{2gy_{max}}$ (found in part (a)) to its value of 0 at the peak height, the corresponding height changes from 0 to y_{max}. That is what we should expect.	

73. A rocket carrying a new 950-kg satellite into orbit misfires and places the satellite in an orbit with an altitude of 125 km, well below its operational altitude in low-Earth orbit.

a. (N) What would be the height of the satellite's orbit if its total energy were 500 MJ greater?

INTERPRET and ANTICIPATE
For a satellite in circular orbit around a planet, we use the orbital properties to determine the energy of the orbit. We can in turn determine how the orbit changes if additional energy is input into the system.

SOLVE The total energy for a satellite in circular orbit depends on the masses of the satellite and planet and their separation, as given by Equation 8.19.	$E_{\text{tot}} = \frac{GmM_E}{2r}$
We can express the change in energy to the change in radius of the orbit.	$\Delta E = \frac{GmM_E}{2}\left(\frac{1}{r_i} - \frac{1}{r_f}\right)$
Solving for the quantity in parentheses, we can insert numerical values.	$\left(\frac{1}{r_i} - \frac{1}{r_f}\right) = \frac{2\Delta E}{GmM_E}$ $\left(\frac{1}{r_i} - \frac{1}{r_f}\right) = \frac{2(500\times10^6\ \text{J})}{\left(6.67\times10^{-11}\,\frac{\text{N}\cdot\text{m}^2}{\text{kg}^2}\right)(950\ \text{kg})\left(5.98\times10^{24}\,\text{kg}\right)}$ $\left(\frac{1}{r_i} - \frac{1}{r_f}\right) = 2.64\times10^{-9}\ \text{m}^{-1}$
The initial orbit is 125 km above the surface of the Earth. The new orbit is an unknown height h_f above the surface of the Earth.	$\left(\frac{1}{(6370+125)\times10^3} - \frac{1}{(6370+h_f)\times10^3}\right) = 2.64\times10^{-9}$ $h_f = 2.38\times10^5\ \text{m} = \boxed{238\ \text{km}}$
CHECK and THINK The final height is above the initial height of 125 km, which we would expect. That is, energy was added to the system and the satellite moved up to a higher potential.	

b. (N) What would be the difference in the system's kinetic energy?
c. (N) What would be the difference in the system's potential energy?

As the satellite is lifted from the lower to the higher orbit, the gravitational potential energy increases. Since energy is added to the system, the total energy must also increase. For a satellite to stay in orbit at larger distance from Earth, it must move slower, so the kinetic energy decreases. To quantitatively determine how all of these quantities relate, we refer to Section 8.9. The potential energy is negative. The total energy is half of the potential energy, $E = \frac{1}{2}U_G$ (Equation 8.21). The kinetic energy is positive and equal in magnitude to the total energy

($E = -K$, see Problem 59). So, numerically, the gravitational energy increases by 1000 MJ, the kinetic energy decreases by 500 MJ, and the total energy increases by 500 MJ. As we would expect, energy is conserved and $\Delta E = \Delta U + \Delta K$ holds.

76. Two 67.0-g arrows are fired in quick succession with an initial speed of 80.0 m/s. The first arrow makes an initial angle of 33.0° with the horizontal, and the second arrow is fired straight upward. Assume an isolated system and choose the reference configuration at the initial position of the arrows.

a. (N) What is the maximum height of each of the arrows?

INTERPRET and ANTICIPATE Both arrows are fired at the same speed and therefore have the same initial kinetic energy. The y velocity decreases to zero at the highest point, while the x velocity remains unchanged. We expect the arrow that is shot straight upwards will travel higher.	
SOLVE In an isolated system, with $y_i = 0$, we can use energy conservation.	$K_i + U_i = K_f + U_f$ $\frac{1}{2}mv_i^2 + mgy_i = \frac{1}{2}mv_f^2 + mgy_f$ $\frac{1}{2}mv_i^2 = \frac{1}{2}mv_f^2 + mgy_f$
By taking the magnitude of the vector $\vec{v} = v_x\hat{i} + v_y\hat{j}$, we find $v^2 = v_x^2 + v_y^2$. The final y velocity is zero at the highest point reached.	$\left(\frac{1}{2}mv_{xi}^2 + \frac{1}{2}mv_{yi}^2\right) = \left(\frac{1}{2}mv_{xf}^2 + 0\right) + mgy_f$
Now use $v_{xi} = v_{xf}$. It is only the vertical component that determines the final height of course, as we expect from previous problems in kinematics.	$\frac{1}{2}mv_{yi}^2 = mgy_f$ $y_f = \frac{v_{yi}^2}{2g}$

Now plugging in numbers for the first arrow…	$y_f = \frac{(80.0\sin 33.0°)^2}{2(9.81\ \text{m/s}^2)} = \boxed{96.8\ \text{m}}$
…and the second arrow.	$y_f = \frac{(80.0\ \text{m/s})^2}{2(9.81\ \text{m/s}^2)} = \boxed{326\ \text{m}}$
CHECK and THINK It is no great surprise that the arrow shot straight up goes significantly higher than the one shot 33 degrees above the horizontal.	

b. (N) What is the total mechanical energy of the arrow–Earth system for each of the arrows at their maximum height?

INTERPRET and ANTICIPATE

The total mechanical energy for each arrow is constant, so we can actually calculate the *initial* mechanical energy, when the arrow has only kinetic energy, which is the same for both arrows. The energy at the peak height must be the same.

SOLVE

$$E_{mech} = K_i + U_i = K_i = \frac{1}{2}(0.067\ \text{kg})(80.0\ \text{m/s})^2 = \boxed{214\ \text{J}}$$

CHECK and THINK

Total mechanical energy is conserved. Since both arrows start at $y = 0$ at the same speed, we need only calculate this initial kinetic energy. *Both* arrows have 214 J of total energy throughout their entire trajectories (at least until they hit the ground and as long as we can ignore forces that dissipate energy, like air resistance).

82. A cluster of grapes is removed from a frictionless, hemispherical bowl 44.0 cm in diameter, leaving behind a single spherical grape of mass 3.00 g initially at rest at the upper edge of the bowl along its horizontal diameter. Choose the bottom of the bowl as the reference configuration where $h = 0$, and answer the following questions as the grape slides to the bottom of the bowl.

a. (N) What is the gravitational potential energy of the grape–Earth system at the grape's initial position?
b. (N) What is the kinetic energy of the grape when it reaches the bottom of the bowl?
c. (N) What is the speed of the grape when it reaches the bottom of the bowl?

d. (N) What are the potential and kinetic energies of the grape when it reaches a point that is a height $h = 15.0$ cm above the bottom of the bowl?

INTERPRET and ANTICIPATE

We present the solutions for all four parts together because it is the same machinery used in each case. Namely, given that there are no forces that lead to energy leaving the system, such as friction, we apply conservation of mechanical energy. We only need to determine the vertical height of the grape above the reference configuration and use the mass provided to determine the gravitational potential energy. At the top of the bowl, the grape has no kinetic energy and all potential energy. As it falls towards the bottom of the bowl, its kinetic energy increases as the potential energy decreases. At the bottom of the bowl (reference configuration, zero potential energy), the grape is moving at its fastest and has only kinetic energy.

SOLVE

Let's first start by drawing a sketch to indicate our notation. We choose point 1 at the initial position at the top of the bowl, point 2 at height h = 15.0 cm above the bottom of the bowl, and point 3 at the bottom of the bowl.

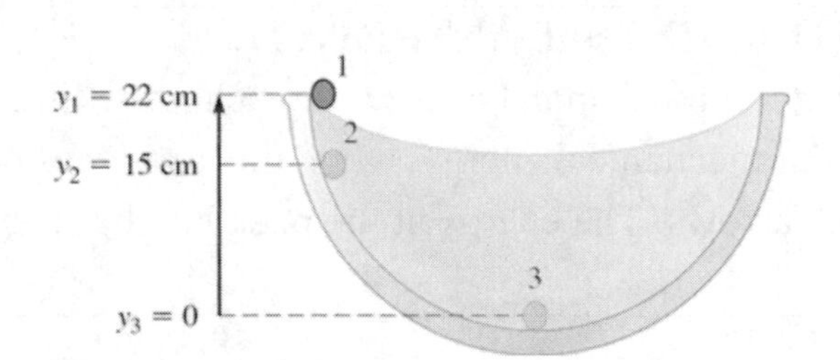

Figure P8.82ANS

a. The radius of the hemispherical bowl, and therefore the initial height of the grape is 44.0 cm / 2 = 22.0 cm = 0.220 m. This allows us to determine the initial potential energy.

$$U_1 = mgh$$
$$U_1 = (0.003 \text{ kg})(9.81 \text{ m/s}^2)(0.220 \text{ m})$$
$$U_1 = \boxed{6.47 \times 10^{-3} \text{ J}}$$

b. Using energy conservation, the potential energy of the grape at the top of the bowl is converted into kinetic energy at the bottom of the bowl.

$$U_1 + K_1 = U_3 + K_3$$
$$U_1 + 0 = 0 + K_3$$
$$K_3 = U_1 = \boxed{6.47 \times 10^{-3} \text{ J}}$$

c. The speed of the grape at the bottom can be found directly using the definition of kinetic energy and the answer to part (b).

$$K_3 = \frac{1}{2} m v_3^2$$
$$v_3 = \sqrt{\frac{2K_3}{m}} = \sqrt{\frac{2(6.47 \times 10^{-3} \text{ J})}{(0.003 \text{ kg})}} = \boxed{2.08 \text{ m/s}}$$

d. We now apply conservation of energy using point 2 on one side of the equation. The *y* position is 15 cm above the bottom of the bowl.	$U_2 = mgh$ $U_2 = (0.003\ \text{kg})(9.80\ \text{m/s}^2)(0.150\ \text{m})$ $U_2 = 4.41\times10^{-3}\ \text{J}$ $U_1 + K_1 = U_2 + K_2$ $K_2 = U_1 - U_2$ $K_2 = 6.47\times10^{-3}\ \text{J} - 4.41\times10^{-3}\ \text{J}$ $K_2 = \boxed{2.06\times10^{-3}\ \text{J}}$
CHECK and THINK The gravitational potential energy depends on the mass and the vertical position relative to the reference configuration.	

9

Energy in Nonisolated Systems

4. During practice, a hockey player passes the puck to his coach who is 6.5 m away as shown in Figure 9.2. The puck, which has a mass of 2.0 kg, was initially at rest. The hockey player exerts a constant 47.4-N force as his stick pushes the puck 0.25 m. Include only the puck in the system and assume the friction between the ice and the puck is negligible.

a. (C) Draw a free-body diagram for the puck while it is in contact with the player's stick.

The puck is resting on the ground and accelerating horizontally. There is a normal force, which balances the weight of the puck, and a net horizontal force due to the stick. 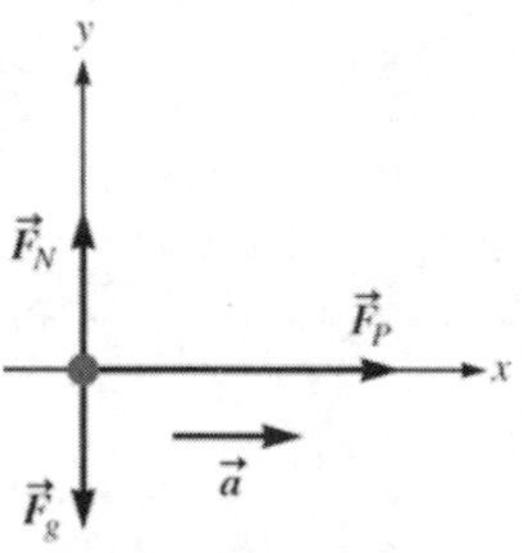 **Figure P9.4ANS**

b. (N) Find the work done by all the forces in your free-body diagram.

INTERPRET and ANTICIPATE According to our sketch, the puck's displacement is to the right in the positive x direction. The normal force and gravity do not do work on the puck because they are perpendicular to the displacement. The player does positive work on the puck.	
SOLVE We can apply Equation 9.1 to determine the work done by the player, assuming that the force is constant over the distance that the puck is accelerated.	$W = F_x \Delta x$ $W_P = F_P \Delta x$ $W_P = (47.4\ \text{N})(0.25\ \text{m})$ $W_P = 11.9\ \text{N}\cdot\text{m} = \boxed{12\ \text{J}}$

For both the normal force and weight, the angle between the force and the displacement is 90 degrees, so the work done is zero according to Equation 9.6.	$W_N = W_g = F\Delta r \cos 90° = 0$

CHECK and THINK

As expected the play does positive work on the puck (which leads to an increase in its speed as seen in part (c)). We can also express the energy in Joules, using $1\ \text{N}\cdot\text{m} = 1\ \text{J}$.

c. (N) What is the speed of the puck as it leaves the player's stick?

INTERPRET and ANTICIPATE

For a particle, we can apply the work-kinetic energy theorem to find the puck's speed. The work done on the puck by the stick is equal to the final kinetic energy of the puck.

SOLVE	
Solve Equation 9.2 for speed. The puck is initially at rest and so it has no kinetic energy. The change in kinetic energy equals the puck's final kinetic energy.	$W_{\text{tot}} = \Delta K$ $W_P = \frac{1}{2}mv^2$ $v = \sqrt{\frac{2W_P}{m}} = \sqrt{\frac{2(11.9\ \text{N}\cdot\text{m})}{2.0\ \text{kg}}}$ $v = \boxed{3.4\ \text{m/s}}$

CHECK and THINK

The positive work done by the player increases the puck's speed.

d. (C) In an all-star hockey game, the puck reaches speeds of 100 mph. Use that information to check your results to parts (a) through (c).

We convert 100 mph to 45 m/s. We find the speed of the puck here is well below the maximum possible speed, so it is a physically sensible answer.

7. (N) Kerry is pulling a 154-kg sled along a snowy, horizontal path with a 615-N force directed at an angle of 30.0° above the ground. If he pulls the sled over a distance of 30.0 m, how much work has Kerry performed on the sled?

INTERPRET and ANTICIPATE The component of the force along the displacement determines the total work Kerry does on the sled. We expect that Kerry does positive work as he exerts a force that has a component in the same direction as the direction the sled moves.	
SOLVE To find the work performed by Kerry, use Eq. 9.6. The angle between the displacement of the sled and his applied force is 30.0°.	$W = F\Delta r\cos\theta$ $W = (615\ \text{N})(30.0\ \text{m})\cos(30.0°)$ $W = \boxed{1.60\times10^4\ \text{J}}$
CHECK and THINK As anticipated, the work is positive and depends on the applied force, the distance over which it's exerted, and the angle between these vectors.	

12. (N) An object is subject to a force $\vec{F} = \left(512\hat{i} - 134\hat{j}\right)$ N such that 10,125 J of work is performed on the object. If the object travels 25.0 m in the positive x direction while this work is performed, what must be the displacement of the object in the y direction?

INTERPRET and ANTICIPATE The total work depends on the applied force, the distance over which it is exerted, and the angle between the force and the displacement. We can write a formula for work using the components of force and displacement and solve for the unknown y component.	
SOLVE Given the force in component form and a single component of the displacement, it is possible to write the work equation using components of force and displacement (Eq. 9.17) and solve for the unknown displacement in the y direction.	$W = F_x\Delta x + F_y\Delta y$
The work, x component of the displacement, and force are given, so we can solve for the y component of the displacement.	$10{,}125\ \text{J} = (512\ \text{N})(25.0\ \text{m}) + (-134\ \text{N})(\Delta y)$ $(134\ \text{N})(\Delta y) = 12{,}800\ \text{J} - 10{,}125\ \text{J}$ $\Delta y = \boxed{20.0\ \text{m}}$
CHECK and THINK We've expressed the work in component form, which allows us to easily relate the work	

with the components of the force and displacement vectors.

16. (N) In Figure P9.16, the magnitude of the vectors are $A = 6.00$ and $B = 3.00$. Find $\vec{A} \bullet \vec{B}$ in each case.

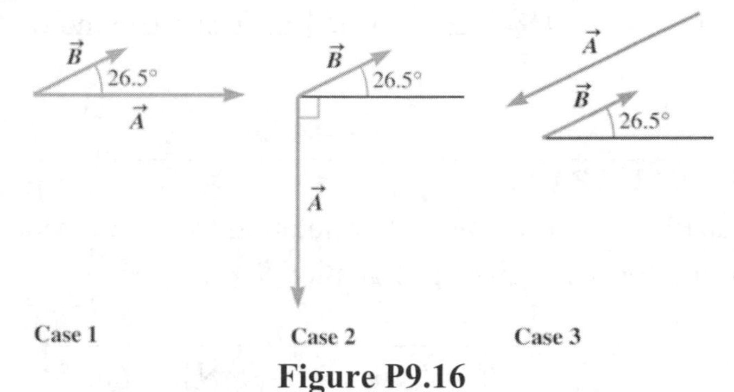

Figure P9.16

INTERPRET and ANTICIPATE For each, we can use the magnitude of the vectors given and the angle between them to calculate the dot product (also called the scalar product).	
SOLVE In each case we apply Equation 9.8.	$D = AB\cos\varphi$
Case 1 The angle is 26.5°.	$D = (6.00)(3.00)\cos 26.5°$ $D = \boxed{16.1}$
Case 2 The total angle between the two vectors is 26.5°+90° = 116.5°. Since there is a component of A in the opposite direction of B, the scalar product is negative.	$D = (6.00)(3.00)\cos 116.5°$ $D = \boxed{-8.03}$
Case 3 The vectors are antiparallel, so the angle is 180°. (There's no need to use the 26.5° directly.) Since they point in the opposite direction, the dot product is negative.	$D = (6.00)(3.00)\cos 180°$ $D = \boxed{-18.0}$
CHECK and THINK In each case, we need only the magnitudes of the vectors and the angle between them. Vectors that have components in opposite directions (that is, the vectors are between 90°	

and 270° apart) will result in a negative scalar product.

23. (N) A constant force of magnitude 4.75 N is exerted on an object. The force's direction is 60.0° counterclockwise from the positive x axis in the xy plane, and the object's displacement is $\Delta\vec{r} = \left(4.2\hat{\boldsymbol{i}} - 2.1\hat{\boldsymbol{j}} + 1.6\hat{\boldsymbol{k}}\right)$ m. Calculate the work done by this force.

INTERPRET and ANTICIPATE Since we have our displacement in unit-vector notation (or component form), we can use the dot product formula for work given in Equation 9.17.	
SOLVE To start, we need to put our force vector into unit-vector notation. In other words, we need to calculate the F_x, F_y, F_z components. The force is in the xy plane, so there is no z component.	$F_x = F\cos\theta = (4.75\text{ N})\cos 60.0°$ $F_y = F\sin\theta = (4.75\text{ N})\sin 60.0°$ $F_z = 0$
We can read off the components of the displacement vector directly.	$\Delta x = 4.2\text{ m}$ $\Delta y = -2.1\text{ m}$ $\Delta z = 1.6\text{ m}$
Now we can use the form for work shown in Equation 9.17 where.	$W = F_x\Delta x + F_y\Delta y + F_z\Delta z$ $W = ((4.75\text{ N})\cos 60.0°)(4.2\text{ m}) + ((4.75\text{ N})\sin 60.0°)(-2.1\text{ m}) + (0)(1.6\text{ m})$ $W = \boxed{1.3\text{ J}}$
CHECK and THINK There is no contribution to the net work from the displacement in the z-direction since the force is constrained to the xy plane.	

25. (N) An object of mass $m = 5.8$ kg moves under the influence of one force. That force causes the object to move along a path given by $x = 6.0 + 5.0t + 2.0t^2$, where x is in meters and t is in seconds. Calculate the work done by the force on the object from $t = 2.0$ s to $t = 7.0$ s.

<table>
<tr><td colspan="2">INTERPRET and ANTICIPATE
We do not know the magnitude of the force acting on the object in this problem so we need to find another way to calculate work. Actually, we could use a kinematic approach and take derivatives of position with respect to time to find velocity and acceleration, calculate the force, integrate with respect to distance to get the work. This would be pretty involved though and there is a simpler way using energy! We know how position changes over time, so we can use the work-kinetic energy theorem to determine the initial and final speeds to calculate the change in kinetic energy, which equals the work done on the object.</td></tr>
<tr><td>SOLVE
The total work done by the force is equal to the change in kinetic energy according to the work-kinetic energy theorem (Eq. 9.5).</td><td>$W_{\text{tot}} = \Delta K$</td></tr>
<tr><td>In order to calculate the change in kinetic energy, we need to calculate the speed from the position function.</td><td>$v = \frac{dx}{dt} = 5.0 + 4.0t$</td></tr>
<tr><td>Use this function to evaluate the speed at $t = 2.0$ and $t = 7.0$ s.</td><td>$v_i = 5.0 + 4.0(2.0\text{ s}) = 13\text{ m/s}$
$v_f = 5.0 + 4.0(7.0\text{ s}) = 33\text{ m/s}$</td></tr>
<tr><td>We can now calculate the work.</td><td>$W = \Delta K = \frac{1}{2}mv_f^2 - \frac{1}{2}mv_i^2$
$W = \frac{1}{2}(5.8\text{ kg})\left[(33\text{ m/s})^2 - (13\text{ m/s})^2\right]$
$W = \boxed{2.7 \times 10^3\text{ J}}$</td></tr>
<tr><td colspan="2">CHECK and THINK
The object's speed increases so the work must be positive as calculated.</td></tr>
</table>

26. (N) A nonconstant force is exerted on a particle as it moves in the positive direction along the x axis. Figure P9.26 shows a graph of this force F_x versus the particle's position x. Find the work done by this force on the particle as the particle moves as follows.

a. From $x_i = 0$ to $x_f = 10.0$ m

b. From $x_i = 10.0$ to $x_f = 20.0$ m
c. From $x_i = 0$ to $x_f = 20.0$ m

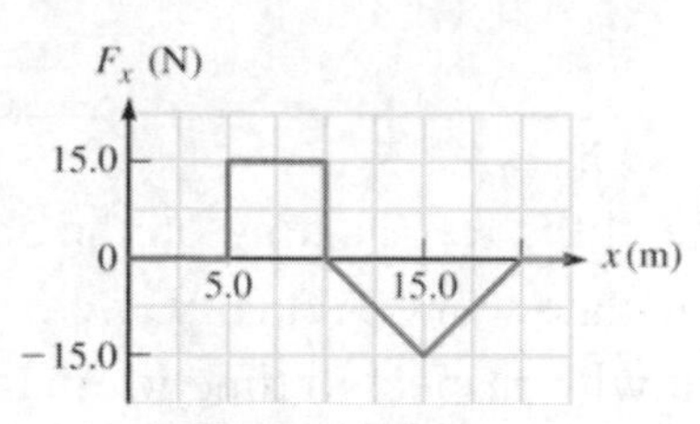

Figure P9.26

<table>
<tr><td colspan="2">INTERPRET and ANTICIPATE
In all three cases, the work equals the area under the force versus position curve. This graphical method is equivalent to saying that the work equals the integral of the force over distance, since an integral on an F vs. x plot is equal to the area under the curve.</td></tr>
<tr><td>SOLVE
a. This is the area of a rectangle. It is positive because the force and displacement both have the same sign, positive (or graphically, because the rectangle is above the x axis). Since the force is measured in Newtons and the distance in meters, the area is in Newton-meters or, equivalently, Joules.</td><td>$W_a = (15.0\ \text{N})(5.00\ \text{m})$
$W_a = \boxed{75.0\ \text{J}}$</td></tr>
<tr><td>b. This is the area of a triangle. It is negative because the force is in the opposite direction as the displacement (or graphically, because the triangle is below the x axis).</td><td>$W_b = \frac{1}{2}(-15.0\ \text{N})(10.0\ \text{m})$
$W_b = \boxed{-75.0\ \text{J}}$</td></tr>
<tr><td>c. This is the sum of the work found in parts (a) and (b), or the area of the rectangle plus the area of the triangle.</td><td>$W_c = W_a + W_b$
$W_c = 75.0\ \text{N}\cdot\text{m} - 75.0\ \text{N}\cdot\text{m}$
$W_c = \boxed{0}$</td></tr>
<tr><td colspan="2">CHECK and THINK
The work can be positive or negative depending on whether the force is applied in the same or opposite direction as the displacement. Graphically, we can calculate this as the</td></tr>
</table>

area under the curve. If the curve is above the x axis, work is positive, and if below, the work is negative. In this case, we happen to have equal and opposite work in parts (a) and (b), so the total work is zero.

30. A particle moves in the xy plane (Fig. P9.30) from the origin to a point having coordinates $x = 7.00$ m and $y = 4.00$ m under the influence of a force given by $\vec{F} = 3y^2\hat{\imath} + x\hat{\jmath}$.

a. (N) What is the work done on the particle by the force F if it moves along path 1 (shown in red)?
b. (N) What is the work done on the particle by the force F if it moves along path 2 (shown in blue)?
c. (N) What is the work done on the particle by the force F if it moves along path 3 (shown in green)?

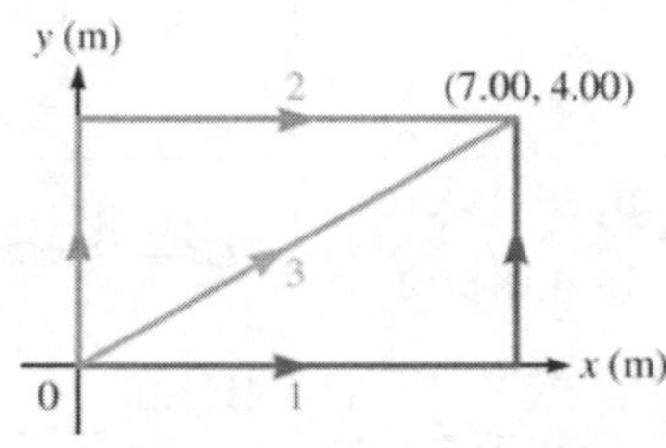

Figure P9.30

INTERPRET and ANTICIPATE	
In all three cases, we can calculate the integral of the force over distance for the specified path.	
SOLVE In each case, the work is found using the integral of force over distance along the path (Equation 9.21).	$W = \int_{r_i}^{r_f} \vec{F} \cdot d\vec{r} = \int_{r_i}^{r_f} \left(F_x dx + F_y dy + F_z dz\right)$
The work done along path 1, we first need to integrate along $d\vec{r} = dx\,\hat{\imath}$ from (0,0) to (7,0) and then along $d\vec{r} = dy\,\hat{\jmath}$ from (7,0) to (7,4).	$W_1 = \int_{x=0;y=0}^{x=7;y=0} \left(3y^2\,\hat{\imath} + x\,\hat{\jmath}\right)\cdot\left(dx\,\hat{\imath}\right) + \int_{x=7;y=0}^{x=7;y=4} \left(3y^2\,\hat{\imath} + x\,\hat{\jmath}\right)\cdot\left(dy\,\hat{\jmath}\right)$
We now perform the dot products.	$W_1 = \int_{x=0;y=0}^{x=7;y=0} 3y^2\,dx + \int_{x=7;y=0}^{x=7;y=4} x\,dy$

Along the first part of this path, y = 0 therefore the first integral equals zero. For the second integral, x is constant ($x = 7$ for the entire integration path), so we can pull it out front of the integral and we are left with an integral of dy.	$W_1 = 0 + \int_{x=7;y=0}^{x=7;y=4} x\,dy$ $W_1 = xy\Big\|_{x=7;y=0}^{x=7;y=4}$ $W_1 = \boxed{28\text{ J}}$
b. The work done along path 2 is along $d\vec{r} = dy\,\hat{j}$ from (0,0) to (0,4) and then along $d\vec{r} = dx\,\hat{i}$ from (0,4) to (7,4).	$W_2 = \int_{x=0;y=0}^{x=0;y=4} \left(3y^2\,\hat{i} + x\,\hat{j}\right)\cdot\left(dy\,\hat{j}\right)$ $+ \int_{x=0;y=4}^{x=7;y=4} \left(3y^2\,\hat{i} + x\,\hat{j}\right)\cdot\left(dx\,\hat{i}\right)$
Now, calculate the dot products.	$W_2 = \int_{x=0;y=0}^{x=0;y=4} x\,dy + \int_{x=0;y=4}^{x=7;y=4} 3y^2\,dx$
Since $x = 0$ along the first integral, it is zero. For the second integral, y is constant along the entire path ($y = 4$), so "$3y^2$" is a constant (equal to 48 along the entire integration path) that we can pull out front and we are left with only an integral of dx.	$W_2 = 0 + 3y^2x\Big\|_{x=0;y=4}^{x=7;y=4}$ $W_2 = \boxed{336\text{ J}}$
c. To find the work along the third path, we first write the expression for the work integral.	$W = \int_{r_i}^{r_f} \vec{F}\bullet d\vec{r} = \int_{r_i}^{r_f}\left(F_x dx + F_y dy + F_z dz\right)$ $W = \int_{r_i}^{r_f}\left(3y^2 dx + x\,dy\right) \quad (1)$
At first glance, this appears to be similar to the first two paths, but there is an important difference. We can't simply integrate $\int x\,dy = xy$, because the value of x changes as we vary y. That is, on this slanted line, x changes as y changes, so x is a function of y – we can't pull it out of the integral since it varies, but we just can't ignore it. In parts (a) and (b), on a straight horizontal or vertical line, *only* x or y changes, which makes life easier. Here, we need to take into account that we're performing the integral on this specific line, along which both x and y	dr dy θ dx **Figure P9.30ANS**

change in proportion to each other. But how? The trick is to write each integral so that the integral depends *only* on one variable, because we know how to perform that integral. One way to do this is to parameterize both x and y in terms of another variable, say t, substitute, and write the entire integral in terms of t. Here, we will take a slightly different approach and relate dx to dy to express each integral only in terms of x or y. We start by sketching a path along the line to see how to relate dx and dy.	
Looking at the figure, we can write dx in terms of dy, and vice versa.	$\tan\theta = \frac{dy}{dx}$ $dy = \tan\theta\, dx \quad \text{and} \quad dx = \frac{dy}{\tan\theta} \qquad (2)$
Now, use equation (2) in (1) to express each integral in terms of only one variable. The angle is constant (on this straight line), so each integral really is just an integral over a single variable, x or y, which we can do!	$W = \int_{x=0;y=0}^{x=7;y=4} 3y^2\, dx + \int_{x=0;y=0}^{x=7;y=4} x\, dy$ $W = \int_{y=0}^{y=4} 3y^2 \frac{dy}{\tan\theta} + \int_{x=0}^{x=7} x \tan\theta\, dx \qquad (3)$
We need to determine the tangent of the angle. This angle is the angle the path makes with the x axis. We don't actually need the angle, though we could take the inverse tangent to find it ($\theta = 29.7$ degrees).	$\tan\theta = \frac{4.00}{7.00} = 0.570 \qquad (4)$
Insert the value of the tangent (equation (4) into (3)) and solve the integrals.	$W = \frac{3}{0.570} \frac{y^3}{3}\Big\|_{y=0}^{y=4} + 0.570 \frac{x^2}{2}\Big\|_{x=0}^{x=7}$ $W = 112 + 14 = \boxed{126\ \text{J}}$

CHECK and THINK

In each case, we integrate the force over distance. The difficult part can be integrating along the specified path. Equation 9.21 works in each case, but the way that we perform the integral depends on the path over which we integrate.

d. (C) Is the force F conservative or nonconservative? Explain.

Since the work done is *not* path-independent, this is a non-conservative force. That is, since the answers for (a), (b), and (c) are different, the integral clearly depends on the path taken and therefore the force is by definition not conservative.

36. A 2.15-g hailstone, which can be modeled as a particle, falls a vertical distance of 145 m at constant speed.
a. (N) What is the work done on the hailstone by gravity?

INTERPRET and ANTICIPATE	
The work done by gravity is found by multiplying the gravitational force (weight = mg) by the distance fallen. Since the force and displacement are in the same direction, the work done by gravity will be positive.	
SOLVE The gravitational force is constant (i.e. the weight of the hailstone, *mg*) and in the same direction as the displacement, so we can just multiply the force times distance (Equation 9.1). We choose downward to be positive so both the force and displacement are positive. (It doesn't matter, of course. If you choose up to be positive, the force and displacement are both negative and the sign of the work is unchanged.)	$W_G = F_y \Delta y$ $W_G = mgh$ $W_G = (2.15 \times 10^{-3}\text{ kg})(9.81\text{ m/s}^2)(145\text{ m})$ $W_G = \boxed{3.06\text{ J}}$
CHECK and THINK When the force acts in the same direction as the displacement, the work is positive and the calculation is straightforward. But if gravity did work on the hailstone, why didn't it speed up? This is the focus of part (b).	

b. (N) What is the work done on the hailstone by air resistance?

INTERPRET and ANTICIPATE	
Since the hailstone falls at constant speed, it is not accelerating and the work done by gravity clearly did not go into kinetic energy. Instead, there is another force in the upward direction (friction/air resistance) which does an equal amount of negative work at is it falls.	
SOLVE Since the hailstone is falling at	$F_R = -mg$ (still assuming downward is positive).

constant speed, air resistance must be balancing the force of gravity. Since we already chose downward as the positive direction in part (a), we will stick with that convention and the upward drag force is negative. Again, if you choose upward to be positive, that works too: the drag force is positive and the displacement negative, so the work is still negative.	$W_R = (-mg)\Delta y$ $W_R = -W_G = \boxed{-3.06\text{ J}}$
CHECK and THINK Friction does negative work on the hailstone, since the force is in the opposite direction of the displacement. The total work (by gravity and friction) is zero, so the kinetic energy of the hailstone is unchanged as it falls.	

39. A shopper weighs 3.00 kg of apples on a supermarket scale whose spring obeys Hooke's law and notes that the spring stretches a distance of 3.00 cm.

a. (N) What will the spring's extension be if 5.00 kg of oranges are weighed instead?

INTERPRET and ANTICIPATE Assuming the spring obeys Hooke's law, we can use the result for 3 kg to determine the spring constant and then use the spring constant to find the extension if 5 kg of apples are weighed. We expect the extension will be proportionally larger for the larger mass.	
SOLVE For the 3.00 kg of apples, apply Hooke's Law.	$F = ky$ $k = \dfrac{Mg}{y} = \dfrac{(3.00\text{ kg})(9.81\text{ m/s}^2)}{0.03\text{ m}} = 9.81\times10^2\text{ N/m}$
Using the same spring constant for 5.00 kg of apples, we can calculate the new extension.	$y = \dfrac{mg}{k}$ $y = \dfrac{(5.00\text{ kg})(9.81\text{ m/s}^2)}{9.81\times10^2\text{ N/m}}$ $y = 0.0500\text{ m} = \boxed{5.00\text{ cm}}$
CHECK and THINK Indeed, when we weigh 5 kg instead of 3 kg, the extension is proportionally larger: 5 cm instead of 3 cm.	

b. (N) What is the total amount of work that the shopper must do to stretch this spring a total distance of 7.00 cm beyond its relaxed position?

INTERPRET and ANTICIPATE

The integral of the force over distance determines the work, as usual. For the spring, since the force is proportional to the distance from equilibrium, the work depends as the square of this distance. Positive work must be done as the shopper pushes downward, in the same direction the scale displaces.

SOLVE	
We could calculate the work by integrating the force (Hooke's law) over distance, as was done to get Equation 9.25. If the initial extension is 0 and the final extension is 0.07 m, we can use the spring force from part (a) to determine the total work.	$W = \frac{1}{2}ky_f^2$ $W = \frac{1}{2}(9.81\times10^2\ \text{N/m})(0.07\ \text{m})^2$ $W = \boxed{2.40\ \text{J}}$

CHECK and THINK

The work is positive, as anticipated. A larger spring constant (stiffer spring) would require more work as well as extending the spring further.

45. (A) A bullet flying horizontally hits a wooden block that is initially at rest on a frictionless, horizontal surface. The bullet gets stuck in the block, and the bullet–block system has a final speed of v_f. Find the final speed of the bullet–block system in terms of the mass of the bullet m_b, the speed of the bullet before the collision v_b, the mass of the block m_{wb}, and the amount of thermal energy generated during the collision E_{th}.

INTERPRET and ANTICIPATE

We can use conservation of energy and start with the work-energy theorem (Eq. 9.31). In this case, there is no potential energy or outside work performed on the system. Initially, the bullet has kinetic energy, while in the final situation, the bullet/wooden-block has kinetic energy and some thermal energy might have been produced. We could sketch a bar chart, though we don't know how much thermal versus kinetic energy there is in the final situation.

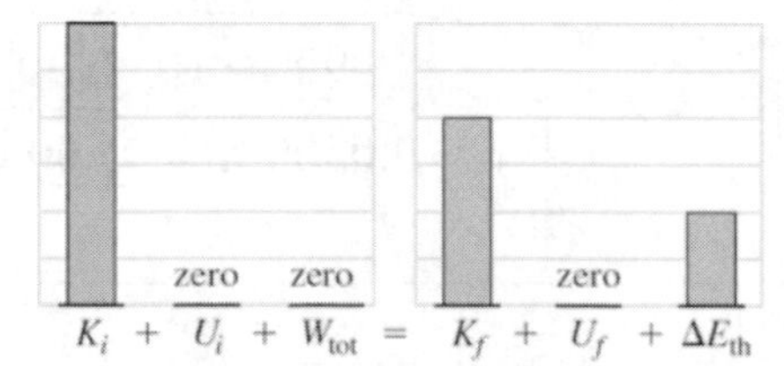

Figure P9.45ANS

SOLVE Start with work-energy theorem (Eq. 9.31). Initially, the bullet has kinetic energy based on its mass m_b and speed v_b. In the final situation, the bullet/wooden-block has kinetic energy (total mass $m_b + m_{wb}$ and speed v_f) and some thermal energy (internal energy) might be produced.	$K_i = K_f + \Delta E_{th}$ $\frac{1}{2}m_b v_b^2 = \frac{1}{2}(m_b + m_{wb})v_f^2 + \Delta E_{th}$
We can solve this expression for the final speed.	$v_f = \sqrt{\frac{m_b v_b^2 - 2\Delta E_{th}}{m_b + m_{wb}}}$

CHECK and THINK
The final velocity is larger with a larger initial bullet velocity and smaller as more of the energy is lost to friction (internal energy). The fraction of energy lost to friction is related to how inelastic versus elastic the collision is. We will encounter these concepts when we learn about momentum and collisions in the next couple chapters.

50. A small 0.65-kg box is launched from rest by a horizontal spring as shown in Figure P9.50. The block slides on a track down a hill and comes to rest at a distance d from the base of the hill. Kinetic friction between the box and the track is negligible on the hill, but the coefficient of kinetic friction between the box and the horizontal parts of track is 0.35. The spring has a spring constant of 345 N/m, and is compressed 30.0 cm with the box attached. The block remains on the track at all times.

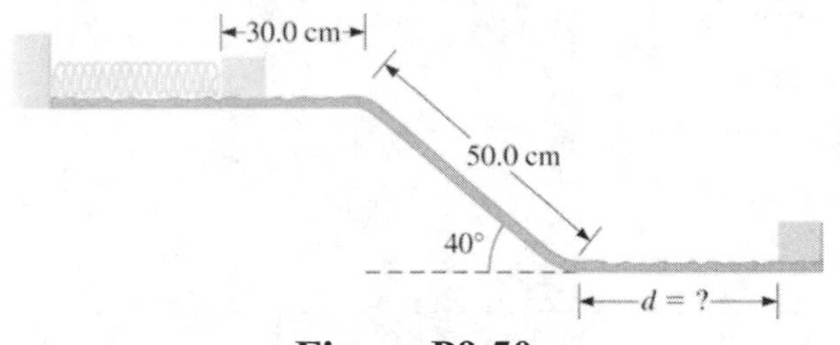

Figure P9.50

a. (C) What would you include in the system? Explain your choice.

It is best to include both the box and the track's surface because of kinetic friction increases both of their thermal energies. Since it's impossible to determine how much of the thermal energy goes into the box versus the track, we should include both of them in the system and all of this thermal energy is "internal." We also include the Earth and the spring, to account for them in terms of changes in gravitational and elastic potential energy, with nothing left outside the system to do work. A bar chart of the initial and final energies can help organize which energies need to be taken into account.

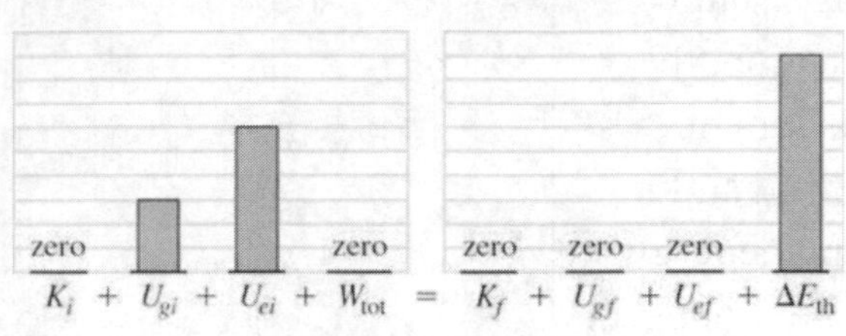

Figure P9.50ANS

b. (N) Calculate *d*.

INTERPRET and ANTICIPATE	
We can use the work-energy theorem. The bar chart from part (a) shows us which terms we must keep and which vanish.	
SOLVE We use the work-energy theorem (Eq. 9.31). The reference configuration for the spring is when it is relaxed and for gravity it is when the box is at the bottom of the ramp. The box is initially at rest $(K_i = 0)$, there are no external forces $(W_{tot} = 0)$, and in the final situation, the box is again at rest at the reference height and the spring is relaxed $(K_f = 0,\ U_{sf} = 0,\ U_{gf} = 0)$.	$K_i + U_{gi} + U_{si} + W_{\text{tot}} = K_f + U_{gf} + U_{sf} + \Delta E_{\text{th}}$ $U_{gi} + U_{si} = \Delta E_{\text{th}}$
Insert formulas for the gravitational potential energy, spring potential energy, and thermal energy (Eq. 9.1, 9.25, 9.29). Kinetic friction acts only on the initial horizontal section (distance $x = 0.3$ meters) and a distance *d* as the box travels on	$mgy + \frac{1}{2}kx^2 = F_k s$ $mgy + \frac{1}{2}kx^2 = F_k(x+d)$ (1)

the lower horizontal section before coming to rest.	
The y displacement as the block slides down the hill is found from trigonometry.	$y = (1.8\text{ m})\sin 40° = 1.16\text{ m}$ (2)
Kinetic friction (only on the horizontal surfaces in this problem) is proportional to the normal force, which equals the weight.	$F_k = \mu_k F_N = \mu_k mg$ (3)
Solve Equations (1) – (3) for distance d.	$d = \frac{y}{\mu_k} + \frac{1}{2}\frac{kx^2}{\mu_k mg} - x$ (4) $d = \frac{1.16\text{ m}}{0.35} + \frac{1}{2}\frac{(345\text{ N/m})(0.30\text{ m})^2}{(0.35)(0.65\text{ kg})(9.81\text{ m/s}^2)} - 0.30\text{ m}$ $d = \boxed{10\text{ m}}$

CHECK and THINK

The block slides for 10 m on the lower surface, which sounds like a possible distance for an object to slide across the floor after sliding down a hill. You might be able to see from the final equation (4) that if the height of the hill y was larger or the spring was compressed further (larger x), the final distance d would be larger. If the friction coefficient were larger, the final distance would be smaller. These all sound physically plausible. So, although equation (4) is fairly complicated, we can verify that it behaves as expected.

53. (N) A box of mass $m = 2.00$ kg is dropped from rest onto a massless, vertical spring with spring constant $k = 2.40 \times 10^2$ N/m that is initially at its natural length. How far is the spring compressed by the box if the initial height of the box is 1.75 m above the top of the spring?

INTERPRET and ANTICIPATE

The initial and final kinetic energies of the box are both zero, since the box is dropped from rest and briefly comes to rest at the point of the maximum compression of the spring. At this point, the gravitational potential energy of the box, having dropped a total distance of $h + x$, is converted to potential energy in the spring.

SOLVE Using energy conservation, we include the box and spring. We choose a reference configuration to be when the spring is relaxed and the mass is just in contact with the relaxed spring. There is only gravitational potential energy initially and spring potential energy and negative gravitational potential at the end (when the spring is compressed and the box is below the reference configuration). In between, of course there is kinetic energy, but we don't need to worry about this part of the motion to answer the question. We can draw a bar chart to represent this.	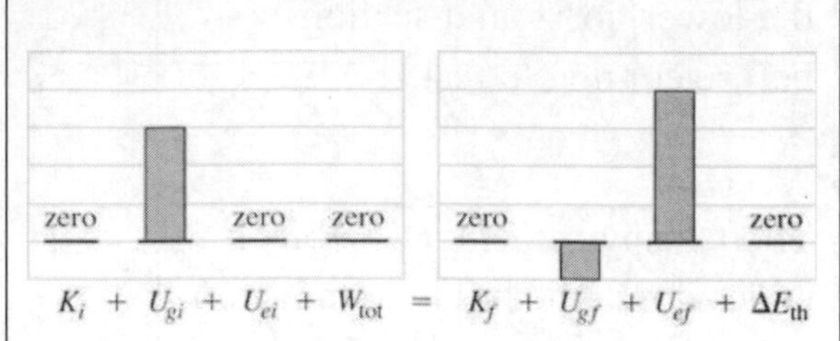 **Figure P9.53ANS**
We can use Eq. 9.26. The initial energy is the gravitational potential energy when the mass is at a height h. In the final situation, the spring is compressed a distance x and the mass is a distance x below the reference configuration.	$\Delta K + \Delta U = 0$ $K_f - K_i + U_f - U_i = 0$ $0 - 0 + \left(\frac{1}{2}kx^2 + mg(-x)\right) - mgh = 0$
This gives a quadratic equation in x.	$\frac{1}{2}kx^2 - mgx - mgh = 0$
Plug in values.	$\frac{2.40\times10^2}{2}x^2 - (2.00\text{ kg})\left(9.81\ \frac{\text{m}}{\text{s}^2}\right)x$ $-(2.00\text{ kg})\left(9.81\ \frac{\text{m}}{\text{s}^2}\right)(1.75\text{ m}) = 0$
We can simplify and solve for x.	$(1.20\times10^2)x^2 - 19.6x - 34.3 = 0$ $x = \frac{19.6 \pm 130}{240}$

The negative root does not correspond to the setup of the problem, so we keep the positive root as the correct answer.	$x = \boxed{0.623 \text{ m}}$
CHECK and THINK We've used energy conservation to determine the final compression of the spring. We chose the initial and final positions so that we didn't have to determine anything about the kinematics as the box fell, only the initial and final potential energy.	

59. (N) Calculate the force required to pull a stuffed toy duck (mass $m = 1.25$ kg) at a constant velocity of 3.6 m/s horizontally across the floor if the string is 50.0° above the horizontal. The coefficient of kinetic friction between the duck and the floor is 0.70.

INTERPRET and ANTICIPATE We choose our system to include the toy duck and the floor. There are several energy terms involved in this scenario and so we need to use the work-kinetic energy theorem in the form of Equation 9.31 to ensure we account for each energy term.	
SOLVE Energy is conserved throughout the motion so our choice of initial and final configurations is somewhat arbitrary. A bar chart helps us keep track of the conversion between the various forms of energy.	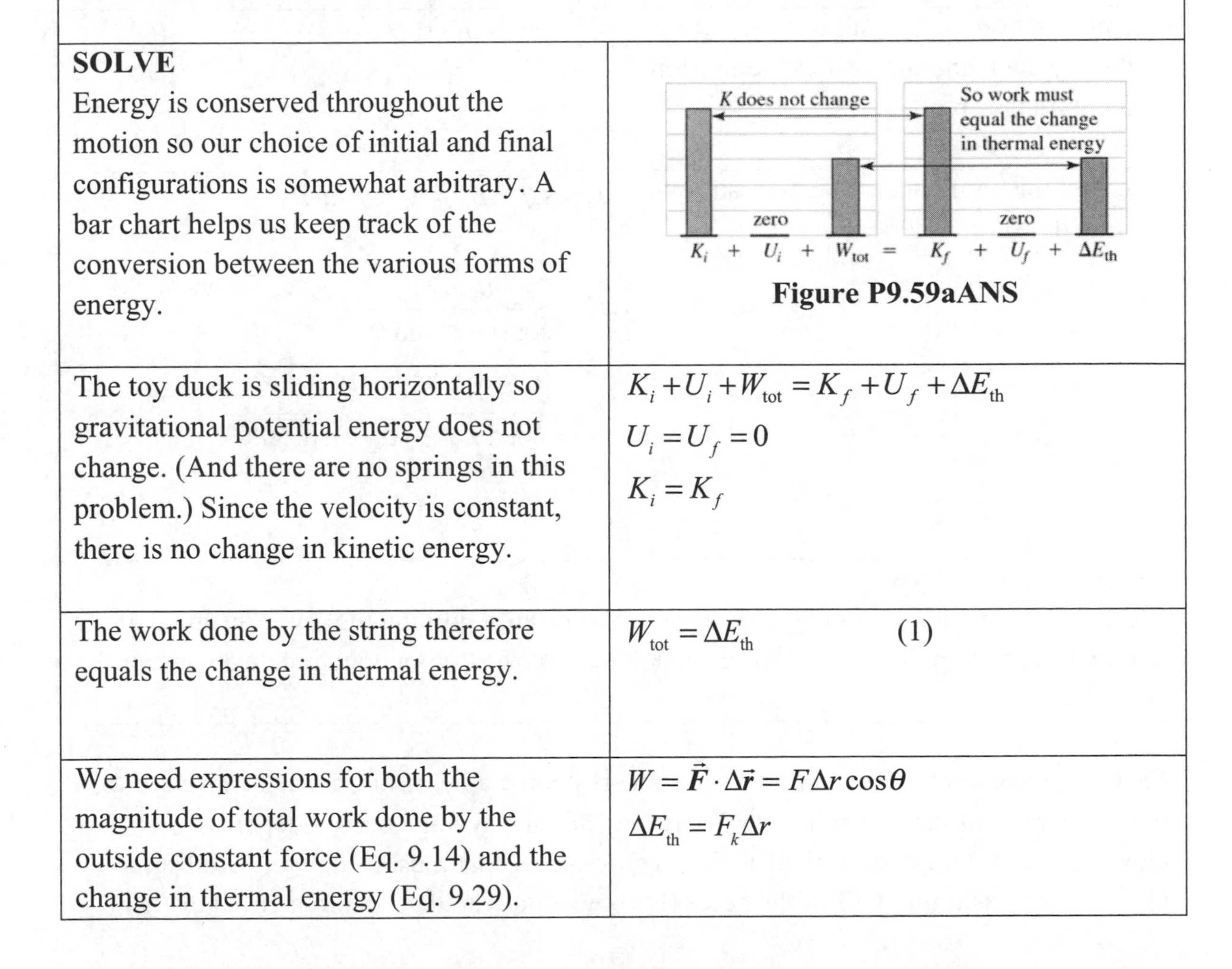 **Figure P9.59aANS**
The toy duck is sliding horizontally so gravitational potential energy does not change. (And there are no springs in this problem.) Since the velocity is constant, there is no change in kinetic energy.	$K_i + U_i + W_{tot} = K_f + U_f + \Delta E_{th}$ $U_i = U_f = 0$ $K_i = K_f$
The work done by the string therefore equals the change in thermal energy.	$W_{tot} = \Delta E_{th}$ (1)
We need expressions for both the magnitude of total work done by the outside constant force (Eq. 9.14) and the change in thermal energy (Eq. 9.29).	$W = \vec{F} \cdot \Delta\vec{r} = F\Delta r \cos\theta$ $\Delta E_{th} = F_k \Delta r$

We now insert these into equation (1).	$F\Delta r\cos\theta = F_k\Delta r$ $F\cos\theta = \mu_k F_N$ (2)
Let's also sketch a free-body diagram.	Figure P9.59bANS
The normal force is the sum of the weight of the toy duck and the vertical component of the pulling force.	$F_N = mg - F\sin\theta$
Inserting this into equation 2, we can solve for F and insert numerical values.	$F\cos\theta = \mu_k(mg - F\sin\theta)$ $F = \dfrac{\mu_k mg}{\cos\theta + \mu_k\sin\theta}$ $F = \dfrac{(0.70)(1.25\text{ kg})(9.81\text{ m/s}^2)}{\cos 50.0° + (0.70)\sin 50.0°}$ $F = \boxed{7.3\text{ N}}$

CHECK and THINK

We can check your formula by considering, for instance, pulling the toy at an angle of 0°, in which case the pulling force would equal the force of friction to maintain a constant velocity.

63. (N) An elevator motor moves a car with six people upward at a constant speed of 2.50 m/s. The mass of the elevator is 8.00×10^2 kg, and the average mass of a person on the elevator is about 80.0 kg. Calculate the electric power that must be delivered to lift the elevator car, assuming half of the necessary power delivered goes into thermal energy.

INTERPRET and ANTICIPATE The elevator motor is doing positive mechanical work by increasing the potential energy of the elevator with people. It exerts a force upwards, in the same direction that the elevator moves. This work per unit time would be, in an ideal case, equal to electrical power consumed. In our case, due to the thermal energy loss, the actual electrical power used will be twice this theoretical value.	
SOLVE Starting with Equation 9.31, we assume that the thermal energy losses are of the same magnitude as the potential energy required to lift the elevator. Assuming constant speed, the change in kinetic energy is zero. The change in potential energy is that of a constant gravitational force (*mg*) exerted over a change in height (Δy), using Eq. 9.1.	$W_{tot} = \Delta U_f + \Delta E_{th}$ $W_{tot} = 2\Delta U_f = 2mg\Delta y$
Average power is equal to the work performed per unit time (Equation 9.34). We can calculate the power by finding the energy required for one second, during which time, the height increases by 2.5 m. The total mass for the elevator and 6 passengers is 800 kg + 6(80 kg) =1280 kg. The units are Joules per second or Watts.	$P = \frac{W}{\Delta t} = \frac{2mg\Delta y}{\Delta t}$ $P = \frac{2(1280\text{ kg})(9.81\text{ m/s}^2)(2.5\text{ m})}{1\text{ s}}$ $P = \boxed{6.28 \times 10^4\text{ W}}$
CHECK and THINK Power is a *rate* of performing work or using energy. In this case, the motor must exert around 63 kW while it's lifting the passengers. In fact, most of the energy is needed to lift the 800 kg elevator rather than the six passengers.	

64. (N) A pail in a water well is hoisted by means of a frictionless winch, which consists of a spool and a hand crank. When Jill turns the winch at her fastest water-fetching rate, she can lift the pail the 25.0 m to the top in 12.2 s. Calculate the average power supplied by Jill's muscles during the upward ascent. Assume the pail of water when full has a mass of 6.82 kg.

INTERPRET and ANTICIPATE

Power is equal to the rate of energy transfer or the rate at which work is performed. For the purpose of answering this question the rate of energy transfer can be calculated by multiplying the average velocity by the force exerted.

SOLVE Power can be calculated as the amount of work performed divided by time. We also learned in the chapter that power can be calculated using the force applied times the velocity of the point at which the force is applied (Eq. 9.38). We can calculate the force of the rope on the pail and then use the fact that Jill ultimately supplies that power. The force due to the rope and the velocity of the pail are both upwards, so the dot product is equal to multiplying the magnitudes (i.e. $\theta = 0$).	$P = \vec{F} \cdot \vec{v} = Fv$
Calculate the average velocity. The velocity of ~2 m/s seems feasible for lifting a pail of about 15 pounds (6.82 kg).	$v_{av} = \frac{\Delta y}{\Delta t} = \frac{25.0 \text{ m}}{12.2 \text{ s}} = 2.05 \text{ m/s}$
We also need the force exerted. Assuming constant velocity, the force is equal to weight of the pail.	$F = mg = (6.82 \text{ kg})(9.81 \text{ m/s}^2) = 66.9 \text{ N}$
Equation 9.38 can now be used to calculate the power.	$P = Fv$ $P = (66.9 \text{ N})(2.05 \text{ m/s})$ $P = \boxed{137 \text{ W}}$

CHECK and THINK

Trying to check this power output is a little tricky. 137 W is a pretty bright light bulb, but how does this relate to Jill? Jill's muscle cells receive the necessary energy to lift the pail through chemical energy in the food she eats. We can detemine Jill's total energy output during lifting the pail by multiplying her power output by time.

$$E = P \cdot t = (137 \text{ W})(12.2 \text{ s}) = 1670 \text{ J}$$

Remembering that food calories are really kilocalories and using the conversion between calories and joules, 1 cal = 4.184 J, this is a pretty small amount of energy of less than half a Calorie! We'll have to lift the pail over 600 times to work off a 300-Calorie slice of pizza.

Is that reasonable? Another way to look at this value is to compare it to energy burning for moderate exercise, which is about 200-600 kcal per hour.

$$\frac{(400{,}000\text{ cal})\left(4.184\dfrac{\text{J}}{\text{cal}}\right)}{3600\text{ sec}}=464.9\ \frac{\text{J}}{\text{s}}$$

Since moderate exercise is the equivalent of a power output of about 465 W, our value of 137 W for lifting a pail actually does sound reasonable.

71. (N) An object is subject to a nonconstant force $\vec{F}=\left(6x^3-2x\right)\hat{i}$ such that the force is in newtons when x is in meters. Determine the work done on the object as a result of this force as the object moves from $x = 0$ to $x = 1.00 \times 10^2$ m.

INTERPRET and ANTICIPATE

The force only has an x component and we are only considering displacement in the x direction. We can integrate the force with respect to distance to calculate the total work.

SOLVE Equation 9.18 can be used to compute the work done by integrating the force times distance as the object is moved from $x = 0$ to $x = 100$ m.	$W=\int_{x_i}^{x_f}F_x dx$ $W=\int_0^{100}\left(6x^3-2x\right)dx$
Now, integrate to find the final answer.	$W=\int_0^{100}6x^3dx-\int_0^{100}2xdx$ $W=\frac{3}{2}x^4\Big\vert_0^{100}-x^2\Big\vert_0^{100}$ $W=\frac{3}{2}(100)^4-(100)^2$ $W=\boxed{1.50\times10^8\text{ J}}$

CHECK and THINK

We don't have a way to interpret the final magnitude, but we've integrated the force over

distance to find the work performed by the force.

74. Kerry is pulling a 154-kg sled along a snowy, horizontal path with a 615-N force directed at an angle of 30.0° above the ground. He pulls the sled over a distance of 30.0 m, and the coefficient of kinetic friction between the sled and the ground is 0.0612.
a. (C) Define the system to be used to account for the change in thermal energy as the sled is moved across the ground.

The system will consist of the sled and the ground. In this way, the energy exchanged between the sled and the ground, as thermal energy, will stay in the system. The total external work is supplied by Kerry. If the sled moves at constant speed on horizontal ground, there is no change in kinetic or potential energy.

b. (N) How much work is performed by Kerry as he pulls the sled over this distance?

INTERPRET and ANTICIPATE We are given the force, displacement, and angle between these vectors. This is sufficient to calculate the work that Kerry performs. Since Kerry's force has a component in the same direction as the displacement, the work should be positive.	
SOLVE Given the force, displacement, and angle between these vectors, we can use Equation 9.6. The angle between the displacement of the sled and his applied force is 30.0°.	$W = F\Delta r\cos\theta$ $W = (615\text{ N})(30.0\text{ m})\cos(30.0°)$ $W = \boxed{1.60\times10^4\text{ J}}$
CHECK and THINK The work is positive as expected. In this case, the work does not cause the sled to speed up, but rather is offset by negative work by friction (a force exerted in the opposite direction that the sled moves).	

c. (N) What is the increase in thermal energy experienced by the system during the motion?

INTERPRET and ANTICIPATE In this case, the change in thermal energy is due to the kinetic friction force between the sled and the ground.	
SOLVE Equation 9.29	$\Delta E_{\text{th}} = F_k s$

allows us to find the change in thermal energy. It is the kinetic friction force times the distance over which the force acts.	
In order to make use of this, however, we need to find the magnitude of the normal force on the sled in order to calculate the friction force (Eq. 5.9).	$F_k = \mu_k F_N$
We sketch the free-body diagram for the sled.	**Figure P9.74ANS**
The sled has zero acceleration and no net force in the y direction.	$\sum F_y = F_N + F_{\text{pull}\,y} - F_g = 0$ $F_N = F_g - F_{\text{pull}\,y}$
Then, using the formula for the gravitational force on Earth, and writing out the y component of Kerry's pulling	$\Delta E_{\text{th}} = \mu_k F_N s$ $\Delta E_{\text{th}} = \mu_k s\left(F_g - F_{\text{pull,y}}\right)$ $\Delta E_{\text{th}} = \mu_k s\left(mg - F_{\text{pull}} \sin\theta\right)$

force using trigonometry, we can determine the thermal energy. In the next step, substitute values and solve.	$\Delta E_{th} = (0.0612)(30.0\text{ m})\left[(154\text{ kg})\left(9.81\ \frac{\text{m}}{\text{s}^2}\right) - (615\text{ N})\sin(30.0°)\right]$ $\Delta E_{th} = \boxed{2.21\times10^3\text{ J}}$

CHECK and THINK

The energy lost to friction as thermal energy is less than the work that Kerry does, which we found in part (b) (it can't be greater, so that's a good sign!). The additional energy Kerry supplies must go somewhere though. This must mean that the sled is actually accelerating, since kinetic energy is really our only option for where the energy ends up if he pulls the sled on flat ground.

77. Maria sets up a simple track for her toy block ($m = 0.25$ kg) as shown in Figure P9.77. She holds the block at the top of the track, 0.54 m above the bottom, and releases it from rest.

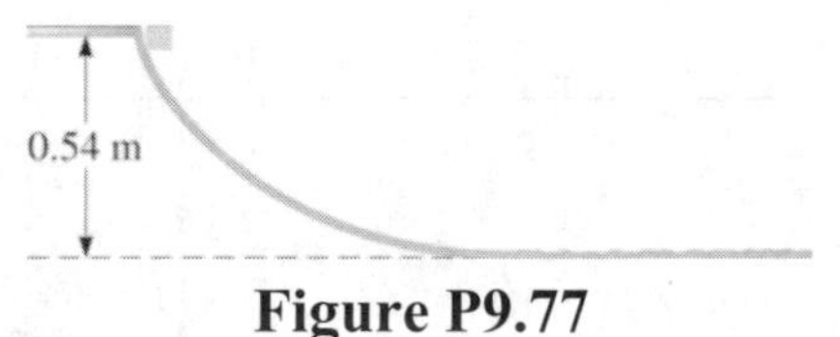

Figure P9.77

a. (N) Neglecting friction, what is the speed of the block when it reaches the bottom of the curve (the beginning of the horizontal section of track)?

INTERPRET and ANTICIPATE

We begin by applying conservation of energy principles, as in Eq. 9.31, to the block-track system from the start to the moment the block reaches the bottom of the track. Because there is no friction during this motion, $\Delta E_{int} = 0$. Also, there are no identifiable forces performing work on the system, other than gravity, so $W_{tot} = 0$.

SOLVE Choosing the bottom of the track to be where $y = 0$ and letting the initial velocity be zero, Eq. 9.31 becomes:	$mg\Delta y = \frac{1}{2}mv_f^2$

We can now solve this equation for the final velocity.	$(9.81\text{ m/s}^2)(0.54\text{ m}) = \frac{1}{2}v_f^2$ $v_f = \boxed{3.3\text{ m/s}}$
CHECK and THINK Using conservation of energy, the calculation is straightforward, even in a case where the block changes direction as it slides down a curved path (that is, where kinematics would be difficult to carry out).	

b. (N) If friction is present on the horizontal section of track and the block comes to a stop after traveling 0.75 m along the bottom, what is the magnitude of the friction force acting on the block?

INTERPRET and ANTICIPATE Again, apply the conservation of energy condition from the moment the block reaches the bottom of the track until it finally stops. While there are no external forces performing work on the block-track system during this motion ($W_{tot} = 0$), friction will cause a rise in internal energy. In this case, the final velocity (and therefore kinetic energy) is zero.	
SOLVE Write out Eq. 9.31 between the initial drop and the final point after energy is dissipated by friction.	$\frac{1}{2}mv_i^2 = \Delta E_{th}$
Inserting numerical values, we determine the total initial energy, which is also the energy dissipated by friction.	$\Delta E_{th} = \frac{1}{2}(0.25\text{ kg})(3.3\text{ m/s})^2$
Then, using Eq. 9.29, we write the thermal energy in terms of the magnitude of the friction force and distance over which friction is in effect.	$\Delta E_{th} = F_k s$
Inserting known values, we can determine the magnitude of the friction force.	$F_k = \Delta E_{th}/s$ $F_k = \left[\frac{1}{2}(0.25\text{ kg})(3.3\text{ m/s})^2\right]\Big/0.75\text{ m} = \boxed{1.8\text{ N}}$

CHECK and THINK
The friction force exerted over a distance is equal to the energy dissipated as thermal energy.

84. Shawn (m = 45.0 kg) rides his skateboard at a local skate park. He starts from rest at the top of the track as seen in Figure P9.84 and begins a descent down the track, always maintaining contact with the surface. The mass of the skateboard is negligible, as is friction except where noted.

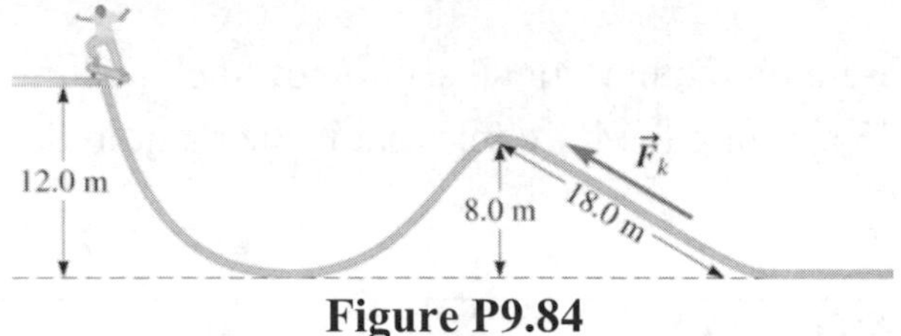

Figure P9.84

a. (N) What is Shawn's speed when he reaches the bottom of the initial dip, 12.0 m below the starting point?

INTERPRET and ANTICIPATE	
We begin by applying conservation of energy principles, as in Eq. 9.31, to the Shawn-skateboard-track system from the start to the moment Shawn reaches the bottom of the track. Because there is no friction during this motion, we expect $\Delta E_{int} = 0$. Also, there are no identifiable forces performing work on the system, other than gravity, so $W_{tot} = 0$.	
SOLVE Choosing the bottom of the track to be where $y = 0$, we can write Eq. 9.31 for this situation. In the absence of friction or external forces, the initial potential energy is equal to the final kinetic energy.	$mgy_i = \frac{1}{2}mv_f^2$
With this, we insert numerical values and solve for the speed.	$(9.81 \text{ m/s}^2)(12.0 \text{ m}) = \frac{1}{2}v_f^2$ $v_f = \boxed{15.3 \text{ m/s}}$
CHECK and THINK Without friction, conservation of energy tells us that the initial potential energy equals the final kinetic energy. Shawn is traveling about 30 mph!	

b. (N) He then ascends the other side of the dip to the top of a hill, 8.0 m above the ground. What is his speed when he reaches this point?

INTERPRET and ANTICIPATE

Again, we apply conservation of energy principles, from the start to the moment Shawn reaches the height of 8.0 m above the lowest point. Because there is no friction during this motion, we expect $\Delta E_{int} = 0$. Also, there are no identifiable forces performing work on the system, other than gravity, so $W_{tot} = 0$. We expect that Shawn will reach top speed at the bottom of the initial hill and travel slower at this point in the motion.

SOLVE

Again choosing the bottom of the track to be where $y = 0$, we write Eq. 9.31. This is similar to part (a), except that at the top of the hill, Shawn has both kinetic and potential energy. We solve the resulting equation for the final speed (final meaning at the top of the hill in this case).	$mgy_i = \frac{1}{2}mv_f^2 + mgy_f$ $m(9.81\text{ m/s}^2)(12.0\text{ m}) = (1/2)mv_f^2 + m(9.81\text{ m/s}^2)(8.0\text{ m})$ $(1/2)v_f^2 = (9.81\text{ m/s}^2)(12.0\text{ m}) - (9.81\text{ m/s}^2)(8.0\text{ m})$ $v_f = \boxed{8.9\text{ m/s}}$

CHECK and THINK

Shawn's speed at the top of the hill is indeed slower than his speed at the lowest point in part (a). Since the hill is lower than his initial height though, he does maintain a non-zero speed and makes it over the hill.

c. (N) As he begins to descend again, down a straight, 18.0-m-long slope, he slows his skateboard down by using friction on the tail of the board. He is able to produce a friction force with a magnitude of 120.0 N. What is the change in thermal energy of the board–rider–track system as he descends the 18.0-m length of track?

INTERPRET and ANTICIPATE

The change in thermal energy is the change in internal energy for the system and is equal to the friction force times the distance over which the force is applied.

SOLVE

The thermal energy can be found using Eq. 9.29.	$\Delta E_{th} = F_f d$

The friction force is given and it is exerted over a distance of 18 m (the total distance of the downhill slope).	$\Delta E_{th} = F_f d = (120\ \text{N})(18\ \text{m}) = \boxed{2.2\times10^3\ \text{J}}$
CHECK and THINK The thermal energy dissipated depends on both the friction force and the distance over which it is applied. Friction is always in the opposite direction of displacement, so we don't need to consider any angles and the thermal energy dissipated (which is the negative of the work done by friction) is always positive.	

d. (N) What is his speed when he reaches the bottom (the end of the 18.0-m length of track)?

INTERPRET and ANTICIPATE Again, we apply conservation of energy principles, from the start to the moment Shawn reaches the bottom at the very end (after the 18.0 m incline). There are no identifiable forces performing work on the system, other than gravity, so $W_{tot} = 0$.	
SOLVE Choosing the bottom of the track to be where $y = 0$, we write Eq. 9.31. The initial potential energy is converted to energy dissipated by friction on the hill (found in part (c)) and Shawn's final kinetic energy.	$mg\Delta y = \frac{1}{2}mv_f^2 + \Delta E_{th}$
Plugging in known values, we can determine the final speed.	$(45.0\ \text{kg})(9.81\ \text{m/s}^2)(12.0\ \text{m}) = \frac{1}{2}(45.0\ \text{kg})v_f^2 + 2.16\times10^3\ \text{J}$ $\frac{1}{2}(45.0\ \text{kg})v_f^2 = (45.0\ \text{kg})(9.81\ \text{m/s}^2)(12.0\ \text{m}) - 2.16\times10^3\ \text{J}$ $v_f = \boxed{11.8\ \text{m/s}}$
CHECK and THINK The final speed is slower than the speed from part (a) since some energy is dissipated by friction (c) on the final hill.	

10

Systems of Particles and Conservation of Momentum

4. A mother pushes her son in a stroller at a constant speed of 1.52 m/s. The boy tosses a 56.7-g tennis ball straight up at 1.75 m/s and catches it. The boy's father sits on a bench and watches.

a. (N) According to the mother, what are the ball's initial and final momenta?

<table>
<tr><td colspan="2">INTERPRET and ANTICIPATE
From the point of view of the mother and son, the ball goes straight up from and back to the son. From the father's point of view, the mother, son, and ball are all moving at a constant speed of 1.52 m/s throughout this process. So, according to the mother, the initial momentum is the in the positive y direction and the final momentum in the negative y direction.</td></tr>
<tr><td>SOLVE
We can sketch the situation from the point of view of the mother and son.</td><td>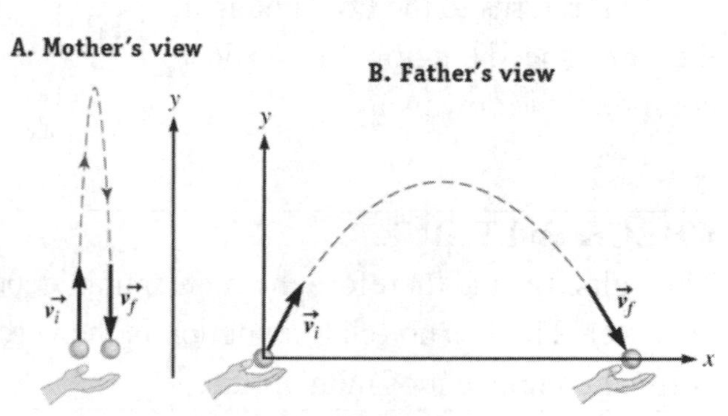

Figure P10.4ANS</td></tr>
<tr><td>Momentum of an object depends on the mass times the velocity (Equation 10.1).</td><td>$\left(\vec{p}_i\right)_{\text{mom}} = m\vec{v}_i$
$\left(\vec{p}_i\right)_{\text{mom}} = \left(56.7\times10^{-3}\text{ kg}\right)\left(1.75\hat{j}\text{ m/s}\right)$
$\left(\vec{p}_i\right)_{\text{mom}} = \boxed{9.92\times10^{-2}\,\hat{j}\text{ kg}\cdot\text{m/s}}$</td></tr>
</table>

The ball returns to the boy's hand at the same speed but now the velocity is downward.	$(\vec{p}_f)_{\text{mom}} = (56.7\times10^{-3}\text{ kg})(-1.75\hat{j}\text{ m/s})$ $(\vec{p}_f)_{\text{mom}} = \boxed{-9.92\times10^{-2}\,\hat{j}\text{ kg}\cdot\text{m/s}}$

CHECK and THINK
Only the direction of the ball's velocity has changed, and so only the direction of the momentum changes.

b. (N) According to the father, what are the ball's initial and final momenta?

INTERPRET and ANTICIPATE
According to the father, in addition to moving up and down under the action of gravity, the ball, mother and son are all moving horizontally at a constant speed. This is actually identical to a projectile, so the father will see the ball travel along a parabolic path. The horizontal velocity equals the velocity of the stroller. The vertical velocities are the same as those observed by the mother.

SOLVE We can refer to the figure in part (a) and again apply the definition of momentum (Eq. 10.1).	$(\vec{p}_i)_{\text{dad}} = m\vec{v}_i$ $(\vec{p}_i)_{\text{dad}} = (56.7\times10^{-3}\text{ kg})(1.52\hat{i}+1.75\hat{j}\text{ m/s})$ $(\vec{p}_i)_{\text{dad}} = \boxed{(8.62\times10^{-2}\hat{i}+9.92\times10^{-2}\hat{j})\text{ kg}\cdot\text{m/s}}$
The ball returns to the boy's hand at the same speed but now the vertical component is downward.	$(\vec{p}_f)_{\text{dad}} = (56.7\times10^{-3}\text{ kg})(1.52\hat{i}-1.75\hat{j}\text{ m/s})$ $(\vec{p}_f)_{\text{dad}} = \boxed{(8.62\times10^{-2}\hat{i}-9.92\times10^{-2}\hat{j})\text{ kg}\cdot\text{m/s}}$

CHECK and THINK
The velocity, and therefore the momentum, depends on the frame of reference of the observer. There is no relative motion in the vertical direction, so they agree on the vertical component of momentum.

c. (C) According to the mother, is the ball's momentum ever zero? If so, when? If not, why not?

The mother will see the ball momentarily stop at the top of its flight. Since its velocity is zero, she would conclude the momentum is zero.

d. (C) According to the father, is the ball's momentum ever zero? If so, when? If not, why not?

From the father's point of view, the ball has a constant x velocity throughout the flight, so it never has a velocity of zero. He will therefore conclude that the momentum is never zero.

10. The velocity of a 10-kg object is given by $\vec{v} = 5t^2\hat{\imath} - (7t + 2t^3)\hat{\jmath}$, where if t is in seconds, then v will be in meters per second.
a. (A) What is the net force on the object as a function of time?

INTERPRET and ANTICIPATE We can use Newton's second law and the fact that acceleration is the rate of change of the velocity.	
SOLVE Using Newton's second law, we can determine the force as mass times acceleration. The acceleration of an object is equal to the derivative of velocity with respect to time.	$\vec{F}_{\text{tot}} = m\dfrac{d\vec{v}}{dt}$
Since we have a formula for velocity, we can determine an expression for the force.	$\vec{F}_{\text{tot}} = (10\text{ kg})\dfrac{d}{dt}\left[5t^2\hat{\imath} - (7t + 2t^3)\hat{\jmath}\right]$ $\vec{F}_{\text{tot}} = \boxed{100t\,\hat{\imath} - (70 + 60t^2)\hat{\jmath}}$
CHECK and THINK For the final expression, time is measured in seconds and force in newtons.	

b. (N) What is the momentum of the object when $t = 15.0$ s?

INTERPRET and ANTICIPATE Momentum is mass times velocity. We have an expression for velocity that we can evaluate at 15.0 seconds, so we can calculate the momentum directly.	
SOLVE We can use the equation for momentum (Eq. 10.1). To find the velocity at 15.0 seconds, we plug this time into the velocity formula.	$\vec{p} = m\vec{v}$ $\vec{p} = (10.0)\left[5(15.0)^2\hat{\imath} - \left(7(15.0) + 2(15.0)^3\right)\hat{\jmath}\right]$ $\vec{p} = \boxed{\left(1.13\times10^4\hat{\imath} - 6.86\times10^4\hat{\jmath}\right)\text{ kg}\cdot\text{m/s}}$

CHECK and THINK

Given, the mass and velocity (or, in this case, an equation for the velocity), determining the momentum is straightforward. Its direction is the same as the velocity at this moment in time.

13. (N) Latoya, sitting on a sled, is being pushed by Dewain on the horizontal surface of a frozen lake. Dewain slips and falls, giving the sled one final push, and the sled comes to rest 9.50 s later. The speed of the sled after the final push is 4.00 m/s, and the combined mass of the sled and Latoya is 32.5 kg. Using a momentum approach, determine the magnitude of the average friction force acting on the sled during this interval.

INTERPRET and ANTICIPATE

The change in momentum of the sled occurs because a horizontal force (friction) acts on it over an interval of time. We can use the momentum change and time to determine the average force.

SOLVE	
The force on an object is equal to the change in momentum in time (Equation 10.2). We can find the average force by finding the total momentum change that occurs in a given time interval.	$F_x = \frac{dp_x}{dt} \quad \rightarrow \quad F_x = \frac{\Delta p_x}{\Delta t}$
The change in momentum is the final momentum of zero minus the initial, which we can calculate.	$F_x = \frac{mv_{xf} - mv_{xi}}{\Delta t}$ $F_x = \frac{0 - mv_{xi}}{\Delta t}$ $F_x = \frac{-(32.5\text{ kg})(4.00\text{ m/s})}{9.50\text{ s}}$ $F_x = -13.7\text{ N}$ $\lvert F_x \rvert = \boxed{13.7\text{ N}}$

CHECK and THINK

The negative friction force means that it is in the opposite direction of the positive velocity.

18. Two metersticks are connected at their ends as shown in Figure P10.18. The center of mass of each individual meterstick is at its midpoint, and the mass of each meterstick is m.

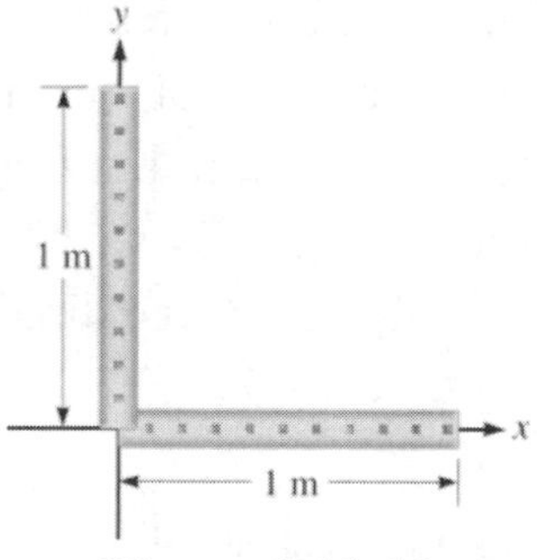

Figure P10.18

a. (N) Where is the center of mass of the two-stick system as depicted in the figure, with the origin located at the intersection of the sticks?

INTERPRET and ANTICIPATE

We are told that the center of mass of each meterstick is in its center. This means we can replace each meterstick by a single point at its center of mass and then find the center of mass of these two points. The center point between the two centers of the metersticks will be somewhere in between them (inside the triangle that they form).

SOLVE	
The center of mass of the stick on the x axis would be at (0.5 m, 0) and the center of mass of the stick on the y axis would be at (0, 0.5 m), assuming the sticks are uniform.	$\vec{r}_{\text{CM1}} = 0.5\hat{\imath}$ m $\vec{r}_{\text{CM2}} = 0.5\hat{\jmath}$ m
We can then use Equations 10.3 to find the x and y coordinates of the center of mass. The mass of each is m, so the total mass is $M = 2m$.	$x_{\text{CM}} = \frac{1}{M}\sum_{j=1}^{n} m_j x_j = \frac{1}{2m}\left[m(0.50\text{ m})\right] = 0.25\text{ m}$ $y_{\text{CM}} = \frac{1}{M}\sum_{j=1}^{n} m_j y_j = \frac{1}{2m}\left[m(0.50\text{ m})\right] = 0.25\text{ m}$
We can now express this center of mass position as a vector.	$\vec{r}_{\text{CM}} = \left(0.25\,\hat{\imath} + 0.25\,\hat{\jmath}\right)\text{m}$ $\vec{r}_{\text{CM}} = \boxed{(0.25\text{ m}, 0.25\text{ m})}$

CHECK and THINK

If you imagine that the metersticks form two sides of a triangle, the center of mass is inside.

b. (C) Can the the two-stick system be balanced on the end of your finger so that it remains lying flat in front of you in the orientation shown? Why or why not?

No. The location of the center of mass is not located on the object, so your finger would not be in contact with the object! To balance an object on a finger, the center of mass of the object needs to be above the point of support, but it is not possible in this case. In a different orientation (perhaps holding it vertically such that one of the metersticks was pointing straight up), balancing by applying a force directly under the center of mass might be possible.

19. A boy of mass 25.0 kg is sitting on one side of a seesaw while his older sister, who has a mass of 35.0 kg, is sitting on the other side. Each child is 1.50 m from the center of the seesaw.

a. (N) Where is the center of mass of the two children relative to the center of the seesaw?

INTERPRET and ANTICIPATE

The center of mass is a weighted average of position. It should be somewhat closer to the older sister, who has a larger mass.

SOLVE

Equation 10.3 can be used to calculate the center of mass. Any coordinate system can be used. Given that the measurements are provided from the center of the seesaw, we'll use that as the origin, with the 25.0 kg child at $x = -1.50$ m and the 35.0 kg older sister at $x = +1.50$ m.

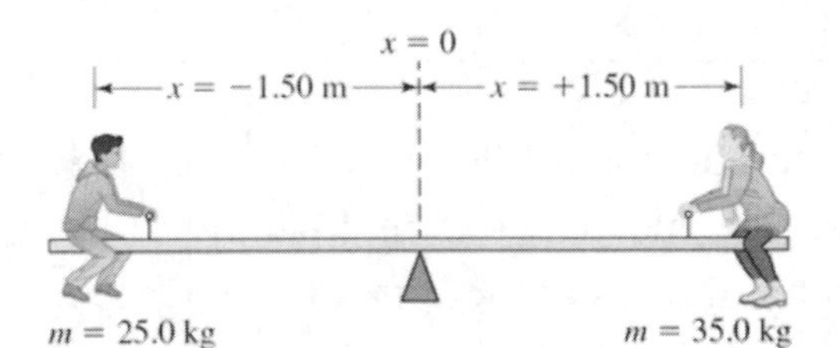

Figure P10.19ANS

Using this coordinate system, we can plug values into Equation 10.3 for the positions of the two children. The total mass of the children M is 60.0 kg.	$M = 25.0\text{ kg} + 35.0\text{ kg} = 60.0\text{ kg}$ $x_{\text{CM}} = \frac{1}{M}\sum_{j=1}^{n} m_j x_j$ $x_{\text{CM}} = \frac{1}{60.0\text{ kg}}\big((25.0\text{ kg})(-1.50\text{ m}) + (35.0\text{ kg})(+1.50\text{ m})\big)$ $x_{\text{CM}} = \boxed{0.25\text{ m}}$

CHECK and THINK

The center of mass position is positive, or to the right of the origin, which is closer to the older sister (the person with the larger mass).

b. (N) The older sister moves so that the center of mass of the two children is directly above the center of the seesaw. How far did she move and in which direction? Did she move closer to or further from the center?

INTERPRET and ANTICIPATE

To move the center of mass towards the left, the older sister should move inwards towards the center of the seesaw (to the left).

SOLVE Using Equation 10.3 again, we want the center of mass to be at $x_{\text{CM}} = 0$ but we will allow the position of the older sister x to vary.	$x_{\text{CM}} = \frac{1}{M}\sum_{j=1}^{n} m_j x_j$ and $x_{\text{CM}} = 0$ $0 = \frac{1}{60.0\text{ kg}}\big((25.0\text{ kg})(-1.50\text{ m}) + (35.0\text{ kg})(x)\big)$
Solving this equation, we determine that she moved from $x = 1.50$ m to $x = 1.07$ m. Therefore, she moved 0.43 m towards the center.	$x = \frac{25.0\text{ kg}}{35.0\text{ kg}} 1.50\text{ m} = 1.07\text{ m}$ $\Delta x = 1.07\text{ m} - 1.50\text{ m} = -0.43\text{ m}$

CHECK and THINK

The older sister moved closer to the center of the seesaw in order for the center of mass to move to the center.

29. (N) Two particles with masses 2.0 kg and 4.0 kg are approaching each other with accelerations of 1.0 m/s^2 and 2.0 m/s^2, respectively, on a smooth, horizontal surface (with negligible friction). Find the magnitude of the acceleration of the center of mass of the system.

INTERPRET and ANTICIPATE	
The system consists of two moving masses of 2.0 kg and 4.0 kg. We use the equation for the center of mass to find the acceleration of the system.	
SOLVE We start with the formula for the center of mass position (Equation 10.5). Since velocity is equal to the derivative of position versus time and acceleration is the derivative of velocity versus time, we can take the time derivative twice to obtain the equation for acceleration of the center of mass.	$\vec{r}_{CM} = \frac{1}{M}(m_1\vec{r}_1 + m_2\vec{r}_2)$ $\vec{v}_{CM} = \frac{d\vec{r}_{CM}}{dt} = \frac{1}{M}(m_1\vec{v}_1 + m_2\vec{v}_2)$ $\vec{a}_{CM} = \frac{d\vec{v}_{CM}}{dt} = \frac{d^2\vec{r}_{CM}}{dt^2} = \frac{1}{M}(m_1\vec{\mathbf{a}}_1 + m_2\vec{\mathbf{a}}_1)$
The total mass is $M = m_1 + m_2$. The two masses are moving towards each other, so their accelerations are antiparallel. We take the 2.0 kg mass to be moving in the positive direction.	$\vec{a}_{CM} = \frac{(2.0\text{ kg})(1.0\text{ m/s}^2) + (4.0\text{ kg})(-2.0\text{ m/s}^2)}{2.0\text{ kg} + 4.0\text{ kg}}$ $\vec{a}_{CM} = -1.0\text{ m/s}^2$ $\vec{a}_{CM} = \boxed{1.0\text{ m/s}^2}$ in the direction the 4.0 kg mass travels
CHECK and THINK Since $m_2\|\vec{a}_2\| > m_1\|\vec{a}_1\|$, the direction of the acceleration of the center of mass is in the direction of $\vec{a}_2$.	

30. (N) A billiard player sends the cue ball toward a group of three balls that are initially at rest and in contact with one another. After the cue balls strikes the group, the four balls scatter, each traveling in a different direction with different speeds as shown in Figure P10.30. If each ball has the same mass, 0.16 kg, determine the total momentum of the system consisting of the four balls immediately after the collision.

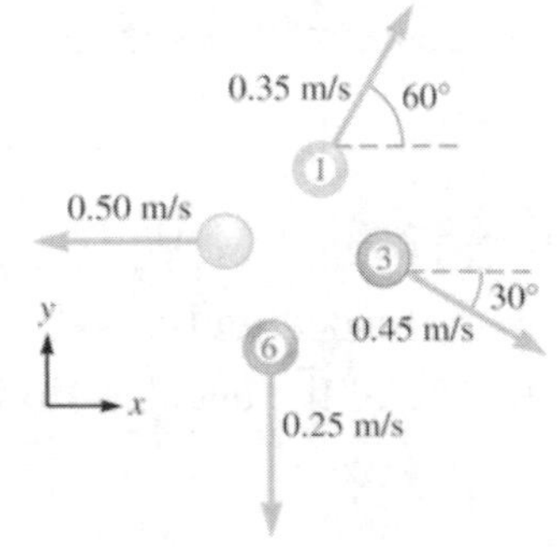

Figure P10.30

INTERPRET and ANTICIPATE The momentum of each ball equals its mass times velocity. The total momentum is the vector sum of the individual momenta.	
SOLVE The total momentum is the vector sum of the individual momenta (Equations 10.7 and 10.1).	$\vec{p}_{\text{tot}} = \sum_{j=1}^{n} \vec{p}_j = \sum_{j=1}^{n} m_j \vec{v}_j$
To begin, write the velocity of each ball in component form, in reference to the coordinate system provided.	$\vec{v}_1 = (0.35 \text{ m/s})\cos(60°)\hat{i} + (0.35 \text{ m/s})\sin(60°)\hat{j} = (0.175\hat{i} + 0.303\hat{j}) \text{ m/s}$ $\vec{v}_3 = (0.45 \text{ m/s})\cos(30°)\hat{i} - (0.45 \text{ m/s})\sin(30°)\hat{j} = (0.390\hat{i} - 0.225\hat{j}) \text{ m/s}$ $\vec{v}_6 = (-0.25\hat{j}) \text{ m/s}$ $\vec{v}_{\text{cue}} = (-0.50\hat{i}) \text{ m/s}$
The momentum of each ball, using Eq. 10.1, can now be calculated.	$\vec{p}_1 = (0.16)(0.175\hat{i} + 0.303\hat{j}) \text{ kg}\cdot\text{m/s} = (0.028\hat{i} + 0.048\hat{j}) \text{ kg}\cdot\text{m/s}$ $\vec{p}_3 = (0.16)(0.390\hat{i} - 0.225\hat{j}) \text{ kg}\cdot\text{m/s} = (0.062\hat{i} - 0.036\hat{j}) \text{ kg}\cdot\text{m/s}$ $\vec{p}_6 = (0.16)(-0.25\hat{j}) \text{ kg}\cdot\text{m/s} = -0.040\hat{j} \text{ kg}\cdot\text{m/s}$ $\vec{p}_{\text{cue}} = (0.16)(-0.50\hat{i}) \text{ kg}\cdot\text{m/s} = -0.080\hat{i} \text{ kg}\cdot\text{m/s}$
Finally, add up the individual momenta as vectors.	$\vec{p}_{\text{tot}} = \sum_{j=1}^{n} \vec{p}_j$ $\vec{p}_{\text{tot}} = \left[(0.028\hat{i} + 0.048\hat{j}) + (0.062\hat{i} - 0.036\hat{j}) - 0.040\hat{j} - 0.080\hat{i} \right] \text{ kg}\cdot\text{m/s}$ $\vec{p}_{\text{tot}} = \boxed{(0.010\hat{i} - 0.028\hat{j})}$
CHECK and THINK Whether we have four objects or two, the procedure is the same: the momentum of each is its mass times its velocity and the total is the vector sum of all the individual momenta. In this case, the resultant momentum points down and to the right.	

33. (N) A particle initially at rest is constrained to move on a smooth, horizontal surface. Another identical particle moving along the surface hits the stationary particle with a speed v. If the particles move together after the collision and their total momentum is the same as the initial momentum of the two-particle system, what is the speed of the combination just after the impact?

INTERPRET and ANTICIPATE The system consists of two identical particles of same masses. There is no net external force on the system of particles parallel to the surface, so we use momentum conservation to solve this problem. Since the initial momentum of the incoming particle now must be shared by the two particles moving together, they must move slower after the collision.	
SOLVE The particles have the same mass m and momentum is conserved (Eq. 10.9). The initial momentum of particle 2 is zero and their final velocities are identical since they stick together.	$m_1v_{1i} + m_2v_{2i} = m_1v_{1f} + m_2v_{2f}$ $0 + mv = 2mv_f$ $v_f = \boxed{\frac{v}{2}}$
CHECK and THINK The particles move slower after the collision as we expect. Though we're only asked for the speed, the direction must be the same as in the initial momentum, which is in the direction of the velocity of the incoming particle.	

34. According to the National Academy of Sciences, the Earth's surface temperature has risen about 1°F since 1900. There is evidence that this *climate change* may be due to human activity. The organizers of World Jump Day argue that if the Earth were in a slightly larger orbit, we could avoid global warming and climate change. They propose that we move the Earth into this new orbit by jumping. The idea is to get people in a particular time zone to jump together. The hope is to have 600 million people jump in a 24-hour period. Let's see if it will work. Consider the Earth and its inhabitants to make up the system.

a. (E) Estimate the number of people in your time zone. Assume they all decide to jump at the same time; estimate the total mass of the jumpers.

The world population is 7 billion people, so the number should be much smaller than that. If we were to talk 1/24th of that, we'd get around 300 million. This neglects the fact that some time zones might be mostly ocean and others more populated, but this is just

meant to be a rough estimate.

We could also start with the US population (300 million), assume that maybe 100 million people live on each of the coasts, and assume that with Canada, South America, etc., we might triple that number and again get to 300 million people. You might even think in terms of large cities in your time zones as a starting point. Again, it's an estimate, so anything in that ballpark is reasonable.

$$N \approx \boxed{3\times10^{8}}$$

b. (C) What is the net external force on the Earth–jumpers system?

Of course, there are many objects exerting gravitational forces on the Earth-jumper system, but these are weak compared to the force exerted by the Sun. So for the purpose of this estimate, we'll assume the Sun's gravity is the only force exerted on the system. This is the force that keeps the Earth in a nearly circular orbit around the Sun. The mass of the people is negligible compared to the mass of the Earth. So the external force on the system is simply the gravitational force exerted by the Sun on the Earth.

$$F_g = G\frac{M_{\odot}M_{\oplus}}{r^2} = \left(6.67\times10^{-11}\ \text{N}\cdot\text{m}^2/\text{kg}^2\right)\frac{\left(1.99\times10^{30}\,\text{kg}\right)\left(5.98\times10^{24}\,\text{kg}\right)}{\left(1.50\times10^{11}\ \text{m}\right)^2}$$

$$F_g = \boxed{3.53\times10^{22}\ \text{N}}$$

c. (N) Assume the jumpers use high-tech Flybar pogo sticks (Fig P8.32), which allow them to jump 6 ft. What is the displacement of the Earth as a result of their jump?

Figure P8.32

INTERPRET and ANTICIPATE
The system consists of the people and the Earth. So when the people exert a force on the Earth to jump, that force is internal to the system. So, the center of mass is not accelerated by that force. (It is only accelerated by external forces, like that due to the

Sun.) Of course, the center of mass continues to move in its circular orbit around the Sun. Since the people move 6 feet off the Earth, in principle the Earth must move in order keep the center of mass on its original circular orbit.	
SOLVE The system consists of the people and the Earth. So, when the people exert a force on the Earth to jump, that force is internal to the system. Since the people move 6 feet off the Earth, in principle the Earth must move in order keep the center of mass on its original circular orbit. Assume that each jumper has a mass of 100 kg (this is probably a conservative overestimate to determine the maximum possible effect).	$M=(100\text{ kg})(300\text{ million people})=3\times10^{10}\text{ kg}$
Say all these people are about 6 feet or 2 m off the Earth and moving in the same direction from the center of mass. We need the Earth to move some distance *d* in the opposite direction to keep the center of mass at the same location. Let's place the origin at the center of mass, so the center of mass position is zero, and multiply both side by the total mass *M*.	$\vec{r}_{\text{CM}}=\frac{1}{M}\sum_{j=1}^{n}m_j\vec{r}_j$ $0=(5.98\times10^{24}\text{ kg})d-(3\times10^{10}\text{ kg})(2\text{ m})$ $d=\frac{(3\times10^{10}\text{ kg})(2\text{ m})}{(5.98\times10^{24}\text{ kg})}$ $d\approx\boxed{10^{-14}\text{ m}}=10\text{ fm}$
CHECK and THINK So, the Earth's displacement is only 10 femtometers (10 fm), about the distance spanned by 10 protons! In other words, even this very large group of very heavy people using high-tech pogo sticks cannot move the Earth an appreciable amount.	

d. (C) What happens to the Earth when the jumpers land?

The center of mass does not leave its circular path as a result of this jump, so the Earth must move back 10 fm when the people return to the surface. Not only an incredibly small displacement, but it quickly disappears when everyone lands!

42. A submarine with a mass of 6.26×10^6 kg contains a torpedo with a mass of 354 kg. The submarine fires the torpedo at an angle of 25° with respect to the horizontal as shown in Figure P10.42.

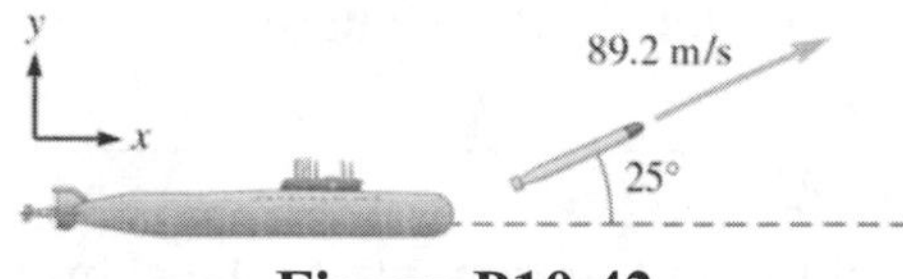

Figure P10.42

a. (N) If the submarine and the torpedo were initially at rest and the torpedo left the submarine with a speed of 89.2 m/s, what is the recoil speed of the submarine?

INTERPRET and ANTICIPATE We assume that there are no external forces (at least initially, as the drag from the water will probably quickly become apparent). That allows us to use momentum conservation.	
SOLVE Model the behavior of the submarine-torpedo system using conservation of momentum (Eq. 10.9). The initial velocity of each object is 0. Begin by finding the components of the final velocity of the torpedo.	$\vec{v}_{f,\text{torp}} = (89.2 \text{ m/s})\cos(25°)\hat{\imath} + (89.2 \text{ m/s})\sin(25°)\hat{\jmath}$ $\vec{v}_{f,\text{torp}} = (80.8\hat{\imath} + 37.7\hat{\jmath}) \text{ m/s}$
Then, apply Eq. 10.10, with zero initial momentum.	$\vec{p}_{i,\text{tot}} = \vec{p}_{f,\text{tot}}$ $0 = m_t\vec{v}_{f,\text{torp}} + m_{\text{sub}}\vec{v}_{f,\text{sub}}$ $0 = (354 \text{ kg})(80.8\hat{\imath} + 37.7\hat{\jmath}) \text{ m/s} + (6.26 \times 10^6 \text{ kg})\vec{v}_{f,\text{sub}}$ $\vec{v}_{f,\text{sub}} = \dfrac{-(354 \text{ kg})(80.8\hat{\imath} + 37.7\hat{\jmath}) \text{ m/s}}{6.26 \times 10^6 \text{ kg}}$ $\vec{v}_{f,\text{sub}} = (-4.57 \times 10^{-3}\hat{\imath} - 2.13 \times 10^{-3}\hat{\jmath}) \text{ m/s}$
We can now find the final speed of the submarine.	$v_{f,\text{sub}} = \sqrt{(-4.57 \times 10^{-3})^2 + (-2.13 \times 10^{-3})^2}$ $v_{f,\text{sub}} = \boxed{5.04 \times 10^{-3} \text{ m/s}}$

CHECK and THINK

The recoil velocity of the submarine is such that the torpedo momentum is equal and magnitude and opposite the submarine momentum.

b. (N) What is the direction of the recoil of the submarine?

INTERPRET and ANTICIPATE

The direction can be found using the components of the submarine's final velocity. Because both components are negative, the submarine will be moving down and to the left in the original picture.

SOLVE	
The angle with respect to the negative x axis can be found using the tangent function.	$\tan\theta = \dfrac{2.13\times 10^{-3}}{4.57\times 10^{-3}}$ $\theta = \tan^{-1}\left[0.466\right] = \boxed{25.0° \text{ below the } -x \text{ axis}}$

CHECK and THINK

The recoil velocity of the submarine is down and to the left, in the opposite direction that the torpedo is fired.

45. (N) A model rocket is shot straight up and explodes at the top of its trajectory into three pieces as viewed from above and shown in Figure P10.44. The masses of the three pieces are $m_A = 100.0$ g, $m_B = 20.0$ g, and $m_C = 30.0$ g. Immediately after the explosion, piece A is traveling at 1.50 m/s, and piece B is traveling at 7.00 m/s in a direction 30° below the negative x axis as shown. What is the velocity of piece C?

INTERPRET and ANTICIPATE

The initial momentum is zero. The velocity of the third piece should be down and to the right as indicated such that its momentum is equal and opposite the total momentum of pieces A and B. The velocities do not all add to zero, but the momenta should.

SOLVE	
Since there are no external forces on the rocket in the horizontal plane, we can use the conservation of momentum as stated in Equation 10.9. (Momentum is not conserved vertically because gravity acts on the rocket, but we are only concerned about the horizontal components.) The total initial momentum is zero, so the sum of all the	$m_A\vec{v}_{Af} + m_B\vec{v}_{Bf} + m_C\vec{v}_{Cf} = 0$

final momenta should be zero.	
The velocities for A and B can be expressed as vectors.	$\vec{v}_{Af} = 1.50\hat{j}$ m/s $\vec{v}_{Bf} = -(7.00\hat{i}$ m/s$)\cos 30° - (7.00\hat{j}$ m/s$)\sin 30°$ $= (-6.06\hat{i} - 3.50\hat{j})$m/s
Using the masses provided, we can now substitute values into the momentum equation. $(100.0\text{ g})(1.50\hat{j}\text{ m/s}) + (20.0\text{ g})(-6.06\hat{i} - 3.50\hat{j})\text{m/s} + (30.0\text{ g})\vec{v}_{Cf} = 0$	
Now, solve for the velocity of piece C.	$\vec{v}_{Cf} = \boxed{(4.04\hat{i} - 2.67\hat{j})\text{m/s}}$
CHECK and THINK The velocity of piece C is down and to the right as expected. The magnitude of the velocity (4.84 m/s) is intermediate between that of A and B, as suggested by the figure.	

48. A model rocket consisting of a 60.7-g chassis and a 30.5-g engine is taken on an interstellar mission and fired in outer space. The engine produces 4.95 N of thrust while expending 13.5 g of fuel at a constant rate during a 2.10-s burn.
a. (N) What is the average exhaust speed of this rocket's engine?

INTERPRET and ANTICIPATE The thrust of the rocket, which is provided, depends on the rate at which material is ejected from the rocket and the exhaust speed.	
SOLVE The force of thrust of the rocket is given by Equation 10.16.	$\vec{F}_{\text{thrust}} = (\vec{v}_E)_R \dfrac{dM_R}{dt}$
Using the amount of fuel and the time for the fuel to burn, we calculate the rate at which fuel burns, converting to metric base units.	$\dfrac{dM_R}{dt} = \dfrac{13.5\text{ g}}{2.10\text{ s}}\left(\dfrac{1\text{ kg}}{10^3\text{ g}}\right)$ $\dfrac{dM_R}{dt} = 6.43\times10^{-3}$ kg/s

Now solve the thrust equation for the exhaust speed.	$(v_E)_R = \frac{F_{\text{thrust}}}{dM_R/dt}$ $(v_E)_R = \frac{4.95 \text{ N}}{6.43\times10^{-3} \text{ kg/s}}$ $(v_E)_R = \boxed{770 \text{ m/s}}$

CHECK and THINK

This is a very large speed, but around the order of magnitude we might expect for a rocket burn.

b. (N) If the rocket starts from rest, what is its final velocity?

INTERPRET and ANTICIPATE

We now use the expression for rocket propulsion with an initial speed of zero, to find the magnitude of the final speed.

SOLVE We can apply the first rocket equation (Equation 10.12), which relates the change of velocity with the exhaust speed and the change of mass of the rocket based on the fuel that's expelled. Let's imagine for the moment that the thrust is in the negative x direction.	$\Delta\vec{v}_R = (\vec{v}_E)_R \ln\left(\frac{M_{Rf}}{M_{Ri}}\right)$
We plug in values, with the initial mass equal to the chassis, engine, and fuel, and the final mass equal to the chassis plus engine mass.	$\vec{v}_f = (-770\hat{i} \text{ m/s})\ln\left(\frac{60.7 \text{ g}+30.5 \text{ g}-13.5 \text{ g}}{60.7 \text{ g}+30.5 \text{ g}}\right)$ $\vec{v}_f = \boxed{123\,\hat{i} \text{ m/s}}$

CHECK and THINK

Given the thrust and initial/final masses, we can calculate the final velocity. The rocket velocity is in the opposite direction of the thrust.

52. (N) A shuttlecraft is initially moving at 94.5 m/s as it travels in deep space. The mass of the shuttle given its current fuel level is approximately 6.33×10^6 kg. When the shuttle fires its rockets to move faster, the speed of the exhaust leaving the rockets is 2750 m/s.

What is the mass of the fuel necessary to get the shuttlecraft up to a final speed of 125 m/s?

INTERPRET and ANTICIPATE
The first rocket equation allows us to determine the change of velocity of the shuttlecraft based on the thrust velocity and the initial and final masses of the rocket. Since we have all of these quantities other than the final mass of the shuttle with fuel, we can calculate the final mass and then the change in mass (which is the amount of fuel expended).

SOLVE We can use the magnitude of Eq. 10.12 to find the mass of the rocket plus the remaining fuel after this change in speed occurs. Solve the equation for this mass.	$\Delta\vec{v}_R = (\vec{v}_E)_R \ln\left(\frac{M_{Rf}}{M_{Ri}}\right) \rightarrow \frac{\Delta\vec{v}_R}{(\vec{v}_E)_R} = \ln\left(\frac{M_{Rf}}{M_{Ri}}\right)$ $\frac{M_{Rf}}{M_{Ri}} = e^{\left(\Delta\vec{v}_R/(\vec{v}_E)_R\right)}$ $M_{Rf} = M_{Ri}e^{\left(\Delta\vec{v}_R/(\vec{v}_E)_R\right)}$
Now, use the values in the problem statement and choosing the velocity of the rocket exhaust to be in the negative x direction, while the rocket moves in the positive x direction.	$M_{Rf} = \left(6.33\times10^6 \text{ kg}\right)e^{(125-94.5)/(-2750)}$ $M_{Rf} = 6.26\times10^6 \text{ kg}$
The difference between this final mass and the original mass will be the mass of the fuel used.	$m_{\text{fuel}} = M_{Ri} - M_{Rf}$ $m_{\text{fuel}} = 6.33\times10^6 \text{ kg} - 6.26\times10^6 \text{ kg}$ $m_{\text{fuel}} = \boxed{7.0\times10^4 \text{ kg}}$

CHECK and THINK
The fuel needed is only about 1% of the mass of the rocket, so this sounds pretty manageable.

56. The cryogenic main stage of a rocket has an exhaust speed of 4.21×10^3 m/s and burns liquid hydrogen and liquid oxygen at a combined rate of 317 kg/s.
a. (N) What is the thrust produced by the rocket's main engine?

INTERPRET and ANTICIPATE
The thrust depends on the exhaust speed and the rate at which mass is ejected from the rocket, both of which we have.

SOLVE The thrust can be calculated using Equation 10.16 (exhaust speed times rate at which mass is ejected). We are given both quantities and can plug them in to find the answer.	$F_{\text{thrust}} = (v_E)_R \dfrac{dM_R}{dt}$ $F_{\text{thrust}} = (4.21\times10^3\ \text{m/s})(317\ \text{kg/s})$ $F_{\text{thrust}} = \boxed{1.33\times10^6\ \text{N}}$
CHECK and THINK This problem is fairly straightforward with the information given. The thrust force is in newtons.	

b. (N) If the initial mass of the rocket is 1.10×10^5 kg, what is the initial acceleration of the rocket upon liftoff from the Earth?

INTERPRET and ANTICIPATE This is actually a Newton's second law problem. With the force and mass, we can calculate the acceleration.	
SOLVE While taking off, the net force upwards is the upwards thrust minus the downward weight of the rocket.	$\sum F_y = F_{\text{thrust}} - Mg = Ma$
Now, plug in values and solve for the acceleration.	$a = \dfrac{F_{\text{thrust}}}{M} - g = \dfrac{1.33\times10^6\ \text{N}}{1.10\times10^5\ \text{kg}} - 9.81\ \text{m/s}^2$ $a = \boxed{2.32\ \text{m/s}^2}$
CHECK and THINK Although the acceleration is relatively low (less than ¼ g), but there is a net force and acceleration upwards allowing the rocket to take off.	

59. (N) A ball of mass $m = 450.0$ g traveling at a speed of 8.00 m/s impacts a vertical wall at an angle of $\theta_i = 45.0°$ below the horizontal (x axis) and bounces away at an angle of $\theta_f = 45.0°$ above the horizontal. What is the average force exerted by the wall on the ball if the ball is in contact with the wall for 250.0 ms?

INTERPRET and ANTICIPATE The net external force acting on an object over a time interval causes its momentum to change. We are given information about the ball's mass and velocity, so we can determine the change in momentum, and using the contact time we can determine the average force on the ball during the collision.	
SOLVE The change in momentum of the ball can be found by applying Equation 10.2 over a finite time interval. The *average* force is equal to the total change in momentum over the total time.	$\vec{F} = \frac{d\vec{p}}{dt}$ $\vec{F} = \frac{\Delta\vec{p}}{\Delta t}$
The change in momentum in the vertical direction is zero.	$\Delta p_y = m(v_{fy} - v_{iy})$ $\Delta p_y = mv\sin 45.0° - mv\sin 45.0°$ $\Delta p_y = 0$
We can calculate the change in momentum in the horizontal direction.	$\Delta p_x = m(v_{fx} - v_{ix})$ $\Delta p_x = m(-v\cos 45.0° - v\cos 45.0°)$ $\Delta p_x = -2mv\cos 45.0°$
Now, plug in values and calculate the force.	$F_{\text{avg}} = \frac{\Delta p_x}{\Delta t}$ $F_{\text{avg}} = \frac{-2mv\text{s}\cos 45.0°}{\Delta t}$ $F_{\text{avg}} = \frac{-2(0.450\text{ kg})(8.00\text{ m/s})\cos 45.0°}{0.250\text{ s}}$ $\lvert F_{\text{avg}}\rvert = \boxed{20.4\text{ N perpendicular to the wall}}$
CHECK and THINK The force is 20.4 N, outward from (and normal to) the wall. The negative sign in the second to last step indicates that if we take the initial velocity to be in the positive direction, then the force is in the negative direction, causing the ball to bounce back.	

65. (N) A racquetball of mass $m = 43.0$ g, initially moving at 30.0 m/s horizontally in the positive x direction, is struck by a racket. After being struck, the ball moves back in the opposite direction at an angle of 30.0° to the horizontal with a speed of 50.0 m/s. What is the average vector force exerted on the racket by the ball if they are in contact for 2.50 ms?

INTERPRET and ANTICIPATE	
Similar to problem 59, the net external force acting on an object over a time interval causes its momentum to change. We are given information about the ball's mass and velocity, so we can determine the change in momentum, and using the contact time we can determine the average force on the ball during the collision.	
SOLVE The change in momentum of the ball can be found by applying Equation 10.2 over a finite time interval. The *average* force is equal to the total change in momentum over the total time.	$\vec{F} = \frac{d\vec{p}}{dt}$ $\vec{F} = \frac{\Delta\vec{p}}{\Delta t} = \frac{m\vec{v}_f - m\vec{v}_i}{\Delta t}$
Choosing the initial direction of the ball's motion as positive x, we can write expressions for the initial and final velocity.	$\vec{v}_i = 30.0\,\hat{i}$ m/s $\vec{v}_f = \left[-(50.0\text{ m/s})\cos 30.0°\right]\hat{i} + \left[(50.0\text{ m/s})\sin 30.0°\right]\hat{j}$ $= \left(-43.3\,\hat{i} + 25.0\,\hat{j}\right)\text{m/s}$ $\Delta t = 2.50\text{ ms} = 2.50\times10^{-3}\text{s}$
Now, plug in values and calculate the force.	$\vec{F}_{\text{on ball}} = \frac{m\vec{v}_f - m\vec{v}_i}{\Delta t}$ $\vec{F}_{\text{on ball}} = (0.0430\text{ kg})\frac{-43.3\hat{i} + 25.0\hat{j} - 30.0\hat{i}}{2.50\times10^{-3}}\frac{\text{m}}{\text{s}^2}$ $\vec{F}_{\text{on ball}} = \left(-1.26\hat{i} + 0.430\hat{j}\right)\times10^3\text{N}$
By Newton's third law, the force on the racket is equal and opposite the force on the ball.	$\vec{F}_{\text{on racquet}} = -\vec{F}_{\text{on ball}} = \left(1.26\hat{i} - 0.430\hat{j}\right)$ kN $= \boxed{\left(1.26\times10^3\hat{i} - 4.30\times10^2\hat{j}\right)\text{ N}}$
CHECK and THINK	
If the ball is initially moving in the positive direction, the racket exerts a force in the negative direction on the ball and the ball exerts and equal and opposite force in the positive direction on the racket. If the ball is initially going to the right, we expect the racket to recoil to the right due to the impact with the ball.	

67. The position of a particle of mass $m_1 = 1.00$ kg is described by the vector $\vec{r}_1 = (2t + 5t^2)\hat{\imath} + 4t\hat{\jmath}$, where t is in seconds and $\vec{r}$ is in meters. The motion of a second particle, of mass $m_2 = 2.50$ kg, is described by the vector $\vec{r}_2 = (3 - 5t)\hat{\imath} + (-t - 2t^2)\hat{\jmath}$.

a. (N) What is the vector position of the center of mass at $t = 3.00$ s?

INTERPRET and ANTICIPATE

The center of mass formula can be used directly to calculate the answer. The fact that the positions of the particles are a function of time just means that the center of mass will depend on time, which we can evaluate at $t = 3$ seconds.

SOLVE

The vector position of the center of mass depends on the mass and position for each particle and is given by Equation 10.5.We write this expression for the case of two particles using the fact that the total mass is $M = m_1 + m_2$.

$$\vec{r}_{CM} = \frac{1}{M}\sum_{j=1}^{n} m_j \vec{r}_j$$

$$\vec{r}_{CM} = \frac{m_1\vec{r}_1 + m_2\vec{r}_2}{m_1 + m_2}$$

We can now plug in the expressions for the positions of the particles.

$$\vec{r}_{CM} = \frac{(1.00\text{ kg})((2t + 5t^2)\hat{\imath} + 4t\,\hat{\jmath}) + (2.50\text{ kg})((3 - 5t)\hat{\imath} + (-t - 2t^2)\hat{\jmath})}{1.00\text{ kg} + 2.50\text{ kg}}$$

$$\vec{r}_{CM} = (1.43t^2 - 3t + 2.14)\hat{\imath} + (-1.43t^2 + 0.429t)\hat{\jmath}$$

Now, evaluate the expression at $t = 3.00$ s.

$$\vec{r}_{CM} = \left(1.43(3.00)^2 - 3(3.00) + 2.14\right)\hat{\imath} + \left(-1.43(3.00)^2 + 0.429(3.00)\right)\hat{\jmath}$$

$$\vec{r}_{CM} = \boxed{\left(6.00\hat{\imath} - 11.6\hat{\jmath}\right)\text{ m}}$$

CHECK and THINK

Given the positions and masses of the particles (even when the positions depend on time), we can apply the center of mass formula to determine the center of mass position.

b. (N) What is the velocity of the center of mass at $t = 3.00$ s?

INTERPRET and ANTICIPATE

The velocity of the center of mass is determined from the time derivative of the position.

SOLVE Since we have an expression of the center of mass position as a function of time from part (a), we need to take the time derivative to find the velocity.	$\vec{v}_{CM} = \frac{d\vec{r}_{CM}}{dt}$
Insert the expression from part (a) and take the derivative.	$\vec{v}_{CM} = \frac{d}{dt}(1.43t^2 - 3t + 2.14)\hat{i} + (-1.43t^2 + 0.429t)\hat{j}$ $\vec{v}_{CM} = (2.86t - 3)\hat{i} + (-2.86t + 0.429)\hat{j}$
Now, evaluate the expression at $t =$ 3.00 s.	$\vec{v}_{CM} = (2.86(3.00) - 3)\hat{i} + (-2.86(3.00) + 0.429)\hat{j}$ $\vec{v}_{CM} = \boxed{\left(5.58\hat{i} - 8.15\hat{j}\right)\text{ m/s}}$
CHECK and THINK With an expression for the center of mass position as a function of time, we are able to take the derivative to find the center of mass velocity.	

c. (N) What is the total linear momentum of the system at $t = 3.00$ s?

INTERPRET and ANTICIPATE The total linear momentum of the system is the total mass times the velocity of the center of mass.	
SOLVE Apply Equation 10.1 to calculate the momentum of a particle. We know the velocity from part (b).	$\vec{p} = M\vec{v}_{CM}$ $\vec{p} = (1.00\text{ kg} + 2.50\text{ kg})\left(\left(5.58\hat{i} - 8.15\hat{j}\right)\text{ m/s}\right)$ $\vec{p} = \boxed{\left(19.5\hat{i} - 28.5\hat{j}\right)\text{ kg}\cdot\text{m/s}}$
CHECK and THINK The momentum is in the same direction as the velocity in part (b).	

70. A pendulum is used to measure the speed of bullets. It comprises a heavy block of wood of mass M suspended by two long cords. A bullet of mass m is fired into the block horizontally. The block, with the bullet embedded in it, swings upward (Fig. P10.70). The center of mass of the combination rises through a vertical distance h before coming to rest momentarily. In a particular experiment, a bullet of mass 40.0 g is fired into a wooden block of mass 10.0 kg. The block–bullet combination is observed to rise to a maximum height of 20.0 cm above the block's initial height.

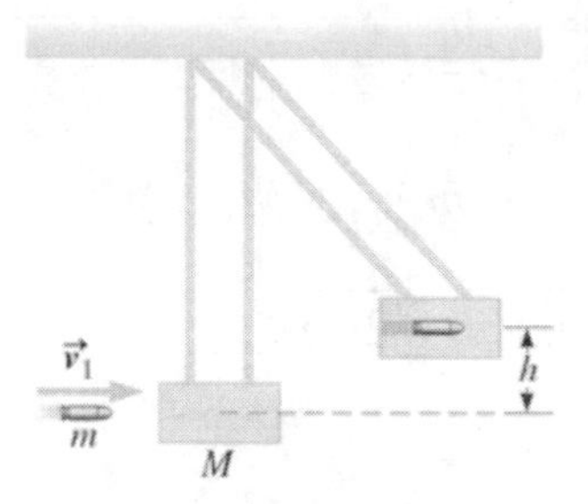

Figure P10.70

a. (N) What is the initial speed of the bullet?

INTERPRET and ANTICIPATE	
We will break the problem in two steps. The first step is the where the bullet hits the block and gets embedded. In this process, momentum is conserved. In the next step, the block and embedded bullet embedded swing upwards to a maximum height h. We will use the conservation of mechanical energy in this step.	
SOLVE ***Step 1:*** Use conservation of momentum for the instant right before and after the bullet hits the block (Equation 10.11). There are no additional external forces in the horizontal direction during this impact, so momentum is conserved. The initial momentum of the system $\vec{p}_{i,\text{tot}}$ is due to the momentum of the bullet. The block is at rest. Momentum right after the collision is $\vec{p}_{f,\text{tot}}$ is the momentum of the block and the bullet. The velocity of the bullet/block immediately after the collision is $\vec{V}$.	$m_{bullet}\vec{v}_{bullet} = (m_{bullet} + M_{block})\vec{V}$ $\vec{v}_{bullet} = \dfrac{m_{bullet} + M_{block}}{m_{bullet}}\vec{V}$ (1)

Step 2: Now, using conservation of energy, the kinetic energy of the bullet/block system immediately after the impact is equal to the gravitational potential energy of the bullet/block when it reaches its highest position. (Since the string is exerting a tension that changes direction as the block swings, momentum of the block is *not* conserved.)	$K_i = U_f$ $\frac{1}{2}(m_{bullet} + M_{block})V^2 = (m_{bullet} + M_{block})gh$ $V = \sqrt{2gh}$ (2)
Now, we take the magnitude of equation (1) and combine it with equation (2).	$v_{bullet} = \frac{m_{bullet} + M_{block}}{m_{bullet}}\sqrt{2gh}$ $= \frac{40.0\times10^{-3}\text{ kg} + 10.0\text{ kg}}{40.0\times10^{-3}\text{ kg}}\sqrt{2\left(9.81\ \frac{\text{m}}{\text{s}^2}\right)\left(20.0\times10^{-2}\text{ m}\right)}$ $v_{bullet} = \boxed{497\text{ m/s}}$

CHECK and THINK

In this case, we used momentum conservation during the collision and energy conservation as the block traveled upwards to a higher potential energy. The bullet speed of about 500 m/s sounds like the right order of magnitude.

b. (N) What is the fraction of initial kinetic energy lost after the bullet is embedded in the block?

INTERPRET and ANTICIPATE

Now that we know the speed of the bullet from part (a), we can determine the initial kinetic energy. Since we know how fast the block is moving after the collision and the height that it reaches, we can calculate either its kinetic energy after the collision or its final potential energy (since, based on energy conservation used above, must be the same magnitude).

SOLVE First, calculate the initial kinetic energy is that of the bullet.	$K_i = \frac{1}{2}m_{bullet}v_1^2$ $K_i = \frac{1}{2}(40.0\times10^{-3}\text{ kg})(497\text{ m/s})^2$ $K_i = 4.94\times10^3\text{ J}$

The final kinetic energy after the impact is also equal to the final potential energy when the block/bullet system reaches its highest position. In fact, we used energy conservation in part (a) to equate these two quantities. The benefit of recognizing this is simply that we have all of the quantities to calculate the final potential energy and it avoids an intermediate step of calculating the speed V. You can calculate $\frac{1}{2}mV^2$ to convince yourself they are equal if you have any doubt though!	$K_f = U_f = (m_{bullet} + M_{block})(gh)$ $K_f = (40.0\times10^{-3}\text{ kg} + 10.0\text{ kg})(9.81\text{ m/s}^2)(20.0\times10^{-2}\text{ m})$ $K_f = 19.7\text{ J}$
We now calculate how much energy was lost.	$\frac{\Delta K}{K_i} = \frac{K_i - K_f}{K_i}$ $\frac{\Delta K}{K_i} = \frac{4.94\times10^3\text{ J} - 19.7\text{ J}}{4.94\times10^3\text{ J}}$ $\frac{\Delta K}{K_i} = 0.996\text{ lost} = \boxed{99.6\%\text{ lost}}$

CHECK and THINK

In this case, almost all of the energy of the bullet is actually lost in the collision! This is an example of an inelastic collision (in which energy is not conserved), which we will encounter in the next couple chapters.

71. (A) Rochelle and Sheldon are two astronauts in space tethered together by a strong rope of length L. Initially, Sheldon is holding a large tool, which Rochelle needs. He then throws it to Rochelle, who catches it. Sheldon, Rochelle, and the tool all have the same mass m. Find an expression for the distance traveled by the tool relative to the center of mass of the system.

INTERPRET and ANTICIPATE There are no external forces on the system composed of the two astronauts and the tool, so momentum is conserved. Initially, nothing is moving and the momentum is zero. When Sheldon throws the tool to Rochelle, he recoils, dragging Rochelle along with him until she catches the tool. After she catches it, all three must be at rest to keep the total momentum zero. It is possible to use momentum conservation, solve for the relative speeds, and determine the distance traveled. However, if there are no external forces and all objects are initially at rest, so the center of mass cannot accelerate and change position (Equation 10.6). We can simply determine the center of mass initially and make sure that it remains unchanged in the final situation.	
SOLVE We sketch the situation and realize that the tool will travel less than the length of the rope.	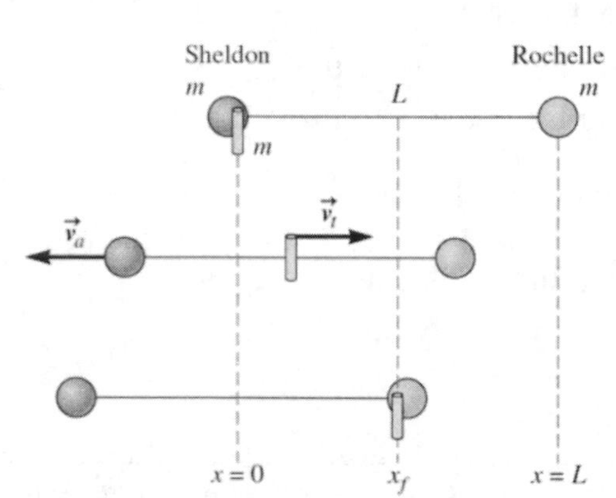 **Figure P10.71ANS**
Since there are no external forces and everything is initially at rest, the center of mass position remains unchanged. Using an axis as which $x = 0$ corresponds to Sheldon and the tool's initial position, with Rochelle at $x = L$, we first calculate the center of mass using Equation 10.3.	$x_{\text{CM}} = \frac{1}{M}\sum_{j=1}^{n} m_j x_j$ $x_{\text{CM}} = \frac{1}{3m}\left[(2m)(0)+(m)(L)\right]$ $x_{\text{CM}} = \frac{L}{3}$
Since the center of mass remains at this point, we need only calculate the final position x_f of Rochelle and the tool, at which point Sheldon has moved a distance $L - x_f$ to the	$x_{\text{CM}} = \frac{L}{3} = \frac{1}{3m}\left[(m)\left(-(L-x_f)\right)+(2m)(x_f)\right]$ $L = -L + 3x_f$ $x_f = \boxed{\frac{2}{3}L}$

left. This final position (relative to the starting position of the tool) is also equal to the distance the tool travels.	

CHECK and THINK

We find that the tool travels less than the distance L, which is what we expected after sketching the situation.

We also said that it would be possible to determine the velocities for the tool and Sheldon/Rochelle and calculate it that way. It's a little more work, but let's sketch this out to convince ourselves that we get the same answer. Let's suppose Sheldon throws the tool at speed v_t. Using momentum conservation while the tool is between the astronauts, we can find the speed of the astronauts v_a.

$$(m)(v_t)+(2m)(v_a)=0 \quad \rightarrow \quad v_a=-\frac{v_t}{2}$$

From the figure, we see that the distance traveled by the tool to the right plus the distance traveled by the astronauts to the left must equal the length of the rope connecting the astronauts.

$$\Delta x_t+\Delta x_a=L \qquad (1)$$

If the tool is in the air for some unknown time Δt, the distance traveled by the tool and the astronauts are:

$$\Delta x_t=v_t\Delta t \qquad (2)$$

$$\Delta x_a=|v_a|\Delta t=\frac{1}{2}v_t\Delta t \quad (3)$$

Using equations (1) – (3),

$$v_t\Delta t+\frac{1}{2}v_t\Delta t=L \quad \rightarrow \quad v_t\Delta t=\frac{2}{3}L$$

From equation (2), this quantity is actually the distance traveled by the tool.

$$\Delta x_t=v_t\Delta t=\frac{2}{3}L$$

So, it does work, but recognizing that the center of mass position remains unchanged is simpler!

78. A light spring is attached to a block of mass $4m$ at rest on a frictionless, horizontal table. A second block of mass m is now placed on the table, in contact with the free end of the spring, and the two blocks are pushed together (Fig. P10.78). When the blocks are released, the more massive block moves to the left at 2.50 m/s.

Figure P10.78

a. (N) What is the speed of the less massive block?

<table>
<tr><td colspan="2">INTERPRET and ANTICIPATE
We take the system to be the two blocks and the spring. Since there are no external forces in the horizontal plane, momentum is conserved during this process. We expect that the more massive block will move slower since their momenta should be the same.</td></tr>
<tr><td>SOLVE
We can use momentum conservation. The initial momentum of the system is zero since everything is at rest.</td><td>$p_i = p_f$
$0 = mv_m + 4mv_{4m}$</td></tr>
<tr><td>Insert values to determine the speed of the smaller mass.</td><td>$v_m = -4v_{4m}$
$v_m = -4(2.50\text{ m/s})$
$v_m = -10.0\text{ m/s}$

The speed is then $\boxed{10.0\text{ m/s.}}$</td></tr>
<tr><td colspan="2">CHECK and THINK
As expected, the smaller mass moves with a larger velocity since it has the same momentum as the more massive block. It also moves in the negative direction compared to the more massive block.</td></tr>
</table>

b. (N) If $m = 1.00$ kg, what is the elastic potential energy of the system before it is released from rest?

<table>
<tr><td colspan="2">INTERPRET and ANTICIPATE
Without friction, energy is conserved, so the initial elastic potential energy of the system is equal to the final kinetic energy of the system.</td></tr>
<tr><td>SOLVE
We use energy conservation. The initial spring elastic energy equals the final kinetic energy, which we can calculate.</td><td>$E_s = \frac{1}{2}kx^2 = \frac{1}{2}mv_m^2 + \frac{1}{2}(4m)v_{4m}^2$
$E_s = \frac{1}{2}(1.00 \text{ kg})(10.0 \text{ m/s})^2 + \frac{1}{2}(4.00 \text{ kg})(2.50 \text{ m/s})^2$
$E_s = \boxed{62.5 \text{ J}}$</td></tr>
<tr><td colspan="2">CHECK and THINK
Using energy conservation, we can determine the spring energy (even though we don't know the spring constant) by calculating the kinetic energy (since we know all of the quantities needed for this calculation).</td></tr>
</table>

11

Collisions

4. (N) A 35.0-kg child steps off a 4.0-ft-high diving board and executes a cannonball jump into a pool. (The child holds her body in a tight ball so that air resistance is negligible as she falls downward.) What is the impulse exerted by the water on the child? In the CHECK and THINK step, explain why the child would be greatly injured if she were to land on the cement edge of the pool instead of in the water.

INTERPRET and ANTICIPATE The impulse exerted on the child is her change in momentum as she stopped by the water. Her initial momentum as she enters the water can be calculated using her free fall data. The final momentum is zero because the water stops her. Throughout this problem we use an upward pointing y axis.	
SOLVE The impulse she experiences equals her change in momentum according to Eq. 11.4.	$\vec{I}_{\text{tot}} = \sum \vec{I} = \Delta\vec{p}$
Since the water stops her, her final momentum is zero. Her initial momentum is mv_i, where the speed v_i is her speed as she enters the water. (So, v_i is her speed after falling 4 feet from the diving board, which we can determine using either kinematics or energy conservation.)	$\vec{I}_{\text{tot}} = \vec{p}_f - \vec{p}_i = 0 - \vec{p}_i$ $\vec{I}_{\text{tot}} = -m\vec{v}_i$
We now calculate v_i, the speed she reaches after falling 4 feet, using conservation of energy. Take upward to be the positive direction to write her velocity as a vector. Of course, we could also use kinematic equations if we wanted, in which case we'd reach the same answer:	$mgh = \frac{1}{2}mv^2$ $v_y = \pm\sqrt{2gh} = \pm\sqrt{2\left(-9.81\,\text{m/s}^2\right)\left(-1.9\text{ m}\right)}$ $v_y = \pm 6.1\text{ m/s}$ $\vec{v}_y = -6.1\hat{j}\,\text{m/s}$

$\Delta\vec{y} = -1.9\hat{j}\,\text{m}$ $\vec{v}_{0y} = 0$ $\vec{a} = -g\hat{j} = -9.81\hat{j}\,\text{m/s}^2$ $\vec{v}_y$ needed $v_y^2 = v_{0y}^2 + 2a_y\Delta y$ $v_y = \pm\sqrt{2gh}$	
Now, calculate the impulse.	$\vec{I}_{\text{tot}} = -m\vec{v}_i = -(35.0\ \text{kg})(-6.1\hat{j}\,\text{m/s})$ $\vec{I}_{\text{tot}} = \boxed{2.1\times10^2\,\hat{j}\ \text{kg}\cdot\text{m/s}}$

CHECK and THINK

Whether she landed in the water or on the cement edge of the pool, her momentum would change from the value just before she lands to zero. So in either case, the water or the cement must exert the *same* impulse. We know from Eq. 11.3 that the impulse can *also* be expressed in terms of a force exerted over time, $\vec{I} = \vec{F}_{\text{av}}\Delta t$. That is, to achieve this impulse, we can exert a smaller force over a longer time or a very large force over a very short time. The water exerts this impulse over a longer period of time, and so the force exerted by the water is smaller. The cement would exert a much larger force over a very short time to achieve the same impulse, which would likely injure the child.

9. (N) A 65.0-kg driver of a vehicle traveling with a velocity of $11.5\hat{i}$ m/s in a parking lot hits a parked car and comes to rest in 0.150 s.

a. Find the impulse experienced by the driver.

b. Find the average force acting on the driver during the collision.

INTERPRET and ANTICIPATE

The impulse exerted on the driver is equal to the change in momentum of the driver (Eq. 11.4, $\vec{I}_{\text{tot}} = \Delta\vec{p}$), which we can calculate with the information given. This impulse is also equal to the average force acting on the driver times the time over which the force acts (Eq. 11.3, $\vec{I}_{\text{tot}} = \vec{F}_{\text{av}}\Delta t$), so we can then find the average force on the driver if we know the duration of the collision.

SOLVE	
The impulse is given by Equation 11.4. We take the driver's initial velocity (about 26 mph) to be in	

the $+\hat{\imath}$ direction. The final momentum, when the car is at rest, is zero.	$\vec{I}_{tot} = \sum \vec{I} = \Delta\vec{p}$ $\vec{I}_{tot} = \vec{p}_f - \vec{p}_i = -\vec{p}_i$ $\vec{I}_{tot} = -m\vec{v}_i = -(65.0\text{ kg})(11.5\hat{\imath}\text{ m/s})$ $\vec{I}_{tot} = \boxed{-748\hat{\imath}\text{ kg}\cdot\text{m/s}}$
The average force can now be calculated using Equation 11.3 and the time over which the collision occurs. We want to find the magnitude of the force, which is 4.98×10^3 N.	$\vec{I}_{tot} = \vec{F}_{av}\Delta t$ $\vec{F}_{av} = \dfrac{\vec{I}_{tot}}{\Delta t}$ $\vec{F}_{av} = \dfrac{\vec{I}_{tot}}{\Delta t} = \dfrac{-(65.0\text{ kg})(11.5\hat{\imath}\text{ m/s})}{0.150\text{ s}} = -4.98\times10^3\hat{\imath}\text{ N}$ $F_{av} = \boxed{4.98\times10^3\text{ N}}$

CHECK and THINK

This force is about 8 times the weight of the driver. The driver will certainly notice this collision, but the force is lower than might be expected for a high-speed collision. If he was traveling at highway speeds, about double the speed of the 26 mph he was traveling at here, the impulse would be twice as large and the force would potentially double if the collision took the same amount of time.

11. (N) An empty bucket is placed on a scale, and the scale is reset to read 0.00 N. Water falls from a faucet, high overhead, into the bucket without splashing at a rate of 350 mL/s. What is the reading on the scale 4.50 s after the water first hits the bottom of the empty bucket if at that time the falling water travels 1.25 m before hitting the water already in the bucket?

INTERPRET and ANTICIPATE

There are two contributions to the weight that the scale reads. One is the weight of the water in the bucket. We can determine how much water has entered the bucket in this 4.5 second period to find this. The other is the force that the jet of water exerts on the scale. Imagine shooting a fire hose at a wall, for instance. There is a contact force with the wall, which must stop the water (or in other words, reduce its momentum to zero). We can determine this force by relating the force over time to the change in momentum of the water jet.

SOLVE	
We first determine the amount of water in the bucket 4.5 seconds after the water first hits the bottom. It	$\left(350\dfrac{\text{mL}}{\text{s}}\right)(4.5\text{ s}) = 1580\text{ mL} = 1.58\text{ L}$

enters at a rate of 350 mL/s, which is a volume per time, so we multiply this rate times the time to get a total volume.	
Now, determine the weight of 1.58 liters of water. One liter of water has a mass of one kilogram (you may also know that 1 mL has a mass of 1 g and 1 L = 1000 mL). The scale would read 15.5 N to support the weight of the water. We could also say that the scale reads 15.5 N, the upward force that it exerts to support the weight of the water.	$(1.58\text{ L})\left(\frac{1\text{ kg}}{1\text{ L}}\right)=1.58\text{ kg}$ $F_{\text{weight}}=mg=(1.58\text{ kg})(9.81\text{ m/s}^2)=15.5\text{ N}$
The scale also exerts an *additional* upward force to stop the downward motion of the water pouring into the bucket. That is, it exerts a force needed to decrease the momentum of the water jet to zero. Using Equations 11.3 $(I=\Delta p)$ and 11.4 $(I=F\Delta t)$, we can relate the "extra" force (in addition to the weight of the water) due to the change in momentum. The water enters the bucket at some impact velocity and its final speed is zero.	$F\Delta t=\Delta p$ $F_{\text{impact}}\Delta t=mv_{\text{impact}}-mv_f=mv_{\text{impact}}-0$
From this, we write an expression that depends on the rate at which mass enters the bucket ($m/\Delta t$) and the impact speed. The scale must exert a larger upward force to cause the water to stop, so this represents an increase in force registered by the scale.	$F_{\text{impact}}=\frac{m}{\Delta t}v_{\text{impact}}$

Water is entering the bucket with an impact speed that we can determine using energy conservation.	$mgy_{top} = \frac{1}{2}mv_{impact}^2$ $v_{impact} = \sqrt{2gy_{top}}$ $v_{impact} = \sqrt{2(9.81\text{ m/s}^2)(1.25\text{ m})}$ $v_{impact} = 4.95\text{ m/s downward}$
The mass per time can be found knowing that water enters at a rate of 350 mL/s and that 1 L = 1 kg. So, $m/\Delta t = 0.350$ kg/s.	$F_{extra} = \left(0.350\ \frac{\text{kg}}{\text{s}}\right)\left(4.95\ \frac{\text{m}}{\text{s}}\right) = 1.73\text{ N}$
In total, the scale must support the weight of the water and the extra force due to the impact, so it will read 17.2 N.	$F_{total} = F_{weight} + F_{impact} = 15.5\text{ N} + 1.73\text{ N} = \boxed{17.2\text{ N}}$

CHECK and THINK

The impact force is smaller, but not negligible. Of course, when the water first hits, there is no water weight to support and the scale reads only the 1.73 N impact force. As the bucket fills, the weight of the water continues to increase while the impact force remains constant.

18. (N) A comet is traveling through space with a speed of 3.33×10^4 m/s when it collides with an asteroid that was at rest. The comet and the asteroid stick together during the collision process. The mass of the comet is 1.11×10^{14} kg and the mass of the asteroid is 6.66×10^{20} kg. For the comet and asteroid,

a. what is the magnitude of the momentum of the system's center of mass before the collision?

b. What is the magnitude of the momentum of the system's center of mass after the collision?

INTERPRET and ANTICIPATE

Because the momentum of the system is conserved, the momentum of the center of mass before the collision must equal the momentum of the center of mass after the collision. This is because the momentum of the center of mass is the same as the total momentum of the system.

SOLVE **a.** The total momentum of the	

system initially is the same as the momentum of the center of mass before the collision. The asteroid's initial momentum is 0, so the total momentum initially is the momentum of the comet.	$p_{CM} = (1.11\times10^{14}\ \text{kg})(3.33\times10^{4}\ \text{m/s})$ $p_{CM} = \boxed{3.70\times10^{18}\ \text{kg}\cdot\text{m/s}}$
b. After the collision, the asteroid and the comet are moving together. As we stated, the total momentum of the system is unchanged, and so the momentum of the center of mass is unchanged as well.	$p_{CM} = \boxed{3.70\times10^{18}\ \text{kg}\cdot\text{m/s}}$

CHECK and THINK

The conservation of momentum of a system of objects can be modeled from the viewpoint of the center of mass, where the momentum and velocity of the center of mass of a system never change as long as the system is not subject to a net external force.

20. (N) A skater of mass 45.0 kg standing on ice throws a stone of mass 7.65 kg with a speed of 20.9 m/s in a horizontal direction. Find the distance over which the skater will move in the opposite direction if the coefficient of kinetic friction between the skater and the ice is 0.03.

INTERPRET and ANTICIPATE

Momentum is conserved in the throwing process, so given the masses and speed of the stone, we can determine the velocity of the skater (who moves in the opposite direction). The skater will slide backward a certain distance before stopping due to the frictional force. The kinetic energy lost by the skater must be equal to the work done by the frictional force.

SOLVE Since this problem is identical to Problem 19, but with specific numbers provided, we'll use the variables from that problem. At the end, we will also write a symbolic solution that is the goal of Problem 19. First, apply momentum conservation. The initial momentum of the skater-stone system (before the skater	$\vec{p}_i = \vec{p}_f$ $0 = m_{\text{ska}}\vec{V}_{\text{ska}} + M_{\text{sto}}\vec{v}_{\text{sto}}$ $\vec{V}_{\text{ska}} = -\dfrac{M_{\text{sto}}}{m_{\text{ska}}}\vec{v}_{\text{sto}}$ Taking the stone to be traveling in the positive direction, insert the values given.

throws the stone) is zero. The final momentum is the momentum of the stone plus the momentum of the skater. Since they travel in opposite directions, they will have opposite signs and add to zero.	$V_{\text{ska}} = -\frac{M_{\text{sto}}}{m_{\text{ska}}} v_{\text{sto}} = -\frac{7.65\text{ kg}}{45.0\text{ kg}}(20.9\text{ m/s}) = -3.55\text{ m/s}$
After initially moving backwards at this speed, the skater loses all of this kinetic energy due to the work done by the frictional force. We first calculate the change in kinetic energy as the skater comes to rest. This is a negative value since energy was lost.	$\Delta K = K_f - K_i$ $\Delta K = 0 - \frac{1}{2} m_{\text{ska}} V_{\text{ska}}^2$ $\Delta K = -\frac{1}{2} m_{\text{ska}} \left(\frac{M_{\text{sto}} v_{\text{sto}}}{m_{\text{ska}}} \right)^2$ $\Delta K = -\frac{M_{\text{sto}}^2 v_{\text{sto}}^2}{2m_{\text{ska}}}$ Numerically, $\Delta K = -\frac{1}{2} m_{\text{ska}} V_{\text{ska}}^2 = -\frac{1}{2}(45.0\text{ kg})(3.55\text{ m/s})^2$ $\Delta K = -284\text{ J}$
Now, equate this to the work done by friction $\left(W_f = \vec{F} \cdot \vec{x} = -\mu_k N x\right)$ to calculate the distance x that the skater slides. The work done by friction is negative since the friction force acts in the opposite direction of the displacement of the sliding skater.	$\Delta K = W_f = -\mu_k N x \quad (1)$ $-\frac{M_{\text{sto}}^2 v_{\text{sto}}^2}{2m_{\text{ska}}} = -\mu_k (m_{\text{ska}} g) x \quad (2)$ Numerically, using Eq. (1), $x = \frac{-\Delta K}{\mu_k m_{\text{ska}} g} = \frac{284\text{ J}}{(0.03)(45.0\text{ kg})(9.81\text{ m/s}^2)} = 21.4\text{ m}$
We can also use Equation (2) to write the symbolic answer asked for in Problem 19. We could verify this by plugging in known values to determine the same final answer as in the last step.	$x = \frac{M_{\text{sto}}^2 v_{\text{sto}}^2}{2m_{\text{ska}}^2 \mu_k g} \quad (3)$ $x = \frac{(7.65\text{ kg})^2 (20.9\text{ m/s})^2}{2(45.0\text{ kg})^2 (0.03)(9.81\text{ m/s}^2)} = \boxed{21.4\text{ m}}$

CHECK and THINK

The skater recoils due to momentum conservation and the skater's kinetic energy is

dissipated by friction during the slide. Given the very small friction coefficient, the skater slides a significant distance (about 70 feet!) before stopping. Notice that the symbolic form of the answer, Eq. (3), behaves as we might expect. If the mass of speed of the stone that is thrown was larger (increasing the numerator), the skater would slide further backwards. Similarly, if the friction coefficient was smaller (decreasing the denominator), the skater would slide further. If the skater weighed more (increased denominator), the skater would not slide as far.

23. (N) Ezra ($m = 25.0$ kg) has a tire swing and wants to swing as high as possible. He thinks that his best option is to run as fast as he can and jump onto the tire at full speed. The tire has a mass of 10.0 kg and hangs 3.75 m straight down from a tree branch. Ezra stands back 10.0 m and accelerates to a speed of 3.50 m/s before jumping onto the tire swing.

a. How fast are Ezra and the tire moving immediately after he jumps onto the swing?

INTERPRET and ANTICIPATE Ezra jumps onto the tire swing and they end up moving together, so this is an inelastic collision. Since Ezra's total momentum just before he lands on the tire swing must be equal to the momentum of the entire system after this inelastic collision, we expect that they will be moving slower than the his initial speed.	
SOLVE Since Ezra holds onto the tire swing, their final speeds are the same. We can use momentum conservation in the horizontal direction to find the final velocity of the combined system. While Ezra eventually swings up, the actual collision is one-dimensional and we can use Equation 11.13 for a completely inelastic collision with a stationary target.	$(\vec{p}_{tot})_i = (\vec{p}_{tot})_f$ $m_1 v_{1_i} \hat{i} = (m_1 + m_2) v_f \hat{i}$
We can assume Ezra is running in the positive direction to write his initial velocity and record their masses.	$\vec{v}_{1i} = 7.50\hat{i}$ m/s $m_1 = 25.0$ kg $m_2 = 10.0$ kg

Filling in these values, we can solve for the final velocity of the combined system.	$(25.0\text{ kg})(3.50\hat{\imath}\text{ m/s}) = (25.0\text{ kg} + 10.0\text{ kg})\vec{v}_f$ $\vec{v}_f = \boxed{2.50\hat{\imath}\text{ m/s}}$
While not asked of us, we can also confirm that energy was lost in this inelastic collision by calculating the initial and final kinetic energy of the system.	$K_{\text{tot},\,i} = \frac{1}{2}mv_i^2 = \frac{1}{2}(25.0\text{ kg})(7.50\text{ m/s})^2 = 703\text{ J}$ $K_{\text{tot, f}} = \frac{1}{2}mv_f^2 = \frac{1}{2}(35.0\text{ kg})(2.50\text{ m/s})^2 = 109\text{ J}$ $K_{\text{tot},\,i} > K_{\text{tot},\,f}$

CHECK and THINK
As expected, the combined system (which has more total mass) moves slower than Ezra's speed when he approached the swing. Energy was also lost in this inelastic collision, which we expect when objects stick together in a collision.

b. How high does the tire travel above its initial height?

INTERPRET and ANTICIPATE
We can use the conservation of energy to determine how high the swing reaches. We know the initial kinetic energy of Ezra and the tire swing (since we know their total mass and speed). We can set this equal to the final potential energy as they swing to the highest point.

SOLVE	
The kinetic energy of the swing and child after the collision was found in part (a). We can use the final velocity of the system after the collision as the initial kinetic energy of the swing as it travels upwards.	$K_{\text{tot},i} = P_{tot,f}$ $\frac{1}{2}mv^2 = mgh$ $\frac{1}{2}(35.0\text{ kg})(2.50\text{ m/s})^2 = (35.0\text{ kg})(9.81\text{ m/s}^2)h$ $h = \boxed{0.319\text{ m}}$

CHECK and THINK
The child swings up about a third of a meter or one foot. Ezra was probably hoping to get a little higher than that!

29. (A) A dart of mass m is fired at and sticks into a block of mass M that is initially at rest on a rough, horizontal surface. The coefficient of kinetic friction between the block and the surface is μ_k. After the collision, the dart and the block slide a distance D before coming to rest. If the dart were fired horizontally, what would its speed be immediately before impact with the block?

INTERPRET and ANTICIPATE The dart impacts and then sticks into the block, so this is an inelastic collision. After the collision, the dart and block are traveling together with some speed and come to rest due to friction with the surface, so their kinetic energy is dissipated due to work done by friction. We would expect that if the dart's initial speed v_{di} was larger, the block would slide further, so our calculated value of v_{di} should increase with D.	
SOLVE The collision between the dart and the metal block is completely inelastic. Momentum is conserved in the collision according to Eq. 11.13. We first find the relationship between the initial speed of the dart (v_{di}) and the speed of the dart+block (v_f) just after impact, taking the initial speed of the block (v_{Bi}) to be zero.	$mv_{di} + Mv_{Bi} = (m+M)v_f$ $v_{di} = \frac{(m+M)}{m} v_f$ (1)
Now use the work-energy theorem to find the relation between the speed v_f just after impact and the distance the block slides before stopping. Kinetic energy is lost and the work done by friction is negative since the friction force points in the opposite direction of the displacement of the block, $W = \vec{f} \cdot \vec{D} = -fD$.	$\Delta K = W$ $0 - \frac{1}{2}(m+M)v_f^2 = -fD$
The friction force can be expressed as the coefficient of friction times the normal force, $f = \mu N = \mu m_{tot} g$.	$-fD = -\mu F_N D = -\mu(m+M)gD$
We can solve this for the final speed of the dart/block after the collision.	$\frac{1}{2}(m+M)v_f^2 = \mu(m+M)gD$ $v_f = \sqrt{2\mu gD}$ (2)

Now plug Eq. (2) into (1) to determine the initial speed of the dart.	$v_{di} = \boxed{\dfrac{(m+M)}{m}\sqrt{2\mu g D}}$

CHECK and THINK
As expected, the greater the distance that they slide D, the larger the initial speed v_{di} must have been. We can also make sense of the dependence on the other variables, but we have to be a little careful. For instance, *given a fixed sliding distance D*, if the mass of the block M was larger, the initial speed o the dart v_{di} must have been larger to cause it to slide that distance. If the friction coefficient was larger, the initial speed also would have had to be larger to make the block slide the same distance.

34. (N) In Examples 8.9 (page 230) and 9.8 (page 267), we found the maximum height of a small ball launched straight up from a springloaded dart gun. In this problem, we find the muzzle speed v_m of the dart gun. The dart gun is held horizontally such that its muzzle is very close to an open container of modeling dough (called *Fun Doh*) so that the ball hits the Fun Doh at the muzzle velocity. The ball of mass $m_1 = 2.7 \times 10^{-2}$ kg is fired into the packed Fun Doh container of total mass $m_2 = 0.15$ kg. The container is suspended from a string so that it can swing as a pendulum bob. The ball becomes embedded in the Fun Doh, and the center of mass of the ball–Fun Doh system rises to a maximum height $h = 0.26$ m. Assume dissipative forces may be ignored. Find the muzzle speed of the dart gun.

INTERPRET and ANTICIPATE
Let's consider the motion of the ball-Fun-Doh system at three key times: just before the collision, just after the collision, and when the ball-Fun-Doh composite comes to rest at the top of its trajectory. The collision is completely inelastic. Momentum is conserved and initially only one object (the ball) is moving. Our task is to find the initial speed of the ball. After the collision we use a conservation of energy to relate the maximum height of the system h to the velocity of the combined system just after the collision v_c.

SOLVE Initially particle 1—the ball, with mass m_1—is in motion. After the collision, the ball and Fun Doh stick together forming a composite system with total mass $M = m_1 + m_2$ moving at velocity v_c. We can use momentum conservation for a	$m_1 v_m = M v_c$ $v_c = \dfrac{m_1}{M} v_m$ (1)

1D inelastic collision, Eq. 11.13.	
Use conservation of energy to relate v_c to the height h. The reference configuration is set to the lowest point in the container's path, so the initial potential energy is zero. The final kinetic energy, at the highest point, is zero.	$K_i + U_i = K_f + U_f$ $\frac{1}{2}Mv_c^2 + 0 = 0 + Mgh$ $v_c^2 = 2gh$
Substitute Equation (1) for v_c. When solving for the muzzle speed v_m, choose the positive root since the initial velocity of the ball is in the positive x direction.	$\left(\frac{m_1}{M}v_m\right)^2 = 2gh$ $v_m^2 = \left(\frac{M}{m_1}\right)^2 (2gh)$ $v_m = \pm\frac{M}{m_1}\sqrt{2gh}$ $v_m = \frac{M}{m_1}\sqrt{2gh}$
Plug in values.	$v_m = \frac{m_1 + m_2}{m_1}\sqrt{2gh}$ $v_m = \frac{(2.7\times10^{-2}\text{ kg} + 0.15\text{ kg})}{2.7\times10^{-2}\text{ kg}}\sqrt{2(9.81\text{ m/s}^2)(0.26\text{ m})}$ $v_m = \boxed{15\text{ m/s}}$

CHECK and THINK

Our result is in the form we expected, but it might seem a bit fast (34 mph). It may help to note that the typical speed of a dart hitting a dart board is about 40 mph (18 m/s) so 15 m/s seems about right. This is actually higher than the freight train's speed in the Case Study, though in that case, the train was pushed backward for 110 m before coming to rest. In the train collision, the moving freight train was much more massive than the passenger train, whereas in this problem the ball is much less massive than the *Fun Doh.* Momentum, not speed, is conserved during a collision, so the masses of both objects matter.

39. (N) Two objects collide head-on (Fig. P11.39). The first object is moving with an initial speed of 8.00 m/s, and the second object is moving with an initial speed of 10.00

m/s. Assuming the collision is elastic, $m_1 = 5.15$ kg, and $m_2 = 6.25$ kg, determine the final velocity of each object.

Figure P11.39

INTERPRET and ANTICIPATE

Because this collision is elastic, both momentum and kinetic energy are conserved in the collision process. We can use Eq. 11.14 and Eq. 11.15 to find the final velocities of each object since they were derived for this kind of collision. Assume that the positive x axis points to the right. Given that the objects have similar masses and are moving at nearly the same speed in opposite directions, we expect that the first object must rebound and end up moving to the left, while the second object must rebound and end up moving to the right. This will ensure both the momentum and kinetic energy are conserved.

SOLVE

Use Eq. 11.14 to find the final velocity of the first object.

$$\vec{v}_{1f} = \frac{m_1 - m_2}{M}\vec{v}_{1i} + \frac{2m_2}{M}\vec{v}_{2i}$$

$$\vec{v}_{1f} = \frac{(5.15\text{ kg} - 6.25\text{ kg})}{(5.15\text{ kg} + 6.25\text{ kg})}(8.00\hat{\imath}\text{ m/s}) + \frac{2(6.25\text{ kg})}{(5.15\text{ kg} + 6.25\text{ kg})}(-10.00\hat{\imath}\text{ m/s})$$

$$\vec{v}_{1f} = \boxed{-11.7\hat{\imath}\text{ m/s}}$$

Or $\boxed{\text{11.7 m/s to the left}}$.

Use Eq. 11.15 to find the final velocity of the second object.

$$\vec{v}_{2f} = \frac{m_2 - m_1}{M}\vec{v}_{2i} + \frac{2m_1}{M}\vec{v}_{1i}$$

$$\vec{v}_{2f} = \frac{(6.25\text{ kg} - 5.15\text{ kg})}{(5.15\text{ kg} + 6.25\text{ kg})}(-10.00\hat{\imath}\text{ m/s}) + \frac{2(5.15\text{ kg})}{(5.15\text{ kg} + 6.25\text{ kg})}(8.00\hat{\imath}\text{ m/s})$$

$$\vec{v}_{2f} = \boxed{6.26\hat{\imath}\text{ m/s}}$$

Or $\boxed{\text{6.26 m/s to the right}}$.

CHECK and THINK

As expected, the first object ended up moving to the left and the second object ended up moving to the right.

42. (N) In an attempt to produce exotic new particles, a proton of mass $m_p = 1.67 \times 10^{-27}$ kg is accelerated to $0.99c$ ($c = 3.00 \times 10^8$ m/s is the speed of light) and crashed into a helium nucleus of mass $m_{He} = 6.64 \times 10^{-27}$ kg initially at rest. The collision is elastic.

a. What is the kinetic energy of the helium nucleus after the collision?
b. What is the kinetic energy of the proton after the collision? (In Chapter 39, we'll learn what Einstein says about making such calculations.)

INTERPRET and ANTICIPATE	
This is a one-dimensional elastic collision with an originally stationary target. Both momentum and kinetic energy are conserved in this process, so we could write equations for moth momentum and energy conservation and solve these. We can also use Equations 11.14 and 11.15, which allow us to determine the final velocities of the two particles in an elastic collision, or Equations 11.16 and 11.17, which are valid specifically for the case of an elastic collision with a stationary target, which we have here.	
SOLVE We call the proton object 1 and the helium nucleus object 2.	$v_{1_i} = v_{p_i} = 0.99c = 2.97 \times 10^8$ m/s $v_{2_i} = v_{He_i} = 0$ $m_1 = m_p = 1.67 \times 10^{-27}$ kg $m_2 = m_{He} = 6.64 \times 10^{-27}$ kg
Use Eq. 11.16 (or Eq. 11.14, which reduces to this equation) to find the final velocity of the proton.	$v_{pf} = \frac{(m_p - m_{He})}{(m_p + m_{He})} v_{pi}$ $v_{pf} = \frac{(1.67 - 6.64) \times 10^{-27}\text{ kg}}{(1.67 + 6.64) \times 10^{-27}\text{ kg}} v_{pi}$ $v_{pf} = -0.598 v_{pi}$
Similarly, use Eq. 11.17 (or Eq. 11.15) to find the final velocity of the helium nucleus.	$v_{Hef} = \frac{(2m_p)}{(m_p + m_{He})} v_{pi}$ $v_{Hef} = \frac{(2 \times 1.67) \times 10^{-27}\text{ kg}}{(1.67 + 6.64) \times 10^{-27}\text{ kg}} v_{pi}$ $v_{Hef} = 0.402 v_{pi}$

a. Now calculate the kinetic energy of the helium nucleus is	$KE_{\text{He}f} = \frac{1}{2} m_{\text{He}} v_{\text{He}f}^2$ $KE_{\text{He}f} = \frac{1}{2}\left(6.64\times10^{-27}\text{ kg}\right)\left(0.402v_{pi}\right)^2$ $KE_{\text{He}f} = \boxed{4.73\times10^{-11}\text{ J}}$
b. And now calculate the kinetic energy of the proton.	$KE_{pf} = \frac{1}{2} m_p v_{pf}^2$ $KE_{pf} = \frac{1}{2}\left(1.67\times10^{-27}\text{ kg}\right)\left(-0.598v_{pi}\right)^2$ $KE_{pf} = \boxed{2.63\times10^{-11}\text{ J}}$

47. (N) A roller-coaster car of mass $m_1 = 8.00 \times 10^2$ kg starts from rest and slides on a frictionless track from point A to point B at a height $h = 10.0$ m below point A, where it collides elastically with a second car of mass $m_2 = 2.00 \times 10^2$ kg, initially at rest. What is the maximum height that the second car rises above point B on the track?

INTERPRET and ANTICIPATE

The first car accelerates as it descends 10 m on the track. We can use conservation of energy to determine how fast it's going when it collides with the second car. The highest that car 2 could possibly go would be if no energy is lost in the collision, so we assume the collision is a 1D elastic collision where the second object is at rest. We can determine how fast the second car is traveling after the collision and then use energy conservation to find how high it climbs after the collision.

SOLVE	
Use conservation of energy to find the speed of the first car at point B just before it collides with the second car. This is the initial velocity of car 1 in the elastic collision.	$K_i + U_i = K_f + U_f$ $0 + m_1 gh = \frac{1}{2} m_1 v_1^2 + 0$ $v_1 = \sqrt{2gh}$ $v_1 = \sqrt{2(9.81\text{ m/s}^2)(10.0\text{ m})} = 14.0\text{ m/s}$
The highest that the second car could go would be if no energy is lost, so we use the elastic collision equation with the second car initially	$v_{2f} = \frac{2m_1}{M} v_{1i} = \left(\frac{2m_1}{m_1 + m_2}\right) v_{1i}$

at rest (Equation 11.17).	$v_{2f}=\left(\frac{2(800)}{800+200}\right)(14.0)=22.4\text{ m/s}$
We can again use conservation of the mechanical energy to find the maximum height the second car will reach.	$K_i+U_i=K_f+U_f$ $\frac{1}{2}m_2v_{2f}^2+0=0+m_2gy_{\max}$ $y_{\max}=\frac{v_{2f}^2}{2g}$ $y_{\max}=\frac{(22.4\text{ m/s})^2}{2(9.81\text{ m/s}^2)}=\boxed{25.6\text{ m}}$

CHECK and THINK

Though it may seem surprising that the second car has traveled higher than the release point of the first car, the first car is four times as massive. The second car has less potential energy at its highest point than the initial potential energy of the first car. This is what we'd expect since this total initial energy must be shared between the two cars after the collision.

52. (N) A proton with an initial speed of 2.00×10^8 m/s in the x direction collides elastically with another proton initially at rest. The first proton's velocity after the collision is 1.64×10^8 m/s at an angle of 35.0° with the horizontal. What is the velocity of the second proton after the collision?

INTERPRET and ANTICIPATE

We first draw a sketch of the collision. For any collision, whether elastic or inelastic, momentum is conserved. Since the first proton scatters at an angle of 35 degrees relative to its initial velocity, this is a 2D collision. Momentum is conserved for both the x direction (which we choose to be along the direction of the first proton) and the y direction independently (Equations 11.21 and 11.22).

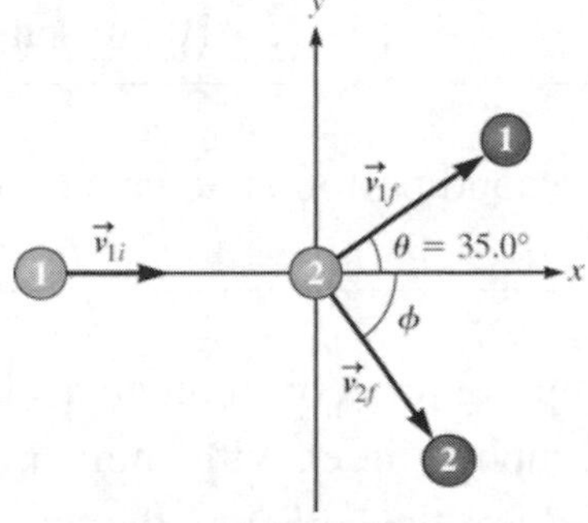

Figure P11.52ANS

SOLVE Use conservation of momentum for the system of the two protons (which have the same mass *m*), first for the *x* direction (Eq. 11.21). Particle 2 is initially at rest, so $v_{2ix}=0$. Proton 1 is initially traveling at $v_{1ix}=2.00\times10^8$ m/s and scatters at an angle of 35° with a final speed $v_{1f}=1.64\times10^8$ m/s. By trigonometry, the *x* component of its final velocity is $v_{1fx}=v_{1f}\cos35°$.	$mv_{1ix}+mv_{2ix}=mv_{1fx}+mv_{2fx}$ $v_{1ix}+0=v_{1f}\cos35°+v_{2fx}$ $\left(2.00\times10^8\text{ m/s}\right)=\left(1.64\times10^8\text{ m/s}\right)\cos35.0°+v_{2fx}$ $v_{2fx}=6.57\times10^7\text{ m/s}$
Similarly, consider the *y* components (Eq. 11.22), for which $v_{2iy}=0$ and $v_{1fy}=v_{1f}\sin35°$. Since proton 1 initially moves only in the *x* direction, $v_{1iy}=0$.	$mv_{1iy}+mv_{2iy}=mv_{1fy}+mv_{2fy}$ $0+0=v_{1f}\sin35°+v_{2fy}$ $0+0=(1.64\times10^8\text{ m/s})\sin35.0°+v_{2fy}$ $v_{2fy}=-9.41\times10^7\text{ m/s}$
We now write the final velocity in component form.	$\vec{v}_{2f}=6.57\times10^7\text{ m/s}\,\hat{\boldsymbol{i}}-9.41\times10^7\text{ m/s}\,\hat{\boldsymbol{j}}$
We can also use trigonometry to calculate the magnitude and direction of the final velocity.	$v=\left(\sqrt{6.57^2+9.41^2}\right)\times10^7\text{ m/s}=1.15\times10^8\text{ m/s}$ $\theta=\tan^{-1}\left(\dfrac{-9.41\times10^7}{6.57\times10^7}\right)=-55.1°$ $\boxed{1.15\times10^8\text{ m/s at }-55.1°}$
CHECK and THINK We see that the velocity of the second proton is down and to the right, which is consistent with our picture above.	

56. (N) A police officer is attempting to reconstruct an accident in which a car traveling southward with a speed of 23.0 mph collided with another car of equal mass traveling eastward at an unknown speed. After the collision, the two cars coupled and slid at an angle of 60.0° south of east. If the speed limit in that neighborhood is 25.0 mph, should the officer cite the second driver for speeding? Explain your answer.

<table>
<tr><td colspan="2">INTERPRET and ANTICIPATE
The cars stick together after the collision, so this is an inelastic collision. Since they are initially traveling perpendicular to each other, this is a 2D collision. We'll apply momentum conservation in both the x and y directions (using either Equations 11.24 and 11.25 or Equations 11.26 and 11.27).</td></tr>
<tr><td>SOLVE
We first sketch the situation described. Use conservation of momentum for the system of two vehicles for both southward and eastward components, to find the original speed of car number 2.</td><td>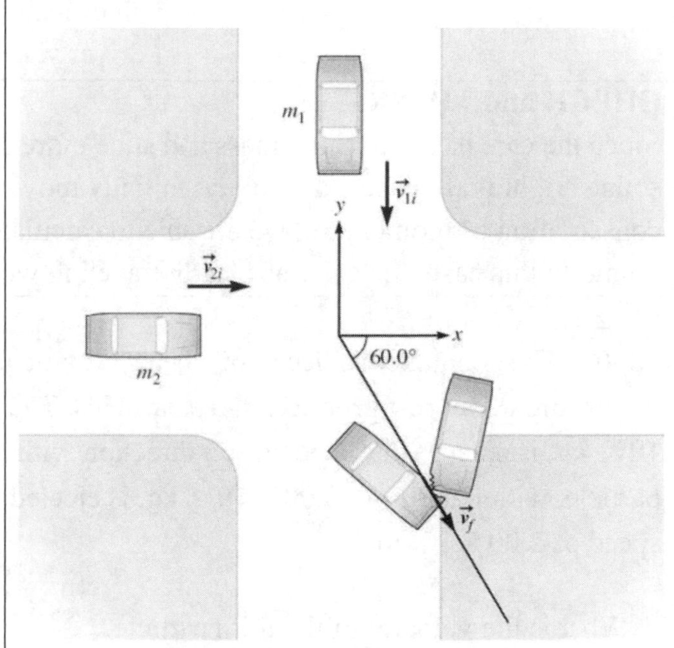

Figure P11.56ANS</td></tr>
<tr><td>Take the x direction to be eastward. Since the car travels due south initially, its initial x velocity is zero. Convert the speed 23.0 mph = 10.3 m/s and use m as the mass of each vehicle and M = 2m the combined final mass. Insert these into Equation 11.26.</td><td>$m_1 v_{1ix} + m_2 v_{2ix} = M v_f \cos\theta$
$m(10.3\text{ m/s}) = (2m) v_f \cos 60.0°$ (1)</td></tr>
<tr><td>Now apply Eq. 11.27, taking south to be the positive direction. (You can also take north to be positive, in which case both terms are negative, which can then be canceled from both sides.)</td><td>$m_1 v_{1i_y} + m_2 v_{2i_y} = M v_f \sin\theta$
$m v_{2i} = (2m) v_f \sin 60.0°$ (2)</td></tr>
</table>

Divide the southward equation (1) by the eastward equation (2) and solve for v_{2i}.	$\frac{m(10.3 \text{ m/s})}{mv_{2i}} = \frac{(2m)v_f \cos 60.0°}{(2m)v_f \sin 60.0°}$ $v_{2i} = (10.3 \text{ m/s}) \tan 60.0°$ $v_{2i} = 17.8 \text{ m/s} = 39.8 \text{ mi/h}$
	The initial speed of the southbound car was significantly larger than 25 mi/h and therefore the driver was speeding.

CHECK and THINK

Since the cars have the same mass and slide more south than east after the collision, it sounds right that the second car was initially moving faster than the first. Using conservation of momentum, we are able to calculate the precise speed it must have been going. In this case, the car was clearly traveling well above the 25 mph speed limit.

58. (N) The spontaneous decay of a heavy atomic nucleus of mass $M = 14.5 \times 10^{-27}$ kg that is initially at rest produces three particles. The first particle, with mass $m_1 = 6.64 \times 10^{-27}$ kg, is ejected in the positive x direction with a speed of 2.50×10^7 m/s. The second particle, with mass $m_2 = 3.00 \times 10^{-27}$ kg, is ejected in the negative y direction with a speed of 2.00×10^7 m/s.

a. What is the velocity of the third particle?

INTERPRET and ANTICIPATE

The heavy nucleus is initially at rest. This decay is like a small explosion that ejects three pieces (somewhat like a collision in reverse) and momentum is conserved. The nucleus initially has a momentum of zero, so the total momentum of the three daughter particles must add up to zero. If we sketch the decay, we might expect that all three pieces will spread apart as shown.

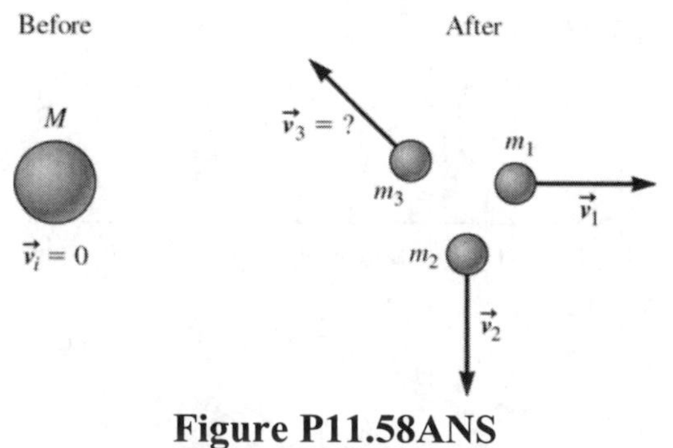

Figure P11.58ANS

SOLVE Let's first keep track of all the variables.	For the parent nucleus (initial): $M = 14.5\times10^{-27}$ kg and $\vec{v}_i = 0$ For the daughter particles (final): $m_1 = 6.64\times10^{-27}$ kg and $\vec{v}_1 = \left(2.50\times10^{7}\text{ m/s}\right)\hat{\imath}$ $m_2 = 3.00\times10^{-27}$ kg and $\vec{v}_2 = -\left(2.00\times10^{7}\text{ m/s}\right)\hat{\jmath}$
The three daughter particles have a total mass that equals the parent particle.	$M = m_1 + m_2 + m_3$ $m_3 = M - m_1 - m_2 = 4.86\times10^{-27}$ kg
Now use momentum conservation.	$\left(\vec{p}_{\text{tot}}\right)_i = \left(\vec{p}_{\text{tot}}\right)_f$ $0 = m_1\vec{v}_1 + m_2\vec{v}_2 + m_3\vec{v}_3$

Insert values from above and solve.

$$\left(6.64\times10^{-27}\text{kg}\right)\left(2.50\times10^{7}\text{m/s}\right)\hat{\imath} - \left(3.00\times10^{-27}\text{kg}\right)\left(2.00\times10^{7}\text{m/s}\right)\hat{\jmath} + \left(4.86\times10^{-27}\text{kg}\right)\vec{v}_3 = 0$$

$$\vec{v}_3 = \boxed{\left(-3.42\times10^{7}\text{ m/s}\right)\hat{\imath} + \left(1.23\times10^{7}\text{ m/s}\right)\hat{\jmath}}$$

with speed $v_3 = 3.63\times10^{7}$ m/s

CHECK and THINK

In this "explosion" momentum is conserved, which allows us to determine the velocity of the third piece. It moves up and too the left, consistent with our sketch above.

b. What is the increase in the kinetic energy of the system after the decay?

INTERPRET and ANTICIPATE

Generally, we think of mechanical energy as either being conserved (elastic collision) or dissipated (inelastic collision). In this case, the nuclear reaction *creates* energy which is converted into kinetic energy. An exploding firecracker is similar in that the chemical energy in some explosive fuel is converted into kinetic energy of the fragments.

SOLVE The initial kinetic energy of the system is zero. Therefore, the increase in kinetic energy is given by the total energy of the three daughter particles that are moving apart from each other.	$E = \frac{1}{2}m_1v_1^2 + \frac{1}{2}m_2v_2^2 + \frac{1}{2}m_3v_3^2$

Plug in the values from part (a).

$$E = \frac{1}{2}\left[\left(6.64\times10^{-27}\right)\left(2.50\times10^7\right)^2 + \left(3.00\times10^{-27}\right)\left(2.00\times10^7\right)^2 + \left(4.86\times10^{-27}\right)\left(3.63\times10^7\right)^2\right]$$

$$E = \boxed{5.88\times10^{-12}\text{ J}}$$

CHECK and THINK

Energy is clearly created in this nuclear reaction. Even though it seems like a small amount of energy, this is the energy produced in a *single* nuclear fission event. If we had a kilogram of the material, with something like an Avogadro's number of nuclei, the energy produced can be significant — in fact, this is the basis for applications such as nuclear power plants!

63. (N) In an experiment designed to determine the velocity of a bullet fired by a certain gun, a wooden block of mass $M = 500.0$ g is supported only by its edges, and a bullet of mass $m = 8.00$ g is fired vertically upward into the block from close range below. After the bullet embeds itself in the block, the block and the bullet are measured to rise to a maximum height of 25.0 cm above the block's original position. What is the speed of the bullet just before impact?

INTERPRET and ANTICIPATE

We're told that the bullet embeds itself in the block, which is a cue that the bullet and block experience a completely inelastic collision and stick together. Since we know their masses and the initial velocity of the bullet, we can calculate the final velocity of the bullet/block. The bullet/block then has some amount of kinetic energy, so we can determine the highest point they reach using energy conservation.

SOLVE Momentum is conserved in this completely inelastic 1D collision. Use Equation 11.13 with v_i as the initial bullet speed and v as the	$m_{bullet}v_i = \left(m_{bullet} + m_{block}\right)v$ $v_i = \frac{\left(m_{bullet} + m_{block}\right)v}{m_{bullet}}$ (1)

final speed of the bullet/block as they move upwards together.	
Using energy conservation, we can relate the speed of the bullet-block v and the height h the block reaches above the initial point. The initial potential energy is zero as is the kinetic energy at the highest point.	$K_i + U_i = K_f + U_f$ $\frac{1}{2}(m+M)v^2 + 0 = 0 + (m+M)gh$ $v = \sqrt{2gh}$ (2)
Now plug equation (2) into (1), solve for v_i, and plug in values.	$v_i = \frac{m_{bullet} + m_{block}}{m_{bullet}}\sqrt{2gh}$ $v_i = \frac{(0.008\text{ kg} + 0.500\text{ kg})}{0.008\text{ kg}}\sqrt{2(9.81\text{ m/s}^2)(0.250\text{ m})}$ $v_i = \boxed{141\text{ m/s}}$

CHECK and THINK

The initial speed of the bullet is fast (over 300 mph), which we would expect for a bullet fired out of a gun.

67. (N) Assume the pucks in Figure P11.66 stick together after their collision at the origin. Puck 2 has four times the mass of puck 1 ($m_2 = 4m_1$). Initially, puck 1's speed is three times puck 2's speed ($v_{1i} = 3v_{2i}$), puck 1's position is $\vec{r}_{1i} = -x_{1i}\hat{i}$, and puck 2's position is $\vec{r}_{2i} = -y_{2i}\hat{j}$.

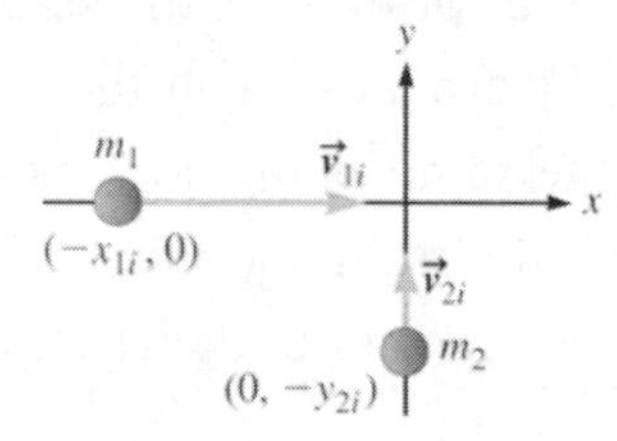

Figure P11.66

a. Find an expression for their velocity after the collision in terms of puck 1's initial velocity.

INTERPRET and ANTICIPATE The collision is completely inelastic, so only momentum is conserved. We must consider each direction separately in this 2D collision.	
SOLVE First consider the momentum in the x direction using Equation 11.24. Before the collision, only puck 1 moves in the x direction, so $v_{2ix} = 0$. Puck 2 is four times the mass of puck 1, $m_2 = 4\,m_1$, so the total mass is $M = 5m_1$.	$m_1 v_{1ix} + m_2 v_{2ix} = M v_{fx}$ $m_1 v_{1i} = 5 m_1 v_{fx}$ $v_{fx} = \frac{1}{5} v_{1i}$ (1)
No apply Eq. 11.25 for the y direction. Only puck 2 moves in the y direction initially, so $v_{1ix} = 0$. We also know that $v_{1i} = 3v_{2i}$, or $v_{2i} = v_{1i}/3$.	$m_1 v_{1iy} + m_2 v_{2iy} = M v_{fy}$ $(4m_1)\frac{v_{1i}}{3} = 5 m_1 v_{fy}$ $v_{fy} = \frac{4}{15} v_{1i}$ (2)
Combine Equations (1) and (2) to write the final velocity.	$\vec{v}_f = v_{fx}\hat{i} + v_{fy}\hat{j}$ $\boxed{\vec{v}_f = \frac{1}{5} v_{1i}\hat{i} + \frac{4}{15} v_{1i}\hat{j}}$
CHECK and THINK After the collision, the system's center of mass is the same as the position of the pucks since they are stuck together, so they are moving at the center of mass velocity. The answer we found here is actually the same as the center of mass velocity found in Problem 66. Since puck 1 moves to the right and puck 2 moves up, when they stick together, the composite object moves up and to the right, as we would expect.	

b. What is the fraction K_f / K_i that remains in the system?

INTERPRET and ANTICIPATE Since the objects stick together, this is a completely inelastic collision and we expect that kinetic energy is lost.

Solve We first write the total initial kinetic energy of the two pucks. As in part (a), we use the facts that $m_2 = 4\,m_1$ and $v_{1i} = 3v_{2i}$.	$K_i = \frac{1}{2}m_1 v_{1i}^2 + \frac{1}{2}m_2 v_{2i}^2$ $K_i = \frac{1}{2}m_1 v_{1i}^2 + \frac{1}{2}(4m_1)\left(\frac{v_{1i}}{3}\right)^2$ $K_i = \frac{13}{18}m_1 v_{1i}^2$ (1)
After the collision, the pucks move together at the velocity given in part (a).	$K_f = \frac{1}{2}Mv_f^2$ $K_f = \frac{1}{2}(5m_1)\left[\frac{1}{25}v_{1i}^2\left(1+\frac{16}{9}\right)\right]$ $K_f = \frac{5}{18}m_1 v_{1i}^2$ (2)
Finally, take the ratio of Eq. (2) to (1).	$\frac{K_f}{K_i} = \left(\frac{\frac{5}{18}m_1 v_{1i}^2}{\frac{13}{18}m_1 v_{1i}^2}\right) = \boxed{\frac{5}{13}}$
CHECK and THINK The ratio is less than 1, so energy is indeed lost in this inelastic collision.	

70. (N) A ball of mass 50.0 g is dropped from a height of 10.0 m. It rebounds after losing 75% of its kinetic energy during the collision process. If the collision with the ground took 0.010 s, find the magnitude of the impulse experienced by the ball.

INTERPRET and ANTICIPATE The impulse-momentum theorem relates the impulse to the change in momentum of the ball. Using energy conservation for the initial drop, we can determine the initial downward velocity. We can then use the fact 75% of the kinetic energy is lost to find the velocity after the collision. From these, we calculate the change in momentum.	
SOLVE According to the impulse-momentum theorem (Eq. 11.4), the impulse equals the change in momentum.	$\vec{I} = \Delta\vec{p}$ $\vec{I} = m(\vec{v}_2 - \vec{v}_1)$

We can find the velocity of the mass just before it lands by using conservation of energy and the initial drop height. We take up to be the positive direction.	$\frac{1}{2}mv_1^2 = mgh$ $v_1 = \sqrt{2gh} = \sqrt{2(9.81\,\text{m/s}^2)(10.0\,\text{m})} = 14.0\,\text{m/s}$ $\vec{v}_1 = -14.0\,\hat{j}\,\text{m/s}$
Now determine the upward (positive) velocity just after the collision using the fact that 75% of the energy is dissipated or, equivalently, that the final kinetic energy is 25% of the initial kinetic energy.	$\frac{1}{2}mv_2^2 = (0.25)\frac{1}{2}mv_1^2$ $v_2 = \frac{v_1}{2} = 7.00\,\text{m/s}$ $\vec{v}_2 = +7.00\,\hat{j}\,\text{m/s}$
Now apply Equation 11.4 and calculate the magnitude of the impulse.	$\lvert\vec{I}\rvert = \lvert m(\vec{v}_2 - \vec{v}_1)\rvert$ $I = (50.0\times10^{-3}\,\text{kg})(7.00\,\text{m/s} + 14.0\,\text{m/s})$ $I = \boxed{1.05\,\text{N}\cdot\text{s}}$

CHECK and THINK

In this case, we are able to determine the change in momentum and therefore the impulse. We don't need to know the average force or contact time, which would be a second way to find impulse (using Eq. 11.3).

73. Three runaway train cars are moving on a frictionless, horizontal track in a railroad yard as shown in Figure P11.73. The first car, with mass $m_1 = 1.50 \times 10^3$ kg, is moving to the right with speed $v_1 = 10.0$ m/s; the second car, with mass $m_2 = 2.50 \times 10^3$ kg, is moving to the left with speed $v_2 = 5.00$ m/s, and the third car, with mass $m_3 = 1.20 \times 10^3$ kg, is moving to the left with speed $v_3 = 8.00$ m/s The three railroad cars collide at the same instant and couple, forming a train of three cars.

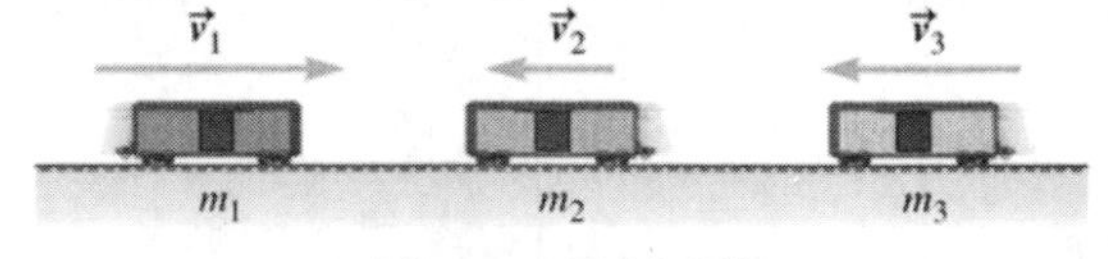

Figure P11.73

a. (N) What is the final velocity of the train cars immediately after the collision?

INTERPRET and ANTICIPATE

The three train cars collide at the same time, so we can treat this as a single collision and use the conservation of momentum.

SOLVE Apply conservation of momentum. Take the first car, traveling to the right, to be moving in the positive direction, so the second and third cars have negative velocities. They join together, so the entire mass $M = m_1 + m_2 + m_3$ moves at the same final velocity v_f.	$\left(\sum \vec{p}\right)_{\text{before}} = \left(\sum \vec{p}\right)_{\text{after}}$ $m_1 v_{1i} + m_2 v_{2i} + m_3 v_{3i} = (m_1 + m_2 + m_3) v_f$

Insert values and solve for v_f. The combined mass is $M = (1500 + 2500 + 1200)$ kg = 5200 kg.

$$(1500\text{ kg})(10.0\text{ m/s}) + (2500\text{ kg})(-5.00\text{ m/s}) + (1200\text{ kg})(-8.00\text{ m/s}) = (5200\text{ kg})v_f$$

$$v_f = \frac{(15000 - 12500 - 9600)\text{kg}\cdot\text{m/s}}{5200\text{ kg}} = \frac{-7100}{5200}\text{m/s} = -1.37\text{ m/s}$$

or $v = \boxed{1.37\text{ m/s toward the left}}$

CHECK and THINK

Even with three objects colliding, this is a straightforward application of momentum conservation. The total initial momentum is towards the left, so the train cars move to the left after they join.

b. (C) Would the answer to part (a) change if the three cars did not collide at the same instant? Explain.

$\boxed{\text{No.}}$ Momentum is conserved, so the total initial momentum equals the total final momentum regardless of the order in which they couple. Therefore the final velocity is the same after they join together.

We can verify this by first imagining that, for instance, cars 1 and 2 join together first. They will have a speed after they couple which we can find from momentum conservation.

$$m_1 v_{1i} + m_2 v_{2i} = (m_1 + m_2) v_{12}$$

$$v_{12} = \frac{m_1 v_{1i} + m_2 v_{2i}}{m_1 + m_2} = \frac{(1500\text{ kg})(10.0\text{ m/s}) + (2500\text{ kg})(-5.00\text{ m/s})}{1500\text{ kg} + 2500\text{ kg}} = 0.625\text{ m/s}$$

Now imagine that this combined object of total mass 4000 kg and velocity 0.625 m/s collides with the third train. Let's calculate the final velocity and compare it to part (a).

$$v_f = \frac{m_{12}v_{12} + m_3 v_{3i}}{m_{12} + m_3} = \frac{(4000\ \text{kg})(0.625\ \text{m/s}) + (1200\ \text{kg})(-8.00\ \text{m/s})}{4000\ \text{kg} + 1200\ \text{kg}} = -1.37\ \text{m/s}$$

This is indeed the same answer we calculated above.

76. A dramatic (and perhaps unexpected) collision occurs if you hold a tennis ball on top of a basketball and drop them at the same time. After impacting the ground, the tennis ball can be launched much higher than the original height of the two balls. Assume this process can be modeled as an elastic collision of the basketball with the ground followed by a second elastic collision of the basketball with the tennis ball. The basketball (with mass m_b) and tennis ball (with mass m_t) each falls a distance h with an acceleration g before the basketball hits the ground.

a. (A) What is the velocity of each ball the instant before the basketball collides with the ground in terms of the variables specified?

INTERPRET and ANTICIPATE

This question concerns the system before the collision, during the period when the ball is in free fall before hitting the ground. Each ball falls the same distance and should have the same velocity which we can determine using energy conservation. We expect an algebraic answer.

SOLVE

The initial velocity of each ball is zero. The change in height of each ball is given as h. Using conservation of energy we express the final velocities of the basketball and tennis ball. The final velocity is pointing downwards, which we call the $-\hat{\boldsymbol{j}}$ direction.

$$mgh = \frac{1}{2}mv^2$$

$$v = \sqrt{2gh} \qquad (1)$$

$$\vec{\boldsymbol{v}}_b = \vec{\boldsymbol{v}}_t = \boxed{-\sqrt{2gh}\,\hat{\boldsymbol{j}}}$$

CHECK and THINK

The final answer indicates a downward velocity and is larger if the initial drop height is larger.

b. (A) Assume the basketball first undergoes an elastic collision with the ground. How fast is the basketball moving after this collision?

INTERPRET and ANTICIPATE	
This is an example where the projectile has a much smaller mass than the target (the target is the Earth in this case!). This is Case B in Figure 11.13 for elastic collisions and we expect the projectile will leave with a velocity equal in magnitude and opposite the direction of its incident velocity.	
SOLVE See Case B in Figure 11.13. For a massive target at rest, the projectile will leave with a velocity equal in magnitude and opposite its incident velocity.	$\vec{v}_{b,f} \simeq -\vec{v}_{b,i}$ $\vec{v}_{b,f} \simeq \boxed{\sqrt{2gh}\hat{j}}$
CHECK and THINK The velocity is equal and opposite as expected for a collision of an elastic object with a much larger object.	

c. (A) The basketball then collides elastically with the tennis ball. What is the final velocity of the tennis ball?

INTERPRET and ANTICIPATE	
While the basketball is somewhat larger than the tennis ball, this is not an example of one of our special cases and we need to use the general equations for a 1D elastic collision. We expect that the tennis ball will leave with a significant velocity based on the description of the experiment above.	
SOLVE The general equations for an elastic collision are Equations 11.14 and 11.15. The tennis ball is traveling downward at the speed given in part (a). The basketball is traveling upwards at the speed determined in part (b). We label the tennis ball as object 1 and the basketball as object 2.	$\vec{v}_{t,f} = \frac{m_t - m_b}{m_t + m_b}\vec{v}_{t,i} + \frac{2m_b}{m_t + m_b}\vec{v}_{b.i}$ $\vec{v}_{t,f} = \frac{m_t - m_b}{m_t + m_b}\left(-\sqrt{2gh}\hat{j}\right) + \frac{2m_b}{m_t + m_b}\left(\sqrt{2gh}\hat{j}\right)$ $\vec{v}_{t,f} = \boxed{\frac{3m_b - m_t}{m_t + m_b}\left(\sqrt{2gh}\hat{j}\right)}$ (2)

CHECK and THINK
The expression depends on the quantities provided. Assuming the basketball has a larger mass than the tennis ball, the final velocity of the tennis ball will be in the $+\hat{j}$ direction.

d. (N) Assume the basketball has a mass that is eight times that of the tennis ball. What is the ratio of the final speed of the tennis ball to the speed of the balls just before impact?

INTERPRET and ANTICIPATE	
We expect that the final speed should be large based on the original description of the experiment.	
SOLVE The relative masses can be expressed since the basketball has a mass of 8 times the mass of the tennis ball.	$m_b = 8m_t$
Equation (2) from part (c) can be simplified using this fact. The speed is the magnitude of the velocity.	$v_{t,f} = \lvert\vec{v}_{t,f}\rvert = \frac{3(8m_t) - m_t}{m_t + 8m_t}\left(\sqrt{2gh}\right)$ $v_{t,f} = \frac{23}{9}\left(\sqrt{2gh}\right)$
Equation (1) from part (a) gives the speed of the balls just before impact. Dividing the calculated final speed with the speed of the balls just before impact gives us the desired ratio.	$\frac{v_{t,f}}{v_{t,i}} = \frac{\frac{23}{9}\left(\sqrt{2gh}\right)}{\sqrt{2gh}} = \boxed{\frac{23}{9}}$
CHECK and THINK The final speed of the tennis ball is about 2.5 times faster than its speed just before impact. For a single ball dropped and an elastic collision, we would expect this ratio to be 1 (same speed after the bounce). Apparently, this arrangement leads to the tennis ball having a much larger upwards velocity after the bounce. In Problem 77, we find that this results in the ball bouncing about 6.5 times higher than the initial drop height! This is actually the somewhat dramatic result that you would observe if you try this.	

82. (A) The force on a particle is given by $\vec{F} = F_{\max}\cos(2\pi t/T)\hat{i}$, where $F_{\max}$ and T are constants. Find expressions for the impulse on the particle during each of the following intervals.

a. $0 < t < T/4$
b. $T/4 < t < 3T/4$
c. $0 < t < T$

<table>
<tr><td colspan="2">INTERPRET and ANTICIPATE
In many cases, we've used the average force times the time to determine the impulse $\left(\vec{I} = \vec{F}_{av}\Delta t\right)$, but in the case of a variable force, we can use this integral form of the equation, $\left(\vec{I} = \int_{t_i}^{t_f} \vec{F}(t)\,dt\right)$, Eq. 11.3. Since we have an expression for the force as a function of time, we can integrate this expression and evaluate it for each of the three time intervals.</td></tr>
<tr><td>SOLVE
Use Equation 11.3 to integrate the force over time to determine the impulse.</td><td>$\vec{I} = \int_{t_i}^{t_f} \vec{F}(t)\,dt = F_{max}\int_{t_i}^{t_f} \cos\left(\frac{2\pi t}{T}\right)dt$</td></tr>
<tr><td>a. First integrate between the limits of $t = 0$ and $t = T/4$. The integral of the cosine function is
$\int_{t_1}^{t_2} \cos(at) = \frac{1}{a}\sin(at)\Big|_{t_1}^{t_2}$
where $a = \frac{2\pi}{T}$ in this case.
Remember that sin(0) = 0 and sin ($\pi/2$) = 1.</td><td>$\vec{I} = F_{max}\left[\int_0^{T/4} \cos\left(\frac{2\pi t}{T}\right)dt\right]\hat{i}$
$\vec{I} = F_{max}\left[\left(\frac{T}{2\pi}\right)\sin\left(\frac{2\pi t}{T}\right)\Big|_0^{T/4}\right]\hat{i}$
$\vec{I} = F_{max}\left(\frac{T}{2\pi}\right)\left[\sin\left(\frac{\pi}{2}\right) - \sin(0)\right]\hat{i}$
$\vec{I} = \boxed{\frac{F_{max}T}{2\pi}\hat{i}}$</td></tr>
<tr><td>b. The integration is the same in this part. We only need to evaluate the integral with the new limits for the time interval $T/4$ to $3T/4$, using sin($3\pi/2$) = –1 and sin ($\pi/2$) = 1.</td><td>$\vec{I} = F_{max}\left[\left(\frac{T}{2\pi}\right)\sin\left(\frac{2\pi t}{T}\right)\Big|_{T/4}^{3T/4}\right]\hat{i}$
$\vec{I} = F_{max}\left(\frac{T}{2\pi}\right)\left[\sin\left(\frac{3\pi}{2}\right) - \sin\left(\frac{\pi}{2}\right)\right]\hat{i}$
$\vec{I} = \boxed{-\frac{F_{max}T}{\pi}\hat{i}}$</td></tr>
</table>

c. Evaluate the integral from 0 to T, with sin(0) = 0 and sin (2π) = 0.	$\vec{I} = F_{\text{max}}\left[\left(\frac{T}{2\pi}\right)\sin\left(\frac{2\pi t}{T}\right)\Big\vert_0^T\right]\hat{i}$ $\vec{I} = F_{\text{max}}\left(\frac{T}{2\pi}\right)\left[\sin(2\pi) - \sin(0)\right]\hat{i}$ $\vec{I} = \boxed{0}$

CHECK and THINK

We've calculated the impulse for each of the three time intervals. To make sense of this, we can also look at the force versus time plot (which is just a cosine function). Notice that between 0 and T/4, the function is positive, so the integral in part (a) is positive. Between T/4 and 3T/4, the force is negative, as is the impulse. Over a full period, there's as much positive as negative (or just as much push as pull) and the integral over one period and the total impulse are both zero.

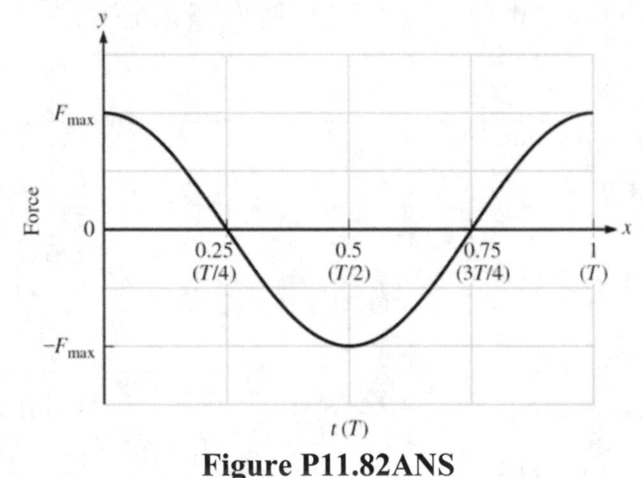

Figure P11.82ANS

12

Rotation I: Kinematics and Dynamics

5. A ceiling fan is rotating counterclockwise with a constant angular acceleration of π rad/s^2 about a fixed axis perpendicular to its plane and through its center. Assume the fan starts from rest.

a. (N) What is the angular velocity of the fan after 4.0 s?

INTERPRET and ANTICIPATE

With the given angular acceleration (π rad/s^2) and time (4.0 s), we can calculate the change in angular velocity using Equation 12.7.

SOLVE	
Similar to linear kinematics, an angular acceleration acting over a time interval leads to a change in angular velocity, as seen in Eq. 12.7.	$\omega = \omega_0 + \alpha t$
The initial velocity ω_0 is zero and the angular acceleration (πrad/s^2) and time (4.0 s) are given.	$\omega = 0 + (\pi \text{ rad/s}^2)(4.0 \text{ s})$ $\omega = 4.0\pi$ rad/s $\omega = \boxed{13 \text{ rad/s}}$

CHECK and THINK

The fan is rotating counterclockwise. With a positive angular acceleration acting over time, the angular velocity increases from zero to a positive final value.

b. (N) What is the angular displacement of the fan after 4.0 s?

INTERPRET and ANTICIPATE

Analogous to kinematic equations for linear motion in Chapter 2, with a constant acceleration acting over time, we can determine the displacement of the fan. (In this case, it is an angular acceleration leading to an angular displacement.) Equation 12.9 relates these variables.

SOLVE We first identify the known and unknown variables.	$\alpha = \pi \text{ rad/s}^2$ $\omega_0 = 0$ $t = 4.0\text{ s}$ $\theta_0 = 0$ $\theta = ?$
Using Eq. 12.9 to relate these variables, we solve for the final angle in radians.	$\theta = \theta_0 + \omega_0 t + \frac{1}{2}\alpha t^2$ $\theta = 0 + 0 + \frac{1}{2}(\pi \text{ rad/s}^2)(4.0\text{ s})^2$ $\theta = 8\pi \text{ rad} = \boxed{25\text{ rad}}$

CHECK and THINK
The fan is rotating counterclockwise. So the angular displacement is positive. In part (c), we will see how many revolutions this corresponds to.

c. (N) How many revolutions has the fan gone through in 4.0 s?

INTERPRET and ANTICIPATE
We use the result from part (b) to find the number of turns. One full revolution equals 360 degrees or 2π radians.

SOLVE From part (b), the angular displacement is 8π rad. One turn equals 2π rad.	$\text{\# revolutions} = \dfrac{8\pi \text{ rad}}{2\pi \text{ rad/rev}} = \boxed{4\text{ rev}}$

CHECK and THINK
The fan makes four turns in 4.0 s. Notice that the angular velocity is not a constant. The fan is still speeding up since it has a positive acceleration.

8. A rotating object's angular position is given by $\theta(t) = \left(1.54t^2 - 7.65t + 2.75\right)$ rad, where t is measured in seconds.

a. (N) When is the object momentarily at rest?

INTERPRET and ANTICIPATE
To find the moment the object is at rest, we take a derivative to find an expression for the angular speed, and set that equal to zero. We expect to find a time in seconds and $\alpha =$ __ rad/s^2. To help check our result we write the expression for angular position as

$\theta(t) = (1.54\ \text{rad/s}^2)t^2 - (7.65\ \text{rad/s})t + 2.75\ \text{rad}$.	
SOLVE The first derivative of $\theta(t)$ is the angular velocity (Eq. 12.4).	$\omega = \frac{d\theta}{dt} = (3.08\ \text{rad/s}^2)t - 7.65\ \text{rad/s}$
Set this equal to zero to find the moment the object is at rest.	$0 = (3.08\ \text{rad/s}^2)t - 7.65\ \text{rad/s}$ $t = \frac{7.65\ \text{rad/s}}{3.08\ \text{rad/s}^2}$ $t = \boxed{2.48\ \text{s}}$
CHECK and THINK The time determined is when the angular velocity (derivative of the angle versus time) is zero.	

b. (N) What is the magnitude of the angular acceleration at that time?

INTERPRET and ANTICIPATE The derivative of the angular velocity versus time (or, equivalently, the second derivative of the angle versus time) equals the angular acceleration. The original expression for the angle is a quadratic function of time, and so we expect the acceleration is constant.	
SOLVE The second derivative (derivative of the angular velocity) with respect to time is the angular acceleration (Eq. 12.6).	$\alpha = \frac{d\omega}{dt} = \boxed{3.08\ \text{rad/s}^2}$
CHECK and THINK The angular acceleration is constant (does not depend on t) in this case. So, the value we found is correct not only at 2.48 s, but at *all* times.	

11. (N) The long hand on a clock is known as the *minute hand*. It completes one clockwise rotation each hour. Choose a coordinate system such that a z axis points out of the clock. Consider a 15-minute time interval.

a. Find the minute hand's angular displacement.

b. Find the minute hand's angular velocity.

c. Find the minute hand's angular acceleration.

INTERPRET and ANTICIPATE In the 15 minute time interval, we know that the minute hand completes a quarter of its rotation. We only need to convert this to radians to find its angular displacement. A working clock does not speed up or slow down. So the angular velocity is constant; we can find the angular velocity at any momentum from the average velocity over the 15-minute interval. The direction comes from the right hand rule. Finally, we expect the angular acceleration is zero because the angular velocity is constant.	
SOLVE **a.** The angular displacement is the ¼ of a full circle. To find the direction, use the right hand rule. Wrap your right fingers clockwise, such that your thumb points into the page in the negative z direction.	$\Delta\theta = \frac{1}{4}(2\pi)$ $\Delta\theta = \frac{\pi}{2}$ rad $\Delta\vec{\theta} = \boxed{-\pi/2\hat{k} \text{ rad}}$
b. To determine the average angular velocity, we calculate the change in angle over time (Equation 12.3).	$\Delta t = 15 \text{ min} = 900 \text{ s}$ $\omega_{av} = \frac{\Delta\theta}{\Delta t} = \frac{\pi/2 \text{ rad}}{900 \text{ s}}$ $\omega = 1.7\times10^{-3}$ rad/s
To find the direction, wrap your right fingers clockwise, and you find your thumb points into the clock. This is the negative z direction $(-\hat{k})$.	$\vec{\omega} = \boxed{-1.7\times10^{-3}\hat{k} \text{ rad/s}}$
c. Since the angular velocity is constant, its derivative and the angular acceleration are zero (Equation 12.6).	$\vec{\alpha} = \frac{d\vec{\omega}}{dt} = \boxed{0}$
CHECK and THINK This problem helps to build our intuition. You have probably very familiar with the motion of a minute hand. Its angular speed is about 10^{-3} rad/s.	

17. Jeff, running outside to play, pushes on a swinging door, causing its motion to be briefly described by $\theta = t^2 + 0.800t + 2.00$, where t is in seconds and θ is in radians.
a. (N) What is the angular position of the door at $t = 0$ and at $t = 1.50$ s?

INTERPRET and ANTICIPATE We are given an equation for $\theta(t)$, so we can plug in the value of time and simply calculate the angle.	
SOLVE First, calculate the angle for $t = 0$.	$\theta(t=0) = (0)^2 + 0.800(0) + 2.00 = \boxed{2.00 \text{ rad}}$
Now, calculate for $t = 1.50$.	$\theta(t=1.50) = (1.50)^2 + 0.800(1.50) + 2.00 = \boxed{5.45 \text{ rad}}$
CHECK and THINK All of the coefficients are positive, so it is no surprise that the angle is increasing in time.	

b. (N) What is the angular speed of the door at $t = 0$ and at $t = 1.50$ s?

INTERPRET and ANTICIPATE The angular speed equals the derivative of the angle with respect to time (Equation 12.4).	
SOLVE We apply Equation 12.4 to calculate the derivative of the angle with respect to time. The resulting expression is $\omega(t)$.	$\omega = \frac{d\theta}{dt}$ $\omega = \frac{d}{dt}\left(t^2 + 0.800t + 2.00\right)$ $\omega = 2t + 0.800$
Using this expression, we plug in the times, 0 and 1.50 s.	$\omega(t=0) = 2(0) + 0.800 = \boxed{0.800 \text{ rad/s}}$ $\omega(t=1.50) = 2(1.50) + 0.800 = \boxed{3.80 \text{ rad/s}}$
CHECK and THINK From our calculations, we see that the door is speeding up during this time interval.	

c. (N) What is the angular acceleration of the door at $t = 0$ and at $t = 1.50$ s?

INTERPRET and ANTICIPATE
The angular acceleration equals the rate of change of the angular velocity (Eq. 12.6). From part (b), we know that the average acceleration during the time interval is positive. This is no guarantee that the acceleration at these moments in time will be positive, but we would expect our expression to be positive for part (or all) of this time interval.

SOLVE Using Equation 12.6, we calculate the derivative of the angular velocity found in part (b) with respect to time.	$\alpha = \dfrac{d\omega}{dt}$ $\alpha = \dfrac{d}{dt}(2t + 0.800)$ $\alpha = \boxed{2.00\ \text{rad/s}^2}$
CHECK and THINK In this case, we find that the angular acceleration is constant and positive. This is consistent with the fact that the door speeds up as we saw in part (b).	

20. A wheel starts from rest and in 12.65 s is rotating with an angular speed of 5.435π rad/s.

a. (N) Find the magnitude of the constant angular acceleration of the wheel.

INTERPRET and ANTICIPATE We use the constant angular acceleration techniques for this problem.	
SOLVE First, list the known variables in SI units.	$t = 12.65\ \text{s}$ $\omega = 5.435\pi\ \text{rad/s}$ $\omega_0 = 0$ $\alpha = ?$
We can use Equation 12.7 to relate these variables.	$\omega = \omega_0 + \alpha t$ $\alpha = \dfrac{\omega - \omega_0}{t}$
Substitute values.	$\alpha = \dfrac{5.435\pi\ \text{rad/s} - 0}{12.65\ \text{s}}$ $\alpha = \boxed{1.350\ \text{rad/s}^2}$
CHECK and THINK The angular acceleration is a positive constant value as expected. It is in the same direction (i.e. has the same sign) as the angular velocity because the wheel is speeding up.	

b. (N) Through what angle does the wheel move in 6.325 s?

INTERPRET and ANTICIPATE From part (b), we know four of the kinematic variables and can use any of the constant acceleration equations with $\Delta\theta$.	
SOLVE We choose Equation 12.9 as one that relates the change in angle to other known values. We want to calculate the position at a time of 6.325 s.	$\Delta\theta = \omega_0 t + \frac{1}{2}\alpha t^2$ $\Delta\theta = 0 + \frac{1}{2}(1.350\ \text{rad/s}^2)(6.325\ \text{s})^2$ $\Delta\theta = \boxed{27.00\ \text{rad}}$
CHECK and THINK Since the angular velocity and acceleration are both positive, the angular displacement is also positive. This corresponds to a bit over four revolutions.	

23. A potter's wheel is rotating with an angular velocity of $\vec{\omega} = 24.2\hat{k}$ rad/s. The wheel is slowing down as it is subject to a constant angular acceleration and eventually comes to a stop.

a. (N) If the wheel goes through 21.0 revolutions as it is slowed to a stop, what is the constant angular acceleration applied to the wheel?

INTERPRET and ANTICIPATE Knowing the angular distance over which the angular speed is reduced from its initial value to zero, we can use Equation 12.11 to find the magnitude of the angular acceleration.	
SOLVE We start by listing the known and desired quantities, taking positive values as indicating motion on the $+\hat{k}$ direction.	$\omega_0 = 24.2$ rad/s $\omega = 0$ $\Delta\theta = 21.0$ rev $\alpha = ?$
We can use Eq. 12.11 to relate the initial and final velocities to the acceleration acting as the wheel rotates a given distance. We will need to convert the angular displacement to radians as we perform the calculation	$\omega^2 = \omega_0^2 + 2\alpha\Delta\theta$

Now, plug in values.	$0=(24.2\text{ rad/s})^2+2\alpha(21.0\text{ rev}\cdot 2\pi\text{ rad}/1\text{ rev})$ $\alpha=-\dfrac{(24.2\text{ rad/s})^2}{2(21\text{ rad})}=-2.22\text{ rad/s}^2$
The negative sign indicates the direction of the angular acceleration is opposite that of the original angular velocity, therefore it's in the $-\hat{\boldsymbol{k}}$ direction.	$\vec{\alpha}=\boxed{-2.22\,\hat{\boldsymbol{k}}\text{ rad/s}^2}$
CHECK and THINK We've calculated the acceleration using rotational kinematic equations. Given that the wheel slows to a stop, the acceleration is in the opposite direction of the initial angular velocity.	

b. (N) How long does it take for the wheel to come to rest?

INTERPRET and ANTICIPATE We now know all of the kinematic quantities other than the time. We can choose any of the kinematic equations and solve for the time for the wheel to come to rest.	
SOLVE In order to find the time over which the wheel stops, we can use Eq. 12.7.	$\omega=\omega_0+\alpha t$
Substitute values and solve for time.	$0=(24.2\text{ rad/s})+(-2.22\text{ rad/s}^2)t$ $t=\dfrac{24.2\text{ rad/s}}{2.22\text{ rad/s}^2}=\boxed{10.9\text{ s}}$
CHECK and THINK The wheel decelerates at a constant rate and slows in 1.7 seconds.	

27. (N) An electric food processor comes with many attachments for blending and slicing food. Assume the motor maintains the same angular speed for all the various attachments.

The largest attachment has a diameter of 12.0 cm, and the smallest has a diameter of 5.00 cm. Find the ratio of the translational speed of the points on the edges of each attachment.

INTERPRET and ANTICIPATE

The angular speed is the same for both attachments. However, the translational speed depends on radius. We expect a point on the edge of larger attachment has faster translational speed.

SOLVE	
Find the radius of each attachment. There is no need to convert to meters in this case, because we are finding a ratio and the cm will cancel.	Small attachment: $r = 2.50$ cm Large attachment: $R = 6.00$ cm
The speed of a point on the edge depends on the angular velocity and radius of the attachment (Equation 12.13). We write a ratio for the two attachments. The angular speed is the same, and so it cancels.	$v = r\omega$ $\frac{v_{\text{large}}}{v_{\text{small}}} = \frac{R\omega}{r\omega} = \frac{R}{r}$ $\frac{v_{\text{large}}}{v_{\text{small}}} = \frac{6.00 \text{ cm}}{2.50 \text{ cm}} = \boxed{2.40}$

CHECK and THINK

The ratio is greater than 1. As expected, the larger one has a greater translational speed.

31. (N) A disk is initially at rest. A penny is placed on it at a distance of 1.0 m from the rotation axis. At time $t = 0$ s, the disk begins to rotate with a constant angular acceleration of 2.0 rad/s^2 around a fixed, vertical axis through its center and perpendicular to its plane. Find the magnitude of the net acceleration of the coin after 1.5 s.

INTERPRET and ANTICIPATE

The net acceleration of the coin is the vector sum of tangential and radial accelerations. The tangential acceleration can be found using the angular acceleration. The radial acceleration for an object in circular motion is the centripetal acceleration.

SOLVE	
The tangential acceleration depends on the angular acceleration and the radius from the rotational axis (Eq. 12.14).	$a_t = r\alpha$ $a_t = (1.0 \text{ m})(2.0 \text{ rad/s}^2) = 2.0 \text{ m/s}^2$

The radial acceleration is the centripetal acceleration, which depends on the tangential or angular velocity and radius of the penny from the rotation axis (Eq. 12.15).	$a_R = a_c = \frac{v^2}{r} = \omega^2 r$
We can calculate the angular velocity at a time of 1.5 seconds, since we know the angular acceleration and the fact that the initial angular velocity was zero (Eq. 12.7).	$\omega = \omega_0 + \alpha t$ $\omega = 0 + 2.0\ \text{rad/s}^2 \times 1.5\ \text{s}$ $\omega = 3.0\ \text{rad/s}$
We now calculate the radial acceleration.	$a_R = \omega^2 r = (3.0\ \text{rad/s})^2 (1.0\ \text{m}) = 9.0\ \text{m/s}^2$
Since the radial and tangential acceleration are always perpendicular, the magnitude of the total acceleration can be found from the Pythagorean theorem.	$a = \sqrt{a_t^2 + a_c^2} = \sqrt{2.0^2 + 9.0^2}\ \text{m/s}^2$ $a = \boxed{9.2\ \text{m/s}^2}$

CHECK and THINK

In this case, the total acceleration is approximately equal in magnitude to the radial acceleration, which is much larger than the tangential acceleration. The total acceleration makes and angle $\tan\theta = \frac{a_R}{a_T} = 4.5$ or $\theta = 77°$ from the tangential direction.

36. Two children, each with a mass of 25.0 kg, are at fixed locations on a merry-go-round (a disk that spins about an axis perpendicular to the disk and through its center; Fig. P12.36). One child is 0.75 m from the center of the merry-go-round, and the other is near the outer edge, 3.00 m from the center. With the merry-go-round rotating at a constant angular speed, the child near the edge is moving with translational speed of 12.5 m/s.

Figure P12.36

a. (N) What is the angular speed of each child?

<table>
<tr><td colspan="2">INTERPRET and ANTICIPATE
For a rigidly rotating object, as we have here, the angular speed is the same for every point on the object. We can calculate the angular velocity of the child at the outer edge with the information given.</td></tr>
<tr><td>SOLVE
Equation 12.13 allows us to calculate ω, which is the same for each child, assuming they stay at one point on the object.</td><td>$v_{outer} = r\omega$
$\omega = \frac{v_{outer}}{r}$
$\omega = \frac{12.5\ \text{m/s}}{3.00\ \text{m}} = \boxed{4.17\ \text{rad/s}}$</td></tr>
<tr><td colspan="2">CHECK and THINK
This rotation rate is a bit more than half a revolution per second ($4.17/2\pi = 0.66$ revolutions per second). The children are both rotating at the same rate.</td></tr>
</table>

b. (N) Through what angular distance does each child move in 5.0 s?

<table>
<tr><td colspan="2">INTERPRET and ANTICIPATE
Since the children both move at the same angular velocity, they traverse the same angular displacement in 5 seconds.</td></tr>
<tr><td>SOLVE
Since the angular speed is constant in this problem, Equation 12.3 may be used to find the angular distance through which both children move in 5.0 s.</td><td>$\omega = \frac{\Delta\theta}{\Delta t}$
$\Delta\theta = \omega\Delta t$
$\Delta\theta = (4.17\ \text{rad/s})(5.0\ \text{s}) = \boxed{21\ \text{rad}}$</td></tr>
<tr><td colspan="2">CHECK and THINK
In part (a), we found that they rotate a bit more than half a revolution per second. After 5 seconds, we find that they have rotated $21/2\pi = 3.3$ times, which sounds reasonable.</td></tr>
</table>

c. (N) Through what distance in meters does each child move in 5.0 s?

<table>
<tr><td colspan="2">INTERPRET and ANTICIPATE
Each child moves through a different distance because they rotate at the same angular velocity and are at different distances from the center of rotation.</td></tr>
<tr><td>SOLVE
Since the children rotate at the same angular velocity, the child at the larger distance from</td><td>$d = r\Delta\theta$
$d = r(21\ \text{rad})$</td></tr>
</table>

the rotation axis sweeps out a larger distance as seen in Equation 12.12.	
Plug in the radius for each.	$d_{outer} = (3.00\text{ m})(21\text{ rad}) = \boxed{63\text{ m}}$ $d_{inner} = (0.75\text{ m})(21\text{ rad}) = \boxed{16\text{ m}}$
CHECK and THINK Since the outermost child is four times away from the axis of rotation, the distance traveled is four times as large.	

d. (N) What is the centripetal force experienced by each child as he or she holds on? Which child has a more difficult time holding on?

INTERPRET and ANTICIPATE If we imagine standing at the center of the merry-go-round, we might get dizzy, but we won't get thrown off like we would if we were at the outside of a quickly-rotating merry-go-round. The child further away from the center should have a higher centripetal acceleration.	
SOLVE For a given *angular* velocity, the centripetal force increases with radius from the rotation axis, as from Equation 12.15 and Newton's second law.	$a_c = \omega^2 r$ $F_c = ma_c = m\omega^2 r$
We now plug in values.	$F_{c,\text{outer}} = (25.0\text{ kg})(4.17\text{ rad/s})^2(3.00\text{ m})$ $F_{c,\text{outer}} = \boxed{1.30\times10^3\text{ N}}$ $F_{c,\text{inner}} = (25.0\text{ kg})(4.17\text{ rad/s})^2(0.75\text{ m})$ $F_{c,\text{inner}} = \boxed{326\text{ N}}$
CHECK and THINK The outer child will have a more difficult time holding on.	

41. (N) A student pulls with a 300.0-N force on the outer edge of a door of width 1.51 m that pivots about point P as shown in Figure P12.41. Find the magnitude of the torque applied to the door in each case, where the force is applied in a different direction as shown.

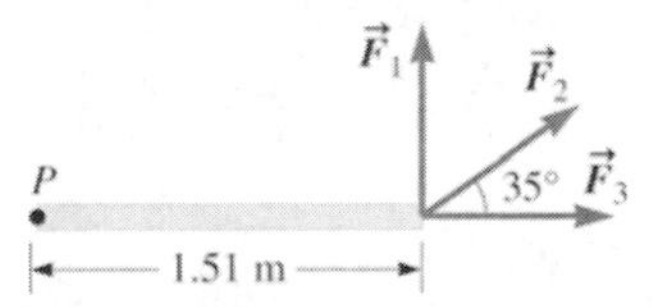

Figure P12.41

<table>
<tr><td colspan="2">INTERPRET and ANTICIPATE
The torque applied depends on the magnitude of the force and the angle at which it's applied. A perpendicular force, such as force 1, will produce a large torque, while a force applied along the radial direction, such as force 3, will produce a zero torque around the pivot point.</td></tr>
<tr><td>SOLVE
In each case, use Equation 12.16 in order to find the magnitude of the torque.</td><td>$\tau = rF\sin\varphi$</td></tr>
<tr><td>The force (300 N) and radial distance (1.51 m) is the same in each case, but the angle between the force and the radial vector is different:

Case 1: $\phi = 90°$
Case 2: $\phi = 35°$
Case 3: $\phi = 0°$</td><td>$\tau_1 = (1.51\text{ m})(300.0\text{ N})\sin(90°) = 453\text{ N}\cdot\text{m} = \boxed{4.53\times10^2\text{ N}\cdot\text{m}}$
$\tau_2 = (1.51\text{ m})(300.0\text{ N})\sin(35°) = \boxed{2.60\times10^2\text{ N}\cdot\text{m}}$
$\tau_3 = (1.51\text{ m})(300.0\text{ N})\sin(0°) = \boxed{0\text{ N}\cdot\text{m}}$</td></tr>
<tr><td colspan="2">CHECK and THINK
As anticipated, the largest torque occurs when the force is applied perpendicular to the radial direction and is zero when the force is applied along the radial direction.</td></tr>
</table>

44. (N) A uniform plank 6.0 m long rests on two supports, 2.5 m apart (Fig. P12.44). The gravitational force on the plank is 100 N. The left end of the plank is 1.5 m to the left of the left support, so the plank is not centered on the supports. A person is standing on the plank half a meter to the right of the right support. The gravitational force on this person is 80.0 N. How far to the right can the person walk before the plank begins to tip?

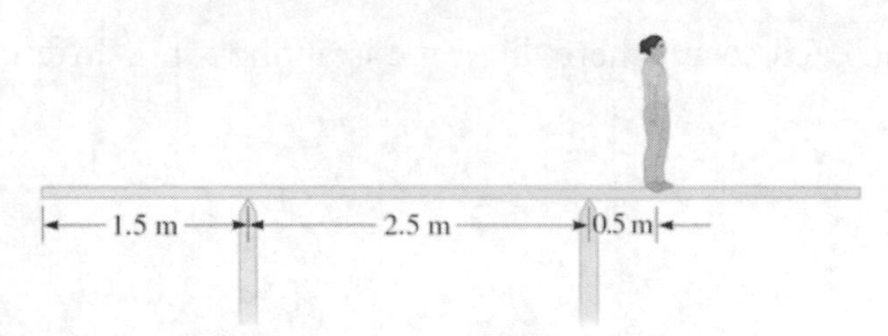

Figure P12.44

INTERPRET and ANTICIPATE

We are asked to find the condition for when the plank tips. We first sketch the situation. As the person begins to walk to the right, the plank remains at rest and the net force and net torque on the plank must be zero. At the moment that the plank just begins to tip, it will begin to pivot around the right support (pivot point P). That is, the net torque becomes non-zero and the plank rotates around point P. We want to find the condition at this moment in time, just before it rotates. At this instant, the contact force between the left support and the plank will become zero as it starts to lift and the only forces which contribute to the torque around the right support are the gravitational forces on the plank and the person. (The force of the right support on the plank does not contribute to the torque *around* this support because the moment arm is zero and the force is exerted at $r = 0$.) Therefore, the condition for the plank to just become unstable is that the torque around the right hand support due to the weight of the person and the weight of the plank must be equal and opposite such that the net torque is zero.

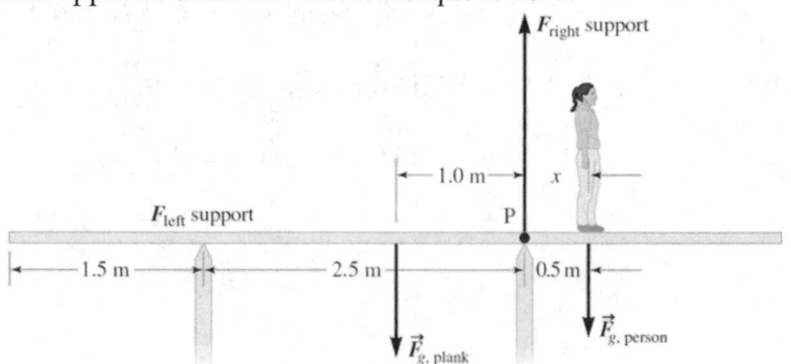

Figure P12.44ANS

SOLVE For the moment in time the plank begins to rotate, the net torque is zero and the net contact force from the support on the left is zero.	$\tau_{\text{tot}} = \tau_{\text{plank}} + \tau_{\text{person}} = 0$

These two torques must have equal magnitudes with opposite signs. The center of mass of the plank is 1.0 meter to the left of the support on the right and we indicate the distance of the person from this support as x.	$(100\text{ N})(1.0\text{ m}) = (80\text{ N})(x\text{ m})$ $x = 1.25\text{ m}$ $x \approx \boxed{1.3\text{ m to the right of the support}}$

CHECK and THINK

The person can walk an additional 0.8 meters from the initial position of 0.5 meters before the plank tips. Notice that if the person weighed less than 80 N, the distance x calculated would be larger. The lighter the person, the further he or she could walk along the plank before it tips, which makes sense.

52. (N) Given a vector $\vec{A} = 4.5\hat{i} + 4.5\hat{j}$ and a vector $\vec{B} = -4.5\hat{i} + 4.5\hat{j}$, determine the magnitude of the cross product of these two vectors, $\vec{A} \times \vec{B}$. *Hint*: Make a sketch of both vectors including a coordinate system.

INTERPRET and ANTICIPATE

We sketch the vectors as recommended and see that there is a 90° angle between the directions defined by these two vectors.

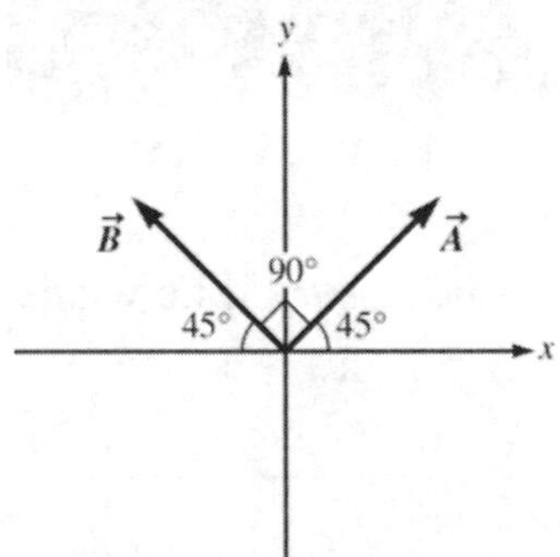

Figure P12.52ANS

SOLVE We can calculate the cross product with Equation 12.22.	$R = AB\sin\varphi$
The magnitude of each vector can be found.	$A = \sqrt{(4.5)^2 + (4.5)^2} = 6.36$ $B = \sqrt{(-4.5)^2 + (4.5)^2} = 6.36$

From our sketch, the angle between the vectors is 90 degrees. We plug values in to calculate the cross product. We keep two significant figures based on the precision of the original values.	$R = (6.36)(6.36)\sin(90°) = \boxed{41}$
CHECK and THINK Given the magnitudes and angle between the vectors, we can determine the cross product. An angle of 90 degrees results in the maximum cross product for two given vectors.	

56. Disc Jockeys (DJs) use a turntable in applying their trade, often using their hand to speed up or slow down a disc record so as to produce a desired change in the sound (Fig. P12.56). Suppose DJ Trick wants to slow down a record initially rotating clockwise (as viewed from above) with an angular speed of 33.0 rpm to an angular speed of 22.0 rpm. The record has a rotational inertia of $0.012\ \text{kg} \cdot \text{m}^2$ and a radius of 0.15 m.

Figure P12.56

a. (N) What angular acceleration is necessary if he wishes to accomplish this feat in exactly 0.65 s with a constant acceleration?

INTERPRET and ANTICIPATE The angular acceleration is equal to the change in angular velocity over time. The direction should be opposite the direction of the angular velocity vector since the record is slowing down.	
SOLVE First convert the velocities to rad/s.	$\omega_i = (33.0\ \text{rev/min}) \cdot \left(\frac{2\pi\ \text{rad}}{1\ \text{rev}}\right) \cdot \left(\frac{1\ \text{min}}{60\ \text{s}}\right) = 3.46\ \text{rad/s}$ $\omega_f = (22.0\ \text{rev/min}) \cdot \left(\frac{2\pi\ \text{rad}}{1\ \text{rev}}\right) \cdot \left(\frac{1\ \text{min}}{60\ \text{s}}\right) = 2.30\ \text{rad/s}$

The angular velocity vectors would point down (viewed from above), since the record is rotating clockwise ($-\hat{k}$).	$\vec{\omega}_i = -3.46\hat{k}$ rad/s $\vec{\omega}_f = -2.30\hat{k}$ rad/s
Now, the angular acceleration can be calculated with Equation 12.5.	$\vec{\alpha}_{av} = \frac{\Delta\vec{\omega}}{\Delta t}$ $\vec{\alpha}_{av} = \frac{(-2.30\hat{k}\ \text{rad/s}) - (-3.46\hat{k}\ \text{rad/s})}{0.65\ \text{s}}$ $\vec{\alpha}_{av} = \boxed{1.8\hat{k}\ \text{rad/s}^2}$

CHECK and THINK
The angular acceleration is in the opposite direction of the angular velocity and found as the change in angular velocity over time.

b. (N) How many revolutions does the record go through during this change in speed?

INTERPRET and ANTICIPATE
The number of revolutions can be found by modeling this scenario as a case of constant angular acceleration. We know the initial and final angular velocities, acceleration, and time, so we can choose any of the constant acceleration formulas that contain the desired quantity, $\Delta\theta$.

SOLVE We can use Equation 12.9 to find the angular displacement given the other kinematic quantities from part (a).	$\Delta\theta = \omega_i t + \frac{1}{2}\alpha t^2$ $\omega_i = -3.46$ rad/s $\alpha = 1.8$ rad/s^2 $t = 0.65$ s
Substitute values to calculate the angular displacement. Because the record is rotating clockwise, we get a negative answer for $\Delta\theta$.	$\Delta\theta = (-3.46\ \text{rad/s})(0.65\ \text{s}) + \frac{1}{2}(1.8\ \text{rad/s}^2)(0.65\ \text{s})^2$ $\Delta\theta = -1.9$ rad
To determine the number of revolutions, we convert the answer and take the absolute value.	$\lvert\Delta\theta\rvert = \lvert -1.9\ \text{rad}\rvert \cdot \left(\frac{1\ \text{rev}}{2\pi\ \text{rad}}\right)$ $\lvert\Delta\theta\rvert = \boxed{0.30\ \text{rev}}$

CHECK and THINK
Since the angular velocity is in the negative direction, the angular displacement is also negative (clockwise when viewed from above). The record rotates about a third of a revolution as it slows, which sounds plausible.

c. (N) If DJ Trick applies a vertical force with his finger to the edge of the record, with what force must he push so as to slow the record in the above time? Assume the coefficient of kinetic friction between his finger and the record is 0.50, and ignore the mass of the finger.

INTERPRET and ANTICIPATE	
The downward force applied by the finger will be the magnitude of the normal force between the finger and the record. This causes a frictional force to be applied to the record, which will cause a torque based on Eq. 12.16, $\tau = rF_f \sin\theta$. This torque causes an angular acceleration based on Eq. 12.24, $\tau = I\alpha$.	
SOLVE Equating Equations 12.16 and 12.24, we can relate the frictional force causing the torque in the record to the angular acceleration. Once we know the frictional force, we'll be able to calculate the normal force that the DJ applies.	$\tau = rF_f \sin\theta \quad \text{and} \quad \tau = I\alpha$ $rF_f \sin\theta = I\alpha$ $F_f = \dfrac{I\alpha}{r\sin\theta}$
The frictional force is related to the magnitude of the normal force as $F_f = \mu_k F_N$, so we can write a final expression for the magnitude of the applied force (normal force).	$\mu_k F_N = \dfrac{I\alpha}{r\sin\theta}$ $F_N = \dfrac{I\alpha}{\mu_k r\sin\theta}$
If we assume the angle θ will be equal to 90° since the frictional force will be tangent to the record, as the record rotates underneath his finger, the magnitude of the applied force (normal force) can then be found.	$F_N = \dfrac{(0.012\ \text{kg/m}^2)(1.8\ \text{rad/s}^2)}{(0.50)(0.15\ \text{m})\sin 90^\circ}$ $F_N = \boxed{0.29\ \text{N}}$
CHECK and THINK The DJ applies a relatively small normal force to slow the record.	

59. A wheel initially rotating at 85.0 rev/min decelerates at the constant rate of 3.50 rad/s^2 when a braking mechanism is engaged.

a. (N) What is the time interval required to bring the wheel to a complete stop?

INTERPRET and ANTICIPATE We are provided the initial angular velocity, the final angular velocity (zero), and the acceleration. Since angular acceleration is change of angular velocity over time, we can calculate the time interval.	
SOLVE We first convert the initial angular velocity to metric units, radians per second.	$\omega_0 = \frac{85.0\text{ rev}}{1.00\text{ min}}\left(\frac{1\text{ min}}{60.0\text{ s}}\right)\left(\frac{2\pi\text{ rad}}{1.00\text{ rev}}\right)$ $\omega_0 = \frac{17\pi}{6}\text{ rad/s}$
The final angular velocity is zero and the angular acceleration given, so we use Equation 12.7.	$\omega = \omega_0 + \alpha t$
Insert the values given.	$t = \frac{\omega - \omega_0}{\alpha}$ $t = \frac{0 - \left(\frac{17\pi}{6}\text{ rad/s}\right)}{-3.50\text{ rad/s}^2}$ $t = \boxed{2.54\text{ s}}$
CHECK and THINK The time depends only on how much the angular velocity must change and the rate at which it changes.	

b. (N) What is the angle in radians through which the wheel rotates in this time interval?

INTERPRET and ANTICIPATE We can use the constant angular acceleration formulas to determine the total angular displacement. We'll list the known and unknown variables and determine which equation relates these variables.	
SOLVE First, list the kinematic variables.	$\omega_0 = \frac{17\pi}{6}\text{ rad/s}$ $\omega = 0$ $\alpha = -3.50\text{ rad/s}^2$ $t = 2.54\text{ s}$ $\Delta\theta = ?$

We have a few options, but we use Equation 12.9 to relate these variables.	$\Delta\theta = \omega_0 t + \frac{1}{2}\alpha t^2$
Plug in values and solve for the angular displacement.	$\Delta\theta = \left(\frac{17\pi}{6}\frac{\text{rad}}{\text{s}}\right)(2.54\text{ s}) + \frac{1}{2}\left(-3.50\frac{\text{rad}}{\text{s}^2}\right)(2.54\text{ s})^2$ $\Delta\theta = \boxed{11.3\text{ rad}}$
CHECK and THINK This is about two revolutions.	

61. A centrifuge used for training astronauts rotating at 0.810 rad/s is spun up to 1.81 rad/s with an angular acceleration of 0.050 rad/s^2.

a. (N) What is the magnitude of the angular displacement that the centrifuge rotates through during this increase in speed?

INTERPRET and ANTICIPATE This is constant angular acceleration motion, so we can use one of the angular kinematic equations. With the angular acceleration and initial and final angular velocity, we can determine the angular displacement.	
SOLVE We first list the kinematic variables.	$\omega = 1.81\text{ rad/s}$ $\omega_0 = 0.81\text{ rad/s}$ $\alpha = 0.050\text{ rad/s}^2$ $\Delta\theta = ?$
Equation 2.11 will allow us to calculate the angular displacement.	$\omega^2 = \omega_0^2 + 2\alpha\Delta\theta$ $\Delta\theta = \frac{\omega^2 - \omega_0^2}{2\alpha}$
Substitute values and solve.	$\Delta\theta = \frac{(1.81\text{ rad/s})^2 - (0.81\text{ rad/s})^2}{2(0.050\text{ rad/s}^2)}$ $\Delta\theta = \boxed{26.2\text{ rad}}$
CHECK and THINK The centrifuge rotates less than two times as it accelerates. Since the angular velocities and accelerations are positive, the angular displacement is also positive.	

b. (N) If the initial and final speeds of the centrifuge were tripled and the angular acceleration remained at 0.050 rad/s^2, what would be the factor by which the result in part (a) would change?

From the equation used in part (a), when the angular acceleration is constant, the displacement is proportional to the difference in the *squares* of the final and initial angular speeds. Therefore, the angular displacement would **increase by a factor of 9** if both of these speeds were tripled. Mathematically, we can triple the angular velocities to reach the same conclusion:

$$\Delta\theta_3 = \frac{(3\omega)^2 - (3\omega_0)^2}{2\alpha} = 9\left(\frac{\omega^2 - \omega_0^2}{2\alpha}\right) = 9\Delta\theta$$

64. (A) A potter's wheel rotates with an angular acceleration $\alpha = 4at^3 - 3bt^2$, where t is the time in seconds, a and b are constants, and α has units of radians per second squared. The initial position of the wheel is θ_0, and the initial angular velocity is ω_0. Obtain expressions for the angular velocity and the angular displacement of the wheel as a function of time.

INTERPRET and ANTICIPATE

The angular acceleration is not a constant for this problem. So we have to use calculus to obtain the angular velocity and angular displacement. The integral of acceleration is the change in velocity and the integral of velocity is displacement.

SOLVE Using the definition of angular acceleration, we can express the change in angular velocity as the integral of angular acceleration over time.	$\alpha = \frac{d\omega}{dt}$ $\int_{\omega_0}^{\omega} d\omega = \int_0^t \alpha dt$ $\omega - \omega_0 = \int_0^t \left(4at^3 - 3bt^2\right) dt$ $\omega - \omega_0 = 4a\frac{t^4}{4} - 3b\frac{t^3}{3}$ $\omega = \boxed{\omega_0 + at^4 - bt^3}$
Similarly, using the definition of angular velocity, we can express the change in angular position as the integral of angular velocity over time.	$\omega = \frac{d\theta}{dt}$ $\int_{\theta_0}^{\theta} d\theta = \int_0^t \omega dt$

	$\theta-\theta_0=\int_0^t\left(\omega_0+at^4-bt^3\right)dt$ $\theta-\theta_0=\omega_0 t+a\frac{t^5}{5}-b\frac{t^4}{4}$ $\theta=\boxed{\theta_0+\omega_0 t+a\frac{t^5}{5}-b\frac{t^4}{4}}$

CHECK and THINK

Though we don't have an easy way to verify the final expressions, we are able to integrate from expressions for acceleration to velocity to displacement. We could also take the first or second derivative of $\theta(t)$ and recover our expressions for $\omega(t)$ and $\alpha(t)$.

68. Lara is running just outside the circumference of a carousel, looking for her favorite horse to ride, with a constant angular speed of 1.00 rad/s. Just as she spots the horse, one-fourth of the circumference ahead of her, the carousel begins to move, accelerating from rest at 0.050 rad/s^2.

a. (N) Taking the time when the carousel begins to move as $t = 0$, when will Lara catch up to the horse?

INTERPRET and ANTICIPATE

Lara is running at a constant speed. The carousel starts from rest and accelerates uniformly. We can write expressions for both Lara and the horse and find when they reach the same position (in this case, the same angle).

SOLVE	
Let's call Lara's initial position zero. She runs at constant angular velocity around the carousel with no acceleration. We could use Equation 12.9 with zero acceleration to express her angular position over time.	$\theta_L(t)=(1.00\text{ rad/s})t$
The horse starts a quarter of the way around the circle ahead of Lara, its initial angular velocity is zero, and it accelerates at the constant angular velocity provided.	$\theta_0=\frac{1}{4}(2\pi\text{ rad})=\frac{\pi}{2}\text{ rad}$ $\omega_0=0$ $\alpha=0.050\text{ rad/s}^2$

Using Equation 12.9, we can express the angular position of the horse as a function of time.	$\theta_H(t) = \theta_0 + \omega_0 t + \frac{1}{2}\alpha t^2$ $\theta_H(t) = \frac{1}{4}(2\pi \text{ rad}) + 0 + \frac{1}{2}(0.050 \text{ rad/s}^2)t^2$
Now, set the expressions for Lara's position and the horse's position equal to each other to determine when they reach the same angular position. The quadratic equation has two solutions. At 1.64 seconds, Lara catches up to the horse. She then passes it (and potentially might even lap it), but since the horse is accelerating (and we are not told what its maximum speed is), our equation says that it will eventually accelerate faster than Lara and pass her again at 38.4 seconds. We should keep in mind though that they are at the same position whenever their positions are the same or different by a multiple of 2π. We've possibly missed additional crossings, which we'll see in part (b), since our expression assumes that angle just increases to infinite and that and angle of 2π is different from an angle of 4π, when these actually correspond to the same location on a circle.	$1.00t = \frac{\pi}{2} + 0.0250t^2$ (1) $0.0250t^2 - 1.00t + \frac{\pi}{2} = 0$ $t = \frac{1.00 \pm \sqrt{1.00^2 - 4(0.0250)1.57}}{0.050}$ $t = \boxed{1.64 \text{ s}}$ or 38.4 s

CHECK and THINK

With Lara running at constant angular velocity around the carousel and the horse accelerating uniformly, we could write equations for each of their positions and find when they are equal. At a time of 1.64 seconds, Lara catches up and is at the same position as the horse.

b. (N) Lara mistakenly passes the horse and keeps running at constant angular speed. If the carousel continues to accelerate at the same rate, when will the horse draw even with Lara again?

<table>
<tr><td colspan="2">INTERPRET and ANTICIPATE
From part (a), we see that the horse will catch up to Lara again at 38.4 seconds, so this is one correct answer for when the horse catches up to Lara. In addition, Lara is actually able to complete a loop around the carousel (while it is still gaining speed) and pass the horse a second time before 38.4 seconds elapses. In this “chase on a circle”, any difference of angle of 2π corresponds to the same position. For instance, if the horse reached an angle of 2π when Lara reaches a position of 4π, they would be at the same position, but Lara will have gone around twice while the horse has gone around once.</td></tr>
<tr><td>SOLVE
We know that Lara has overshot the horse. Using equation (1) from part (a), we look for a solution for which Lara’s position is exactly 2π greater than the horse’s position (meaning she has gone around 1 full loop more than the horse), since they will still be at the same position on the circle if their angular positions are different by any multiple of 2π.</td><td>$$1.00t = \frac{\pi}{2} + 0.0250t^2 + 2\pi \qquad (2)$$</td></tr>
<tr><td>Solving this quadratic equation, we find two more times when Lara and the horse are at the same position. So, Lara will actually pass the horse again at a time of 10.7 seconds.</td><td>$$t = \frac{1.00 \pm \sqrt{1.00^2 - 4(0.0250)7.85}}{0.050}$$
$$t = \boxed{10.7\text{ s}} \text{ or } 29.3\text{ s}$$</td></tr>
<tr><td colspan="2">CHECK and THINK
We’ve now actually found four times for which the horse and Lara will be at the same position on the circle (1.64 s, 10.7 s, 29.3 s, and 38.4 s) and there are surely others that we could find using equation (2) with other multiples of 2π. Since the motion occurs on the circle, Lara will apparently lap the horse at least once and the horse will accelerate faster than Lara and begin to lap her.</td></tr>
</table>

71. (A) Three forces are exerted on the disk shown in Figure P12.71, and their magnitudes are $F_3 = 2F_2 = 2F_1$. The disk’s outer rim has a radius of R, and the inner rim has a radius of $R/2$. As shown in the figure, $\vec{F}_1$ and $\vec{F}_3$ are tangent to the outer rim of the disk, and $\vec{F}_2$ is tangent to the inner rim. $\vec{F}_3$ is parallel to the x axis, $\vec{F}_2$ is parallel to the y

axis, and $\vec{F}_1$ makes a 45° with the negative x axis. Find expressions for the magnitude of each torque exerted around the center of the disk in terms of R and F_1.

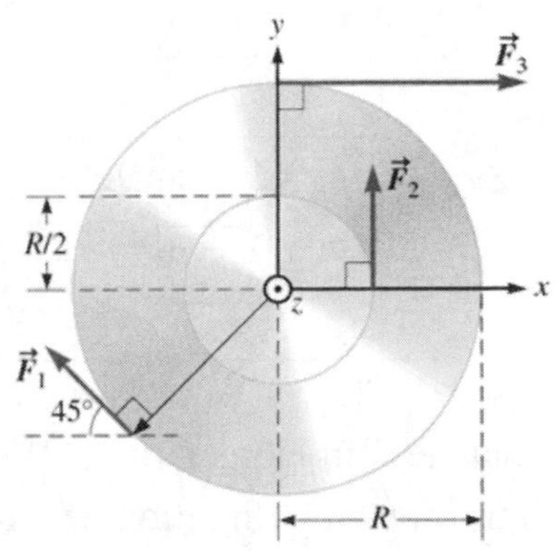

Figure P12.71

INTERPRET and ANTICIPATE This problem has a bit of a trick. All three forces are tangent to the disk, and so all three forces make a right angle with the moment arm. Don't be fooled into thinking that $\vec{F}_1$ makes a 45° with the moment arm just because an angle is specified. We expect that since $\vec{F}_3$ has the greatest magnitude and is exerted at the farthest distance from the center, it will exert the greatest torque.	
SOLVE The torque for each force depends on the force, the radial distance from the pivot point, and the angle between these two vectors (Equation 12.16).	$\tau = Fr\sin\varphi$
The angle is 90° for each force. We are also told that $F_3 = 2F_2 = 2F_1$, so we can express each force in terms of F_1.	$\varphi_1 = \varphi_2 = \varphi_3 = 90°$ $F_3 = 2F_1$ and $F_2 = F_1$
Torque due to force 1:	$\tau_1 = F_1 R\sin 90°$ $\tau_1 = \boxed{F_1 R}$

Torque due to force 2:	$\tau_2 = F_2 \frac{R}{2} \sin 90°$ $\tau_2 = \boxed{\frac{1}{2} F_1 R}$
Torque due to force 3:	$\tau_3 = F_3 R \sin 90°$ $\tau_3 = \boxed{2F_1 R}$

CHECK and THINK

We just found $\tau_3 > \tau_1 > \tau_2$; this matches our expectations. We are only asked for magnitudes, but torque 2 acts in the opposite direction of torques 1 and 3. If we were asked to write these as vectors, we would use the right hand rule to find that torque 2 is in the positive z direction while the others are in the negative z direction.

76. Suppose the angular displacement of a rotating object is given by the equation $\Delta\theta(t) = (-5.3 \text{ rad/s}^3)t^3 + (7.0 \text{ rad/s}^2)t^2 + (8.1 \text{ rad/s})t$.

a. (N) Find a time when the object is momentarily not rotating.

INTERPRET and ANTICIPATE

We are given an expression for an angular position and basically asked when the angular velocity is zero (not rotating). This might occur "momentarily" because the acceleration might be non-zero, even if the angular velocity is zero (similar to the way a ball stops momentarily at the top of its trajectory after being tossed upwards). We can find an expression for angular velocity by taking the derivative of the position versus time and then set it equal to zero to find times for which the object is not rotating.

SOLVE First, take the time derivative of the angular displacement (Equation 12.4).	$\omega = \frac{d\theta}{dt}$ $\omega = \frac{d}{dt}\left[(-5.3 \text{ rad/s}^3)t^3 + (7.0 \text{ rad/s}^2)t^2 + (8.1 \text{ rad/s})t\right]$ $\omega = (-16 \text{ rad/s}^3)t^2 + (14 \text{ rad/s}^2)t + (8.1 \text{ rad/s})$
Now, we set this equal to zero and solve for the times when the angular velocity is zero. We are only asked to find one time for which the velocity is zero, so we could choose the	$\omega = (-16 \text{ rad/s}^3)t^2 + (14 \text{ rad/s}^2)t + (8.1 \text{ rad/s}) = 0$ $t = \frac{-14 \pm \sqrt{(14)^2 - (4)(-16)(8.1)}}{2(-16)}$ $t = -0.40 \text{ s}$ and $\boxed{1.3 \text{ s}}$

positive root, $t = 1.3$ s, though either satisfies the goal of producing a zero angular velocity.	

CHECK and THINK

We've successfully determined the angular velocity as the derivative of the angle versus time and set it equal to zero to find times when the object is not rotating. As typical when we have a quadratic equation, we have two possible answers. Frequently, we have problems in which we consider $t = 0$ as the starting point, in which case a time of –0.40 s would not be physically meaningful. However, we are not told that, for instance, that the objects starts to move at $t = 0$, so –0.40 s would also be a possible answer.

b. (C) Determine whether the object is beginning to turn clockwise or counterclockwise. (Let clockwise be indicated by a negative angular velocity.)

One approach would be to determine the angular velocity just after the time calculated in part (a) and see which way it begins moving. For instance, if you choose a time of 1.4 seconds and insert it into the expression from part (a), you would calculate a negative angular velocity:

$$\omega = \left(-16\ \text{rad/s}^3\right)\left(1.4\ \text{s}\right)^2 + \left(14\ \text{rad/s}^2\right)\left(1.4\ \text{s}\right) + \left(8.1\ \text{rad/s}\right) = -3.7\ \text{rad/s}$$

which means that the wheel has gone from rest (at 1.3 seconds) to clockwise (at 1.4 seconds). Therefore, it is beginning to turn clockwise. Of course, in general we need to be careful that the time we choose really is "just after" it stops and that it has not changed directions a couple times in the meantime, but since we found in part (a) the only two times when the object is at rest, we should be able to choose *any* time after $t = 1.3$ seconds and reach the same conclusion.

Another way to approach this, since we want to know how it's changing *at that moment* in time, is to take one more derivative and calculate the angular acceleration, which tells us how the angular velocity is changing. So, using Equation 12.6, we can find an equation for the angular acceleration as a function of time.

$$\alpha = \frac{d\omega}{dt} = \frac{d}{dt}\left[\left(-16\ \text{rad/s}^3\right)t^2 + \left(14\ \text{rad/s}^2\right)t + \left(8.1\ \text{rad/s}\right)\right]$$

$$\alpha = \left(-32\ \text{rad/s}^3\right)t + \left(14\ \text{rad/s}^2\right)$$

Now we plug in the time from part (a) to see what the angular acceleration is at this time when the angular speed is 0.

$$\alpha = (-32 \text{ rad/s}^3)(1.3 \text{ s}) + (14 \text{ rad/s}^2)$$
$$\alpha = -28 \text{ rad/s}^2$$

Since the angular acceleration is negative, the object is "accelerating clockwise," meaning that the object will begin to rotate clockwise immediately after briefly stopping. This is consistent with what we found above.

81. (N) A uniform 4.55-kg horizontal rod is fixed on one end as shown in Figure P12.79. The rod is 1.75 m long. A rope is attached to the opposite end, and it makes a 30.0° angle with respect to the rod. If the rod is in equilibrium, what is the tension in the rope at the instant shown?

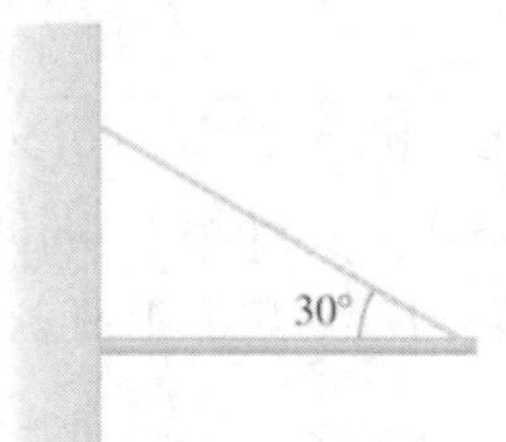

Figure P12.79

INTERPRET and ANTICIPATE There are two forces exerted on the rod that contribute to the torque about the point where the rod is fixed to the wall: the tension force and the gravitational force. (Though the wall at the fixed end clearly exerts a force, this force acts at the pivot point, so $r = 0$ and it does not contribute to the torque around this point.) We must find the torque exerted by each of these forces using Equation 12.16 and use the fact that the net torque on the rod must be 0 since the rod is in equilibrium.	
SOLVE We first express the torque due to the tension of the rope (Eq. 12.16). Using the right hand rule, we find the direction as the positive z direction out of the page.	$\tau_T = RF_T \sin(180° - 30°)$ $\tau_T = (1.75 \text{ m})(F_T)\sin 150°$ $\vec{\tau}_T = (1.75 \text{ m})(F_T)\sin 150° \hat{k}$

Now, we calculate the torque due to gravity acting at the center of mass (acting at a distance from the wall of half the length of the rod). By the right hand rule, this torque points into the page in the negative z direction.	$\tau_g = r_{\text{CM}} F_g \sin 90° = r_{\text{CM}} mg$ $\tau_g = \left(\frac{1.75\ \text{m}}{2}\right)(4.55\ \text{kg})(9.81\ \text{m/s}^2) = 39.1\ \text{N}\cdot\text{m}$ $\vec{\tau}_g = -39.1\,\hat{k}\ \text{N}\cdot\text{m}$
The total torque is the vector sum of these two torques and must be equal to 0 due to the equilibrium condition. We can then solve for the tension in the rope.	$\vec{\tau}_{\text{total}} = \vec{\tau}_T + \vec{\tau}_g = 0$ $\vec{\tau}_{\text{total}} = (1.75\ \text{m})(F_T)\sin 30°\hat{k} - 39.1\,\hat{k}\ \text{N}\cdot\text{m} = 0$ $F_T = \frac{39.1\ \text{N}\cdot\text{m}}{(1.75\ \text{m})\sin 30°} = \boxed{44.6\ \text{N}}$
CHECK and THINK Now that the tension is known, the hinge force (the wall pushing on the rod) between the rod and the wall could be found using the fact that the sum of the forces in each direction must be 0. Note that because the rod is in equilibrium, the total, or net, torque about any point on the rod must be equal to 0. This means we did not have to write the torque about the point where the rod is attached to the wall. However, if we were to choose to use a different point on the rod, we would have to include the torque due to the hinge force about this new point. It is advantageous to write the torque about a point where one of the forces is unknown because it will then not appear in the equilibrium condition for torque.	

13

Rotation II: A Conservation Approach

5. (N) A system consists of four particles connected by very lightweight, stiff rods (Fig. P13.5). The system rotates around the z axis, which points out of the page. Each particle has a mass of 5.00 kg. The distances from the z axis to each particle are $r_1 = 32.0$ cm, $r_2 = 16.0$ cm, $r_3 = 17.0$ cm, and $r_4 = 34.0$ cm. Find the rotational inertia of the system around the z axis.

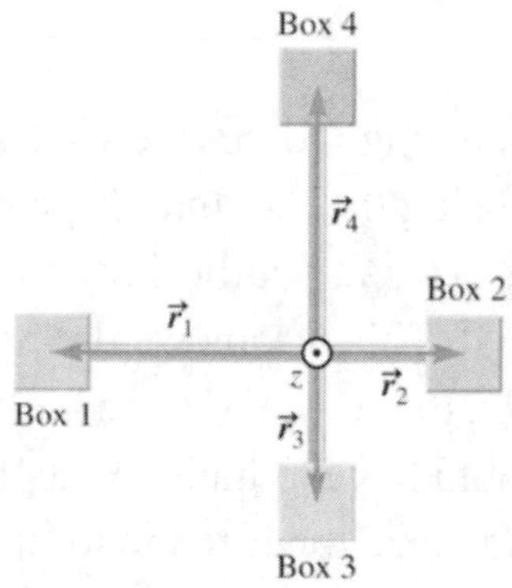

Figure P13.5

INTERPRET and ANTICIPATE	
The rotational inertia depends on the distribution of the mass around the rotation axis, so by knowing the positions of all the masses, it's straightforward to calculate. The problem provides us with a sketch and coordinate system. We expect to find a numerical answer in the form _____ kg·m^2.	
SOLVE We model the system as four particles, and find the rotational inertia using Equation 13.4.	$I = \sum_{i=1}^{n} m_i r_i^2$ $I = m_1 r_1^2 + m_2 r_2^2 + m_3 r_3^2 + m_4 r_4^2$
All the particles have the same mass.	$m_1 = m_2 = m_3 = m_4 \equiv m$

So the expression for I is simplified and we can calculate the answer.	$I = m\left(r_1^2 + r_2^2 + r_3^2 + r_4^2\right)$ $I = (5.00\text{ kg})\left((0.320\text{ m})^2 + (0.160\text{ m})^2 + (0.170\text{ m})^2 + (0.340\text{ m})^2\right)$ $I = \boxed{1.36\text{ kg}\cdot\text{m}^2}$
CHECK and THINK Our answer is in the form we expected.	

6. (N) Use the information in Problem 5 to find the rotational inertia of the system around particle 1.

INTERPRET and ANTICIPATE We can follow the same steps as in Problem 5, but the distances from the rotation axis change. In the previous problem, the axis of rotation is closer to the center of mass than it is in this problem. So in this case, the average location of the system's mass is farther from the axis of rotation and we expect the rotational inertia to be larger.	
SOLVE We model the system as four particles, and find the rotational inertia using Equation 13.4. All the particles have the same mass.	$I = \sum_{i=1}^{n} m_i r_i^2$ $m_1 = m_2 = m_3 = m_4 \equiv m$ $I = m\sum_{i=1}^{n} r_i^2$
Here the distance r must be calculated for each mass from the information in the figure. Mass 1 is at the rotation axis ($r = 0$), mass 2 is at a distance $r = r_1 + r_2$, mass 3 and mass 4 are at distances where we need to use the Pythagorean theorem and are $r = \sqrt{r_1^2 + r_3^2}$ and $r = \sqrt{r_1^2 + r_4^2}$ respectively.	$I = m\left(0 + (r_1 + r_2)^2 + \left(r_1^2 + r_3^2\right) + \left(r_1^2 + r_4^2\right)\right)$ $I = (5.00\text{ kg})\left((0.480\text{ m})^2 + (0.320\text{ m})^2 + (0.170\text{ m})^2 + (0.320\text{ m})^2 + (0.340\text{ m})^2\right)$ $I = \boxed{2.90\text{ kg}\cdot\text{m}^2}$
CHECK and THINK Our answer is in the form we expected. The system has a larger rotational inertia when it is rotated around particle 1 than it had when it was rotated around the z axis.	

10. (A) Suppose a disk having mass M_{tot} and radius R is broken into four equal parts (Fig. P13.10). What is the rotational inertia of one-fourth of the disk around the z axis shown?

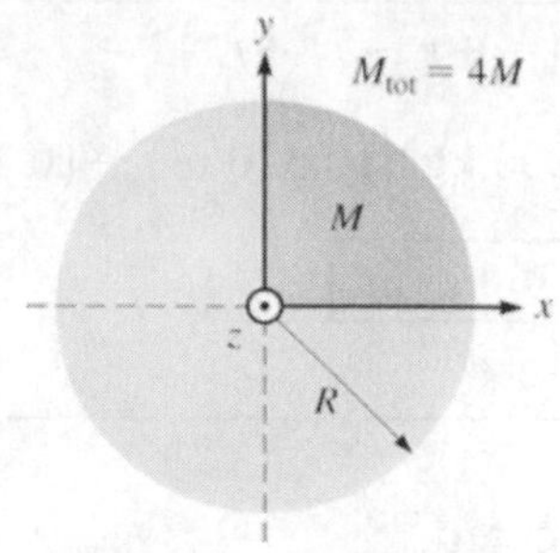

Figure P13.10

INTERPRET and ANTICIPATE We will use symmetry and the moment of inertial of a full disk about the center of mass to solve this problem. The four pieces are equivalent and must add up to the total for a disk.	
SOLVE We first write the moment of inertia of a full disk of mass M_{tot} about the z axis passing through its center of mass using Table 13.1.	$I_{disk} = \frac{1}{2} M_{tot} R^2$
Because of symmetry, each quarter will have same moment of inertia $I_{quarter}$ which add up to the total value. The mass of each piece (M) is also one quarter of the total mass ($4M$).	$I_{quarter} = \frac{1}{4} I_{disk} = \frac{1}{4}\left(\frac{1}{2} M_{tot} R^2\right) = \frac{1}{8} M_{tot} R^2$ $I_{quarter} = \frac{1}{8}(4M)R^2 = \boxed{\frac{1}{2} MR^2}$
CHECK and THINK In this case, we used the symmetry to easily find the moment of inertia of the quarter disk. In fact, it turns out to have the same form as that for the disk, but the mass in the formula is the mass of one quarter of the disk. That makes sense actually because only the distribution of mass *from the rotation axis* matters for the rotational inertia and not how it's spread *around* the rotation axis. That is, a disk or half a disk or quarter of disk, rotated around the center of the original disk, has the same radial distribution of mass, so only the total mass matters.	

16. The net total torque of 50.0 N·m on a wheel rotating around an axis through its center is due to an applied force and a frictional torque at the axle. Starting from rest, the wheel reaches an angular speed of 12.0 rad/s in 5.00 s. At $t = 5.00$ s, the applied force is removed, and the frictional torque brings the wheel to a stop in 30.0 s.

a. (N) What is the rotational inertia of the wheel?

INTERPRET and ANTICIPATE

The rotational inertia of the wheel depends on the distribution of mass around the rotation axis, but we're not given information about this. Another way is to use the fact that the rotational inertia relates the torque on an object to its angular acceleration ($\tau = I\alpha$ for rotational motion is similar to $F = ma$ for linear motion). We can find the angular acceleration and torque and solve for the rotational inertia.

SOLVE	
The net torque is related to the angular acceleration by the moment of inertia (Eq. 12.24). The angular acceleration depends on the change in angular velocity in time, which is information that's provided.	$\sum \tau = I\alpha$ $\alpha = \dfrac{\Delta\omega}{\Delta t}$
Using these relationships, we can calculate the moment of inertia.	$I = \dfrac{\sum \tau}{\alpha} = \dfrac{\sum \tau}{\Delta\omega}\Delta t$ $I = \dfrac{50.0\ \text{N}\cdot\text{m}}{12.0\ \text{rad/s}}(5.00\ \text{s}) = \boxed{20.8\ \text{kg}\cdot\text{m}^2}$

CHECK and THINK

Here we have another way to find the rotational inertia, by considering the dynamics of the wheel.

b. (N) What is the magnitude of the frictional torque acting on the wheel?

INTERPRET and ANTICIPATE

We now have the moment of inertia from part (a) and the deceleration, so we can calculate the torque that caused the deceleration.

SOLVE	
Again using Eq.12.24, we can determine the magnitude of the torque due to friction that results in the angular acceleration of this moment of inertia.	$\lvert\tau_f\rvert = \lvert I\alpha\rvert = \left\lvert I\dfrac{\Delta\omega}{\Delta t}\right\rvert =$ $\lvert\tau_f\rvert = \left\lvert(20.8\text{kg}\cdot\text{m}^2)\dfrac{-12.0\ \text{rad/s}}{30.0\text{s}}\right\rvert$ $\lvert\tau_f\rvert = \boxed{8.33\text{N}\cdot\text{m}}$

CHECK and THINK
The answer has the right units and is fairly straightforward after completing part (a).

c. (N) What is the total number of revolutions the wheel undergoes during this 35.0-s interval?

INTERPRET and ANTICIPATE	
Just like with linear kinematics, where we might specify the acceleration of an object, it's velocity, and time and add for the total displacement, we can perform the same time of calculation for rotational motion. Since the deceleration takes longer, we might expect that the displacement as the wheel slows down is larger than that when it speeds up.	
SOLVE Since the wheel accelerates at a constant rate, the total angular displacement is the average velocity times the time that it accelerates.	$\Delta\theta = \omega_{avg}\Delta t$ $\Delta\theta = \left(\frac{\omega_i + \omega_f}{2}\right)\Delta t$ $\Delta\theta = \left(\frac{0 + 12.0\text{ rad/s}}{2}\right)(5.00\text{ s})$ $\Delta\theta = 30.0\text{ rad}$
We can repeat this calculation to determine the rotation angle for the period of time when the wheel decelerates to a stop.	$\Delta\theta = \omega_{avg}\Delta t$ $\Delta\theta = \left(\frac{\omega_i + \omega_f}{2}\right)\Delta t$ $\Delta\theta = \left(\frac{12.0\text{ rad/s} + 0}{2}\right)(30.0\text{ s})$ $\Delta\theta = 180\text{ rad}$
The total angular displacement for both intervals is the sum, or 210 rad, which we can convert to revolutions.	$\Delta\theta_{tot} = 30.0\text{ rad} + 180\text{ rad} = 210\text{ rad}$ $210\text{ rad}\left(\frac{1\text{ rev}}{2\pi\text{ rad}}\right) = \boxed{33.4\text{ rev}}$
CHECK and THINK It covers less angular displacement while speeding up than it does when slowing down.	

19. (N) A 10.0-kg disk of radius 2.0 m rotates from rest as a result of a 20.0-N tangential force applied at the edge of the disk. What is the kinetic energy of the disk 4.00 s after the force is applied?

INTERPRET and ANTICIPATE	
To calculate kinetic energy, we can use the moment of inertia and angular velocity as seen in Eq. 13.10, $K=\frac{1}{2}I\omega^2$. The moment of inertia can be calculated using the mass and radius of the disk while the angular velocity can be calculated by figuring out how quickly the force accelerates the disk in the given time.	
SOLVE The constant torque depends on the force and distance from the rotation axis as $\tau = FR$ and produces an angular acceleration that depends on the moment of inertia according to Eq. 12.24.	$\sum\tau = I\alpha \quad \rightarrow \quad \alpha = \frac{FR}{I}$
We can use the moment of inertia for a disk (Table 13.1).	$I = \frac{1}{2}MR^2$ $\alpha = \frac{FR}{\frac{1}{2}MR^2} = \frac{2F}{MR}$
If the disk starts from rest, the angular velocity increases linearly with time.	$\omega = \omega_i + \alpha t = \alpha t$
We now apply the definition of rotational kinetic energy (Equation 13.10).	$K = \frac{1}{2}I\omega^2$ $K = \frac{1}{2}I\alpha^2t^2$ $K = \frac{1}{2}\left(\frac{1}{2}MR^2\right)\left(\frac{2F}{MR}\right)^2 t^2$ $K = \frac{F^2t^2}{M}$
Finally, inserting numerical values, we determine the kinetic energy.	$K = \frac{(20.0\ \text{N})^2(4.00\ \text{s})^2}{(10.0\ \text{kg})} = \boxed{6.40\times10^2\ \text{J}}$
CHECK and THINK Similar to linear kinetic energy, which depends on the mass and velocity squared, the rotational kinetic energy depends on the moment of inertia and angular velocity squared, which are quantities we can determine. The final answer is in Joules, as we expect.	

24. (N) When a pitcher throws a baseball or softball, the rotation speed is important when it comes to making the ball "break," or curve away from its expected motion. A baseball that has a mass of 0.143 kg and a radius of 3.65 cm can rotate at a rate of 1800 rev/min on its way to home plate, causing the ball to curve! Suppose this baseball is rotating around an axis through its center at this extreme rate. What is the rotational kinetic energy of the baseball as it travels toward home plate?

INTERPRET and ANTICIPATE	
The rotational kinetic energy is given by Eq. 13.10. In order to make use of it though, we must find the rotational inertia of the ball and its angular speed. The angular speed is given, but must be converted to rad/s. The radius of the ball is 0.0365 m in SI units. The rotational inertia of a solid sphere is $I_{sphere} = \frac{2}{5}MR^2$, as shown in Table 13.1.	
SOLVE First, convert the angular speed to rad/s.	$\omega = 1.80\times10^3 \text{ rpm}\times\left(\frac{2\pi \text{ rad}}{1 \text{ rev}}\right)\times\left(\frac{1 \text{ min}}{60 \text{ s}}\right)$ $\omega = 60.0\pi \text{ rad/s}$
Then, write Eq. 13.10, using the rotational inertia for the solid sphere.	$K_r = \frac{1}{2}I\omega^2 = \frac{1}{2}\left(\frac{2}{5}MR^2\right)\omega^2 = \frac{1}{5}MR^2\omega^2$
Now, substitute the values and calculate the rotational kinetic energy.	$K_r = \frac{1}{5}(0.143 \text{ kg})(0.0365 \text{ m})^2(60.0\pi \text{ rad/s})^2$ $K_r = \boxed{1.35 \text{ J}}$
CHECK and THINK The total kinetic energy of the ball is the sum of the rotational kinetic energy, found here, and the translational kinetic energy of the ball, $K_t = \frac{1}{2}mv^2$. An object's rotational kinetic energy depends on the distribution of the mass about the object's axis of rotation, whereas the translational kinetic energy does not.	

25. (N) In the fall of 2003, Sanyo announced that it could make 10 compact disks (CDs) from one ear of corn. Each CD has a mass of 16.3 g and a radius $R = 6.00$ cm. From Problem 12.82 (page 359), when information is being read off the innermost ring of a CD, its angular speed is $\omega_0 = 52.4 \text{ rad/s}$. The CD slows down within the player so that when information is read off the outermost ring, $\omega = 20.9 \text{ rad/s}$. Assume a Sanyo corn CD can be modeled as a solid disk rotating around its central axis. Find the change in its rotational kinetic energy between the beginning and end of its play cycle.

INTERPRET and ANTICIPATE	
Our job is to find a change in kinetic energy, which depends on the moment of inertia and angular velocity. Since the disk slows down, we expect the change in kinetic energy should be negative.	
SOLVE The rotational kinetic energy can be calculated using Equation 13.10 using the moment of inertia and angular velocity.	$K_r = \frac{1}{2}I\omega^2$
We start by finding the CD's rotational inertia from Table 13.1.	$I = \frac{1}{2}MR^2$ $I = \frac{1}{2}(16.3\times10^{-3}\text{ kg})(6.00\times10^{-2}\text{ m})^2$
Now find the initial and final kinetic energies.	$K_i = \frac{1}{2}I\omega_0^2$ $K_i = \frac{1}{2}\left[\frac{1}{2}(16.3\times10^{-3}\text{ kg})(6.00\times10^{-2}\text{ m})^2\right](52.4\text{ rad/s})^2$ $K_i = 4.03\times10^{-2}\text{ J}$ $K_f = \frac{1}{2}I\omega^2$ $K_f = \frac{1}{2}\left[\frac{1}{2}(16.3\times10^{-3}\text{ kg})(6.00\times10^{-2}\text{ m})^2\right](20.9\text{ rad/s})^2$ $K_f = 6.41\times10^{-3}\text{ J}$
Subtract to find the change in rotational kinetic energy.	$\Delta K = K_f - K_i = 6.41\times10^{-3}\text{ J} - 4.03\times10^{-2}\text{ J}$ $\Delta K = \boxed{-3.39\times10^{-2}\text{ J}}$
CHECK and THINK As expected, the CD loses kinetic energy as it plays.	

30. A bicycle and its rider have a mass of 85.6 kg. The bike's tires have a radius of 0.382 m. The mass of each tire is 0.980 kg and is concentrated near each tire's rim. The rider is riding at a constant speed of 5.40 m/s.

a. (N) Find the kinetic energy of the center of mass.

b. (N) Find the rotational kinetic energy of the tires.

c. (C) Compare and comment on your results.

Chapter 13 – Rotation II: A Conservation Approach

<table>
<tr><td colspan="2">INTERPRET and ANTICIPATE
Finding the kinetic energy of the center of mass and of the two tires is fairly straightforward, and we expect to find two numerical results in joules. Comparing the results is interesting. Often we ignore the rotation energy of the wheels of objects such as cars, bikes and other vehicles. This problem gives us a chance to see if that assumption is reasonable.</td></tr>
<tr><td>SOLVE
a. The kinetic energy comes from the system's mass and center of mass speed.</td><td>$K_{CM} = \frac{1}{2}mv_{CM}^2$
$K_{CM} = \frac{1}{2}(85.6 \text{ kg})(5.40 \text{ m/s})^2$
$K_{CM} = \boxed{1.25\times10^3 \text{ J}}$</td></tr>
<tr><td>b. The rotational kinetic energy is given by Equation 13.10. Since there are two tires, we multiply by two.</td><td>$K_r = \frac{1}{2}I\omega^2$
$K_{tires} = 2K_r = I\omega^2$</td></tr>
<tr><td>The tires are modeled as hoops, and their rotational inertia comes from Table 13.1. Since they roll with outside slipping, we can relate the angular and linear velocity with Equation 13.12.</td><td>$I = MR^2$
$v_{CM} = \omega R$</td></tr>
<tr><td>Now the kinetic energy of the tires can be written in terms of their center of mass speed and calculated.</td><td>$K_{tires} = I\omega^2 = (MR^2)\left(\frac{v_{CM}}{R}\right)^2 = Mv_{CM}^2$
$K_{tires} = (0.980 \text{ kg})(5.40 \text{ m/s})^2 = \boxed{28.6 \text{ J}}$</td></tr>
<tr><td>c. Such a comparison is perhaps best made by finding a ratio. Here we find the ratio of the kinetic energy of the tires to the kinetic energy of the center of mass. The kinetic energy in the tires is only 2.3% of the center of mass kinetic energy. It seems safe to ignore this energy when we ignore the rotation energy of the wheels of objects such as cars, bikes and other vehicles in problems.</td><td>$\frac{K_{tires}}{K_{CM}} = \frac{28.6 \text{ J}}{1.25\times10^3 \text{ J}} = 2.3\times10^{-2} = \boxed{0.023}$</td></tr>
</table>

CHECK and THINK
This makes sense since we often do ignore the rotational energy of tires on bicycles and cars. If we found that the kinetic energy of the tires was significant, we'd have to worry about making this approximation!

33. A spring with spring constant 25 N/m is compressed a distance of 7.0 cm by a ball with a mass of 202.5 g (Fig. P13.33). The ball is then released and rolls without slipping along a horizontal surface, leaving the spring at point *A*. The process is repeated, using a block instead, with a mass identical to that of the ball. The block compresses the spring by 7.0 cm and is also released, leaving the spring at point *A*. Assume the ball rolls, but ignore other effects of friction.

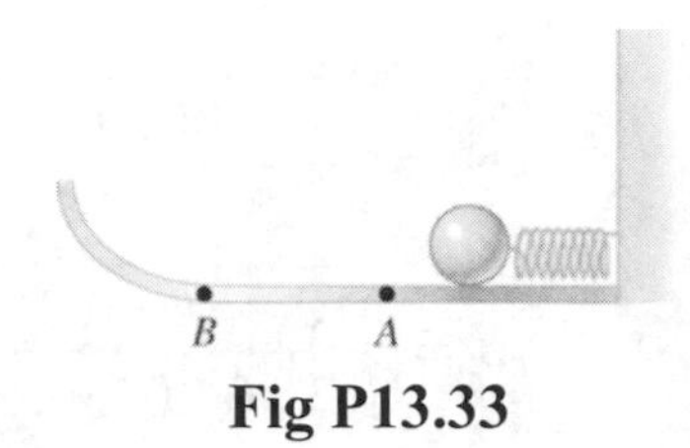

Fig P13.33

a. (N) What is the speed of the ball at point *B*?

INTERPRET and ANTICIPATE	
Because a spring is included in the problem, we should include it in our system along with the ball. Energy is conserved in the system, because the only external force is friction and it is negligible. Using a conservation approach, the total energy of the system is initially only spring potential energy that is converted to translational *and rotational* kinetic energy. This is not a very stiff spring, nor is it compressed by a large amount, so we should expect that the final linear speed of the ball will be relatively small. We are also not given the radius of the ball, so the answer must not actually depend on the size of the ball. (In truth, the ball won't actually start rolling if there was *no* friction, but we assume it's negligible as indicated in the problem.)	
SOLVE First we find the total energy of the system at point *A* by calculating the spring potential energy. This is equal to the energy of the ball at point *B*.	$E_A = \frac{1}{2}kx^2$ $E_A = \frac{1}{2}(25 \text{ N/m})(0.07 \text{ m})^2$ $E_A = 0.0061 \text{ J} = E_B$

At point B, the system has rotational and translational energies. The rotational inertia for a sphere is found from Table 13.1, and we also find use the rolling condition, Equation 13.12.	$E_B = \frac{1}{2}mv^2 + \frac{1}{2}I\omega^2$ $I = \frac{2}{5}mr^2$ $\omega = \frac{v}{r}$
Now, find an expression for the energy at point B and set it equal to point A.	$E_B = \frac{1}{2}mv^2 + \frac{1}{2}\left(\frac{2}{5}mr^2\right)\left(\frac{v}{r}\right)^2$ $E_B = \frac{7}{10}mv^2 = 0.061\text{ J}$
Now, solve for the linear speed.	$v^2 = \frac{10}{7}\cdot\frac{0.061\text{ J}}{m}$ $v = \sqrt{\frac{10}{7}\cdot\frac{0.061\text{ J}}{0.2025\text{ kg}}}$ $v = \boxed{0.66\text{ m/s}}$

CHECK and THINK

The linear speed of the ball is not very great. Part of the energy imparted to the ball by the spring has been transferred to rotational kinetic energy and part to translational kinetic energy.

b. (N) What is the speed of the block at point B?

INTERPRET and ANTICIPATE

Replacing the ball with the block will not change the amount of energy stored in the spring, nor will it alter the fact that mechanical energy is conserved (since friction is being neglected). At point B, however, the system now only has translational motion. Because none of the energy has been transferred to rotational energy, we should expect that the linear speed of the block will be larger than that of the ball.

SOLVE The total energy is initially all spring potential energy, which we found in part (a).	$E_A = 0.061\text{ J} = E_B$

The total energy at point A is purely translational kinetic energy.	$E_B = \frac{1}{2}mv^2 = 0.061\ \text{J}$
Once again we solve for the linear speed.	$v = \sqrt{2\frac{0.061\ \text{J}}{0.2025\ \text{kg}}}$ $v = \boxed{0.78\ \text{m/s}}$
CHECK and THINK The ball has a lower translational velocity in part (a), because the spring energy is split between rotational and translational energy, and a higher translational velocity in part (b), since all of the spring energy goes into translational energy for the block.	

38. A merry-go-round at a park is subject to a constant torque with a magnitude of 645 N · m as a parent pushes the ride.

a. (N) How much work is performed by the torque as the merry-go-round rotates through 1.75 revolutions?

INTERPRET and ANTICIPATE When a torque acts on an object through an angular distance, work is performed on the object, given by $W = \int \tau\, d\theta$. When the torque is constant, the work performed should end up being equal to the product of the torque and the angular distance through which the torque is applied.	
SOLVE To begin, we first convert the number of revolutions to radians.	$1.75\ \text{rev} \times \left(\frac{2\pi\ \text{rad}}{1\ \text{rev}}\right) = 3.5\pi\ \text{rad}$
Then, integrate the torque over the range of the angular distance from $\theta = 0$ to $\theta = 3.5\pi$.	$W = \int \tau\, d\theta = \tau \int_0^{\theta_f} d\theta = \tau\theta_f$ $W = (645\ \text{N} \cdot \text{m})(3.5\pi\ \text{rad}) = \boxed{7.09 \times 10^3\ \text{J}}$
CHECK and THINK If the torque is not constant, it may depend on angular position, which would make the integration more difficult.	

b. (N) If it takes the merry-go-round 4.51 s to go through the 1.75 revolutions, what is the power transferred by the parent to the ride?

INTERPRET and ANTICIPATE Knowing the work from part (a) and the time over which it is performed, we can find the power by using the definition of power: The work performed per unit time.	
SOLVE The power should equal the work performed divided by the time over which it occurred.	$P = W/\Delta t = \left(7.09\times10^3\ \text{J}\right)/\left(4.51\ \text{s}\right) = \boxed{1.57\times10^3\ \text{W}}$
CHECK and THINK While there are formulae for finding the power, it is helpful to remember that fundamentally it is the work performed per unit time. If the parents pushed with a greater torque, they would perform more work over the same angular distance and their power would be greater.	

44. (A) An object of mass M is thrown with a velocity v_0 at an angle θ with respect to the horizontal (Fig. P13.44). Find the angular momentum of the object around the origin when the object is at the highest point of its trajectory.

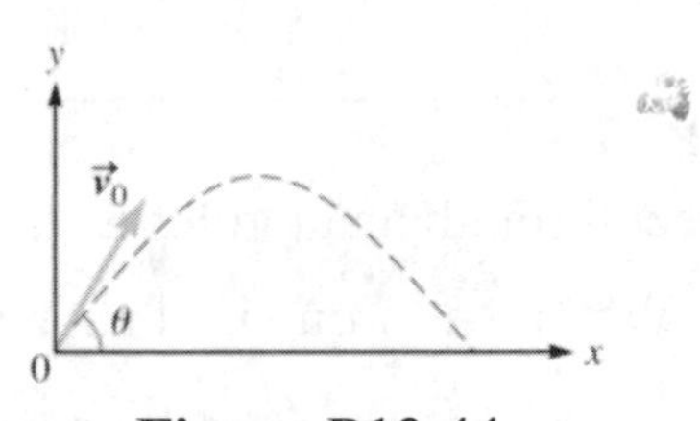

Figure P13.44

INTERPRET and ANTICIPATE

To calculate the angular momentum, we need three things: the mass of the object, radial vector from the rotation axis, and its velocity. We'll have to use kinematic equations to determine the last two. Since, like torque, angular momentum is found as the cross products of these quantities, we only need to determine the components of the radial vector and velocity that are perpendicular.

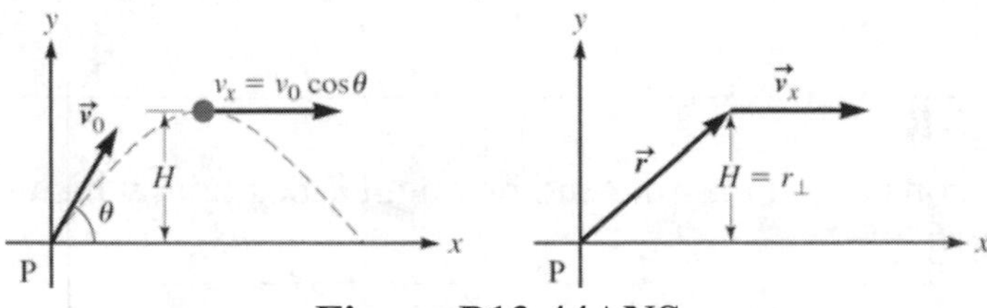

Figure P13.44ANS

SOLVE At the highest point, the particle has only horizontal velocity, equal to its initial horizontal velocity.	$v_x = v_0 \cos\theta$
The length of the perpendicular to the horizontal velocity is equal to the height *H*, which we can determine referring back to our equations for projectile motion in Chapter 4, using the fact that the final *y* velocity is zero, the initial *y* velocity is $v_0 \sin\theta$ and acceleration is due to gravity.	$v_{y,f}^2 = v_{y,i}^2 + 2ad$ $0 = (v_0 \sin\theta)^2 + 2(-g)H$ $H = \frac{v_0^2 \sin^2\theta}{2g}$
Angular momentum is the vector product of linear momentum of the particle and the length of the perpendicular to the horizontal velocity (Eq. 13.24). The magnitude of the angular momentum can also be calculated using Eq. 13.25 and the direction using the right hand rule to determine that the angular momentum is directed along the negative *z*-axis.	$\vec{L} = \vec{r} \times \vec{p} = (H\,\hat{j}) \times (p_x\,\hat{i})$ $\vec{L} = -H(mv_x)\,\hat{k} = -\frac{mv_0^3 \sin^2\theta\cos\theta}{2g}\hat{k}$ or $L = r_\perp p = H(mv_x)$ $L = \boxed{\frac{mv_0^3 \sin^2\theta\cos\theta}{2g} \text{ (in the } -\hat{k} \text{ direction)}}$

CHECK and THINK

At the highest point the velocity of the particle is horizontal, so it is easy to find the length of the perpendicular line which in this case is the maximum height reached by the particle. We used the result for the maximum height derived for the two dimensional projectile to find the angular momentum. The resulting formula is not entirely transparent, but we can find limits that make sense: if $\theta = 0$ (shot straight into the ground), the angular momentum is zero, if $\theta = 90$ (shot straight up), the angular momentum around the rotation point is zero, with a larger mass or speed, the angular momentum goes up. Those all sound reasonable.

47. (N) A cylinder of length 2.65 m, radius 0.35 m, and mass 13.7 kg is rotated at an angular speed of 3.89 rad/s around an axis parallel to the length of the cylinder and through its center. Find the magnitude of the cylinder's angular momentum.

INTERPRET and ANTICIPATE	
Finding the angular momentum involves first finding the rotational inertia. Once that is known, it is straightforward to find a numerical answer for the angular momentum of this rotating rod using the angular velocity (Eq. 13.26, $L = I\omega$).	
SOLVE The rod's rotational inertia is found in Table 13.1.	$I_{CM} = \frac{1}{12} ML^2$
Its angular momentum is given by Equation13.27.	$L = I\omega$ $L = \frac{1}{2} MR^2\omega$ $L = \frac{1}{2}(13.7\text{ kg})(0.35\text{ m})^2(3.89\text{ rad/s})$ $L = \boxed{3.26\ \frac{\text{kg}\cdot\text{m}^2}{\text{s}}}$
CHECK and THINK We probably don't have a lot of intuition about this value, but the units are correct for angular momentum.	

53. Two children (m = 30.0 kg each) stand opposite each other on the edge of a merry-go-round. The merry-go round, which has a mass of 1.80×10^2 kg and a radius of 1.5 m, is spinning at a constant rate of 0.50 rev/s. Treat the two children and the merry-go-round as a system.

a. (N) Calculate the angular momentum of the system, treating each child as a particle.

INTERPRET and ANTICIPATE	
First we need to calculate the rotational inertia for the merry-go-round and each child and add them together. We expect that the total angular momentum of the system should be a fairly large quantity compared to many other problems because, even though its not spinning particularly fast, the mass of the merry-go-round is large and the children are standing as far out as they can on the merry-go-round. (Of course, this tends to be the most fun point on a merry-go-round, so who can blame them!)	
SOLVE The angular momentum equals the moment of inertia of the system	$L = I\omega$

times the angular velocity (Eq. 13.26).	
The moment of inertia is the sum of the merry-go-round plus the two children. We calculate the rotational inertia for each child using Eq. 13.4 and the merry-go-round treating at a disk and using Table 13.1.	$I_{\text{child}} = mr^2$ $I_{\text{child}} = (30.0 \text{ kg})(1.5 \text{ m})^2 = 68 \text{ kg}\cdot\text{m}^2$ $I_{\text{mgr}} = \frac{1}{2}mr^2$ $I_{\text{mgr}} = \frac{1}{2}(180 \text{ kg})(1.5 \text{ m})^2 = 200 \text{ kg}\cdot\text{m}^2$
The total rotational inertia is the sum of the 2 children plus that of the merry go round.	$I_{\text{system}} = 2I_{\text{child}} + I_{\text{mgr}}$ $I_{\text{system}} = 2(68 \text{ kg}\cdot\text{m}^2) + 200 \text{ kg}\cdot\text{m}^2$ $I_{\text{system}} = 340 \text{ kg}\cdot\text{m}^2$
Now, express the angular velocity in radians per second and apply Eq. 13.26.	$\omega = \frac{1 \text{ rev}}{2 \text{ s}}\frac{2\pi \text{ rad}}{1 \text{ rev}} = \pi \text{ rad/s}$ $L = I\omega$ $L = (340 \text{ kg}\cdot\text{m}^2)(\pi \text{ rad/s})$ $L = \boxed{1.1\times10^3 \text{ kg}\cdot\text{m}^2/\text{s}}$

CHECK and THINK
As expected, our answer is has relatively large value for the angular momentum compared to some of the smaller scale problems we've seen.

b. (N) Calculate the total kinetic energy of the system.

INTERPRET and ANTICIPATE
Since the center of mass of the system is not moving, then the total mechanical energy of the system is only rotational energy, which we can calculate using values from part (a).

SOLVE The rotational kinetic energy depends on the moment of inertia and angular frequency according to Equation 13.10. We found this information in part (a).	$K_r = \frac{1}{2}I\omega^2$ $K_r = \frac{1}{2}(340 \text{ kg}\cdot\text{m}^2)(\pi \text{ rad/s})^2$ $K_r = \boxed{1.7\times10^3 \text{ J}}$

CHECK and THINK
With the values calculated in part (a), finding the kinetic energy follows pretty easily.

c. (N) Both children walk half the distance toward the center of the merry-go-round. Calculate the final angular speed of the system.

INTERPRET and ANTICIPATE	
Since we aren't given a lot of information other than a change of position of the chidren, which will change the moment of inertia of the system, this is similar to an ice skater pulling his arms in while spinning and speeding up. This is a conservation of angular momentum problem. We have actually just calculated two values that *could* be conserved (we have seen conservation of mechanical energy in example 13.7 and conservation of angular momentum in example 13.13). In general, energy is conserved if no work is done on the system and angular momentum is conserved if there is no net torque on the system. (A free body diagram of the system, including all forces may be useful for identifying all possible forces that could either do work on the system or apply a torque.) Energy is not conserved, because the children must do work to walk to the center. However, angular momentum is conserved, because there is no net torque on the system as the children move. Just like the ice skater, we expect that the final angular speed of the system *increases*, because the rotational inertia of the system *decreases*.	
SOLVE Angular momentum is conserved since there is no net torque on the system (Eq. 13.30).	$I_i\omega_i = I_f\omega_f \quad \rightarrow \quad \omega_f = \frac{I_i\omega_i}{I_f}$
The original moment of inertia and angular momentum were found in part (a). The final moment of inertia (after the children move) is the moment of inertia of each child (which is different than part (a) since their distance from the rotation axis is different) and the moment of inertia of the merry-go-round (which is the same as in part (a)).	$I_f = 2m_{child}r^2 + I_{mgr}$ $I_f = \frac{1}{2}2(30.0\text{ kg})(0.75\text{ m})^2 + 200\text{ kg}\cdot\text{m}^2$

Now determine the final angular speed of the system.	$\omega_f = \frac{I_i \omega_i}{I_f}$ $\omega_f = \frac{\left(2(68\ \text{kg}\cdot\text{m}^2) + 200\ \text{kg}\cdot\text{m}^2\right)(\pi\ \text{rad/s})}{\frac{1}{2}2(30.0\ \text{kg})(0.75\ \text{m})^2 + 200\ \text{kg}\cdot\text{m}^2}$ $\omega_f = \boxed{4.9\ \text{rad/s}}$

CHECK and THINK

Our final angular velocity is indeed larger than the initial angular velocity. If we decrease the rotational inertia of an object (without applying an external torque) then the angular velocity of the system increases…just like the ice skater speeding up as he pulls his arms in.

57. (A) The angular momentum of a sphere is given by $\vec{L} = \left(-4.59t^3\right)\hat{\imath} + \left(6.01 - 1.19t^2\right)\hat{\jmath} + \left(6.26t\right)\hat{k}$, where L has units of kg · m²/s when t is in seconds. What is the net torque on the sphere as a function of time?

INTERPRET and ANTICIPATE

The net torque on an object is equal to the derivative of the angular momentum, as a function of time.

SOLVE Using the relationship given in Eq. 13.28, we take the derivative of each component of the angular momentum with respect to time. The result will be the net torque on the sphere.	$\vec{\tau} = \frac{d\vec{L}}{dt} = \frac{d}{dt}\left[\left(-4.59t^3\right)\hat{\imath} + \left(6.01 - 1.19t^2\right)\hat{\jmath} + \left(6.26t\right)\hat{k}\right]$ $\vec{\tau} = \boxed{\left(-13.8t^2\right)\hat{\imath} + \left(-2.38t\right)\hat{\jmath} + \left(6.26\right)\hat{k}}$

CHECK and THINK

Note that if the angular momentum is constant, then there must be no net torque. This means there would be no angular acceleration, as well. In cases like that seen here, the angular momentum is changing as time goes on, so there must be a net torque on the object. Here, the net torque is also changing as time goes on.

62. (A) A rigid rod of mass M and length L is pivoted around point P at one of its ends (Fig. P13.62). Find the speed of the center of mass of the rod when it is vertical if it is released from its horizontal position.

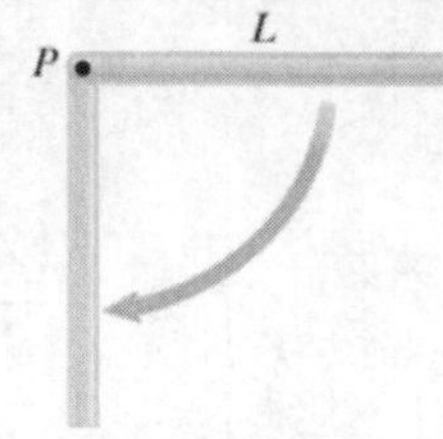

Figure P13.62

<table>
<tr><td colspan="2">INTERPRET and ANTICIPATE
We use the principle of mechanical energy conservation of the center of mass of the rod. As the center of mass of the rod falls, its gravitational potential energy is converted into kinetic energy.</td></tr>
<tr><td>SOLVE
We first find the change in potential energy as its center of mass falls through a vertical distance of L/2.</td><td>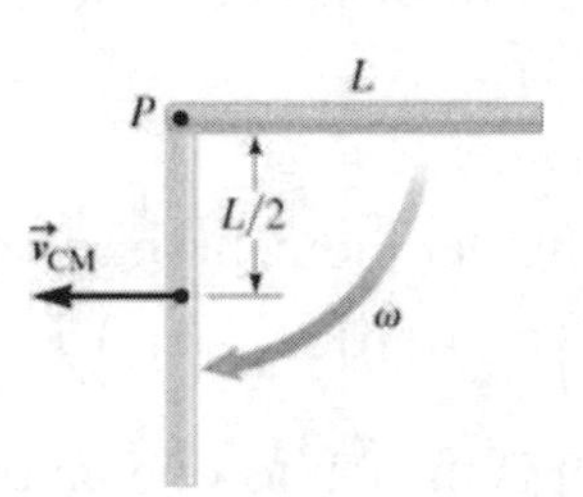

Figure P13.62ANS
$$\Delta U = -Mg\frac{L}{2}$$</td></tr>
<tr><td>The rod's kinetic energy increases from zero. When it falls to the vertical position, it has rotational kinetic energy (Eq. 13.10).</td><td>$$K_R = \frac{1}{2}I_P\omega^2$$</td></tr>
<tr><td>I_P is the moment of inertia of the rod about its end P, which we can determine using the moment of inertia about the center (Table 13.1) and the parallel axis theorem (Equation 13.9).</td><td>$$I_P = I_{CM} + M\left(\frac{L}{2}\right)^2$$
$$I_P = \frac{ML^2}{12} + \frac{ML^2}{4} = \frac{ML^2}{3}$$</td></tr>
<tr><td>We now write an expression for the rotational kinetic energy.</td><td>$$K_R = \frac{1}{2}I_P\omega^2$$
$$K_R = \frac{1}{2}\frac{ML^2}{3}\left(\frac{2v_{CM}}{L}\right)^2$$
$$K_R = \frac{2}{3}Mv_{CM}{}^2$$</td></tr>
</table>

Finally, we use conservation of mechanical energy.	$\Delta K = -\Delta U$ $\frac{2}{3}Mv_{CM}^2 = Mg\frac{L}{2}$ $v_{CM} = \boxed{\frac{\sqrt{3gL}}{2}}$
CHECK and THINK We see from our result that a longer rod (or going to a planet with a larger gravitational field) would result in a larger velocity.	

63. A uniform cylinder of radius $r = 10.0$ cm and mass $m = 2.00$ kg is rolling without slipping on a horizontal tabletop. The cylinder's center of mass is observed to have a speed of 5.00 m/s at a given instant.

a. (N) What is the translational kinetic energy of the cylinder at that instant?
b. (N) What is the rotational kinetic energy of the cylinder around its center of mass at that instant?
c. (N) What is the total kinetic energy of the cylinder at that instant?

INTERPRET and ANTICIPATE The translational kinetic energy depends on the mass and center of mass velocity, which are given. The rotational velocity depends on the moment of inertia and angular velocity, which we can calculate. The total kinetic energy is the sum of these two quantities.	
SOLVE **a.** First, we calculate the translational kinetic energy.	$K_{trans} = \frac{1}{2}mv^2$ $K_{trans} = \frac{1}{2}(2.00\text{ kg})(5.00\text{ m/s})^2$ $K_{trans} = \boxed{25.0\text{ J}}$
b. Using Equation 13.10 and the moment of inertia from Table 13.1, we find that the radius of the cylinder cancels out and is not necessary to obtain its rotational kinetic energy. In fact, we find that for a cylinder, the rotational kinetic energy is half the translational kinetic energy.	$K_{rot} = \frac{1}{2}I\omega^2$ $K_{rot} = \frac{1}{2}\left(\frac{1}{2}mr^2\right)\left(\frac{v^2}{r^2}\right)$ $K_{rot} = \frac{1}{4}mv^2$ $K_{rot} = \frac{1}{4}(2.00\text{ kg})(5.00\text{ m/s})^2$ $K_{rot} = \boxed{12.5\text{ J}}$

c. The total is the sum of the values from parts (a) and (b).	$K_{\text{total}} = K_{\text{trans}} + K_{\text{rot}}$ $K_{\text{total}} = 25.0\text{ J} + 12.5\text{ J}$ $K_{\text{total}} = \boxed{37.5\text{ J}}$

CHECK and THINK
In this case, for a cylinder, the rotational kinetic energy is half the translational kinetic energy. Given the shape, mass, and center of mass velocity, we are able to calculate the magnitude of both types of kinetic energy.

69. (N) The velocity of a particle of mass $m = 2.00$ kg is given by $\vec{v} = -5.10\hat{i} + 2.40\hat{j}$ m/s. What is the angular momentum of the particle around the origin when it is located at $\vec{r} = 8.60\hat{i} - 3.70\hat{j}$ m?

INTERPRET and ANTICIPATE
The angular momentum of a moving object around a particular point is the cross product of the radial vector from the point with the velocity of the particle. The direction of the resulting vector is perpendicular to each. Since both vectors given are in the *xy* plane, we expect the angular momentum will be in the $\pm z$ direction.

SOLVE Use Equation 13.24 to calculate the angular momentum.	$\vec{L} = \vec{r} \times \vec{p}$ $\vec{L} = \left(-8.60\hat{i} - 3.70\hat{j}\right)\text{ m} \times \left(2.00\text{ kg}\right)\left(-5.10\hat{i} + 2.40\hat{j}\right)\text{ m/s}$ $\vec{L} = \left(-41.28\hat{k} - 37.74\hat{k}\right)\text{ kg}\cdot\text{m}^2/\text{s}$ $\vec{L} = \boxed{\left(-79.0\text{ kg}\cdot\text{m}^2/\text{s}\right)\hat{k}}$

CHECK and THINK
The angular momentum is indeed along the *z* axis.

70. A ball of mass $M = 5.00$ kg and radius $r = 5.00$ cm is attached to one end of a thin, cylindrical rod of length $L = 15.0$ cm and mass $m = 0.600$ kg. The ball and rod, initially at rest in a vertical position and free to rotate around the axis shown in Figure P13.70, are nudged into motion.

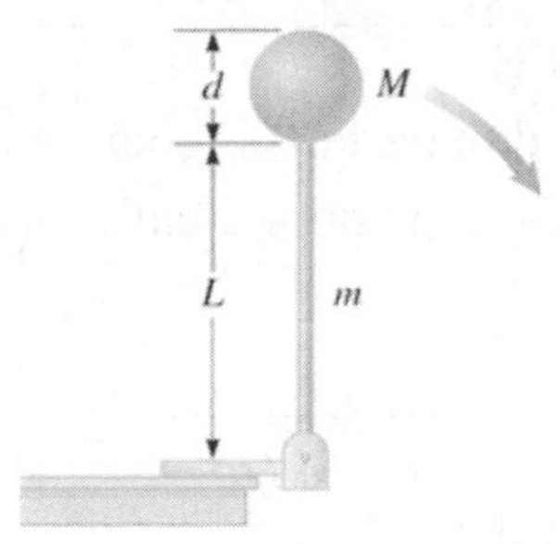

Figure P13.70

a. (N) What is the rotational kinetic energy of the system when the ball and rod reach a horizontal position?

INTERPRET and ANTICIPATE For the isolated rod-ball-Earth system with no friction, mechanical energy is conserved. As the ball rotates from vertical to horizontal, the potential energy decreases and the rotational kinetic energy increases.	
SOLVE We apply conservation of energy. The final energy equals the change in potential energy of both the rod and the ball. We use $y = 0$ to represent the height of the center of mass when it is horizontal, so that $U_f = 0$.	$\Delta K + \Delta U = 0$ $K_f = U_i$ $K_f = m_{\text{rod}} g y_{\text{CM, rod}} + m_{\text{ball}} g y_{\text{CM, ball}}$
Now, insert values and calculate the kinetic energy.	$K_f = \left(m_{\text{rod}} y_{\text{CM, rod}} + m_{\text{ball}} y_{\text{CM, ball}}\right) g$ $K_f = \left(\left(0.600\ \text{kg}\right)\left(0.075\ \text{m}\right) + \left(5.00\ \text{kg}\right)\left(0.200\ \text{m}\right)\right)\left(9.81\ \frac{\text{m}}{\text{s}^2}\right)$ $K_f = \boxed{10.3\ \text{J}}$
CHECK and THINK Energy conservation provides a straightforward way to determine the final kinetic energy.	

b. (N) What is the angular speed of the ball and rod when they reach a horizontal position?

<table>
<tr><td colspan="2">INTERPRET and ANTICIPATE
The final kinetic energy calculated in part (a) is rotational kinetic energy, which depends on the angular velocity (of both the rod and the ball) that we want to find and the moment of inertia.</td></tr>
<tr><td>SOLVE
The rotational kinetic energy found in part (a) depends on the angular velocity and moment of inertia according to Equation 13.10.</td><td>$K = \frac{1}{2} I\omega^2$</td></tr>
<tr><td>For the compound object, the total moment of inertia is the sum of the individual parts (in this case, the rod and the ball). For the rod, we assume it is a thin rod and use the moment of inertia from Table 12.1 (or use the moment of inertia for a rod about the center of mass and the parallel axis theorem, Eq. 13.9). The ball has a relatively small diameter compared to the distance from the rotation axis, so one might treat it as a point particle at a distance $h = l + d/2 = 20\text{ cm}$ from the rotation axis</td><td>$I = \frac{1}{3} M_{\text{rod}} L^2 + m_{\text{ball}} h^2$
$I = \frac{1}{3}(0.600\text{ kg})(0.150\text{ m})^2 + (5.00\text{ kg})(0.200\text{ m})^2$
$I = 0.205\text{ kg}\cdot\text{m}^2$</td></tr>
<tr><td>Now solve Eq. 13.10 for the angular velocity.</td><td>$K_f = \frac{1}{2} I\omega^2$
$\omega = \sqrt{\frac{2K_f}{I}}$
$\omega = \sqrt{\frac{2(10.3\text{ J})}{0.205\text{ kg}\cdot\text{m}^2}}$
$\omega = \boxed{10.0\text{ rad/s}}$</td></tr>
<tr><td colspan="2">CHECK and THINK
To test our approximation of the ball as a point particle, we can also treat it more precisely as a sphere at a distance h from the rotation axis using the parallel axis theorem.</td></tr>
</table>

$$I = \frac{1}{3}M_{\text{rod}}L^2 + \left[\frac{2}{5}m_{\text{ball}}R^2 + m_{\text{ball}}h^2\right]$$

$$I = \frac{1}{3}(0.600 \text{ kg})(0.150 \text{ m})^2 + \frac{2}{5}(5.00 \text{ kg})(5.00\times10^{-2} \text{ m})^2 + (5.00 \text{ kg})(0.200 \text{ m})^2 = 0.210 \text{ kg}\cdot\text{m}^2$$

$$\omega = \sqrt{\frac{2K_f}{I}} = \sqrt{\frac{2(10.3 \text{ J})}{0.210 \text{ kg}\cdot\text{m}^2}} = 9.90 \text{ rad/s}$$

We reach nearly the same answer, indicating that approximating the ball as a point particle in this case is reasonable.

c. (N) What is the linear speed of the center of mass of the ball when the ball and rod reach a horizontal position?

INTERPRET and ANTICIPATE

For an object in circular motion, the linear speed is related to the angular speed by the radius of the orbit.

SOLVE	
The linear speed can be calculated from the angular speed and radius of the circular motion.	$v = r\omega$ $v = (0.200 \text{ m})(9.90 \text{ rad/s})$ $v = \boxed{1.98 \text{ m/s}}$

CHECK and THINK

This sounds like a reasonable speed for this problem.

d. (N) What is the ratio of the speed found in part (c) to the speed of a ball that falls freely through the same distance?

INTERPRET and ANTICIPATE

Referring back to our kinematic equations from Chapter 2, we can apply conservation of energy and determine the speed of a ball falling the same distance.

SOLVE	
Using constant acceleration kinematic equations, we can determine the speed if the ball falls freely.	$v_f^2 = v_i^2 + 2a(y_f - y_i)$

	$v_f = \sqrt{0+2\left(9.81\text{ m/s}^2\right)(0.200\text{ m})}$ $v = 1.98\text{ m/s}$
We take the ratio to compare the speeds.	$\dfrac{v_{\text{swing}}}{v_{\text{fall}}} = \dfrac{1.98}{1.98} = \boxed{1.00}$
CHECK and THINK The resulting speed is the same. In both cases, energy conservation relates the initial potential energy to the final kinetic energy.	

72. (A) A solid sphere and a hollow cylinder of the same mass and radius have a rolling race down an incline as in Example 13.9 (page 372). They start at rest on an incline at a height h above a horizontal plane. The race then continues along the horizontal plane. The coefficient of rolling friction between each rolling object and the surface is the same. Which object rolls the farthest? (Justify your answer with an algebraic expression.)

INTERPRET and ANTICIPATE Conservation of energy provides a very simple approach to this problem. Each object starts at rest on the incline, and each object stops on the horizontal surface. Along the way there is an increase in thermal energy between the surface and the object. Let's include the Earth, the rolling object and the surface in the system. We set the reference configuration to the horizontal surface.	
SOLVE We include a bar chart for this problem which shows: 1. There is initially positive gravitational potential energy. There is no kinetic energy and no work is done on the system. 2. Finally, the gravitational potential energy and the kinetic energy is zero. The thermal energy has increased.	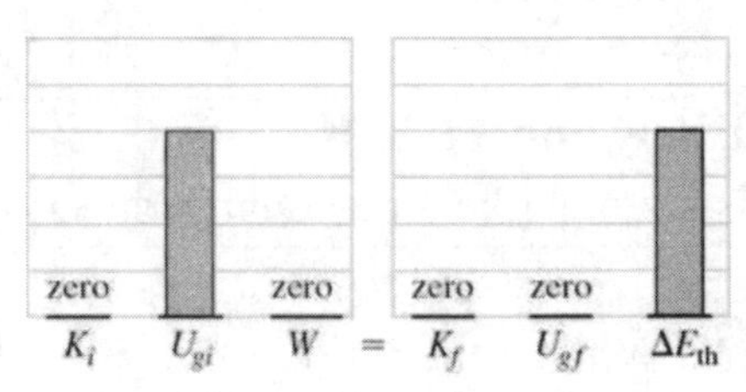 **Figure P13.72ANS**
We can use this to write an equation for the conservation of energy.	$U_{gi} = \Delta E_{th}$

Write expressions for the initial potential energy. The increase in total thermal energy equals the work done by the friction force ($F_f = \mu F_N$) acting over a distance *S*.	$mgh = \mu_r F_N S$
Solve for *S*.	$S = \dfrac{mgh}{\mu_r F_N}$
Each object has the same mass *m*, is released from the same height *h* and has the same coefficient of rolling friction μ_r with the surface. The normal force exerted by the surfaces on the each object is also the same since the objects have the same mass and are on the same inclined and horizontal surfaces. So *S* is the same for both objects. In other words, they travel the same distance from the starting point.	*S* is the same for both objects.

CHECK and THINK
This result may be surprising, but is it not a race in the traditional sense. We didn't ask which object arrived at the finish line first. Instead, we asked where the finish line is. The sphere gets there sooner because it has a small rotational inertia so it rolls down the incline at a higher speed.

77. (N) A rod of length 2.0 m and of negligible mass is constrained to move in a vertical plane (Fig. P13.77). The ends of the rod always touch the *x* and *y* axes, respectively, as the rod slides downward. Find the coordinates of the instantaneous axis of rotation of the rod when it makes an angle of 60° with the horizontal.

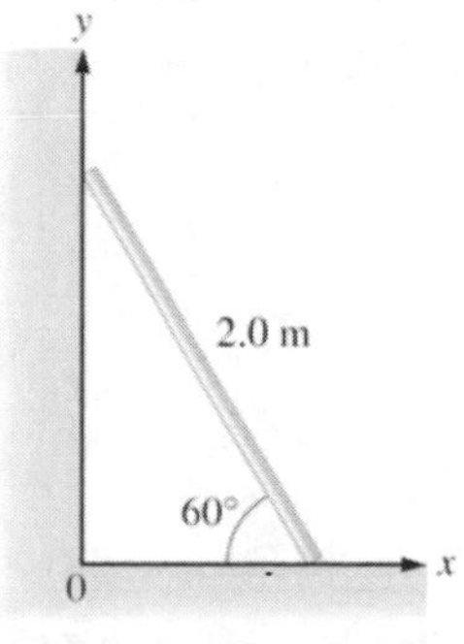

Figure P13.77

INTERPRET and ANTICIPATE

The rod tends to slide down, so the top end of the rod has a velocity along the negative y axis and the bottom end has a velocity along the positive x axis. So this does look like rotation, as the rod is rotating counter-clockwise as it falls, but what determines the axis of rotation? For something in solid body rotation, we expect that all parts of the object are moving in a circular orbit around the rotation axis. Therefore, we'll try to find a point at this instant in time for which the ends of the rod are both moving in a circular orbit or, in other words, are moving perpendicular to a radial vector from the rotation axis.

SOLVE

The top end of the rod has a velocity $\vec{v}_1 = v_1(-\hat{j})$ and the bottom end has a velocity $\vec{v}_2 = v_2\hat{i}$. If we draw perpendicular lines (representing possible radial vectors) we see that they both intersect at point O, so both ends move perpendicularly around this point. This is what we would expect if O is the axis of rotation. Therefore, O is the instantaneous point of rotation and the axis of rotation passes through it perpendicular to the plane of rotation.

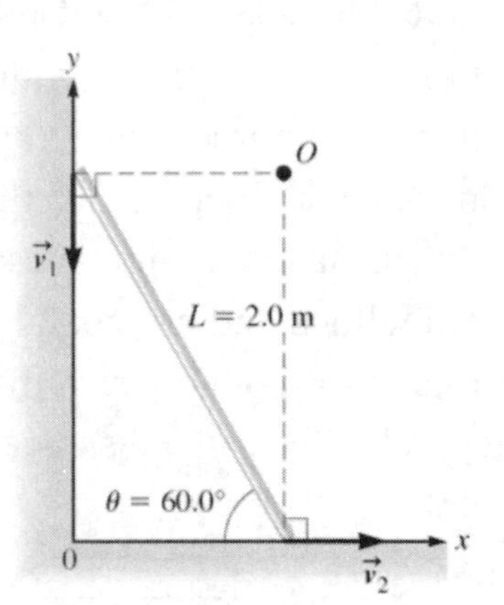

Figure P13.77ANS

That was the hard part! It's now straightforward to calculate the coordinates of point O using the sketch.

$$(L\cos\theta, L\sin\theta) = (2.0\cos 60°\ \text{m}, 2.0\sin 60°\ \text{m})$$

$$(L\cos\theta, L\sin\theta) = \boxed{(1.0\ \text{m}, 1.7\ \text{m})}$$

CHECK and THINK

In this case, we had to think carefully about what determines a rotation axis. It is instantaneously rotating counter-clockwise around point O… part of the difficulty though is that this instantaneous rotation axis is changing as the bar falls.

14

Static Equilibrium, Elasticity, and Fracture

9. (A) The keystone of an arch is the stone at the top (Fig. P14.9). It is supported by forces from its two neighbors, blocks A and B. Each block has mass m and approximate length L. What can you conclude about the force exerted by each block, $\vec{F}_A$ and $\vec{F}_B$, for the keystone to remain in static equilibrium? That is, show that the equilibrium conditions are satisfied by the components of the forces F_{Ax}, F_{Ay}, F_{Bx}, and F_{By}. Assume the arch is symmetric.

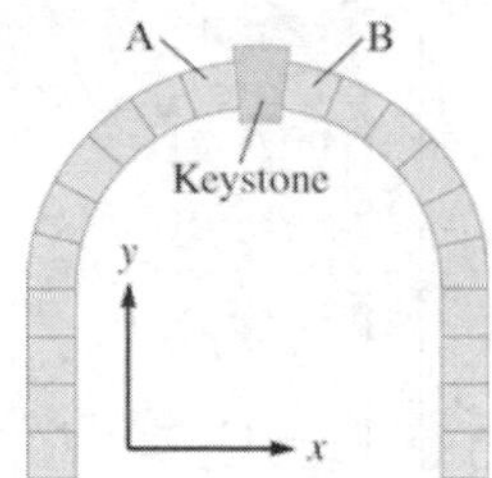

Figure P14.9

INTERPRET and ANTICIPATE Since the keystone is in static equilibrium, both the net force and the net torque must be equal to zero. Since the geometry of the arch is symmetric, we expect that the forces will be symmetric on either side of the keystone as well.	
SOLVE We can write down the equations for force conservation for the x and y components of force, Equation 14.3a and 14.3b.	$\sum F_x = 0$ and $\sum F_y = 0$
There are three forces acting on the keystone. One from each neighbor and the force of gravity, which points downward.	$\vec{F}_A = F_{Ax}\hat{i} + F_{Ay}\hat{j}$ $\vec{F}_B = F_{Bx}\hat{i} + F_{By}\hat{j}$ $\vec{F}_g = -mg\hat{j}$

For the x component of the force we determine that $F_{Ay} = -F_{By}$, which we expect for a symmetric arch.	$F_{Ax} + F_{Bx} = 0$ $F_{Ax} = -F_{Bx}$
For the y component of the force, we know that the keystone has a mass m, which has a weight of magnitude mg. If the arch is symmetric, $F_{Ay} = F_{By}$ and each supports half the weight of the keystone.	$F_{Ay} + F_{By} - mg = 0$ $2F_{Ay} = mg$ $F_{Ay} = F_{By} = \frac{1}{2}mg$
To determine the torques, we can calculate the torques relative to the center of the keystone, in which case the distance at which the force is applied is $l/2$ in each case. First we can calculate the force due to block A relative to the center of the block.	$\vec{\tau}_A = \vec{r}_A \times \vec{F}_A$ $\vec{r}_A = -\frac{L}{2}\hat{i}$ $\vec{F}_A = F_{Ax}\hat{i} + F_{Ay}\hat{j} = F_{Ax}\hat{i} + \frac{1}{2}mg\hat{j}$ $\vec{\tau}_A = \vec{r}_A \times \vec{F}_A$ $\vec{\tau}_A = (r_yF_z - r_zF_y)\hat{i} + (r_zF_x - r_xF_z)\hat{j} + (r_xF_y - r_yF_x)\hat{k}$ $\vec{\tau}_A = (-\frac{L}{2} \cdot \frac{1}{2}mg)\hat{k} = -\frac{1}{4}Lmg\hat{k}$
And now for the torque due to block B.	$\vec{r}_B = \frac{L}{2}\hat{i}$ $\vec{F}_B = F_{Bx}\hat{i} + F_{By}\hat{j} = F_{Bx}\hat{i} + \frac{1}{2}mg\hat{j}$ $\vec{\tau}_B = \vec{r}_B \times \vec{F}_B$ $\vec{\tau}_B = (r_yF_z - r_zF_y)\hat{i} + (r_zF_x - r_xF_z)\hat{j} + (r_xF_y - r_yF_x)\hat{k}$ $\vec{\tau}_B = \left(\frac{L}{2} \cdot \frac{1}{2}mg\right)\hat{k} = \frac{1}{4}Lmg\hat{k}$
Both torques are in the $\hat{k}$ direction and it is clear that torque balance is satisfied.	$\vec{\tau}_A = \vec{\tau}_A + \vec{\tau}_B = -\frac{1}{4}Lmg\ \hat{k} + \frac{1}{4}Lmg\ \hat{k} = 0$

CHECK and THINK

The net force and net torque are zero if $\boxed{F_{Ay} = F_{By} = \frac{1}{2}mg}$ and $\boxed{F_{Ax} = -F_{Bx}}$. This makes sense for the symmetric arch. Each of the neighboring blocks supports half the weight of the keystone and the forces in the x-direction are equal and opposite.

14. (A) The uniform block of width W and height H in Figure P14.13 rests on an inclined plane in static equilibrium. The plane is slowly raised until the block begins to tip over. If the block tips before it slides, determine the minimum coefficient of static friction.

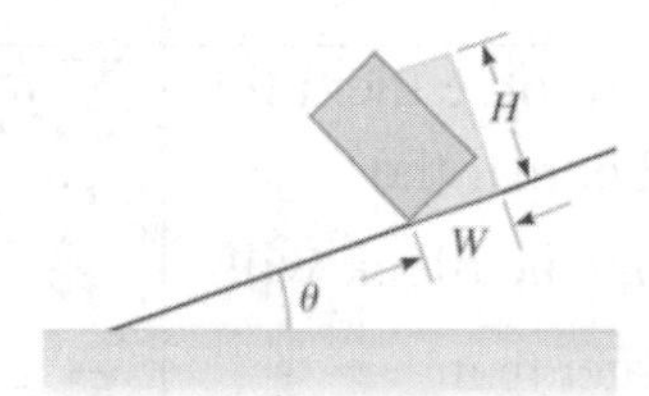

Figure P14.13

INTERPRET and ANTICIPATE

Since the block does not slide, it obeys the conditions for both force and torque balance until the moment it tips. The verge of tipping occurs when the net torque on the block is zero, but any further increase in θ will produce a non-zero net torque and hence, rotation. (This is the focus of Problem 13.) To find the minimum coefficient of friction needed to prevent the block from sliding when it is at the verge of tipping, we need to determine the friction force needed to keep the block in static equilibrium and the corresponding coefficient of friction needed to produce this force. Any larger coefficient of friction would work as well, since this would also produce a large enough friction force to keep the block from sliding.

We know that a tall, skinny block, for instance, will tip easily, whereas a short, wide box won't tip easily, so we might expect the friction force needed to somehow depend on this. We'll check this in the Check and Think step.

SOLVE

The figure shows a free-body diagram of the situation at the verge of slipping. In this case, the normal force acts at the front edge of the block. (Imagine that the block just starts to tip… at that moment, only the front edge of the block remains in contact with the incline as the block tilts forward.)

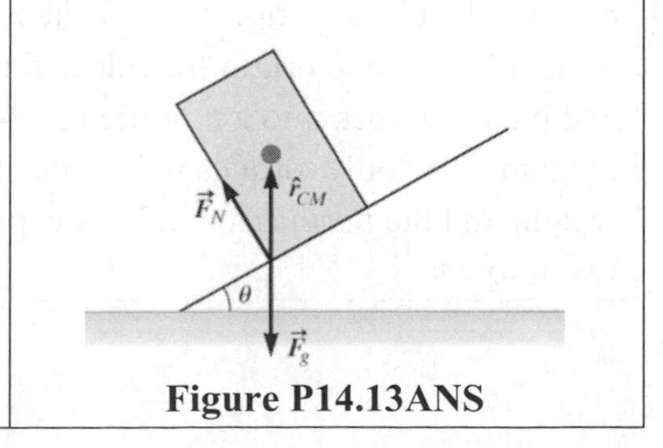

Figure P14.13ANS

To determine what friction coefficient is needed, we only need to consider force balance. The maximum frictional force must be large enough to support the net downward force at the moment the block tips. The frictional force just at the verge of sliding will correspond to the smallest coefficient of friction possible — any smaller and the block will slide before it starts to tip.	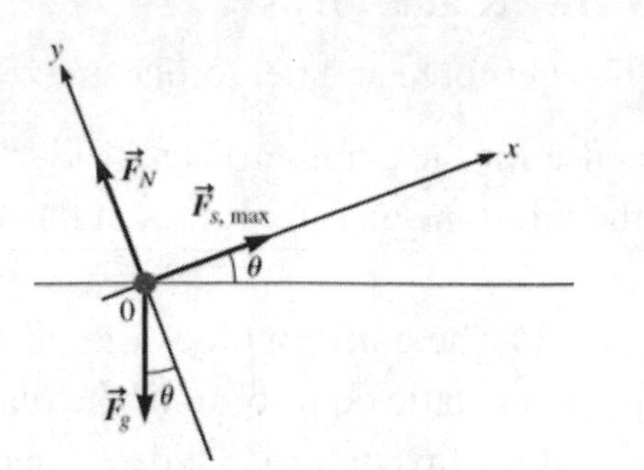 **Figure P14.14ANS**
Use Equations 14.3*x* and14.3*y* and solve them for the angle of the incline. Remember that the friction force is $F_s \le \mu F_N$ and that the block will slip when the friction force is larger than $F_{s,\max} = \mu F_N$.	$\sum F_x = F_{s,\max} - F_g \sin\theta = 0$ $F_{s,\max} = F_g \sin\theta$ $\sum F_y = F_N - F_g \cos\theta = 0$ $F_N = F_g \cos\theta$
After some algebra, we get an expression for θ.	$F_{s,\max} = F_g \sin\theta = \mu_s F_N = \mu_s F_g \sin\theta$ $\mu_s = \tan\theta$ $\theta = \tan^{-1}\mu_s \qquad (1)$
The result of Problem 13 was that $\theta = \tan^{-1}\dfrac{W}{H}$. Let's quickly go through that solution. To determine when the block tips, we need to use the force diagram for the extended object in Figure P14.13 above. The block obeys force balance until the moment it begins to slide. Tipping occurs when the net torque just equals zero, when any further increase in θ will lead to a net torque causing the block to tip. We pick the lower edge of the block as the origin for calculating torques and put the forces into a modified free-body diagram. The only torques will be due to the weight and the normal force. Now express the two torques.	$\vec{\boldsymbol{\tau}}_g = \vec{\boldsymbol{r}}_{cm} \times \vec{\boldsymbol{F}}_g$ $\vec{\boldsymbol{\tau}}_N = \vec{\boldsymbol{r}}_N \times \vec{\boldsymbol{F}}_N$

When the block is on the verge of tipping, the normal force produces no torque. For rotational equilibrium, the torque due to the weight must then also be zero. This will occur when the center of mass is directly above the origin, so that $\vec{r}_{cm}$ and $\vec{F}_g$ are anti-parallel. We can now calculate the angle θ at which this occurs. We use the triangle formed by the line of action of the weight vector and the sides of the block. A smaller angle is needed for a tall, narrow object, as we would expect.	$\theta = \tan^{-1}\dfrac{W}{H}$ (2)
With the result of Problem 13, we can now relate the coefficient of friction to the width and height using Equations 1 and 2.	$\theta = \tan^{-1}\mu_s = \tan^{-1}\dfrac{W}{H}$ $\mu_s = \boxed{\dfrac{W}{H}}$

CHECK and THINK

A short, fat block ($W >> H$) won't tip until θ is very large, so more friction is needed to prevent it from sliding before it tips. By the same reasoning, we would expect a tall, skinny block ($H >> W$) to tip over easily (small θ), and thus the coefficient of friction needed to prevent sliding can be smaller. (In other words, very little friction is needed because the block will tip before much of a friction force builds up.) One special case is $W = H$, for which $\theta = 45°$ and $\mu_s = 1$. Most pairs of materials don't have this high a coefficient of static friction, so slipping would typically occur first for this shape.

19. (N) A force $\vec{F} = (65.8\hat{i} + 96.3\hat{k})$ N is exerted on an object, and the torque that results is given by $\vec{\tau} = (57.1\hat{j})$ N·m. Find $\vec{r} = r_z\hat{k}$.

INTERPRET and ANTICIPATE

This is an application of the torque formula. Torque is the cross product of the force and the radial vector from the rotation axis. Given the first two quantities, we can calculate the third.

SOLVE To relate the force, distance, and torque, use Equation 14.7. Each is given in the problem statement.	$\vec{\tau} = \vec{r} \times \vec{F}$ $\vec{\tau} = (r_yF_z - r_zF_y)\hat{i} + (r_zF_x - r_xF_z)\hat{j} + (r_xF_y - r_yF_x)\hat{k}$

We write the quantities in component form.	$\vec{F} = (65.8\hat{i} + 96.3\hat{k})$ N $\rightarrow$ $F_x = 65.8,\ F_z = 96.3$ $\vec{\tau} = 57.1\hat{j}$ N·m $\rightarrow$ $\tau_y = 57.1$ $\vec{r} = r_z\ \hat{k}$ m $\rightarrow$ r_z?
Now, substitute values into the torque equation and solve for r_z. Then write the final answer as a vector.	$57.1\hat{j}\ \text{N}\cdot\text{m} = (r_z(65.8\ \text{N}) - (0)(96.3))\hat{j}$ $r_z = \dfrac{57.1\ \text{N}\cdot\text{m}}{65.8\ \text{N}} = 0.868\ \text{m}$ $\vec{r} = \boxed{0.868\hat{k}\ \text{m}}$

CHECK and THINK

Given the formula for the cross-product and expressions for the torque and force, we can determine the radial distance to the rotation point.

22. The inner planets of our solar system are represented on a mobile constructed from drinking straws and light strings for a school project (Fig. P14.22). The mass of the piece representing the Earth is 25.0 g, and the mass of the straws can be ignored. (The lengths shown are not proportional to actual distances in the solar system.)

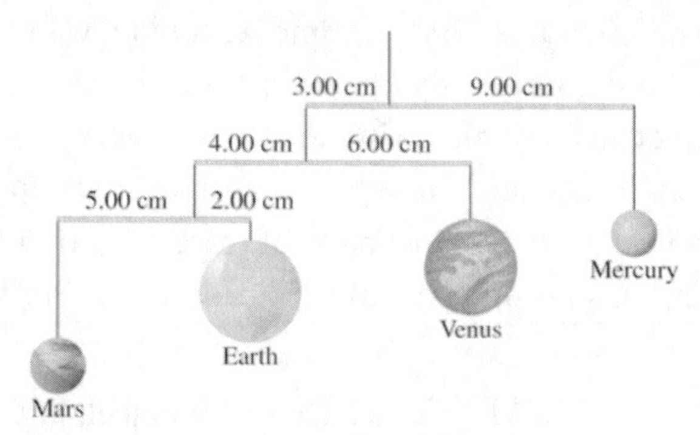

Figure P14.22

a. (N) What is the mass of the piece representing Mars?
b. (N) What is the mass of the piece representing Venus?
c. (N) What is the mass of the piece representing Mercury?

INTERPRET and ANTICIPATE

The important thing about a mobile is that it each of the horizontal bars is balanced. If the lengths or masses are chosen poorly, some of the straws will tip. "Balanced" in this case means that the net torque of each straw is zero around the point where the string is holding it. While we could actually calculate the net torque on a straw around any point, but by choosing the support point, the unknown upward force due to the string acts at $r = 0$ and contributes no torque.

SOLVE For rotational equilibrium, the net torque on each of the rods about its point of support is zero (Eq. 14.2). The torque due to each planet is its weight (*mg*) times its distance from the support point.	$\sum \tau = 0$
a. For the rod with Earth and Mars (m_1):	$+(25.0\text{ g})g(2.0\text{ cm}) - m_1 g(5.00\text{ cm}) = 0$ $\boxed{m_1 = 10.0\text{ g} = 0.0100\text{ kg}}$
b. For the middle rod, Venus (m_2) must support the weight of both Mars and Earth:	$m_2 g(6.00\text{ cm}) - (10.0\text{ g} + 25.0\text{ g})g(4.00\text{ cm}) = 0$ $\boxed{m_2 = 23.3\text{ g} = 0.0233\text{ kg}}$
c. For the top rod, with m_3 for Mercury (which must balance the other three planets):	$m_3 g(9.00\text{ cm}) - (10.0\text{ g} + 25.0\text{ g} + 23.3\text{ g})g(3.00\text{ cm}) = 0$ $\boxed{m_3 = 19.4\text{ g} = 0.0194\text{ kg}}$
CHECK and THINK Each straw must be balanced. By calculating the torque of each relative to its support point, we are able to determine the masses for each. A smaller mass is able to balance a larger mass if its distance from the rotation axis is larger.	

27. (N) Children playing pirate have suspended a uniform wooden plank with mass 15.0 kg and length 2.50 m as shown in Figure P14.27. What is the tension in each of the three ropes when Sophia, with a mass of 23.0 kg, is made to "walk the plank" and is 1.50 m from reaching the end of the plank?

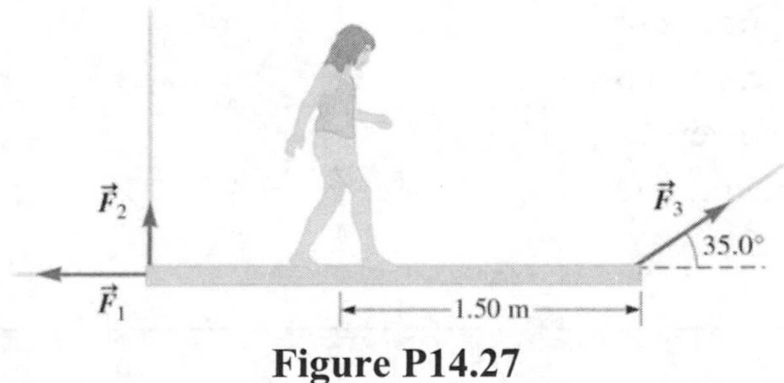

Figure P14.27

INTERPRET and ANTICIPATE

At equilibrium, the sum of forces and sum of torques must each be zero (Eqs. 14.1 and 14.2). We can write equations for both force and torque balance and solve for the unknown forces.

SOLVE

Consider the torques about an axis perpendicular to the page and through the left end of the plank, which we sketch. Forces 1 and 2 contribute no torque around this axis, so we need only consider the weight of Sophia and the plank and force 3. (The net torque around *any* axis on the plank must be zero, so we choose an axis that makes our calculation easiest.)

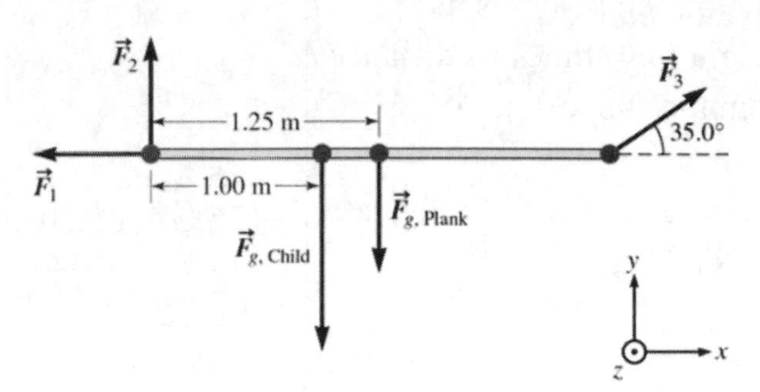

Figure P14.27ANS

Each torque can be expressed as the radial distance from the pivot point times the perpendicular component of the force. For Sophia and the plank, their weight exerts a downward force (*mg*) perpendicular to the radial distance from the rotation axis. Sophia's weight is mg = (23.0 kg)(9.81 m/s^2) and acts 1.0 meter away from our chosen rotation axis. The plank's weight is mg = (15.0 kg)(9.81 m/s^2) and acts at its center, 1.25 meters from our rotation point. For F_3, we can use trigonometry to find the vertical (perpendicular component) and solve for its magnitude.

$$\sum \tau = 0, \quad \tau = rF_{\perp}$$

$$-(23.0\text{ kg})(9.81\text{ m/s}^2)(1.00\text{ m}) - (15.0\text{ kg})(9.81\text{ m/s}^2)(1.25\text{ m}) + (F_3 \sin 35.0°)(2.50\text{ m}) = 0$$

$$F_3 = \boxed{286\text{ N}}$$

Now consider force balance in the x direction. We again use trigonometry to find the x component of F_3.	$\Sigma F_x = 0$ $-F_1 + F_3 \cos 35.0° = 0$ $F_1 = (286\text{ N})\cos 35.0° = \boxed{234\text{ N}}$
Finally, consider force balance in the y direction.	$\Sigma F_y = 0$ $F_2 - (23.0\text{ kg})(9.81\text{ m/s}^2) - (15.0\text{ kg})(9.81\text{ m/s}^2) + F_3 \sin 35.0° = 0$ $F_2 = (23.0\text{ kg})(9.81\text{ m/s}^2) + (15.0\text{ kg})(9.81\text{ m/s}^2) - (286\text{ N})\sin 35.0° = \boxed{209\text{ N}}$

CHECK and THINK

The forces calculated are all positive, which we expect for a force exerted by the tension of each rope.

30. (N) A 5.45-N beam of uniform density is 1.60 m long. The beam is supported at an angle of 35.0° by a cable attached to one end. There is a pin through the other end of the beam (Fig. P14.30). Use the values given in the figure to find the tension in the cable.

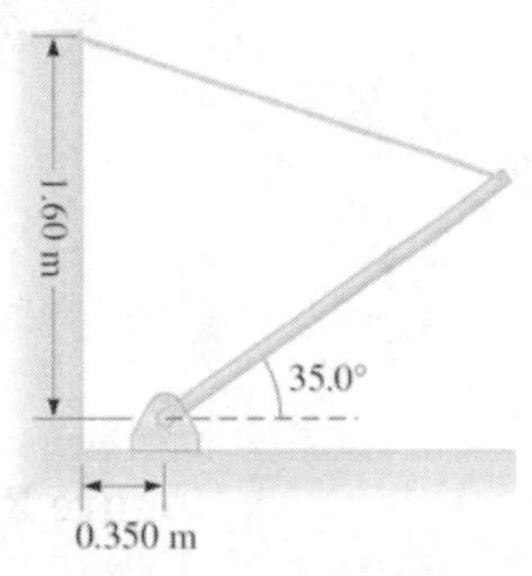

Figure P14.30

INTERPRET and ANTICIPATE

The beam is in static equilibrium, which means that the net force and net torque on it must be zero (Eqs. 14.1 and 14.2). We can write expressions for torque balance and force balance and solve for the tension.

SOLVE Since the beam is in equilibrium, the net torque around the point in contact with the ground is zero.	$\vec{\tau}_g + \vec{\tau}_T = 0$ $\vec{\tau}_g = \vec{r} \times \vec{F}$

The net torque around *any* point on the beam must be zero, but we don't know what the force of the hinge on the beam is, so by choosing this rotation axis, $r = 0$ and the unknown force does not produce a torque around this point. The torques we must consider are those due to the weight of the beam and the tension due to the rope.	
We sketch the force vectors and radial vectors from the pivot point (part A). The gravitational force acts straight down at a point that is halfway up the beam.	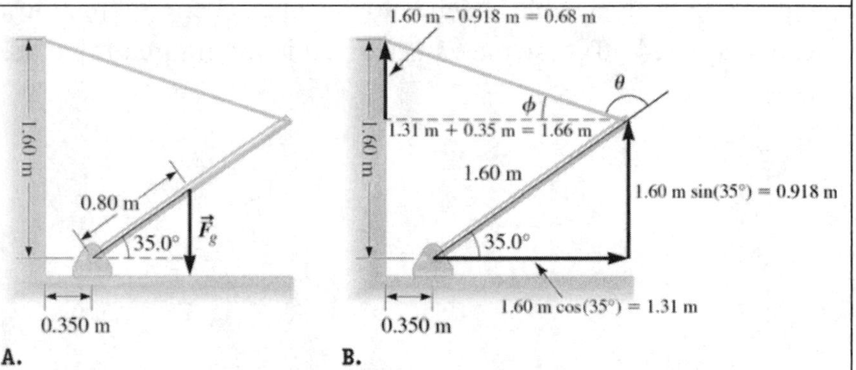 A. B. **Figure P14.30ANS**
First, calculate the torque due to gravity.	$\vec{\tau}_g = \vec{r} \times \vec{F}$ $\vec{\tau}_g = \left(0.8\cos 35°\text{ m}\,\hat{i} + 0.8\sin 35°\text{ m}\,\hat{j}\right) \times \left(-5.45\text{ N}\,\hat{j}\right)$ $\vec{\tau}_g = -3.57\hat{k}\text{ N}\cdot\text{m}$
For the tension force, we need to determine the angle the tension force makes relative to the bar. Using geometry and the	$\phi = \tan^{-1}\left(\dfrac{0.68}{1.66}\right) = 22.3°$

values given in the problem, we determine first the angle of the rope above the horizontal axis by first determining the dimensions of the triangle formed by the rope as shown in part B of the figure.	
This allows us to calculate the angle between the rope and the beam, 35.0° + 22.3° = 57.3°, and the angle between the radial vector (outward from the pivot point) and the tension force, θ=180° – 57.3° = 122.7°.	$\theta = 122.7°$
Now, equate the magnitudes of the torque.	$rF_T \sin\theta = 3.57\ \text{N}\cdot\text{m}$ $F_T = \dfrac{3.57\ \text{N}\cdot\text{m}}{(1.60\ \text{m})\sin 122.7°}$ $F_T = \boxed{2.65\ \text{N}}$

CHECK and THINK

Notice that we didn't need to use force balance. While the net force is also zero, by using the net torque on the beam around the hinge, we were able to avoid using the unknown force of the hinge on the beam.

32. A 215-kg robotic arm at an assembly plant is extended horizontally (Fig. P14.32). The massless support rope attached at point *B* makes an angle of 15.0° with the horizontal, and the center of mass of the arm is at point *C*.

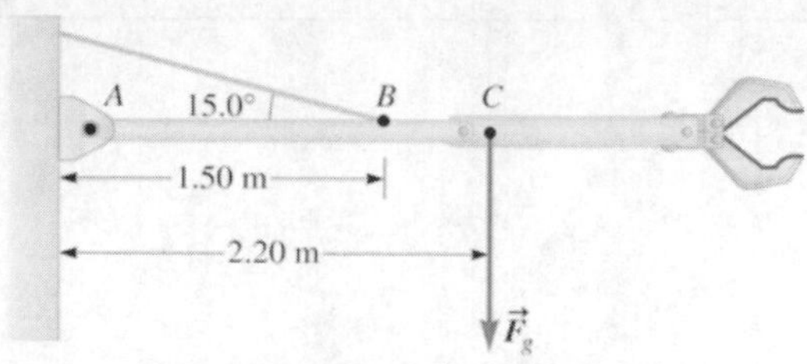

Figure P14.32

a. (N) What is the tension in the support rope?

INTERPRET and ANTICIPATE We first sketch the robotic arm. At equilibrium, we require that the net torque is zero using the shoulder joint at point *A* as a pivot. 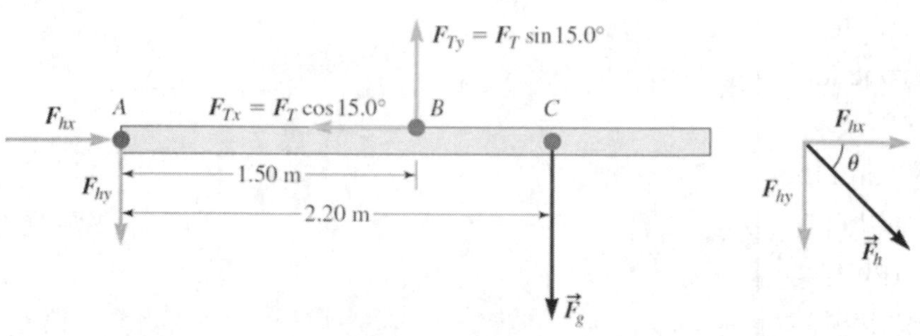 Figure P14.32ANS	
SOLVE Use the fact that net torque is zero (Eq. 14.2) with the shoulder joint at point A as the rotation axis. The two forces producing torque are the tension in the rope and the weight of the arm. The force of the hinge on the arm acts at $r = 0$, so it does not produce a torque.	$\sum \vec{\tau} = \sum \vec{F} \times \vec{r} = 0$ $(F_T \sin 15.0°)(1.50 \text{ m}) - (215 \text{ kg})(9.81 \text{ m/s}^2)(2.20 \text{ m}) = 0$ $F_T = \boxed{1.20 \times 10^4 \text{ N}}$
CHECK and THINK By applying torque balance, we find the unknown tension force.	

b. (N) What are the magnitude and direction of the force exerted by the hinge *A* on the robotic arm to keep the arm in the horizontal position?

INTERPRET and ANTICIPATE	
We know that the net torque on the arm is zero. To keep the arm in a fixed position, the net force on the arm must also be zero.	
SOLVE Use force balance to determine the force due to the hinge F_h (Eq. 14.1). First, we consider the y components.	$\Sigma F_y = 0$ $-F_{hy} + (F_T)(\sin 15.0°) - (215\text{ kg})(9.81\text{ m/s}^2) = 0$ $F_{hy} = \dfrac{(215\text{ kg})(9.81\text{ m/s}^2)(2.20\text{ m})}{(1.50\text{ m})(\sin 15.0°)}(\sin 15.0°) - (215\text{ kg})(9.81\text{ m/s}^2)$ $F_{hy} = 984\text{ N}$
Now, consider the x components.	$\Sigma F_x = 0$ $F_{hx} - \dfrac{(215\text{ kg})(9.81\text{ m/s}^2)(2.20\text{ m})}{(1.50\text{ m})(\sin 15.0°)}(\cos 15.0°) = 0$ $F_{hx} = 1.15\times 10^4\text{ N}$
With the two components of this vector, we can determine its magnitude and angle with respect to the horizontal.	$F_h = \sqrt{F_{hx}^2 + F_{hy}^2}$ $F_h = \sqrt{(1.15\times 10^4\text{ N})^2 + (984\text{ N})^2} = \boxed{1.16\times 10^4\text{ N}}$ $\theta = \tan^{-1}\left\|\dfrac{F_{hy}}{F_{hx}}\right\| = \tan^{-1}\left\|\dfrac{984}{1.15\times 10^4}\right\| = \boxed{4.89°}$
CHECK and THINK We found the force due to the hinge that produces a net force on the hinge. Since we know that the net torque is also zero from part (a), the robotic arm is in static equilibrium.	

35. (N) The owner of the Galaxy Café wishes to suspend a fictional spacecraft outside the front door (Fig. P14.35). The spacecraft weighs 1000 lb. The triangular structure is much lighter than the spacecraft, and its weight can be neglected. Use the values for length given in the figure and find the normal forces exerted by the pins at points A and B.

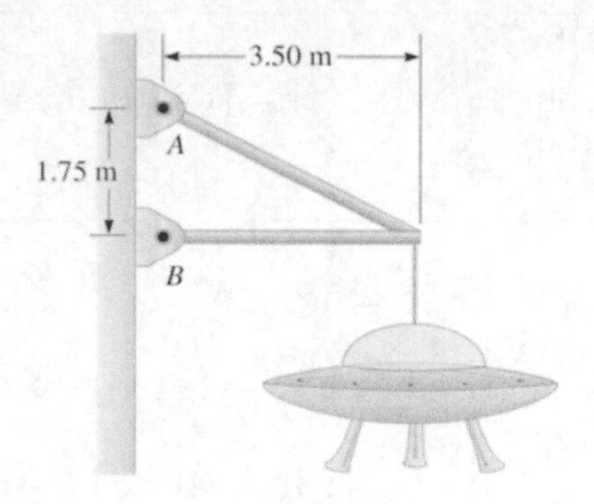

Figure P14.35

INTERPRET and ANTICIPATE	
Since the spacecraftis in static equilibrium, the net force and net torque on it are zero. We can start by sketching the spacecraft and labeling all the forces. We sketch a coordinate system, assuming that the wall will exert an outward force on the rod at point *B* and an inward force on the rod at point *A*. 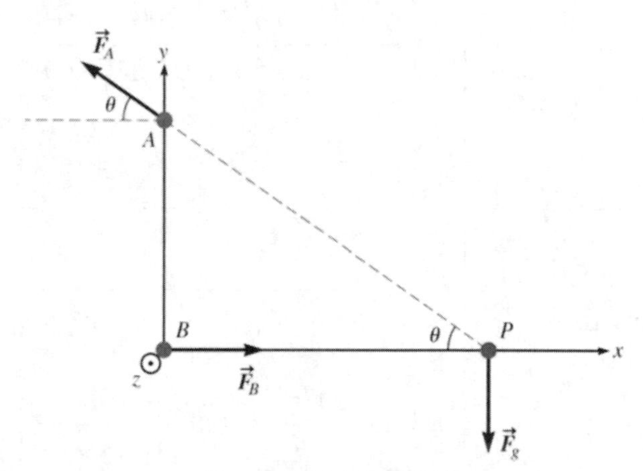 Figure P14.35ANS	
SOLVE Using trigonometry, we determine the angle indicated to find components of the force.	$\theta = \tan^{-1}\left(\frac{1.75}{3.50}\right)$
The spacecraft is in static equilibrium, so the torque acting at point *P* around point *B* is zero (Eq. 14.2). The two contributions to the torque are due to the weight of the sign acting straight down at a distance 3.50 m from the pivot point and the force up towards point *A*. (By using the pivot point at point *B*, the force F_B acts at $r = 0$ and does not produce a torque.) Set the magnitudes of these torques equal to each other	$\tau_g = \tau_A$ $F_g x = F_{A,\perp} x = F_A \sin\theta x$ $F_A = \frac{4448 \text{ N}}{\sin\left[\tan^{-1}\left(\frac{1.75}{3.50}\right)\right]} = \boxed{9.95\times10^3 \text{ N}}$

and convert the 1000 pounds (gravitational force) to 4448 Newtons.	
Now, we can use force balance in the x direction to determine the force at point B (Eq. 14.1). The only contributions are due to the rightward pointing force B and the leftward component of A.	$F_B = F_A \cos\theta$ $F_B = (9.95\times10^3\text{ N})\cos\left[\tan^{-1}\left(\frac{1.75}{3.50}\right)\right]$ $F_B = \boxed{8.90\times10^3\text{ N}}$

CHECK and THINK

For an object in static equilibrium, the net torque and net force must be zero, which allows us to relate the forces acting on the spacecraft.

38. At a museum, a 1300-kg model aircraft is hung from a lightweight beam of length 12.0 m that is free to pivot about its base and is supported by a massless cable (Fig. P14.38). Ignore the mass of the beam.

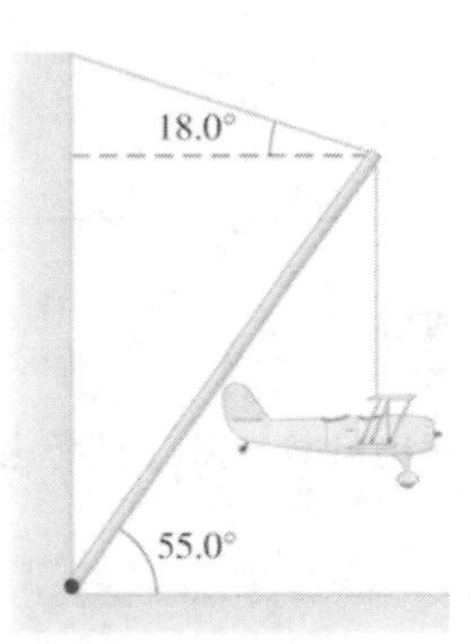

Figure P14.38

a. (N) What is the tension in the section of the cable between the beam and the wall?

INTERPRET and ANTICIPATE

The aircraft is in static equilibrium, which means we can use the fact that the net torque and the net force on the model aircraft is zero. We first sketch a diagram and label the forces.

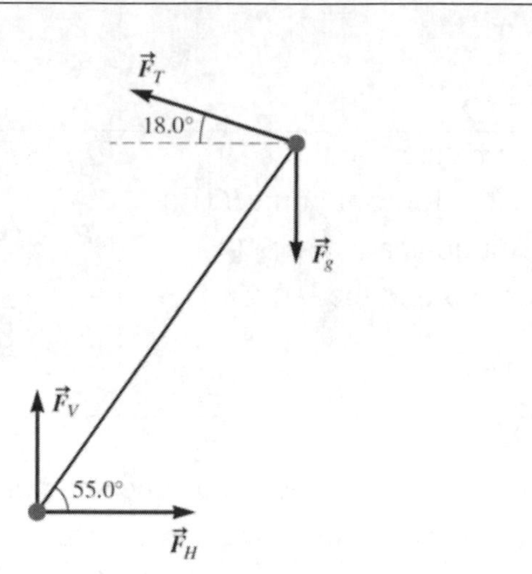

Figure P14.38ANS

SOLVE From the free-body diagram, the angle that the string tension makes with the beam can be found.	$\theta = 55.0° + 18.0° = 73.0°$
To determine the tension we can apply torque balance using the base as the rotation axis. This will allow us to relate the tension with the weight and disregard the forces on the base, for which $r = 0$ and therefore no torque is produced around this pivot. The perpendicular component of the string tension can be determined from the diagram as $F_T \sin 73.0°$. Now, add the torques around the base of the rod (Eq. 14.2).	$\sum \tau = 0$ $-(12.0\text{ m})(1300\text{ kg})(9.81\text{ m/s}^2)\cos 55.0° + F_T(12.0\text{ m})\sin 73.0° = 0$ $F_T = \dfrac{(12.0\text{ m})(1300\text{ kg})(9.81\text{ m/s}^2)\cos 55.0°}{(12.0\text{ m})\sin 73.0°}$ $F_T = \boxed{7.65 \times 10^3\text{ N}}$

CHECK and THINK

For an object in static equilibrium, the net torque and net force are zero. Here, we applied torque balance to solve for the unknown tension force.

b. (N) What are the horizontal and vertical forces that the pivot exerts on the beam?

<table>
<tr><td colspan="2">INTERPRET and ANTICIPATE
In part (a), we applied torque balance (Eq. 14.2) to determine the tension. We can also use the fact that the net force is zero (Eq. 14.1) to determine the unknown forces at the pivot.</td></tr>
<tr><td>SOLVE
First, use the fact that the net force in the x direction is zero.</td><td>$\sum F_x = 0$
$F_H - F_T \cos 18.0° = 0$
$F_H = F_T \cos 18.0°$
$F_H = \left[\frac{(12.0 \text{ m})(1300 \text{ kg})(9.81 \text{ m/s}^2)\cos 55.0°}{(12.0 \text{ m})\sin 73.0°}\right]\cos 18.0°$
$F_H = \boxed{7.27 \times 10^3 \text{ N}}$</td></tr>
<tr><td>Now, use the y components of the forces.</td><td>$\sum F_y = 0$
$F_V - F_T \sin 18.0° - (1300 \text{ kg})(9.81 \text{ m/s}^2) = 0$
$F_V = F_T \sin 18.0° + (1300 \text{ kg})g$
$F_V = \left[\frac{(12.0 \text{ m})(1300 \text{ kg})(9.81 \text{ m/s}^2)\cos 55.0°}{(12.0 \text{ m})\sin 73.0°}\right]\sin 18.0° + (1300 \text{ kg})(9.81 \text{ m/s}^2)$
$F_V = \boxed{1.51 \times 10^4 \text{ N}}$</td></tr>
<tr><td colspan="2">CHECK and THINK
In part (b), we applied the second equation at our disposal for static equilibrium, that the net force is zero, to determine the unknown force. With zero net force and net torque, there is no acceleration and the system is at rest.</td></tr>
</table>

41. (N) CASE STUDY In Example 14.3, we found that one of the steel cables supporting an airplane at the Udvar-Hazy Center was under a tension of 9.30×10^3 N. Assume the cable has a diameter of 2.30 cm and an initial length of 8.00 m before the plane is suspended on the cable. How much longer is the cable when the plane is suspended on it?

INTERPRET and ANTICIPATE
The weight of the plane exerts a tension on the cable, which causes it to stretch in response. The Young's modulus relates the tensile stress exerted on an object to the strain, or fractional elongation.

SOLVE Equation 14.11 relates the stress and strain. The Young's modulus for steel appears in Table 14.1.	$\sigma = Y\varepsilon$ $Y = 210\times10^9\ \text{N/m}^2$
The stress can be calculated with the force exerted on the cable divided by the cross-sectional area (Eq. 14.9). The radius of the cable is half the diameter or 0.0115 m.	$\sigma = \dfrac{9.30\times10^3\ \text{N}}{\pi(0.0115\ \text{m})^2} = 2.24\times10^7\ \dfrac{\text{N}}{\text{m}^2}$
Now substitute into the stress-strain relationship (Eq. 14.11) to solve for the strain.	$\varepsilon = \dfrac{\sigma}{Y} = \dfrac{2.24\times10^7\ \frac{\text{N}}{\text{m}^2}}{2.10\times10^{11}\ \frac{\text{N}}{\text{m}^2}} = 1.07\times10^{-4}$
The strain is equal to the elongation divided by the original length, which we can now solve for.	$\varepsilon = \dfrac{\Delta L}{L} = \dfrac{\Delta L}{8\ \text{m}} = 1.07\times10^{-4}$ $\Delta L = (1.07\times10^{-4})(8\ \text{m}) = \boxed{8.53\times10^{-4}\ \text{m}}$
CHECK and THINK The 8-meter long cable stretches by a little less than a millimeter. We'd be surprised if the steel cable stretched appreciably, so this sounds reasonable.	

44. (N) Consider a nanotube with a Young's modulus of 2.130×10^{12} N/m^2 that experiences a tensile stress of 5.3×10^{10} N/m^2. Steel has a Young's modulus of about 2.000×10^{11} Pa. How much stress would cause a piece of steel to experience the same strain as the nanotube?

INTERPRET and ANTICIPATE The strain experienced by an object can be expressed as $\varepsilon = \sigma/Y$. Since the strain is supposed to be the same for both the nanotube and the steel, we can write $\sigma_{\text{tube}}/Y_{\text{tube}} = \sigma_{\text{steel}}/Y_{\text{steel}}$.	
SOLVE Using the values given in the problem statement, we equate the strains and solve for the necessary stress on the steel.	$\sigma_{\text{tube}}/Y_{\text{tube}} = \sigma_{\text{steel}}/Y_{\text{steel}}$ $\sigma_{\text{steel}} = \left(\sigma_{\text{tube}}/Y_{\text{tube}}\right)Y_{\text{steel}}$ $\sigma_{\text{steel}} = \left[\left(5.3\times10^{10}\ \text{Pa}\right)/\left(2.130\times10^{12}\ \text{Pa}\right)\right]\left(2.000\times10^{11}\ \text{Pa}\right)$ $\sigma_{\text{steel}} = \boxed{5.0\times10^9\ \text{Pa}}$

CHECK and THINK
The nanotube is tough! It takes about ten times more stress to cause the equivalent strain in the nanotube.

49. (N) A tubular steel support is 3.0 m tall and shortens 0.60 mm under a compressive force of 70 kN. If the inner radius of the tube is 0.80 times the outer radius, what is the outer radius? Young's modulus for steel is 2.0×10^{11} Pa .

INTERPRET and ANTICIPATE	
We are provided information associated with the stress (force over area) and strain (change in length over length) for a compressed sample. Stress and strain are related by the Young's modulus, which is provided. We can calculate the strain, find the stress, and use the fact that stress is force per area to calculate the cross-sectional area of the tube. Knowing the ratio of the radii, we can solve for the outer radius.	
SOLVE Calculate the strain (change in length over length) and then use it with Equation 14.11 to find the stress, inserting the proper Young's modulus for steel.	$\varepsilon=\frac{\Delta L}{L}=\frac{0.0006\text{ m}}{3.0\text{ m}}=2.0\times10^{-4}$ $\sigma=Y\varepsilon=(2.0\times10^{11}\text{ N/m}^2)(2.0\times10^{-4})$ $\sigma=4.0\times10^{7}\text{ N/m}^2$
From the definition of stress (force over area), find the area of the tube.	$A=\frac{F}{\sigma}=\frac{70{,}000\text{ N}}{4.0\times10^{7}\text{ N/m}^2}$ $A=1.75\times10^{-3}\text{ m}^2$
Letting r be the *outer* radius, we write the cross-sectional area of the tube. We are told that the inner radius is 0.80 times r.	$A=\pi(r^2-(0.80r)^2)$ $A=0.36\pi r^2$
Solve for r and insert the numerical value of A.	$r=\sqrt{\frac{A}{0.36\pi}}$ $r=\sqrt{\frac{1.75\times10^{-3}\text{ m}^2}{0.36\pi}}=\boxed{0.039\text{ m}}$

CHECK and THINK

The tube has a diameter of 7.8 cm or just over 3", and this is reasonable for a strong material like steel. Tubular structures are favorable because they are strong to bending in addition to compression, but relatively lightweight compared to a solid rod.

53. (N) A steel rod has a radius of 15.0 cm and a length of 1.25 m. The rod is held firmly in place, and a 32.5-kN force is applied parallel to one face of the rod so that it is sheared. What is the (shear) strain of the rod? (If you worked Problem 50, compare your answers.)

INTERPRET and ANTICIPATE

This problem describes a shear experiment. The shear strain, the distance that the rod is flexed relative to its length when pushed perpendicular to the end, depends on the shear stress, how much force per cross-sectional area is exerted, and the strength of the materials (the shear modulus). If we can determine the stress and shear modulus, we can calculate the shear strain.

SOLVE	
We can use the definition of shear stress (Eq. 14.12) and shear modulus (Eq. 14.13) to calculate the shear strain in terms of the applied force, the area over which the force is applied, and the shear modulus. The area is the cross-sectional area of the rod.	$\gamma = \frac{\tau}{G}$ $\gamma = \frac{F}{GA}$ $\gamma = \frac{F}{G\pi r^2}$
We find the shear modulus from Table 14.1 and insert values. We keep three significant figures based on the values in the problem statement.	$\gamma = \frac{32.5\times10^3\ \text{N}}{\left(80\times10^9\ \frac{\text{N}}{\text{m}^2}\right)\pi(0.150\ \text{m})^2}$ $\gamma = \boxed{5.75\times10^{-6}}$

CHECK and THINK

This is about 2.6 times the value found in Problem 50, so the shear strain is larger than the compressive strain. That is, the rod would deflect more than it compresses, which probably agrees with your intuition.

56. (A) A metal wire of length L is stretched an amount ΔL by a weight F_g. What is the fractional change in the wire's volume, $\frac{\Delta V}{V}$?

INTERPRET and ANTICIPATE If we assume that the cross-sectional area of the wire stays constant, we can then determine the fractional change in volume. This is not necessarily the case, but we are not provided any additional information. (Another possibility might be that the volume of the wire does not change, in which case the cross-sectional area would decrease as we elongate the wire.) Assuming the cross-sectional area remains constant, we imagine that only the bonds between atoms along the wire are stretched slightly due to the tension.	
SOLVE If we were to assume that the cross-sectional area of the wire doesn't change and express the volume in terms of length and area as $V = AL$, we can calculate the result.	$\frac{\Delta V}{V} = \frac{A\Delta L}{AL} = \boxed{\frac{\Delta L}{L}}$
CHECK and THINK With out assumption that the cross-sectional area remains unchanged, the change in volume is directly related to the change in length.	

60. Bruce Lee was famous for breaking concrete blocks with a single karate chop. From slow-motion video, the speed of his 1.50-kg hand descending on a block was estimated to be 15.0 m/s, which decreased to a speed of 0.500 m/s in the 2.50×10^{-3} s during which his hand made contact with and broke through the block. The maximum shear stress a concrete block can be subjected to before breaking is 9.50×10^5 N/m^2.

a. (N) What was the force exerted by Lee's hand on the block?

INTERPRET and ANTICIPATE We are given the initial and final velocity and time of contact for the collision. Therefore, using Newton's second law, we can calculate the acceleration and then the force. From this, we can calculate the stress and compare to the maximum shear stress. I would bet that Bruce Lee breaks the board, but we'll see!	
SOLVE Apply Newton's second law.	$F = m\left(\frac{\Delta v}{\Delta t}\right)$ $F = (1.50 \text{ kg})\frac{(15.0 - 0.500)\text{ m/s}}{0.0025 \text{ s}}$ $F = \boxed{8.7 \times 10^3 \text{ N}}$
CHECK and THINK This is pretty significant – about 2000 pounds, though it's exerted over a very short time!	

b. (N) If a typical concrete block broken by Lee was 3.00 cm thick and 15.2 cm wide, what is the shear stress experienced by the concrete block?

INTERPRET and ANTICIPATE The stress is force over area cross-sectional area, which we can calculate with the information given.	
SOLVE Use the definition of shear stress as force over area (Eq. 14.12).	$\tau = \frac{F}{A} = \frac{8.7\times10^{3}\ \text{N}}{(0.030\ \text{m})(0.152\ \text{m})} = \boxed{1.91\times10^{6}\ \text{N/m}^2}$
CHECK and THINK We will compare this to the shear strength in part (c).	

c. (N) Will the concrete block succumb to Lee's karate chop?

$\boxed{\text{Yes}}$, the stress found in part (b) is larger than the maximum shear stress and more than sufficient to break the board. Bruce Lee successfully breaks through the block!

67. (N) While working out at the gym, a Marine finds that the fan (Fig. 14.16, page 402) is too low to keep her cool. She gets the idea to hang the fan from two cables. She removes its pedestal, and the fan's remaining mass is 10.3 kg. When the fan is off, the cables are vertical (Fig. P14.67A). When the fan is on, the cables make an angle θ with the vertical (Fig. P14.67B). She attaches the cables to the fan using pins that allow the fan to remain in the same orientation when it is operating. Assume the fan is exerting the maximum thrust (21.0 N). Use any of the necessary values given and derived in Example 14.5 to find θ.

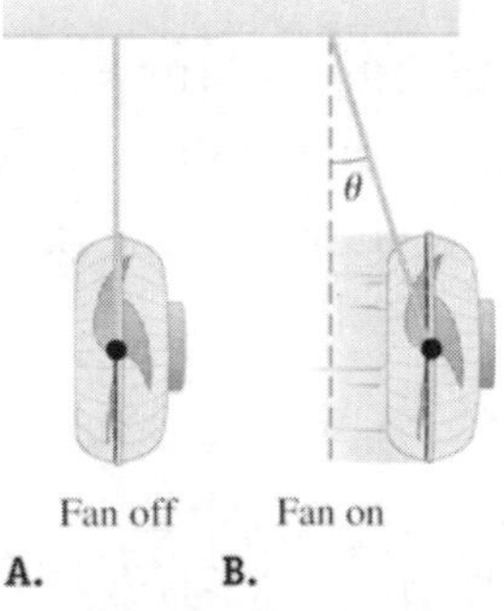

Figure P14.67

INTERPRET and ANTICIPATE The fan is in equilibrium, so we can draw a force diagram and require that the net force on the fan is zero.

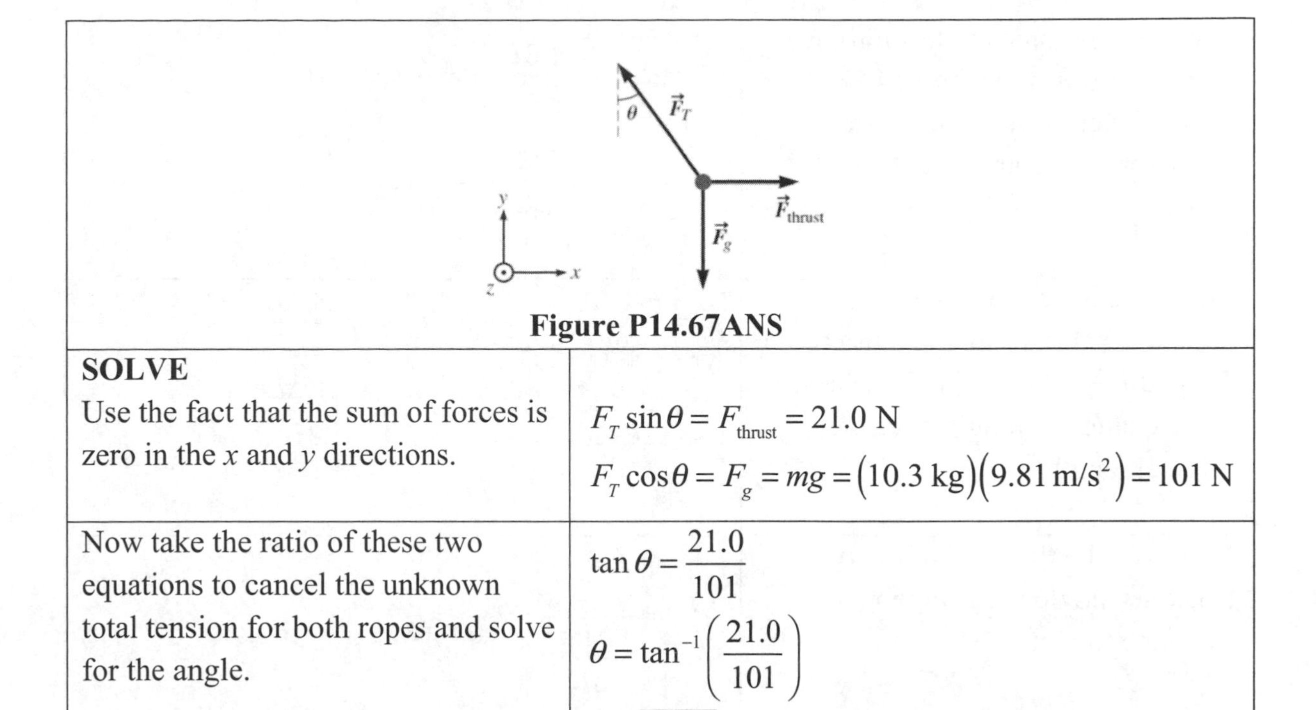

Figure P14.67ANS

SOLVE Use the fact that the sum of forces is zero in the x and y directions.	$F_T \sin\theta = F_{thrust} = 21.0\text{ N}$ $F_T \cos\theta = F_g = mg = (10.3\text{ kg})(9.81\text{ m/s}^2) = 101\text{ N}$
Now take the ratio of these two equations to cancel the unknown total tension for both ropes and solve for the angle.	$\tan\theta = \frac{21.0}{101}$ $\theta = \tan^{-1}\left(\frac{21.0}{101}\right)$ $\theta = \boxed{11.7^\circ}$

CHECK and THINK

The angle at which the fan is stable is the one for which the net force on the fan is zero. An angle of around 12 degrees sounds reasonable. We see that if we were to increase or decrease the thrust of the fan (by turning it to a higher or lower setting maybe), the angle would also increase or decrease respectively, as we'd expect.

70. (N) Two metal wires of the same material—lengths L and $3L$, respectively, and diameters $3D$ and D, respectively—are under different tensions that stretch each by the same amount ΔL. What is the ratio of the two tensions in the wires?

INTERPRET and ANTICIPATE

The wires are made of same material, so they have the same Young's modulus, which relates the stress (tension force over cross-sectional area) to the strain (fractional change in length). We would guess that the short, thick wire would resist stretching more than the long, thin wire, so we expect that it has a larger tension on it to produce the same change in length.

SOLVE Equation 14.11 relates stress and strain with the Young's modulus.	$\sigma = Y\varepsilon$

The stress is force (or tension) over cross-sectional area (Eq. 14.9), which depends on the diameter. We first express the areas of the two wires.	$A_1 = \pi r_1^2 = \pi\left(\frac{3D}{2}\right)^2 = \frac{9}{4}\pi D^2$ $A_2 = \pi r_2^2 = \pi\left(\frac{D}{2}\right)^2 = \frac{1}{4}\pi D^2$
The wires have the same Young's modulus Y. Let's say the tensions are F_{T1} and F_{T2}. The strain in each case is the change in length over the original length.	$Y = \frac{F_{T1}/A_1}{\Delta L/L} \rightarrow F_{T1} = Y\frac{A_1\Delta L}{L}$ $Y = \frac{F_{T2}/A_2}{\Delta L/3L} \rightarrow F_{T2} = Y\frac{A_2\Delta L}{3L}$
Now we take the ratio of the two tensions and do some algebra to calculate the ratio.	$\frac{F_{T1}}{F_{T2}} = \frac{\frac{Y\Delta LA_1}{L}}{\frac{Y\Delta LA_2}{3L}} = 3\frac{A_1}{A_2} = \boxed{27}$ or $F_{T1} = 27F_{T2}$

CHECK and THINK

As predicted, the shorter/thicker wire (wire 1) must have more tension on it in order to produce the same change in length… in this case, wire 1 must have a tension 27 times larger than for wire 2!

73. An aluminum rod has a radius of 15.0 cm and a length of 1.25 m. The rod is held firmly in place.

a. (N) What is the maximum compressive force that can be applied to one face?

b. (N) What is the maximum tensile force that can be applied to one face? (If you worked Problem 72, compare your answers.)

INTERPRET and ANTICIPATE

The maximum force is determined by the maximum tensile/compressive stress and the cross-sectional area according to Eq. 14.9, $F = \sigma A$. The maximum tensile and compressive stresses can be found from Table 14.1, from which we can determine the maximum force.

SOLVE The area in both cases is the same.	$A = \pi r^2$ $A = (3.14)(0.15\text{ m})^2$ $A = 0.0707\text{ m}^2$

a., b. Apply Eq. 14.9. The maximum compressive stress and maximum tensile stress are actually the *same* for both cases (Table 14.1), so the maximum force is also the same in both cases.	$F_{max} = \sigma A$ $F_{max} = \left(110\times10^6\ \text{N/m}^2\right)(\pi)(0.15\ \text{m})^2$ $F_{max} = \boxed{7.78\times10^6\ \text{N}}$

CHECK and THINK

Compared to steel, aluminum withstands about half as much force under compression and a third as much for tensile forces. (For steel, the compressive and tensile strengths are different.)

76. (N) An object is being weighed using an unequal-arm balance (Fig. P14.76). When the object is in the left pan, a downward force of 3.0 N must be exerted on the right pan to balance the gravitational force on the object. When the object is in the right pan, a downward force of 2.0 N must be exerted on the left pan to balance the gravitational force on the object. Determine the magnitude of the gravitational force on the object.

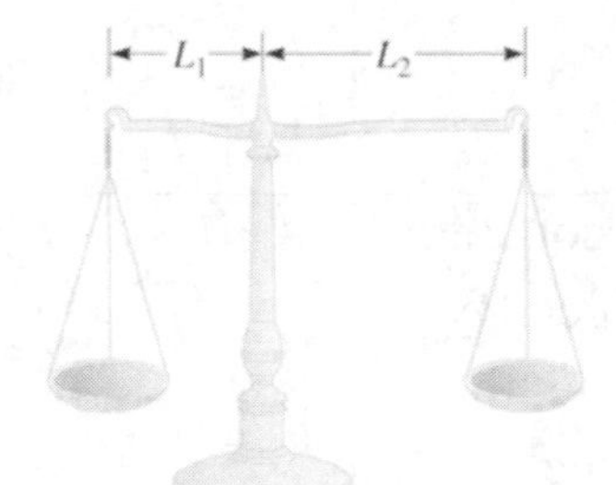

Figure P14.76

INTERPRET and ANTICIPATE

The object is first weighed on one pan and then on the other pan in this unequal-arm balance. When the balance is "balanced" it is not rotating and the net torque must be zero. We can set the net torque equal to zero for each case and determine the actual weight of the object.

SOLVE Let's say the gravitational force of the object is F_g and the left and right arms of the balance are L_1 and L_2 respectively. The object is weighed by putting it in the left pan as shown and applying a force F_1 = 3.0 N on the right.	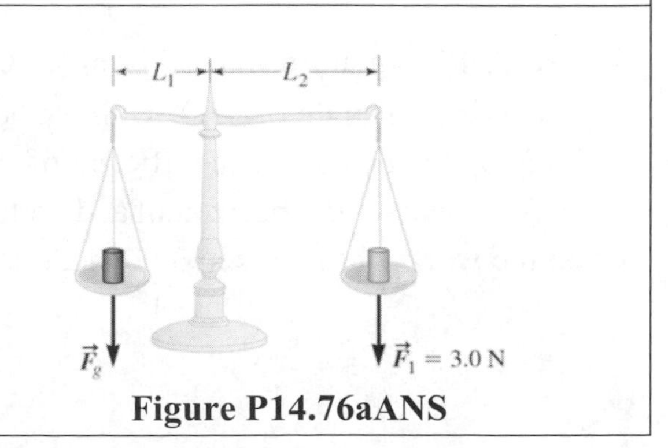 **Figure P14.76aANS**

The weight and the applied force produce torques in the opposite direction and the net torque must be zero for equilibrium.	$\tau_1 + \tau_2 = F_g L_1 - F_1 L_2 = 0$ $\frac{L_1}{L_2} = \frac{F_1}{F_g}$
Next, the object is weighed by putting it in the right pan and applying a force F_2 = 2.0 N on the left.	Figure P14.76bANS $\tau_1 + \tau_2 = F_2 L_1 - F_g L_2 = 0$ $\frac{L_1}{L_2} = \frac{F_g}{F_2}$
Now we use both equations and solve for the gravitational force of the object. By doing both experiments, it turns out that we don't actually need to know the lengths of the two arms.	$\frac{L_1}{L_2} = \frac{F_1}{F_g} = \frac{F_g}{F_2}$ $F_g = \sqrt{F_1 F_2}$ $F_g = \sqrt{(3.0\text{ N})(2.0\text{ N})}$ $F_g = \boxed{2.4\text{ N}}$
CHECK and THINK It's in the ballpark of the two forces we used to balance it, so it seems like the right order of magnitude. If the arms were equal, we'd expect the mass would balance with the same force in both cases, something between 2 and 3 Newtons. The key is that the net torque in this stationary balance is zero.	

83. (N) A 185-kg uniform steel beam 8.00 m in length rests against a frictionless vertical wall, making an angle of 55.0° with the horizontal. A 75.0-kg construction worker begins walking up the beam. The coefficient of static friction between the beam and the ground is 0.750. What are the horizontal and vertical forces exerted on the beam by the ground when the worker has walked 3.00 m along the beam?

INTERPRET and ANTICIPATE

We first sketch a force diagram. At equilibrium, the net force and net torque on the beam are zero, so let's write expressions for each and see if we can solve for the two forces.

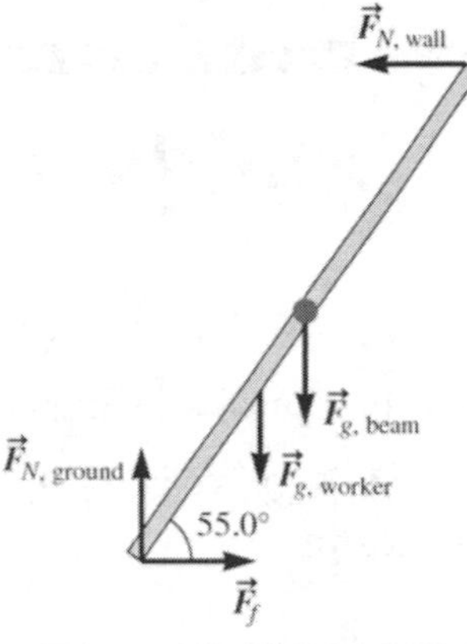

Figure P14.83ANS

SOLVE Considering the forces first (Eq. 14.1):	$\sum F_x = F_f - F_{N,\text{wall}} = 0$ $\sum F_y = F_{N,\text{ground}} - (185\text{ kg})g - (75.0\text{ kg})g = 0$
Using the y component, we solve for the normal force due to the ground.	$F_{N,\text{ground}} = \boxed{2.55\times10^3\text{ N}}$ (upwards).

Consider the torques about an axis at the foot of the beam and use the fact that the net torque is zero (Eq. 14.2).

$$\sum \tau = 0$$

$$-(185\text{ kg})g(4.00\text{ m})\sin35° - (75.0\text{ kg})g(3.00\text{ m})\sin35° + F_{N,\text{wall}}(8.00\text{ m})\cos 35° = 0$$

Solve the torque equation for $F_{N,\text{wall}}$.

$$F_{N,\text{wall}} = \frac{(185\text{ kg})\left(9.81\text{ m/s}^2\right)(4.00\text{ m})\sin35° + (75.0\text{ kg})\left(9.81\text{ m/s}^2\right)(3.00\text{ m})\sin35°}{(8.00\text{ m})\cos35°} = 829\text{ N}$$

Next substitute this value into the F_x equation.	$F_f = F_{N,\text{wall}} = 829\text{ N} = \boxed{8.29\times10^2\text{ N}}$ (the friction force is directed toward the wall)

CHECK and THINK

As with many of the problems in this chapter, if we can sketch a force diagram and use the facts that the net force and torque are zero for an object in static equilibrium. It is then a matter of algebra to determine the unknown forces.

15

Fluids

3. (N) Dry air is primarily composed of nitrogen. In a classroom demonstration, a physics instructor pours 2.00 L of liquid nitrogen into a beaker. After the nitrogen evaporates, how much volume does it occupy if its density is equal to that of the dry air at sea level? Liquid nitrogen has a density of 808 kg/m^3.

INTERPRET and ANTICIPATE We expect that the nitrogen will occupy a much larger volume than the 2 liters it occupies as a liquid. We can determine the mass of liquid nitrogen and use the density of dry air to determine the volume that this mass of nitrogen would occupy.	
SOLVE Using the density of liquid nitrogen, we can determine the mass of liquid nitrogen in 2 liters.	$\rho = \frac{m}{V}$ $m = \rho V$
A liter is a cubic decimeter or 0.001 m^3 and we're given the density.	$2l = 0.002 \text{ m}^3$ $m = \rho V = (808 \text{ kg/m}^3)(0.00200 \text{ m}^3)$
Using the density of dry air from Table 15.1, we can determine the volume for this mass of nitrogen.	$\rho_{\text{air}} = 1.29 \text{ kg/m}^3$ $V = \frac{m}{\rho_{\text{air}}} = \frac{(808 \text{ kg/m}^3)(0.00200 \text{ m}^3)}{1.29 \text{ kg/m}^3} = \boxed{1.25 \text{ m}^3}$
CHECK and THINK While the nitrogen expands by more than a factor of 600 in volume, it's still only about a cubic meter. Large spills of liquid nitrogen could be dangerous if it displaces sufficient oxygen in the room, but this seems like a relatively small amount of nitrogen compared to the volume of a classroom.	

8. One study found that the dives of emperor penguins ranged from 45 m to 265 m below the surface of the ocean.

a. (N) Using the density of seawater, what is the range of pressures experienced by the penguins between these two depths?

INTERPRET and ANTICIPATE We can use Eq. 15.6 to find the pressure at each depth. The pressure at the surface is the atmospheric pressure, 1.01325×10^5 Pa. We can then write the range of the pressures experienced.	
SOLVE First, find the pressure when the penguin is at a depth of 45 m.	$P_{45} = P_0 + \rho g y$ $P_{45} = 1.01325 \times 10^5 \text{ Pa} + (1025.18 \text{ kg/m}^3)(9.81 \text{ m/s}^2)(45 \text{ m})$ $P_{45} = 5.5 \times 10^5 \text{ Pa}$
Then, find the pressure when the penguin is at a depth of 265 m.	$P_{265} = P_0 + \rho g y$ $P_{265} = 1.01325 \times 10^5 \text{ Pa} + (1025.18 \text{ kg/m}^3)(9.81 \text{ m/s}^2)(265 \text{ m})$ $P_{265} = 2.8 \times 10^6 \text{ Pa}$
Then, the range of pressures can be expressed.	$\boxed{5.5 \times 10^5 \text{ Pa} \rightarrow 2.8 \times 10^6 \text{ Pa}}$
CHECK and THINK Emperor penguins dive to depths where the pressure is more than 27× atmospheric pressure!	

b. (N) At what depth below the surface is the pressure 10 times the normal atmospheric pressure?

INTERPRET and ANTICIPATE Again, use Eq. 15.6 to find the depth where the pressure is $10P_0$, or 10× the atmospheric pressure. The pressure at the surface is the atmospheric pressure, 1.01325×10^5 Pa. Given the answer to part (a), we expect this result to be between 45 m and 265 m.	
SOLVE Using Eq. 15.6, we can solve for the depth, y.	$P = 10P_0 = P_0 + \rho g y$ $\rho g y = 9P_0$ $y = 9P_0 / \rho g$
Substituting the known values, we find the depth.	$y = 9P_0 / \rho g$ $y = \left[9(1.01325 \times 10^5 \text{ Pa})\right] / \left[(1025.18 \text{ kg/m}^3)(9.81 \text{ m/s}^2)\right]$ $y = \boxed{90.7 \text{ m}}$

CHECK and THINK
As expected, the depth was between 45 m and 265 m. As long as we know the density of a fluid and the pressure at its surface, we can find the pressure at any depth in that fluid.

13. (N) Imagine a planet that has the same sea-level atmospheric pressure as the Earth, but whose gravitational acceleration is only 0.5*g*. What is the pressure difference between the bottom and the top of a 2.10-m deep swimming pool of water on this planet?

INTERPRET and ANTICIPATE	
The pressure at depth is proportional to the acceleration due to gravity, so we expect the pressure at the bottom of this pool would be half that for the same pool on Earth.	
SOLVE The pressure at depth *y* comes from Equation 15.6 with the acceleration due to gravity replaced by *g*/2. We solve this for the difference in pressure between the bottom and the top of the pool.	$P = \rho \frac{g}{2} y + P_0$ $P - P_0 = \rho \frac{g}{2} y$
Substitute values.	$P - P_0 = \left(1000\ \text{kg/m}^3\right)\left(\frac{9.81\,\text{m/s}^2}{2}\right)\left(2.10\ \text{m}\right)$ $P - P_0 = \boxed{1.03\times10^4\ \text{Pa}}$
CHECK and THINK We check our results by comparing this to the same pool's pressure difference on Earth. $P - P_0 = \rho g y$ $P - P_0 = \left(1000\ \text{kg/m}^3\right)\left(9.81\,\text{m/s}^2\right)\left(2.10\ \text{m}\right)$ $P - P_0 = 2.06\times10^4\ \text{Pa}$ As expected the pressure difference for the pool on the extraterrestrial planet is half the value on Earth.	

17. The dolphin tank at an amusement park is rectangular in shape with a length of 40.0 m, a width of 15.0 m, and a depth of 7.50 m. The tank is filled to the brim to provide maximum splash during dolphin shows.

a. (N) What is the total amount of force exerted by the water on the bottom of the tank?
b. (N) What is the total amount of force exerted by the water on the longer wall of the tank?

c. (N) What is the total amount of force exerted by the water on the shorter wall of the tank?

<table>
<tr><td colspan="2">INTERPRET and ANTICIPATE
The hydrostatic pressure increases with depth under the surface of the water. For the bottom surface (which is all at the same depth of course), we can use the fact that pressure is force over area to determine the force. For the long and short side walls, the pressure changes with depth, so we need to add up all the contributions (in other words, integrate) to find the total force.</td></tr>
<tr><td>SOLVE
a. The hydrostatic pressure (i.e. the pressure beyond atmospheric) on the bottom due to the water can be calculated using Equation 15.6.</td><td>$P_b = \rho g y = (1025.18\ \text{kg/m}^3)(9.81\ \text{m/s}^2)(7.50\ \text{m})$</td></tr>
<tr><td>The force depends on the pressure and area of the bottom surface (Eq. 15.1).</td><td>$F_b = P_b A$
$F_b = (1025.18\ \text{kg/m}^3)(9.81\ \text{m/s}^2)(7.50\ \text{m})(40.0\ \text{m})(15.0\ \text{m})$
$F_b = \boxed{4.53\times 10^7\ \text{N down}}$</td></tr>
<tr><td>b. For the side and end walls, the pressure varies with depth, so we need to add up, or integrate, the contributions to the total force at different depths. We integrate along the vertical direction since the pressure at each depth z is the same. On a strip of height dz and length L, we can write an expression for the small force dF caused by the water pressure over a height dz.</td><td>$dF = PdA = PLdz = \rho gzLdz$</td></tr>
</table>

Integrate from the top $z = 0$ of the swimming pool to the depth $z = h$ to find the total force.	$F = \int_0^h \rho g z L dz$ $F = \frac{1}{2}\rho g L h^2$ $F = \left(\frac{1}{2}\rho g h\right) L h \qquad (1)$
We have $\rho g y = (1025.18 \text{ kg/m}^3) \times (9.81 \text{ m/s}^2) \times (7.50 \text{ m})$ from part (a), so we can plug in values to get the total force on the longer side of the pool.	$F = P_{\text{average}} A$ $F = \frac{1}{2}(1025.18 \text{ kg/m}^3)(9.81 \text{ m/s}^2)(7.50 \text{ m})(40.0 \text{ m})(7.50 \text{ m})$ $F = \boxed{1.13 \times 10^7 \text{ N outward}}$
c. Use equation (1) from part (b) again to find the total force on the short side of the pool.	$F = P_{\text{average}} A$ $F = \frac{1}{2}(1025.18 \text{ kg/m}^3)(9.81 \text{ m/s}^2)(7.50 \text{ m})(15.0 \text{ m})(7.50 \text{ m})$ $F = \boxed{4.24 \times 10^6 \text{ N outward}}$

CHECK and THINK

The total force on the long side of the pool is larger than the force on the short side, which makes sense. The force on the bottom, which has a larger area and is at the highest pressure (being at the greatest depth), is even greater.

22. A spherical submersible 2.00 m in radius, armed with multiple cameras, descends under water in a region of the Atlantic Ocean known for shipwrecks and finds its first shipwreck at a depth of 1.75×10^3 m. Seawater has density 1.03×10^3 kg/m^3, and the air pressure at the ocean's surface is 1.013×10^5 Pa.

a. (N) What is the absolute pressure at the depth of the shipwreck?

INTERPRET and ANTICIPATE

The pressure at a given depth in the water is the hydrostatic pressure plus atmospheric pressure, the pressure at the surface of the ocean.

SOLVE The absolute pressure can be calculated with Equation	

15.6. We have everything need to plug in values and solve the equation.	$P = P_0 + \rho gh$ $P = 1.013\times10^5\ \text{Pa} + (1030\ \text{kg/m}^3)(9.81\ \text{m/s}^2)(1750\ \text{m})$ $P = \boxed{1.78\times10^7\ \text{Pa}}$
CHECK and THINK The pressure is significant – about 100 times larger than atmospheric pressure!	

b. (N) What is the buoyant force on the submersible at the depth of the shipwreck?

INTERPRET and ANTICIPATE The buoyant force equals the weight of the water displaced so it is independent of depth, assuming that the volume of the submersible does not change. Using this volume and the density of the water, we can determine this buoyant force.	
SOLVE Using Eq. 15.8, we calculate the buoyant force.	$F_B = \rho_f V_{\text{disp}} g$ $F_B = \rho \frac{4}{3}\pi r^3 g$ $F_B = (1030\ \text{kg/m}^3)\frac{4}{3}\pi(2.00\ \text{m})^3(9.81\ \text{m/s}^2)$ $F_B = \boxed{3.39\times10^5\ \text{N}}$
CHECK and THINK This is a large buoyant force (about 76,000 pounds!). It does not depend on the pressure of the fluid.	

25. A hollow copper ($\rho_{\text{Cu}} = 8.92 \times 10^3\ \text{kg/m}^3$) spherical shell of mass $m = 0.950$ kg floats on water with its entire volume below the surface.

a. (N) What is the radius of the sphere?

INTERPRET and ANTICIPATE Since the sphere is in equilibrium, the weight of the spherical shell (mg) must be equal to the buoyant force of the water. The displaced volume is the entire volume of the sphere that is submerged just below the surface.	
SOLVE The weight of the spherical shell (mg) must be equal to the buoyant force of the water (Eq. 15.8). The displaced volume is the entire	$F_g = F_B$ $mg = \rho_{\text{water}} \frac{4}{3}\pi r_{\text{outer}}^3 g$

volume of the sphere that is submerged, which depends on the outer radius of the sphere.	
Solve for the radius and insert values.	$r_{outer} = \left(\frac{3m}{4\pi\, \rho_{water}}\right)^{1/3}$ $r_{outer} = \left(\frac{3\times 0.950 \text{ kg}}{4\pi\; 1\,000 \text{ kg/m}^3}\right)^{1/3}$ $r_{outer} = 0.0610 \text{ m} = \boxed{6.10 \text{ cm}}$

CHECK and THINK

We don't have an expectation of what the value is, but this radius leads to the buoyant force equaling the weight of the sphere, which is what's needed for it to float just under the surface of the water.

b. (N) What is the thickness of the shell wall?

INTERPRET and ANTICIPATE

The mass of the ball depends on the density of copper times the volume. We have the mass and density, so we just need to find the inner radius needed for the copper shell to have the total volume that corresponds to the given mass.

SOLVE Density equals mass over volume. The volume of the spherical shell can be expressed as a spherical volume with an outer radius r_o found in part (a) minus a sphere with inner radius r_i. (If one assumes the shell is thin, one could also write the volume as the thickness of the shell times the surface area, $V \approx 4\pi r^2 \Delta r$.)	$m = \rho_{Cu} V$ $V = \left(\frac{4}{3}\pi r_o^3 - \frac{4}{3}\pi r_i^3\right)$

Substitute values and solve for the inner radius.	$m=\rho_{Cu}\left(\frac{4}{3}\pi r_o^3-\frac{4}{3}\pi r_i^3\right)$ $0.950\text{ kg}=8920\frac{\text{kg}}{\text{m}^3}\left(\frac{4}{3}\pi\right)\left((0.0610\text{ m})^3-r_i^3\right)$ $2.54\times10^{-5}\text{ m}^3=2.27\times10^{-4}\text{ m}^3-r_i^3$ $r_i=\left(2.01\times10^{-4}\text{ m}^3\right)^{1/3}=5.86\text{ cm}$
The thickness of the shell wall can now be found.	$\Delta r=r_0-r_i$ $\Delta r=6.10\text{ cm}-5.86\text{ cm}$ $\Delta r=0.237\text{ cm}=\boxed{2.37\text{ mm}}$
CHECK and THINK The shell is quite thin, but this seems believable for a metal shell that floats in water.	

30. Imagine a planet that has an ocean with the same seawater density as that of the Earth, but whose gravitational acceleration is only $g/2$. A 67.5-kg person is submerged in the planet's ocean. Assume the person's volume is the same as it is on the Earth, 0.57 m^3.

a. (N) What is the person's weight on the alien planet?
b. (N) What is the buoyant force exerted on the person?
c. (N) What is the person's acceleration, ignoring any drag forces on the person?

INTERPRET and ANTICIPATE In many ways this problem (especially part (c)) is like Example 15.1. So we'll use it as a guide. For part (a) we expect the person weighs less on the planet since the acceleration due to gravity is weaker. For part (c) we expect the person's acceleration depends on his or her density compared to seawater as described in Example 15.1.	
SOLVE **a.** The person's weight on the planet is given by mass times the gravitational acceleration. As expected this is half of the person's weight on Earth.	$F_g=m\left(\frac{g}{2}\right)=\frac{1}{2}mg$ $F_g=\frac{1}{2}(67.5\text{ kg})(9.81\text{ m/s}^2)$ $F_g=\boxed{331\text{ N}}$

b. The buoyant force equals the weight of the water displaced (Eq. 15.8). Remember that on this planet the gravitational acceleration is $g/2$.	$F_B = \rho_{\text{water}} V_{\text{person}} \left(\frac{g}{2}\right)$ $F_B = \frac{1}{2}(1025.18\text{ kg/m}^3)(0.57\text{ m}^3)(9.81\text{ m/s}^2)$ $F_B = \boxed{2.9\times10^3\text{ N}}$
c. We apply Newton's second law to find the acceleration as in Equation (1) in Example 15.1. Using an upward pointing y axis we write our final answer in component form.	$\sum F_y = F_B - F_g = m_{\text{person}} a_y$ $a_y = \frac{F_B - F_g}{m_{\text{person}}}$ $a_y = \frac{2.9\times10^3\text{ N} - 331\text{ N}}{67.5\text{kg}}$ $a_y = 38\text{ m/s}^2$ $\vec{a} = \boxed{38\hat{j}\text{ m/s}^2}$

CHECK and THINK

We already found that the person weighs less on the planet as expected. To check the direction of the acceleration, we find the person's density and compare it to seawater.

$$\rho_{\text{person}} = \frac{m_{\text{person}}}{V_{\text{person}}} = \frac{67.5\text{ kg}}{0.57\text{ m}^3}$$

$$\rho_{\text{person}} = 120\text{ kg/m}^3 < \rho_{\text{water}}$$

Since the person is less dense than seawater, we expect the acceleration to be upward as we found.

33. A rectangular block of Styrofoam 25.0 cm in length, 15.0 cm in width, and 12.0 cm in height is placed in a large tub of water. Assume the density of Styrofoam is 3.00×10^2 kg/m^3.

a. (N) What volume of the block is submerged?

INTERPRET and ANTICIPATE

Since the Styrofoam floats, its mass must be supported by the buoyant force, which depends on the volume of the block that's submerged.

SOLVE The block has length L, width W, and height H. The buoyant force (Eq. 15.8)	$F_B = F_g$ $F_B = \rho_f V_{\text{disp}} g$ $F_g = \rho_{\text{styrofoam}} V_{\text{block}} g$

supports the weight of the block at equilibrium. The volume of the object submerged is equal to the volume of the water displaced.	$\rho_{water} V_{disp} g = \rho_{styrofoam} V_{block} g$
Using $V_{object} = LWH$, solve for the volume displaced.	$V_{submerged} = V_{disp} = \frac{\rho_{styrofoam}}{\rho_{water}}(LWH)$ $V_{submerged} = (0.300)(25.0 \times 15.0 \times 12.0 \text{ cm}^3)$ $V_{submerged} = 1350 \text{ cm}^3 = \boxed{1.35 \times 10^{-3} \text{m}^3}$
CHECK and THINK 30% of the block is submerged in this case, which is plausible for a Styrofoam block. It is definitely floating!	

b. (N) A copper block is now placed atop the Styrofoam block so that the top of the Styrofoam block is level with the surface of the water. What is the mass of the copper block?

INTERPRET and ANTICIPATE The buoyant force (with the displaced volume equal to the total volume of the Styrofoam block) supports the weight of both blocks.	
SOLVE For the system to be in equilibrium, the buoyant force (now due to the entirely submerged Styrofoam block) must equal the weight of both the Styrofoam and copper blocks.	$F_B = M_{styrofoam} g + M_{copper} g$ $\rho_{water} V_{block} g = \rho_{styrofoam} V_{block} g + M_{copper} g$
Solve for the mass of the copper block and insert values.	$M_{copper} = (\rho_{water} - \rho_{styrofoam}) V_{block}$ $M_{copper} = \left((1.00 \times 10^3 - 3.00 \times 10^2) \frac{\text{kg}}{\text{m}^3}\right)(0.250 \times 0.150 \times 0.120 \text{ m}^3)$ $M_{copper} = \boxed{3.15 \text{ kg}}$

CHECK and THINK
Using the fact that the net force equals zero, we are able to determine when the total mass is balanced by the buoyant force.

38. The gauge pressure of an *empty* scuba tank is 500 psi, whereas a full tank is at 3000 psi. The tank is connected to an open-tube manometer.

a. (N) What is the height of the mercury if the tank is empty?
b. (N) What is the height of the mercury if the tank is full?
c. (C) CHECK and THINK step: Is it practical to use a manometer or a barometer on a scuba tank? Explain.

INTERPRET and ANTICIPATE The key to this problem is knowing that the manometer measures the gauge pressure.	
SOLVE First convert to SI pressure units.	500 psi = 3.45×10^6 Pa 3000 psi = 2.07×10^7 Pa
a. The gauge pressure is related to the height by Equation 15.19. When the tank is empty the gauge pressure is 3.45×10^6 Pa.	$P_{\text{gauge}} = \rho g y$ $y_{\text{empty}} = \dfrac{P_{\text{gauge}}}{\rho g} = \dfrac{3.45\times10^6 \text{ Pa}}{\left(13{,}600 \text{ kg/m}^3\right)\left(9.81 \text{ m/s}^2\right)} = \boxed{25.9 \text{ m}}$
b. The gauge pressure is related to the height by Equation 15.19. When the tank is full the gauge pressure is 2.07×10^7 Pa.	$y = \dfrac{P_{\text{gauge}}}{\rho g} = \dfrac{2.07\times10^7 \text{ Pa}}{\left(13{,}600 \text{ kg/m}^3\right)\left(9.81 \text{ m/s}^2\right)} = \boxed{155 \text{ m}}$
CHECK and THINK **c.** It is possible to easily measure the difference between 25.9 m and 155 m and so a manometer could be used on a SCUBA tank (though it would be a very large gauge!). The barometer measures atmospheric pressure not gauge pressure. It would not help you to find the pressure in the tank.	

41. (N) A hurricane cannot develop unless the height of the column of mercury in a barometer falls below 749 mm. During Hurricane Katrina in 2005, drops in the mercury level of barometers as large as 75.0 mm from the normal level of 760 mm were recorded. If normal atmospheric pressure is 1.013×10^5 Pa, what was the lowest atmospheric pressure recorded during this hurricane?

INTERPRET and ANTICIPATE Normal atmospheric pressure corresponds to 760 mm of mercury. A drop of 75.0 mm of mercury is a pressure change related to the change in height of mercury. We can find this change in pressure and determine the absolute pressure.	
SOLVE The change in pressure registered by the change in height of a barometer is given by Equation 15.6. We can insert values and solve for the change in pressure, which is negative since the mercury level has dropped.	$\Delta P = \rho g \Delta h$ $\Delta P = \left(13.6\times 10^3 \text{ kg/m}^3\right)\left(9.81 \text{ m/s}^2\right)\left(-75.0\times 10^{-3} \text{ m}\right)$ $\Delta P = -1.00\times 10^4 \text{ Pa}$
The total pressure is standard atmospheric pressure plus the change recorded by the barometer.	$P = P_0 + \Delta P = \left(1.013 - 0.100\right)\times 10^5 \text{ Pa}$ $P = \boxed{9.13\times 10^4 \text{ Pa}}$
CHECK and THINK This drop in pressure is about 10%, which is pretty significant.	

44. (N) Water enters a smooth, horizontal tube with a speed of 2.0 m/s and emerges out of the tube with a speed of 8.0 m/s. Each end of the tube has a different cross-sectional radius. Find the ratio of the entrance radius to the exit radius.

INTERPRET and ANTICIPATE The continuity equation tells us that if the water speed increased, it's because the cross-sectional area decreased. We can solve for this ratio given the speeds at both ends of the tube.	
SOLVE We use the continuity equation (Equation 15.21) to relate the fluid speed and cross-sectional area ($A = \pi r^2$ for a circle) for both sides of the tube.	$v_1 A_1 = v_2 A_2$ $\frac{v_2}{v_1} = \frac{A_1}{A_2} = \frac{\pi R_1^2}{\pi R_2^2}$

Solve for the ratio of the radii and insert values assuming side 1 is the inlet and side 2 the outlet.	$\frac{R_1}{R_2} = \sqrt{\frac{v_2}{v_1}}$ $\frac{R_1}{R_2} = \sqrt{\frac{8.0\text{ m/s}}{2.0\text{ m/s}}} = \boxed{2}$
CHECK and THINK As expected, the inlet radius is larger (twice as large) as the outlet radius, leading to the increase in water speed. This is similar to covering most of the end of a hose with your finger, making a small outlet opening and producing a fast water jet shooting out.	

47. When connected to a standpipe, a fire hose 5.08 cm in diameter must deliver a minimum flow rate of 1.00×10^3 L/min

a. (N) At this rate, with what speed does water leave the fire hose?
b. (N) A standard fire hose nozzle has an opening 0.950 cm in radius. What is the speed of the water leaving the fire hose with the nozzle attached?

INTERPRET and ANTICIPATE The volume flow rate equals the area of the hose times the fluid velocity. We can use the first two to find the third.	
SOLVE Volume flow rate equals the cross-sectional area times the speed of water flow.	$R = Av = \frac{1.00 \times 10^3\text{ L}}{1.00\text{ min}}$
a. The cross-sectional area of the hose is $A = \pi r^2 = \pi(2.54\text{ cm})^2$. Solve for v.	$v = \left(\frac{1.00 \times 10^3\text{ L}}{1.00\text{ min}}\right)\left[\frac{1}{\pi(2.54)^2\text{ cm}^2}\right]\left(\frac{1\text{ min}}{60\text{ s}}\right)\left(\frac{10^3\text{ cm}^3}{1\text{ L}}\right)$ $v = (822\text{ cm/s})\left(\frac{1\text{ m}}{10^2\text{ cm}}\right)$ $v = \boxed{8.22\text{ m/s}}$

b. We can use the volume flow rate equation again as we did in part (a)	$v=\left(\frac{1.00\times10^{3}\text{ L}}{1.00\text{ min}}\right)\left[\frac{1}{\pi(0.950)^{2}\text{ cm}^{2}}\right]\left(\frac{1\text{ min}}{60\text{ s}}\right)\left(\frac{10^{3}\text{ cm}^{3}}{1\text{ L}}\right)$ $v=(5880\text{ cm/s})\left(\frac{1\text{ m}}{10^{2}\text{ cm}}\right)$ $v=\boxed{58.8\text{ m/s}}$

CHECK and THINK

The volume flow rate depends on the cross-sectional area and water speed. When the cross-sectional area in decreased in part (a), the flow speed goes up. An in Problem 44, this is similar to covering most of the end of a hose with your finger, making a small outlet opening and producing a fast water jet shooting out.

We can also find the answer to part (b) in a different way, using the continuity equation (Eq. 15.21) to relate the speed and cross-sectional area in part (a) to the new speed occurring for a smaller cross sectional area.

$$v_2=\left(\frac{A_1}{A_2}\right)v_1$$

$$\frac{A_1}{A_2}=\left(\frac{\pi r_1^2}{\pi r_2^2}\right)=\left(\frac{r_1}{r_2}\right)^2=\left(\frac{2.54}{0.950}\right)^2=7.15$$

$$v_2=\left(\frac{A_1}{A_2}\right)v_1=(7.15)(8.22\text{ m/s})=58.8\text{ m/s}$$

We find the same answer as we'd expect.

55. Water is flowing in the pipe shown in Figure P15.55, with the 8.00-cm diameter at point 1 tapering to 3.50 cm at point 2, located y = 12.0 cm below point 1. The water pressure at point 1 is 3.20×10^4 Pa and decreases by 50% at point 2. Assume steady, ideal flow.

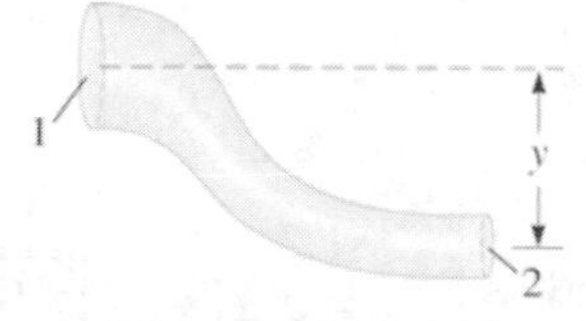

Figure P15.55

a. (N) What is the speed of the water at point 1?
b. (N) What is the speed of the water at point 2?

<table>
<tr><td colspan="2">INTERPRET and ANTICIPATE
We can apply both the continuity equation and Bernoulli's equation to relate the cross-sectional area of the pipe, water speed, height, and pressure.</td></tr>
<tr><td>SOLVE
a. Use the continuity equations (Eq. 15.21) to relate the water speeds at points 1 and 2.</td><td>$A_1v_1 = A_2v_2$
$\pi r_1^2 v_1 = \pi r_2^2 v_2$
$(4.00\text{ cm})^2 v_1 = (1.75\text{ cm})^2 v_2$
$v_2 = 5.22v_1$</td></tr>
<tr><td>For ideal flow, we can then apply Bernoulli's equation (Eq. 15.30).</td><td>$P_1 + \rho g y_1 + \frac{1}{2}\rho v_1^2 = P_2 + \rho g y_2 + \frac{1}{2}\rho v_2^2$</td></tr>
<tr><td colspan="2">For ideal flow, we can then apply Bernoulli's equation (Eq. 15.30), retaining all digits until applying the significant figure rules to the final answer.

$P_1 + \rho g y_1 + \frac{1}{2}\rho v_1^2 = P_2 + \rho g y_2 + \frac{1}{2}\rho v_2^2$
$3.20\times10^4 + (1.00\times10^3)(9.81)(0.120) + \frac{1}{2}(1.00\times10^3)(v_1)^2 =$
$1.60\times10^4 + \frac{1}{2}(1.00\times10^3)(5.22v_1)^2$
$\frac{1}{2}(1.00\times10^3)(5.22v_1)^2 - \frac{1}{2}(1.00\times10^3)(v_1)^2 =$
$3.20\times10^4 + (1.00\times10^3)(9.81)(0.120) - 1.60\times10^4$
$v_1^2 = \frac{1.71772\times10^4}{5.00\times10^2(5.22^2 - 1)}$
$v_1 = \boxed{1.14\text{ m/s}}$</td></tr>
<tr><td>b. From part (a), we have an expression for v_2 in terms of v_1 and we know the value of v_1. We need to retain all digits from the prior calculation and not apply the significant figure rules until this final calculation to get the answer shown here.</td><td>$v_2 = 5.22v_1 = \boxed{5.97\text{ m/s}}$</td></tr>
</table>

CHECK and THINK
Although the pressure decreases at point 2, the decrease in radius is the biggest factor in leading to a larger fluid velocity at point 2.

58. (N) Air flows horizontally with a speed of 108 km/h over a house that has a flat roof of area 20.0 m^2. Find the magnitude of the net force on the roof due to the air inside and outside the house. The density of air is 1.30 kg/m^3, and the thickness of the roof is negligible.

INTERPRET and ANTICIPATE	
There is no air flow just below the roof in the room and air flows at the given speed just above the roof. We should apply Bernoulli's theorem at points just below and just above the roof to determine the difference in pressure caused by the airflow.	
SOLVE Use Bernoulli's equation (Eq. 15.30) at points just below and just above the roof. There is no air flow just below the roof in the room ($v_1 = 0$) and air flows at the given speed just above the roof ($v_2 = v = 108$ km/h = 30.0 m/s). ρ is the density of air. h_1 and h_2 are the heights just below and above the roof – so $h_1 \approx h_2 = h$.	$0 + \rho g h + P_1 = \frac{1}{2}\rho v^2 + \rho g h + P_2$ $P_1 - P_2 = \Delta P = \frac{1}{2}\rho v^2$
The area of the roof is A. So we can find the aerodynamic force exerted on the roof.	$F = (\Delta P)A$ $F = \frac{1}{2}\rho v^2 A$ $F = \frac{1}{2}(1.30\ \text{kg/m}^3)(30.0\ \text{m/s})^2(20.0\ \text{m}^2)$ $F = \boxed{1.17 \times 10^4\ \text{N}}$
CHECK and THINK This is a substantial force, though probably not enough to damage our roof.	

62. (N) A spherical beach ball is 7.00 cm in radius and has a mass of 50.0 g. What is the force that must be exerted on the ball to keep it completely submerged in a swimming pool?

INTERPRET and ANTICIPATE At equilibrium $\sum F = 0$, so the applied force and weight of the ball must equal the magnitude of the upward buoyant force, which we can calculate.	
SOLVE The applied force plus the weight of the ball (both downwards) must exactly balance the upward buoyant force.	$F_{\text{app}} + mg = F_B$
The buoyant force equals the weight of the water displaced (Eq. 15.8).	$F_B = \rho_f V_{\text{disp}} g$
Solve for the applied force. $F_{\text{app}} = \rho_f V_{\text{disp}} g - mg = \rho_{\text{water}} \frac{4}{3}\pi r^3 g - mg$ $F_{\text{app}} = (1.00\times10^3 \text{ kg/m}^3)\frac{4}{3}\pi(0.0700 \text{ m})^3(9.81 \text{ m/s}^2) - (0.050 \text{ kg})(9.81 \text{ m/s}^2)$ $F_{\text{app}} = \boxed{13.6 \text{ N down}}$	
CHECK and THINK The applied force must balance the buoyant force. One tricky part is to remember to include the weight of the ball, which also contributes to the total downward force.	

66. (N) Imagine a planet that has a sea with the same water density (1250 kg/m^3) as the Dead Sea on the Earth, but whose gravitational acceleration is only *g*/2. A 67.5-kg person floats in the planet's sea. What fraction of the person is above the water's surface? Think about your answer by comparing it with what you would find if the person were on the Earth.

INTERPRET and ANTICIPATE In many ways this problem is like Example 15.2, so we'll use it as a guide. We expect that the fraction of the person above the water is the same on Earth as on the planet because in Example 15.2 we found that the answer on depends on the relative densities of the fluid, which is the same on both planets, and not the acceleration due to gravity.	
SOLVE This problem is similar to Example 15.2. The person is not accelerating so the	$F_B = F_g$

buoyant force must equal the weight.	
Let's find the person's weight on the planet.	$F_g = m\left(\frac{g}{2}\right) = \frac{1}{2}\rho_{\text{person}} V_{\text{person}} g$
The buoyant force equals the weight of the water displaced. The volume displaced is the person's total volume minus the volume above the surface.	$F_B = \rho_{\text{water}} V_{\text{disp}} \left(\frac{g}{2}\right)$ $F_B = \frac{1}{2}\rho_{\text{water}}\left(V_{\text{person}} - V_{\text{above}}\right) g$
Equate these forces and solve for the fraction of the body below the surface of the water. We assume the density of the person is around 1000 kg/m^3. This is similar to Equation 15.10, but here we find the fraction above the water instead of below.	$\frac{1}{2}\rho_{\text{water}}\left(V_{\text{person}} - V_{\text{above}}\right) g = \frac{1}{2}\rho_{\text{person}} V_{\text{person}} g$ $\frac{\rho_{\text{person}}}{\rho_{\text{water}}} = \frac{\left(V_{\text{person}} - V_{\text{above}}\right)}{V_{\text{person}}}$ $1 - \frac{V_{\text{above}}}{V_{\text{person}}} = \frac{\rho_{\text{person}}}{\rho_{\text{water}}}$ $\frac{V_{\text{above}}}{V_{\text{person}}} = 1 - \frac{1000\ \text{kg/m}^3}{1250\ \text{kg/m}^3} = \boxed{0.20}$

CHECK and THINK

The person is 80% submerged. This is the same as we find in Concept Exercise 15.5. We see that the expression does not depend on *g*, so it actually identical to Equation 15.10.

69. (N) A student is designing a piece of equipment that holds a camera and remains neutrally buoyant in water. That is, the device will remain at the same height at which it is placed under water. This device has a mass of 25 kg and a volume of 0.025 m^3. If it is instead brought to the ocean and released under water, what would be the acceleration experienced by the device? Would it accelerate up or down?

INTERPRET and ANTICIPATE

The acceleration of an object depends on the density of the object and the fluid. Since the piece of equipment is designed to have the same density of fresh water, which is smaller than the density of seawater, the object should accelerate upwards when submerged in the ocean.

SOLVE Equation 15.9 is an expression for the acceleration of an object	$a_y = \left(\frac{\rho_f - \rho_{\text{obj}}}{\rho_{\text{obj}}}\right) g$

submerged in a fluid.	
The density of seawater is given in Table 15.1. The density of the object is the same as fresh water, which we can confirm since the mass and volume are given.	$\rho_f = 1025.18\ \text{kg/m}^3$ $\rho_o = \dfrac{25\ \text{kg}}{0.025\ \text{m}^3} = 1.0\times10^3\ \text{kg/m}^3$
Solve for acceleration.	$a_y = \left(\dfrac{1025.18 - 1.0\times10^3}{1.0\times10^3}\right) 9.81\ \text{m/s}^2 = \boxed{0.25\ \text{m/s}^2}$

CHECK and THINK
As expected, the acceleration is positive (upwards). The relatively small difference in densities leads to a relatively small acceleration.

71. (N) The density of air in the Earth's atmosphere decreases according to the function $\rho = \rho_0 e^{-h/h_0}$, where $\rho_0 = 1.20\ \text{kg/m}^3$ is the density of air at sea level and h_0 is the scale height of the atmosphere, with an average value of 7640 m. What is the maximum payload that a balloon filled with $2.50 \times 10^3\ \text{m}^3$ of helium ($\rho_{\text{He}} = 0.179\ \text{kg/m}^3$) can lift to an altitude of 10.0 km?

INTERPRET and ANTICIPATE
As the balloon rises, the density of the atmosphere decreases, therefore the buoyant force (which is the weight of the fluid, in this case air, that is displaced) also decreases. Its weight remains unchanged. The balloon stops rising when the net force is zero.

SOLVE The forces acting on the balloon are the upward buoyant force given by Equation 15.8, the downward weight of the helium gas, and the downward weight of the payload of mass *M*. The displaced volume is the volume of the balloon and the mass of helium can be	$F_B - F_{g,\text{He}} - Mg = 0$ $\rho_{\text{air}} V_{\text{bal}} g - \rho_{He} V_{\text{bal}} g - Mg = 0$

expressed as the density of helium times the volume of the balloon.	
Solve for M.	$M = (\rho_{\text{air}} - \rho_{\text{He}})V$ $M = \left[(1.20\ \text{kg/m}^3)e^{-10000/7640} - 0.179\ \text{kg/m}^3\right](2.50\times10^3\ \text{m}^3)$ $M = \boxed{363\ \text{kg}}$

CHECK and THINK
The payload sounds possible (around 800 pounds) for a balloon, though this is the total mass of whatever it carries *and* the mass of the balloon itself.

76. (N) A mother is administering cold medicine with the density of water using an oral syringe to her child. The medicine is pushed by a plunger from the barrel with a cross-sectional area of 1.25 cm^2 into the opening with a cross-sectional radius of 1.20 mm. The syringe is held horizontally, and it is a distance of 7.00 cm in the horizontal direction and 5.00 cm in the vertical direction from the child's mouth. Assume atmospheric pressure is 1.00 atm and neglect air resistance.

a. What is the time of flight of the medicine from the syringe into the child's mouth?
b. With what speed must the medicine leave the opening of the syringe to reach the child's mouth?
c. What is the speed with which the mother must push the plunger for the medicine to reach the child's mouth?
d. What is the pressure of the medicine in the opening of the syringe?
e. What is the pressure in the barrel of the syringe?

INTERPRET and ANTICIPATE
It sounds like this problem will use many of our fluid equations. In particular, we'll likely have to consider the continuity equation and Bernoulli's equation, determine known values, and attempt to solve for unknown values. We expect that the fluid flow in the larger barrel of the syringe will be slower than at the narrow outlet.

SOLVE	
a. Since the syringe is fired horizontally, the emerging water stream has initial velocity components of ($v_{0x} = v_{\text{opening}}, v_{0y} = 0$). It travels at a constant speed horizontally and falls due to gravity. It must fall 5 cm in order	$\Delta y = v_{0y}t + \frac{1}{2}a_y t^2$

to reach the child's mouth. So, using a kinematic equation and $a_y = -g$ we can determine the time for the fluid to fall 5 cm.	$t = \sqrt{\frac{2(\Delta y)}{a_y}}$ $t = \sqrt{\frac{2(-0.0500\text{ m})}{-9.81\text{ m/s}^2}}$ $t = \boxed{0.101\text{ s}}$
b. Since $a_x = 0$, and $v_{0x} = v_{\text{opening}}$, the horizontal range of the emergent stream is $\Delta x = v_{\text{nozzle}} t$ where t is the time of flight from part (a). Thus, the speed of the water emerging from the opening can be found.	$v_{\text{opening}} = \frac{\Delta x}{t}$ $v_{\text{opening}} = \frac{0.0700\text{ m}}{0.101\text{ s}}$ $v_{\text{opening}} = \boxed{0.693\text{ m/s}}$
c. From the continuity equation (Eq. 15.21), we can calculate the speed of the water in the larger cylinder.	$A_1 v_1 = A_2 v_2$ $v_1 = \left(\frac{\pi r_2^2}{\pi r_1^2}\right) v_{\text{opening}}$ $v_1 = \left(\frac{\pi (0.120\text{ cm})^2}{1.25\text{ cm}^2}\right)(0.693\text{ m/s})$ $v_1 = \boxed{0.0251\text{ m/s}}$
d. The pressure at the nozzle is 1.000 atm, or atmospheric pressure to four significant digits.	$P_2 = \boxed{1.013\times10^5\text{ Pa}}$
e. We apply Bernoulli's equation on the barrel and at the opening of the syringe. With the barrel and opening both horizontal, $y_1 = y_2$ the gravity terms from Bernoulli's equation (Eq. 15.30) can be neglected, leaving only the pressure and kinetic terms.	$P_1 + \frac{1}{2}\rho_w v_1^2 = P_2 + \frac{1}{2}\rho_w v_2^2$
Now, we can solve for the pressure in the large cylinder. $P_1 = P_2 + \frac{\rho_w}{2}\left(v_2^2 - v_1^2\right)$ $P_1 = 1.013\times10^5\text{ Pa} + \frac{1.00\times10^3\text{ kg/m}^3}{2}\left[(0.693\text{ m/s})^2 - (0.0251\text{ m/s})^2\right]$ $P_1 = \boxed{1.015\times10^5\text{ Pa}}$	
CHECK and THINK The first couple parts were based on our old friends, the kinematic equations. We see that	

inside the barrel of the syringe, the fluid velocity is lower and the pressure is slightly higher than at the narrower outlet.

80. Carolyn, living on the fourth floor of an apartment building, opens a faucet tap 1.25 cm in radius. The tap is 12.0 m above the main water supply pipe to the building. The main water pipe is 4.00 cm in radius and has a volume flow rate of 1.50 L per second.

a. (N) What is the speed of the water exiting Carolyn's faucet?

INTERPRET and ANTICIPATE We are given the flow rate, which depends on the cross-sectional area and fluid velocity. Given the first two, we can calculate the third.	
SOLVE The flow rate of the water is the same throughout the pipe and depends on the area of the pipe and fluid velocity.	$R = Av = 1.50\text{ L/s} = 1.50\times10^{3}\text{ cm}^{3}/\text{s}$
The area of the faucet tap is $A = \pi(1.25\text{ cm})^2 = 4.91\text{ cm}^2$, so we can find the velocity.	$v = \dfrac{R}{A}$ $v = \dfrac{1.50\times10^{3}\text{ cm}^{3}/\text{s}}{4.91\text{ cm}^{2}}$ $v = 306\text{ cm/s} = \boxed{3.06\text{ m/s}}$
CHECK and THINK The speed sounds like a plausible value.	

b. (N) If no other taps are open in the building, what is the gauge pressure in the main water supply pipe?

INTERPRET and ANTICIPATE Since we are asked to find the pressure in a pipe in which the height is changing, we should immediately think of Bernoulli's equation.	
SOLVE We choose point 1 to be in the entrance pipe and point 2 to be at the faucet tap. Apply Bernoulli's	$P_1 - P_2 = \dfrac{1}{2}\rho\left(v_2^2 - v_1^2\right) + \rho g\left(y_2 - y_1\right)$

equation (Eq. 15.30).	
We use the continuity equation (Eq. 15.21), $A_1v_1 = A_2v_2$, to relate the velocity of the water at points 1 and 2.	$v_1 = \left(\frac{A_2}{A_1}\right)v_2$ $v_1 = \left(\frac{\pi(1.25\text{ cm})^2}{\pi(4.00\text{ cm})^2}\right)(3.06\text{ m/s})$ $v_1 = 0.298\text{ m/s}$

Inserting this into Bernoulli's equation, we can find the pressure difference, which is the gauge pressure.

$$P_1 - P_2 = \frac{1}{2}\left(10^3\text{kg/m}^3\right)\left[\left(3.06\text{ m/s}\right)^2 - \left(0.298\text{ m/s}\right)^2\right] + \left(10^3\text{kg/m}^3\right)\left(9.81\text{ m/s}^2\right)\left(12.00\text{ m}\right)$$

$$P_{\text{gauge}} = P_1 - P_2 = \boxed{1.22\times10^5\text{ Pa}}$$

CHECK and THINK

Though we might not have an intuition about what to expect, the gauge pressure is larger than atmospheric and drives the fluid up to the faucet.

16

Oscillations

4. (N) A simple harmonic oscillator's position is given by $y(t) = (0.850 \text{ m})\cos(10.4t - 5.20)$. Find the oscillator's position, velocity, and acceleration at each of the following times.

a. $t = 0$

b. $t = 0.500$ s

c. $t = 2.00$ s

INTERPRET and ANTICIPATE	
The oscillator has a position given by a cosine function (as in Equation 16.3), which includes the amplitude, angular frequency, and phase. The velocity is the derivative of this expression with respect to time (a sine function, given by Equation 16.6) and the acceleration is the derivative of velocity with respect to time (Equation 16.12). Using these expressions, we can determine the necessary information from the position function and insert this along with the desired time in the expressions for velocity and acceleration.	
SOLVE Using Eq. 16.3, we can determine the amplitude, angular frequency, and phase by matching terms.	$y(t) = (0.850 \text{ m})\cos(10.4t - 5.20)$ $y(t) = y_{max}\cos(\omega t + \varphi)$ $y_{max} = 0.850$ m $\omega = 10.4$ rad/s $\varphi = -5.20$ rad
Insert these into Eq. 16.6 to find an expression for velocity that depends on time.	$v_y(t) = -y_{max}\omega\sin(\omega t + \varphi)$ $v_y(t) = -(0.850 \text{ m})(10.4 \text{ rad/s})\sin(10.4t - 5.20)$ (1)
Similarly, apply Eq. 16.12 to find the acceleration.	$a_y(t) = -\omega^2 x$ $a_y = -(10.4 \text{ rad/s})^2(0.850 \text{ m})\cos(10.4t - 5.20)$ (2)

Substitute times into the given equation for position and equations (1) and (2).

Time (s)	y (m)	v_y (m/s)	a_y (m/s^2)
0	0.398	−7.81	−43.1
0.500	0.850	0	−91.9
2.00	−0.845	0.953	91.4

CHECK and THINK

Given the expression for position, we can determine other kinematic properties, like velocity and acceleration. The acceleration and the position have opposite signs as expected for simple harmonic motion.

7. (N) The equation of motion of a simple harmonic oscillator is given by $x(t) = (8.0\text{ cm})\cos(10\pi t) - (6.0\text{ cm})\sin(10\pi t)$, where t is in seconds.

a. Find the amplitude.

b. Determine the period.

c. Determine the initial phase.

INTERPRET and ANTICIPATE

Since the given expression includes a cosine plus sine term, we can expand our usual formula to try to match terms. In particular, we can use a trigonometric expression, $\cos(A+B) = \cos A \cos B - \sin A \sin B$.

SOLVE Using a trigonometric identity, we expand the expression for the position of an oscillator (Eq. 16.3).	$\cos(A+B) = \cos A \cos B - \sin A \sin B$ $x(t) = x_{\text{max}} \cos(\omega t + \phi)$ $x(t) = x_{\text{max}} \cos(\omega t)\cos\phi - x_{\text{max}} \sin(\omega t)\sin\phi$
a. Now match the amplitude of each term to write two equations that contain two unknown values (x_{max} and ϕ).	$x_{\text{max}} \cos\phi = 8.0\text{ cm}$ and $x_{\text{max}} \sin\phi = 6.0\text{ cm}$
We have two equations and two unknowns. There are a few ways to solve this, but we can isolate the amplitude by squaring both equations and adding them to use the identity $\sin^2\theta + \cos^2\theta = 1$. As an alternative, we can solve for the phase angle (done in part (c) and substitute it into one of the equations to solve for x_{max}.)	$x_{\text{max}}^2 \cos^2\phi + x_{\text{max}}^2 \sin^2\phi = (8.0\text{ cm})^2 + (6.0\text{ cm})^2$ $x_{\text{max}}^2 \left(\cos^2\phi + \sin^2\phi\right) = 100\text{ cm}^2$ $x_{\text{max}}^2 = 100\text{ cm}^2$ $x_{\text{max}} = 10\text{ cm} = \boxed{0.10\text{ m}}$

b. This matches the equation provided only if $\omega = 10\pi$ rad/s. Use the definition of the angular frequency to determine the period.	$T = \frac{2\pi}{\omega} = \frac{2\pi}{10\pi}$ $T = \boxed{0.20\text{ s}}$
c. Finally, the phase angle. For instance, divide the expressions in part (a) to cancel out the x_{max}.	$\frac{x_{max}\sin\phi}{x_{max}\cos\phi} = \frac{6.0}{8.0}$ $\tan\phi = 0.75$ $\phi = \boxed{37°}$

CHECK and THINK

Given an expression for the oscillator, we can write it in a form where we can identify terms. In this case, it required a trigonometric identity, but we are able to isolate and solve for each value.

11. A 1.50-kg mass is attached to a spring with spring constant 33.0 N/m on a frictionless, horizontal table. The spring–mass system is stretched to 4.00 cm beyond the equilibrium position of the spring and is released from rest at $t = 0$.

a. (N) What is the maximum speed of the 1.50-kg mass?

INTERPRET and ANTICIPATE

The maximum speed can be determined using the maximum amplitude and the angular frequency using Equation 16.7, $v_{max} = y_{max}\omega$.

SOLVE We are given the amplitude of motion, y_{max} = 4.00 cm. The mass oscillates at an angular frequency given by Equation 16.26.	$\omega_s = \sqrt{\frac{k}{m}} = \sqrt{\frac{33.0\text{ N/m}}{1.50\text{ kg}}} = 4.69\text{ rad/s}$
Now, apply Eq. 16.7.	$v_{max} = y_{max}\omega$ $v_{max} = (4.00\times10^{-2}\text{ m})(4.69\text{ rad/s})$ $v_{max} = \boxed{0.188\text{ m/s}}$

CHECK and THINK

The maximum speed can be found with the amplitude and angular velocity.

b. (N) What is the maximum acceleration of the 1.50-kg mass?

INTERPRET and ANTICIPATE	
The maximum acceleration can now be found easily using Eq. 16.10, $a_{max} = y_{max}\omega^2$.	
SOLVE Apply Eq. 16.10.	$a_{max} = y_{max}\omega^2$ $a_{max} = \left(4.00\times10^{-2}\text{ m}\right)\left(4.69\text{ rad/s}\right)^2$ $a_{max} = \boxed{0.880\text{ m/s}^2}$
CHECK and THINK As in part (a), with the given quantities, the calculation is straightforward and sounds like a plausible value.	

c. (A) What are the position, velocity, and acceleration of the 1.50-kg mass as functions of time?

INTERPRET and ANTICIPATE	
We now want to write an expression for position, similar to Equation 16.3. $y(t) = y_{max}\cos(\omega t + \varphi)$, with which we can determine the velocity and acceleration. We have the maximum position and angular velocity already, so we just need to determine the phase.	
SOLVE Because y_{max} = 4.00 cm and v = 0 at t = 0, the required solution is $y(t) = y_{max}\cos\omega t$. In other words, the phase must be zero ($\varphi = 0$) so that the maximum in position occurs at t = 0.	$y(t) = y_{max}\cos\omega t$ $\boxed{y(t) = 4.00\cos(4.69t)}$ (position in cm and time in s)
The velocity is the derivative of position with respect to time. For a harmonic oscillator, we can also substitute the values we've determined into the velocity equation (Eq. 16.6).	$v_y(t) = \dfrac{dy}{dt}$ $v_y(t) = 4.00\left[-4.69\sin(4.69t)\right]$ $\boxed{v_y(t) = -18.8\sin(4.69t)}$ or $v_y(t) = -y_{max}\omega\sin(\omega t + \varphi)$ $v_y(t) = -(4.00)(4.69)\sin(4.69t + 0)$ $v_y(t) = -18.8\sin(4.69t)$

	(speed in cm/s and time in s)
The acceleration is the derivative of velocity with respect to time. For a harmonic oscillator, we can also substitute the values we've determined into the acceleration equation (Eq. 16.12).	$a_y(t) = \frac{dv}{dt}$ $a_y(t) = -18.8[4.69\cos(4.69t)]$ $\boxed{a_y(t) = -88.0\cos(4.69t)}$ or $a_y(t) = -\omega^2 y(t)$ $a_y(t) = -(4.69)^2 4.00\cos(4.69t)$ $a_y(t) = -88.0\cos(4.69t)$ (acceleration in cm/s^2 and time in s)
CHECK and THINK In the case of velocity and acceleration, we can either take derivatives of the position as a function of time or use the equations from the text given quantities like the angular frequency and amplitude. Both give the same expressions, as we would expect.	

15. (N) A point on the edge of a child's pinwheel is in uniform circular motion as the wheel spins counterclockwise with a frequency of 1.53 Hz. The point is at the location $x = 30.00$ cm and $y = 0$ when a stopwatch is started to track the motion (Fig. P16.15).

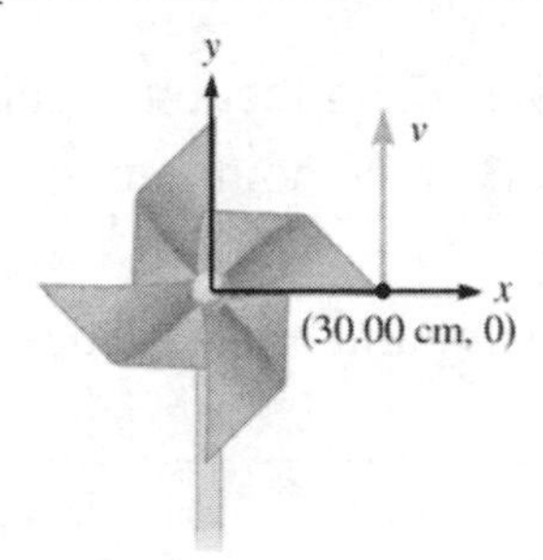

Figure P16.15

a. What is the period of the circular motion?

INTERPRET and ANTICIPATE Knowing the frequency of the motion, we can find the period using Eq. 16.1.	
SOLVE Use Eq. 16.1, and solve for the period.	$T = 1/f = 1/1.53\text{ Hz} = \boxed{0.654\text{ s}}$

CHECK and THINK
Anytime we know the frequency, angular frequency, or period, we can quickly find the other two quantities using Eq. 16.1 and Eq. 16.2.

b. What is the velocity of the point at the instant described?

INTERPRET and ANTICIPATE	
At the location, the velocity is directed in the positive y direction, as indicated by the yellow arrow in Figure P16.15. Since the point is in uniform circular motion, the magnitude of the velocity, the speed, is given by the circumference divided by the period. We will finish by writing the final vector for the velocity. The radius of the path is 0.3000 m in SI units.	
SOLVE Use the circumference of the path and the period to find the speed.	$v = 2\pi r/T = 2\pi(0.3000 \text{ m})/(1/1.53 \text{ Hz}) = 2.88 \text{ m/s}$
Now, because the velocity points in the positive y direction at the point in question, we write the velocity of the point at this moment in time.	$\vec{v} = \boxed{2.88\hat{j} \text{ m/s}}$
CHECK and THINK As time goes on and the point moves, the direction of the velocity will change, always pointing tangent to the circular path, as the point moves counterclockwise.	

c. What is the acceleration of the point at the instant described?

INTERPRET and ANTICIPATE	
Since the point is in uniform circular motion, it is subject only to centripetal acceleration, directed toward the center of motion. At the point under consideration, the centripetal acceleration will point in the negative x direction. We will finish by writing the final vector for the acceleration. The radius of the path is 0.3000 m in SI units.	
SOLVE Solve for the centripetal acceleration of the point, using the speed found in part (b).	$a_c = v^2/r$ $a_c = [2\pi(0.3000 \text{ m})/(1/1.53 \text{ Hz})]^2/(0.3000 \text{ m})$ $a_c = 27.7 \text{ m/s}^2$

Now, because the acceleration points in the negative x direction at the point in question, we write the acceleration of the point at this moment in time.	$\vec{a} = \boxed{-27.7\hat{\imath} \text{ m/s}^2}$

CHECK and THINK
As the point moves in uniform circular motion, the acceleration will always point towards the center of the motion. Thus, it will change direction constantly, as the point moves. Its magnitude, however, will remain constant (27.7 m/s^2).

19. (C, N) A uniform plank of length L and mass M is balanced on a fixed, semicircular bowl of radius R (Fig. P16.19). If the plank is tilted slightly from its equilibrium position and released, will it execute simple harmonic motion? If so, obtain the period of its oscillation.

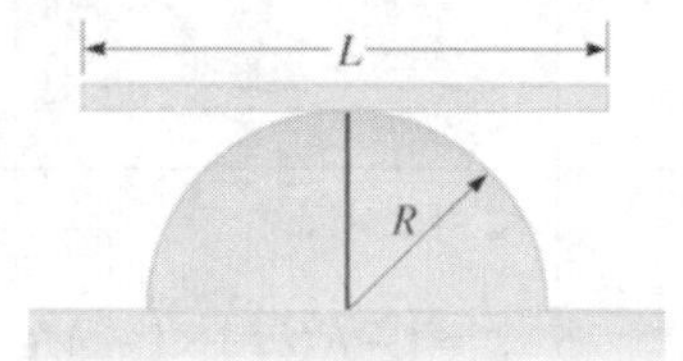

Figure P16.19

INTERPRET and ANTICIPATE
We start by sketching the situation. The key is that as it tilts, the point of contact moves to point P. The force on the board, which acts at its center of mass, actually exerts a torque on the board around point P causing it to return to a horizontal position. With the torque and rotational inertia of the board around the pivot point, we can determine the oscillation frequency.

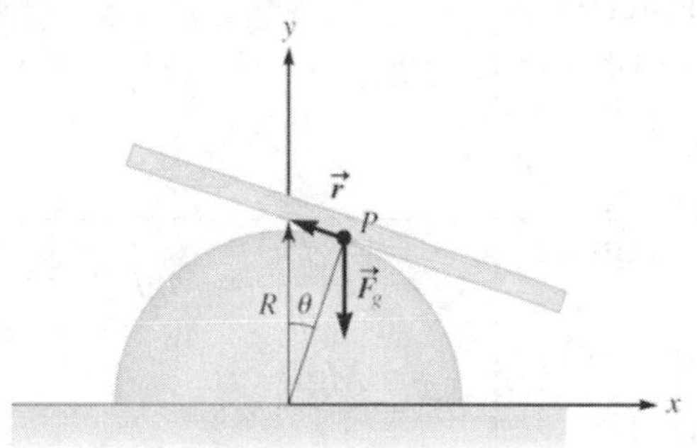

Figure P16.19ANS

SOLVE To determine the torque, we need the expressions for the	$\vec{F}_g = -Mg\,\hat{\jmath}$

gravitational force and the vector $\vec{r}$. The force acting on the board is the gravitational force on the plank.	
The position vector from point P to the center of the board is approximately $R\theta$ in length (using the arc length of a circular section) and is inclined at an angle θ. We use the small angle approxomation.	Small angle approximation: $\cos\theta \approx 1, \quad \sin\theta \approx \theta, \quad \theta \gg \theta^2$
Now, we can write an expression for $\vec{r}$.	$\vec{r} = -R\theta\cos\theta\,\hat{i} + R\theta\sin\theta\,\hat{j} \approx -R\theta\,\hat{i} + R\theta^2\,\hat{j} \approx -R\theta\,\hat{i}$
Find the torque acting on the board. The positive torque does indeed cause the board to rotate counter-clockwise and acts as a restoring force and it will approximately exhibit simple harmonic motion.	$\vec{\tau} = \vec{r} \times \vec{F}_g$ $\vec{\tau} \approx \left(-R\theta\,\hat{i}\right) \times \left(-Mg\,\hat{j}\right)$ $\vec{\tau} \approx MgR\theta\,\hat{k}$
To find the oscillation frequency, use Newton's second law for rotation. I is the rotational inertia of the plank about its center of mass.	$I = \frac{1}{12}ML^2$
The motion is SHM as the angular acceleration $\alpha = \frac{d^2\theta}{dt^2}$ is proportional to the angular displacement θ and we expect, from the hallmark of simple harmonic motion (Eq. 16.12), that they will obey a relationship like	$\tau_z = -I\frac{d^2\theta}{dt^2} = MgR\theta$ $\omega^2 = \frac{MgR}{I}$

$\frac{d^2\theta}{dt^2} = -\omega^2\theta$. This results in Equation 16.32 and 16.33, which can be used to find the angular frequency and then the period.	$\omega^2 = \frac{MgR}{\frac{1}{12}ML^2}$ $\omega = \sqrt{\frac{12gR}{L^2}}$ $T = \frac{2\pi}{\omega} = \boxed{\frac{\pi L}{\sqrt{3gR}}}$
CHECK and THINK An oscillation occurs when the plank is displaced from balanced position and experiences a torque to bring it back to equilibrium. While we can't tell if the expression is right by looking at it, it makes sense that a longer board would oscillate more slowly $(T \propto L)$ and, similar to a pendulum, if we could somehow increase gravity, it would increase the restoring force and decrease the period of oscillation $\left(T \propto 1/\sqrt{g}\right)$.	

23. (N) It is important for astronauts in space to monitor their body weight. In Earth orbit, a simple scale only reads an apparent weight of zero, so another method is needed. NASA developed the body mass measuring device (BMMD) for Skylab astronauts. The BMMD is a spring-mounted chair that oscillates in simple harmonic motion (Fig. P16.23). From the period of the motion, the mass of the astronaut can be calculated. In a typical system, the chair has a period of oscillation of 0.901 s when empty. The spring constant is 606 N/m. When a certain astronaut sits in the chair, the period of oscillation increases to 2.37 s. Determine the mass of the astronaut.

Figure P16.23

INTERPRET and ANTICIPATE   We first write an expression for the period of the oscillation, which is related to the spring constant and total mass that is oscillating. From the data given, the effective mass of the BMMD chair can be determined. The new period will allow calculation of the total mass, from which the astronaut's mass can be found.

SOLVE Combine equation 16.2 and 16.27 to get an expression for the period of the oscillation. Solve this for the mass.	$T = \frac{2\pi}{\omega} = 2\pi\sqrt{\frac{m}{k}}$ $m = \frac{kT^2}{4\pi^2}$
Insert the numerical values to find the mass of the chair alone.	$m_{\text{chair}} = \frac{(606)(0.901^2)}{4\pi^2}$
Now put in the numerical values for the astronaut/chair to find the total mass.	$m_{\text{tot}} = \frac{(606)(2.37^2)}{4\pi^2}$
Subtract the mass of the chair to find the mass of the astronaut.	$m_{\text{astronaut}} = \frac{(606)(2.37^2)}{4\pi^2} - \frac{(606)(0.901^2)}{4\pi^2} = \boxed{73.8 \text{ kg}}$

CHECK and THINK

This mass corresponds to a weight on earth of 723 N or 162 lb, which is quite reasonable for an astronaut. The BMMD actually uses an electronic timer to determine the period and is calibrated to calculate the astronaut's mass automatically.

26. (N) In an undergraduate physics lab, a simple pendulum is observed to swing through 75 complete oscillations in a time period of 2.25 min.

a. What is the period of the pendulum?

The period is the time per oscillation and we are given the information to calculate this directly.

$$T = \frac{(2.25 \text{ min})}{75 \text{ oscillations}}\left(\frac{60 \text{ s}}{1 \text{ min}}\right) = \boxed{1.8 \text{ s}}$$

b. What is the length of the pendulum?

INTERPRET and ANTICIPATE

The period of a pendulum depends on its length and the acceleration due to gravity.

SOLVE The length of the pendulum and acceleration due to gravity determine the period according to Eq. 16.2 and 16.29.	$\omega_{\text{smp}} = \frac{2\pi}{T} = \sqrt{\frac{g}{\ell}}$
Using this expression, we solve for the length using the period from part (a).	$\ell = g\left(\frac{T^2}{4\pi^2}\right)$ $\ell = \left(9.81\text{ m/s}^2\right)\left(\frac{(1.8\text{ s})^2}{4\pi^2}\right)$ $\ell = \boxed{0.81\text{ m}}$
CHECK and THINK A period of a couple seconds for a meter long pendulum sounds reasonable.	

32. (N) A simple pendulum is constructed from a bob of mass m = 150 g and a lightweight string of length ℓ = 1.50 m.

a. What is the period of oscillation for this pendulum in a physics lab at sea level?

b. What is the period of oscillation for this pendulum in an elevator accelerating upward at 2.00 m/s^2?

c. What is the period of oscillation for this pendulum in an elevator accelerating downward at 2.00 m/s^2?

d. What is the period of oscillation for this pendulum in a school bus accelerating horizontally at 2.00 m/s^2?

INTERPRET and ANTICIPATE The period of a pendulum depends on its length and the acceleration due to gravity. In this case, the length of the pendulum remains constant while the gravitational acceleration changes. We'll find a general expression for the period and determine the correct acceleration for each case.	
SOLVE The period of a pendulum depends on its length and the acceleration due to gravity according to Eq. 16.2 and 16.29, but instead of using the usual g for gravity on earth, we consider an "effective gravitational	$T = 2\pi\sqrt{\frac{\ell}{g_{\text{eff}}}}$

acceleration" g_{eff}, which can be different.	
a. At sea level, the acceleration *is* just the acceleration due to gravity, 9.81 m/s^2.	$T = 2\pi\sqrt{\frac{1.50\ \text{m}}{9.81\ \text{m/s}^2}}$ $T = \boxed{2.46\ \text{s}}$
b. The string tension must support the weight of the bob *and* accelerate it upwards, just as if the elevator were at rest in a larger gravity field of (9.81 + 2.00) m/s^2.	$T = 2\pi\sqrt{\frac{1.50\ \text{m}}{(9.81+\ 2.00)\ \text{m/s}^2}}$ $T = \boxed{2.24\ \text{s}}$
c. Similar to part (b), except the effective gravitational field is smaller. (This is perhaps easier to picture. If the elevator was in free fall, there would be no restoring force to drive the pendulum at all and the period becomes infinite.)	$T = 2\pi\sqrt{\frac{1.50\ \text{m}}{\left(9.81\ \text{m/s}^2 - 2.00\ \text{m/s}^2\right)}}$ $T = \boxed{2.75\ \text{s}}$
d. We again need the effective acceleration, which includes both a vertical (gravity) and horizontal (bus) component. The net effect due to both the gravitational field and the acceleration of the car is found using the Pythagorean theorem and can then be used to calculate the period. In this case, since the horizontal acceleration is still relatively small compared to *g*, the answer turns out to be very close to the value without horizontal acceleration.	$g_{\text{eff}} = \sqrt{\left(9.81\ \text{m/s}^2\right)^2 + \left(2.00\ \text{m/s}^2\right)^2} = 10.0\ \text{m/s}^2$ $T = 2\pi\sqrt{\frac{1.50\ \text{m}}{10.0\ \text{m/s}^2}}$ $T = \boxed{2.43\ \text{s}}$

CHECK and THINK

In each case, the oscillation period depends on the (constant) length of the pendulum and the total acceleration (effective gravitational acceleration). A larger net acceleration leads

to a smaller period and vice versa.

37. (N) Three thin sticks of equal mass, each of length 20.0 cm, are connected at the ends to form an equilateral triangle. When the triangle is pivoted around an axis perpendicular to the plane of the triangle through one of the vertices, what is the period of the triangle's oscillation as a physical pendulum?

INTERPRET and ANTICIPATE We follow Example 16.5 and find the rotational inertia of the triangle about the pivot. Based on our experience with other pendulums, we expect the result to be independent of the mass.	
SOLVE We start by drawing a sketch.	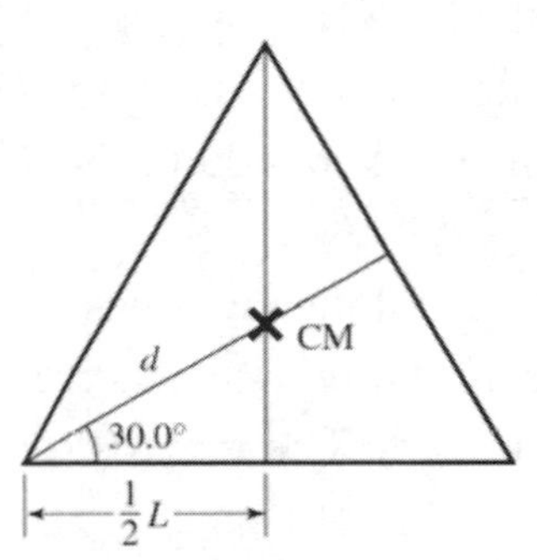 **Figure P16.37ANS**
Referring to the sketch, two sides (A and B) are thin rods of mass *m* and length $L = 20$ cm that rotate about their ends.	$I_A = I_B = \frac{1}{3}mL^2$
For the third side, we use the parallel axis theorem for a rod rotating about its center of mass with the center of mass a distance $r_{cm} = L \sin 60° = \frac{\sqrt{3}}{2}L$ from the pivot point.	$I_C = I_{cm} + mr_{cm}^2 = \frac{1}{12}mL^2 + \frac{3}{4}mL^2 = \frac{5}{6}mL^2$
The rotational inertia for a complex object is equal to the sum of the individual pieces. Combining all three sides give the total rotational inertia about the pivot.	$I = \frac{2}{3}mL^2 + \frac{5}{6}mL^2 = \frac{3}{2}mL^2$

Lastly, we need the distance d from the center of mass of the triangle to the pivot, which is half the distance r_{cm}. From Example 16.5 we get the period of the pendulum.	$T = 2\pi\sqrt{\frac{I}{mgd}}$ $T = 2\pi\sqrt{\frac{\frac{3}{2}mL^2}{mg\frac{L}{\sqrt{3}}}}$ $T = 2\pi\sqrt{\frac{3\sqrt{3}L}{2g}}$ $T = 2\pi\sqrt{\frac{3\sqrt{3}(0.200\text{ m})}{2(9.81\text{ m/s}^2)}}$ $T = \boxed{1.45\text{ s}}$

CHECK and THINK

The final result just depends on the length of the side of the triangle, as expected. For comparison, a simple pendulum with the same period would be 52 cm long.

41. (N) A restaurant manager has decorated his retro diner by hanging (scratched) vinyl LP records from thin wires. The records have a mass of 180 g, a diameter of 12 in., and negligible thickness. The records oscillate as torsion pendulums.

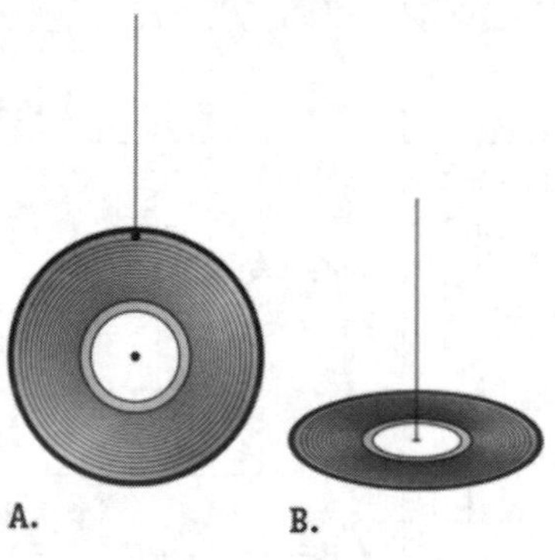

Figure P16.41

a. Records hung from a small hole near their rims have a period of roughly 3.5 s (Fig. P16.41A). What is the torsion spring constant of the wire?

INTERPRET and ANTICIPATE

The angular frequency of a torsion pendulum only depends of the torsion spring constant and the rotational inertia. We can use the period information to find the angular frequency. We can find the rotational inertia, and then find the spring constant.

SOLVE The angular frequency of a torsion pendulum depends on the torsional spring constant and the rotational inertia (Equation 16.36).	$\omega_{tor} = \sqrt{\frac{\kappa}{I}}$
The rotational inertia can be found starting with Table 13.1 for a cylinder rotated around an axis perpendicular to it and through its center of mass. In the case of a vinyl LP, the length is nearly zero (i.e. it is a thin disc).	$I_{CM} = \frac{1}{4}MR^2 + \frac{1}{12}M\ell^2$ $I_{CM} \approx \frac{1}{4}MR^2$
The angular frequency depends on the period (Eq. 16.2).	$T = \frac{2\pi}{\omega}$
Equation 16.36 can be solved for the torsion spring constant.	$\omega_{tor} = \sqrt{\frac{\kappa}{I}}$ $\frac{2\pi}{T} = \sqrt{\frac{4\kappa}{MR^2}}$ $\kappa = \left(\frac{2\pi}{T}\right)^2 \frac{MR^2}{4} = \frac{\pi^2 MR^2}{T^2}$
Now, substitute numerical values after converting to SI units (12 in = 0.305 m, 1000 g = 1 kg).	$\kappa = \frac{\pi^2 (0.180\ \text{kg})(0.305\ \text{m})^2}{(3.5\ \text{s})^2}$ $\kappa = \boxed{1.3\times10^{-2}\ \text{N}\cdot\text{m}}$
CHECK and THINK While it is difficult to have an intuition about the torsion spring constant, it seems reassuring that the value we found here is between the watch and the millennium clock (Example 16.6).	

b. If a record is hung from its center hole using a wire of the same torsion spring constant (Fig. P16.41B), what is its period of oscillation?

INTERPRET and ANTICIPATE Changing the way the wire is attached changes the rotational inertia.	
SOLVE The rotational inertia can be found starting with Table 13.1 for a cylinder or disk rotated around an axis parallel to it and through its center of mass.	$I_{CM} = \frac{1}{2}MR^2$
Equation 16.36 can then be applied. Solve for the period.	$\omega_{tor} = \sqrt{\frac{\kappa}{I}}$ $\frac{2\pi}{T} = \sqrt{\frac{2\kappa}{MR^2}}$ $T = 2\pi\sqrt{\frac{MR^2}{2\kappa}}$
As in part (a), substitute metric values.	$T = 2\pi\sqrt{\frac{(0.180\text{ kg})(0.305\text{ m})^2}{2(1.3\times10^{-2}\text{ N}\cdot\text{m})}}$ $T = \boxed{4.9\text{ s}}$
CHECK and THINK In this configuration, the pendulum has more rotational inertia that in the previous part. The greater rotational inertia means a slower angular frequency and a longer period.	

44. (N) A box of mass 0.900 kg is attached to a spring with $k = 125$ N/m and set into simple harmonic motion on a frictionless, horizontal table. The amplitude of motion is 5.00 cm.

a. What is the total energy of the box–spring system?

INTERPRET and ANTICIPATE
The mechanical energy (potential plus kinetic) is constant as the box oscillates. When the box is at the maximum amplitude, the energy is entirely potential energy, so we can simply solve for this value to find the total energy for any point in the motion.

SOLVE The energy is initially entirely potential energy, so we use Eq. 16.44.	$E = \frac{1}{2}ky_{\max}^2$ $E = \frac{1}{2}(125 \text{ N/m})(5.00\times10^{-2} \text{ m})^2$ $E = \boxed{0.156 \text{ J}}$
CHECK and THINK Given the maximum amplitude and spring constant, we easily calculate the spring potential energy in Joules.	

b. What is the speed of the box when the spring is compressed by 2.00 cm?

INTERPRET and ANTICIPATE The total mechanical energy is constant, which we know from part (a). So, we'll equate this value to the value at the given position and solve for speed.	
SOLVE At any point in the oscillatory motion, the total energy (kinetic plus potential) equals 0.156 J (part (a)).	$E = \frac{1}{2}ky^2 + \frac{1}{2}mv^2 = 0.156 \text{ J}$
Solve for the speed.	$v = \sqrt{\frac{2(0.156 \text{ J}) - ky^2}{m}}$
Insert numerical values.	$v = \sqrt{\frac{2(0.156 \text{ J}) - (125 \text{ N/m})(2.00\times10^{-2} \text{ m})^2}{0.900 \text{ kg}}}$ $v = \boxed{0.540 \text{ m/s}}$
CHECK and THINK The speed sounds plausible for an oscillating mass on a spring.	

c. What is the kinetic energy of the box at this position?
d. What is the potential energy of the spring–box system at this position?

INTERPRET and ANTICIPATE
The kinetic energy depends on the mass and speed, both of which we know. The potential plus kinetic energy must equal the total energy from part (a).

SOLVE **c.** First, determine the kinetic energy.	$K=\frac{1}{2}mv^2$ $K=\frac{1}{2}(0.900\text{ kg})(0.540\text{ m/s})^2$ $K=\boxed{0.131\text{ J}}$
d. The remaining energy must be potential energy.	$U=E-K$ $U=0.156\text{ J}-0.131\text{ J}$ $U=\boxed{0.025\text{ J}}$
CHECK and THINK After parts (a) and (b), it is straightforward to calculate the kinetic and potential energies, which add up to the total mechanical energy.	

46. (N) A block of mass m = 1.23 kg is attached to the end of a spring with a spring constant of 565 N/m. The block rests on a frictionless surface, is pulled to the right, and is held there. When released, the block undergoes simple harmonic motion. The block's maximum speed is 7.12 m/s. There is an instant in the motion when the potential energy of the system is equal to the kinetic energy of the block. At what position x does this situation occur?

INTERPRET and ANTICIPATE When the potential energy is equal to the kinetic energy, the potential energy is equal to half of the total energy in the object-spring system. By expressing half of the total energy at the equilibrium position (when all the energy is kinetic), we can equate this to the potential energy at the point of interest and then solve for the unknown position.	
SOLVE First, express the total energy at the equilibrium position.	$E=\frac{1}{2}mv_{\text{max}}^2=\frac{1}{2}(1.23\text{ kg})(7.12\text{ m/s})^2$
Then, express the potential energy at the point of interest and equate it to one half of the total energy. Solve for the unknown position	$U=\frac{1}{2}E$ $\frac{1}{2}kx^2=\frac{1}{2}\left(\frac{1}{2}mv_{\text{max}}^2\right)$

	$\frac{1}{2}(565\text{ N/m})x^2 = \frac{1}{2}\left[\frac{1}{2}(1.23\text{ kg})(7.12\text{ m/s})^2\right]$ $x = \sqrt{\frac{\frac{1}{2}(1.23\text{ kg})(7.12\text{ m/s})^2}{(565\text{ N/m})}} = \boxed{\pm 0.235\text{ m}}$
CHECK and THINK The block could be on either side of the equilibrium position, 0.235 m away. Note that this is not one half of the amplitude of the motion, $A = 0.332$ m. We could quickly confirm that the object has less potential energy, and thus more kinetic energy, when $x = (1/2)A$.	

50. (N) The bob of a simple pendulum is displaced by an initial angle of 22.0° and released from rest. Because of friction, the amplitude of oscillation is observed to be half the initial value after 765 s. What is the value of the time constant τ for damping for this pendulum?

INTERPRET and ANTICIPATE With friction, the pendulum slows such that the peak amplitude decreases exponentially with a characteristic damping time. We can use Equation 16.53 to represent the amplitude of the damped oscillator and Eq. 16.52 as the definition of the time constant.	
SOLVE Considering Eq. 16.52 and 16.53. Since the information provided is for the amplitude, it is only the factor in front of the cosine term that we need to be concerned with.	$\theta(t) = \theta_{\text{max}} e^{-bt/2m}$
We are told that $\theta(t=0) = 22.0°$ and $\theta(t=765) = 11.0°$.	$\frac{\theta_{765}}{\theta_0} = \frac{Ae^{-bt/2m}}{A} = \frac{11.0}{22.0} = e^{-b(765)/2m}$
Now, solve for the damping time.	$\ln\left(\frac{11.0}{22.0}\right) = -0.693 = \frac{-b(765)}{2m}$ $\tau = \frac{2m}{b} = \frac{765\text{ s}}{0.693} = \boxed{1.10\times10^3\text{ s}}$

CHECK and THINK
The damping time is the amount of time for the amplitude to fall to 1/e, or about 37%, of its initial value. Since the amplitude decreases to half in 765 seconds, we expect the damping time to be a little larger, but on this order of magnitude, so 1100 seconds sounds reasonable.

57. (N) To demonstrate the concept of resonance to your son and his friends, you suspend your smartphone by a lightweight string of length *L* and set the phone on vibrate. Using your landline phone, you call the cell phone, which vibrates with a frequency of 0.900 Hz, causing your makeshift pendulum to oscillate at very large amplitude. What is the length *L* of the string you used in this experiment?

INTERPRET and ANTICIPATE
The pendulum is resonating with the vibration motor of the cell phone. The phone must vibrate at the frequency of a simple pendulum of frequency 0.900 Hz.

SOLVE Using Eq. 16.29, we can relate the oscillation frequency to the length of the pendulum.	$\omega = 2\pi f = \sqrt{\frac{g}{\ell}}$
Solve for the length and insert numerical values.	$\ell = \frac{g}{(2\pi f)^2}$ $\ell = \frac{9.81\ \text{m/s}^2}{\left[2\pi(0.900\ \text{Hz})\right]^2}$ $\ell = \boxed{0.307\ \text{m}}$

CHECK and THINK
The length of a third of a meter seems reasonable for an oscillation period of nearly one second.

62. (A) Use the data in Table P16.59. Write an expression for the magnitude of the block's
a. position,
b. velocity and
c. acceleration.
Assume the maximum position observed is the amplitude.

INTERPRET and ANTICIPATE The block is a simple harmonic oscillator, and so we know the form that we expect for the position, velocity, and acceleration (Equations 16.3, 16.6 and 16.12). The key is to find the angular frequency, amplitude and initial phase from the data provided.	
SOLVE First, we must find the angular frequency from the period. Using the fact that the block starts at a peak value of 5.0 m and returns to this position in 2.00 s, we know that this is the period.	$\omega = \frac{2\pi}{T}$ $\omega = \frac{2\pi}{2.00\text{ s}} = \pi\text{ rad/s} = 3.14\text{ rad/s}$
The amplitude is the maximum observed position.	$x_{max} = 5.0\text{ m}$
The initial phase is found from the initial position (Eq. 16.4). Initially, the block is at its maximum amplitude. Therefore, the phase is zero such that the maximum amplitude occurs at a time of $t = 0$.	$\varphi = \cos^{-1}\frac{x_i}{x_{max}}$ $\varphi = \cos^{-1}\frac{x_{max}}{x_{max}} = \cos^{-1}1 = 0$
a. The position is given by Equation 16.3 with these specific parameters. All values are in SI units.	$x(t) = x_{max}\cos(\omega t + \varphi)$ $\boxed{x(t) = 5.00\cos(3.14t)}$
b. The velocity is given by Equation 16.6 with these specific parameters.	$v_x(t) = -v_{max}\sin(\omega t + \varphi) = -y_{max}\omega\sin(\omega t + \varphi)$ $v_x(t) = -(5.00\text{ m})(3.14\text{ rad/s})\sin(3.14t)$ $\boxed{v_x(t) = -15.7\sin(3.14t)}$
c. Next apply the hallmark of SHM (Eq. 16.12).	$a_x(t) = -\omega^2 x$ $a_x(t) = -(3.14\text{ rad/s})^2\,5.00\cos(3.14t)$ $\boxed{a_x(t) = -49.3\cos(3.14t)}$
CHECK and THINK Compare these expressions to the plots found in Problem 16.60 to verify that they are consistent.	

67. A particle initially located at the origin undergoes simple harmonic motion, moving first in the positive z direction, with a frequency of 3.20 Hz and an amplitude of 1.40 m. The particle oscillates between $z = 1.40$ m and $z = -1.40$ m.
a. (A) What is the equation describing the particle's position as a function of time?

INTERPRET and ANTICIPATE
We want to write an expression of the form given by Eq. 16.3, $z(t) = z_{max}\cos(\omega t + \varphi)$. We need to determine each of the parameters and insert them into the equation.

SOLVE	
The angular frequency can be determined from the given frequency.	$\omega = 2\pi f = 6.40\pi$
At $t = 0$, $z = 0$ and v is positive. Therefore, this situation corresponds to $z = +A\sin\omega t$ and $v = +v_i\cos\omega t$. We can substitute the amplitude and angular frequency.	$z(t) = A\sin\omega t$ $z(t) = 1.40\sin 6.40\pi t$
The sine function is related to the cosine function by a $\pi/2$ phase shift, so we can express the position in a form similar to Eq. 16.3 where z is in centimeters and t is in seconds.	$\boxed{z(t) = 1.40\cos(6.40\pi t - \pi/2)}$

CHECK and THINK
We are able to write the position in a form that looks like what we expect for an oscillator.

b. (N) What is the maximum speed of the particle?
c. (N) What is the maximum acceleration of the particle?

INTERPRET and ANTICIPATE
Once we have the equation for position, finding quantities related to the velocity and acceleration is relatively straightforward.

SOLVE **b.** The maximum speed can be found using the maximum amplitude and angular velocity (Eq. 16.7).	$v_{max} = z_{max}\omega$ $v_{max} = 1.40\text{ m}(6.40\pi\text{ Hz}) = 8.96\pi\text{ m/s}$ $v_{max} = \boxed{28.1\text{ m/s}}$
c. The maximum acceleration can be determined from the maximum amplitude and angular frequency (Eq. 16.10).	$a_{max} = z_{max}\omega^2$ $a_{max} = 1.40\text{ m}(6.40\pi\text{ Hz})^2 = 57.3\pi^2\text{ m/s}^2$ $a_{max} = \boxed{566\text{ m/s}^2}$
CHECK and THINK Given the quantities determined in part (a), we can calculate the maximum velocity and acceleration.	

d. (N) What is the total distance covered by the particle in the first 2.50 s of this motion?

INTERPRET and ANTICIPATE We can determine the period and the number of cycles traversed in this amount of time, which will allow us to determine the number of times the particle moves back and forth and the total distance traversed.	
SOLVE The period and frequency are inversely related.	$T = \frac{1}{f} = \frac{1}{3.20\text{ Hz}} = 0.313\text{ s}$
Since the period is 0.313 s, in 2.50 s, the particle will complete 2.50 s/0.313 s = 8.00 cycles.	$\#\text{ cycles} = \frac{2.50\text{ s}}{0.313\text{ s}} = 8.00\text{ cycles}$
Since the amplitude is 1.40 m, the particle will travel 4 × 1.40 m = 5.60 m in one cycle (i.e. from the center point to +1.40 m, back to center, out to −1.40 m, and back to center).	$\text{distance per cycle} = 4\times 1.40\text{ m} = 5.60\text{ m}$
Therefore, the particle travels the distance of a full cycle 8 times.	$\text{total distance} = \#\text{ cycles} \times \text{distance/cycle}$ $\text{total distance} = 8.00\times 5.60\text{ m} = \boxed{44.8\text{ m}}$
CHECK and THINK With a maximum speed from part (b) of 28 m/s, a total distance of 45 m in 2.5 seconds seems like the right order of magnitude.	

71. (C) A hollow, metal sphere is filled with water, and a small hole is made at the bottom. The sphere hangs by a long thread and is made to oscillate. How will the period of oscillation change over time if water is allowed to flow through the hole until the sphere is empty?

The period of a simple pendulum is given by Equations 16.2 and 16.29, $T = \frac{2\pi}{\omega_{\text{smp}}} = 2\pi\sqrt{\frac{\ell}{g}}$. The mass does not enter, so the change in mass (which is the most obvious change) does not matter at all. There is actually another way that the period can change though, which is a bit more subtle. The period is also proportional to the square root of the effective length of the pendulum. Initially the center of mass of the water-filled sphere is at its center. As the water drains, the center of mass of the sphere shifts downward slightly. Thus, the effective length of the pendulum increases, so the time period of the pendulum increases. Eventually, as the water drains out, most of the mass of the pendulum bob is in the hollow sphere and the center of mass of the bob returns to the center of the (now hollow) sphere and the period returns to the original period.

76. (N) The frequency of a physical pendulum comprising a nonuniform rod of mass 1.25 kg pivoted at one end is observed to be 0.667 Hz. The center of mass of the rod is 40.0 cm below the pivot point. What is the rotational inertia of the pendulum around its pivot point?

INTERPRET and ANTICIPATE The physical pendulum is governed by Eq. 16.33, which relates the angular frequency to the mass, distance to the center of mass, and the rotational inertia of the pendulum. We can rearrange this expression to solve for the rotational inertia.	
SOLVE We start with Equation 16.33 relating the angular frequency of a physical pendulum to quantities we're given and solve for the rotational inertia.	$\omega_{\text{phy}} = 2\pi f = \sqrt{\frac{mgr_{\text{CM}}}{I}}$ $I = \frac{mgr_{\text{CM}}}{4\pi^2 f^2}$

Substitute numerical values.	$I = \dfrac{(1.25\text{ kg})(9.81\text{ m/s}^2)(0.400\text{ m})}{4\pi^2(0.667\text{ Hz})^2}$ $I = \boxed{0.279\text{ kg}\cdot\text{m}^2}$
CHECK and THINK It's unlikely that you have a physical intuition about the magnitude of this number, but we've identified an equation describing a physical pendulum and used the quantities provided.	

79. (N) Air resistance in a lab causes the motion of a 5.00-kg disk attached to a vertical spring with spring constant $k = 5.000 \times 10^3$ N/m to be damped at a rate given by a damping coefficient $b = 4.50$ N · s/m.

a. What is the frequency of the damped oscillation of the system?

INTERPRET and ANTICIPATE The frequency of a damped oscillator is related to the frequency of an undamped oscillator, but is shifted to a smaller frequency due to the damping (Eq. 16.54).	
SOLVE With damping, the angular frequency shifts lower according to Eq. 16.54.	$\omega_D = \sqrt{\omega^2 - \dfrac{b^2}{4m^2}}$
The frequency if undamped depends on the spring constant and mass according to Eq. 16.26.	$\omega = \sqrt{\dfrac{k}{m}}$ $\omega = \sqrt{\dfrac{5.000\times10^3\text{ N/m}}{5.00\text{ kg}}}$ $\omega = 31.6\text{ s}^{-1}$
We now substitute values into Eq. 16.54. In this case, the damping is a minor effect and the angular frequency is unchanged (within significant figures) compared to the undamped case.	$\omega_D = \sqrt{(31.6\text{ s}^{-1})^2 - \left(\dfrac{4.50\text{ N}\cdot\text{s/m}}{2\times5.00\text{ kg}}\right)^2}$ $\omega_D = \sqrt{1000 - 0.203} = 31.6\text{ Hz}$

From the damped angular frequency, we can determine the frequency.	$f = \frac{\omega}{2\pi} = \frac{31.6\text{ s}}{2\pi\text{ s}}$ $f = \boxed{5.03\text{ Hz}}$
CHECK and THINK In general, damping decreases the frequency of an oscillator. We see in this case, the effect is quite small, such that the frequency is nearly identical to the undamped case.	

b. What is the percentage by which the amplitude of motion decreases after each cycle?

INTERPRET and ANTICIPATE With damping, the peak position decreases exponentially according to Eq. 16.53. With the time and damping coefficient, we can determine the amplitude.	
SOLVE The position of the mass is described by Eq. 16.53.	$y(t) = y_{\max} e^{-bt/2m} \cos(\omega_D t + \varphi)$
Over one cycle, a time $T = 2\pi / \omega$, the amplitude changes from $y_{\max}$ to $y_{\max} e^{-b2\pi/2m\omega}$, from which we can calculate the fractional decrease.	$\frac{\Delta y}{y} = \frac{y_{\max} - y_{\max} e^{-\pi b/m\omega}}{y_{\max}}$ $\frac{\Delta y}{y} = 1 - e^{-\pi(4.50)/(5.00 \cdot 31.6)}$ $\frac{\Delta y}{y} = 1 - e^{-0.0894} = 1 - 0.91446$ $\frac{\Delta y}{y} = 0.0855 = \boxed{8.55\%}$
CHECK and THINK After each cycle, the amplitude decreases by 8.55%.	

80. (A) Two springs, with spring constants k_1 and k_2, are connected to a block of mass m on a frictionless, horizontal table (Fig. P16.80). The block is extended a distance x from equilibrium and released from rest. Show that the block executes simple harmonic motion with a period given by

$$T = 2\pi \sqrt{\frac{m(k_1 + k_2)}{k_1 k_2}}$$

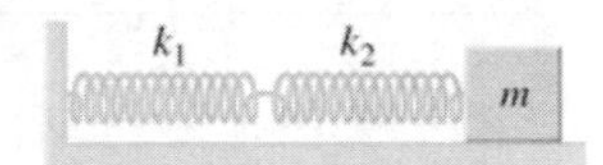

Figure P16.80

INTERPRET and ANTICIPATE
When the mass is displaced, each spring will be stretched. We can determine the total force on the mass to determine its acceleration and then find the oscillation period in the same way we did for a single spring.

SOLVE	
Imagine the mass is displaced a distance x from equilibrium and held in place. Let's suppose spring 1 is stretched a distance x_1 and spring 2 is stretched a distance x_2. Applying Newton's third law at the point where the two springs meet, we expect that the force exerted by each spring must be equal, such that this point remains in equilibrium.	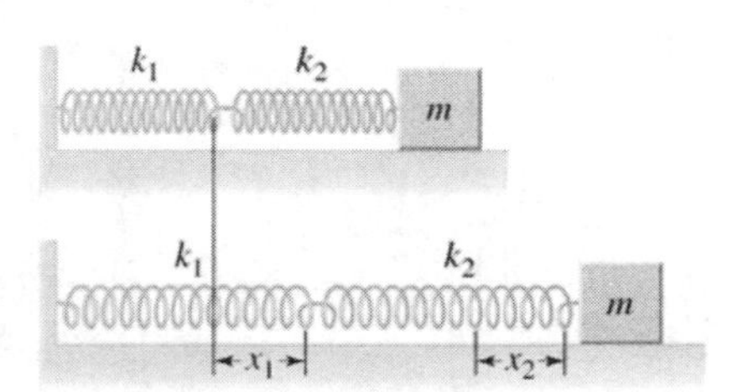 **Figure P16.80ANS** $k_1x_1 = k_2x_2$
When this is combined with the requirement that these compressions add up to the total compression, $x = x_1 + x_2$, we find an expression for x_1 in terms of x.	$x_1 = \left[\dfrac{k_2}{k_1+k_2}\right]x$
Applying, $F_1 = k_1x_1$ or $F_2 = k_2x_2$ the tension force on either spring can be calculated. These forces are equal and also equal the force of the second spring on the mass.	$F_1 = k_1x_1 = k_1\left[\dfrac{k_2}{k_1+k_2}\right]x$ $F = \left[\dfrac{k_1k_2}{k_1+k_2}\right]x$
This force causes the mass to accelerate.	$F = \left[\dfrac{k_1k_2}{k_1+k_2}\right]x = ma$

This is in the form $F = k_{\text{eff}} x = ma$. The period of oscillation is then given by Eq. 16.26 (using Eq. 16.2 to relate angular frequency to the period).	$T = 2\pi\sqrt{\frac{m}{k_{\text{eff}}}}$ $\boxed{T = 2\pi\sqrt{\frac{m(k_1 + k_2)}{k_1 k_2}}}$
CHECK and THINK We confirmed the relationship we were asked to find.	

17

Traveling Waves

7. The wave function of a pulse is

$$y(x,t) = \frac{7.5}{(x+3.5t)^2 + 0.5}$$

where the values are in the appropriate SI units.

a. (N) What are the speed and direction of the pulse? *Hint*: Sketch the wave pulse at $t = 2$, 4, and 6 s as in Figure 17.4.

We first plot the pulse at the three times as indicated (see Problem 6 for a similar example). Our sketch shows the pulses move in the negative x direction. Using the peak of the pulses to find the speed, the pulse moves from $x = -7.0$ m to $x = -21.0$ m in 4 s.

$$\vec{v} = \frac{((-21.0)-(-7.0))\text{ m}}{4\text{ s}}\hat{i} = \boxed{-3.5\,\hat{i}\text{ m/s}}$$

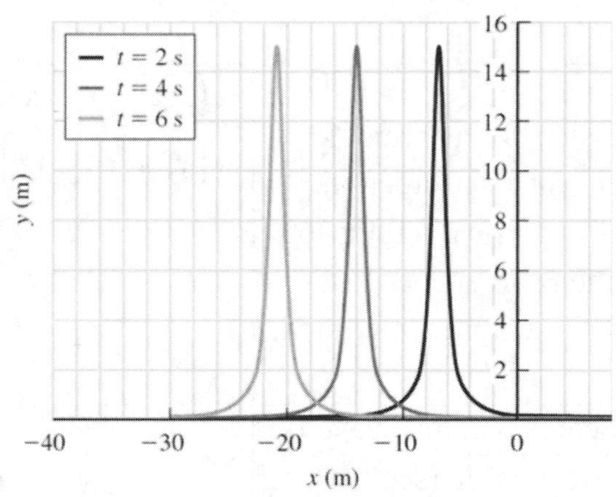

Figure P17.7ANS

b. (A) Write the wave function for a pulse with the same profile and speed but moving in the opposite direction.

The direction of the wave is determined by the sign in $x + 3.5t$. In general, we see that the formula for a pulse (Eq. 17.2) includes the expression $(x - v_x t)$ and it is the sign of the velocity that determines the direction of the pulse (for a pulse traveling in the positive direction, v_x is positive and the expression is $x - v_x$). So, to reverse the direction, replace the plus sign in $(x + 3.5t)$ with a negative sign.

$$\boxed{y(x,t) = \frac{7.5}{(x - 3.5t)^2 + 0.5}}$$

12. The equation of a harmonic wave propagating along a stretched string is represented by $y(x,t) = 4.0\sin(1.5x - 45t)$, where x and y are in meters and the time t is in seconds.

a. (C) In what direction is the wave propagating?
b. (N) What is the amplitude of the wave?
c. (N) What is the wavelength of the wave?
d. (N) What is the frequency of the wave?
e. (N) What is the propagation speed of the wave?

INTERPRET and ANTICIPATE
First we write the general wave equation for a harmonic wave propagating in the positive x direction and compare this to the given equation to determine the various quantities.

SOLVE First, consider the expression for a harmonic wave propagating in the positive x direction, given by Equation 17.4. By comparing this to the expression given in the problem, we can match terms to identify the variables requested.	$y(x,t) = y_{max}\sin(kx - \omega t)$ $y(x,t) = 4.0\sin(1.5x - 45t)$
a. The negative sign indicates that this wave is traveling in the $\boxed{\text{positive } x \text{ direction}}$ (see Problem 7(b) for a similar example).	
b. The amplitude is y_{max}, the factor in front of the sine function.	$y_{max} = \boxed{4.0\,\text{m}}$

c. Use the definition of wave number to determine the wavelength (Eq. 17.5).	$k = \frac{2\pi}{\lambda} = 1.5\,\text{m}^{-1}$ $\lambda = \boxed{4.2\,\text{m}}$
d. Use the definition of angular frequency to determine the frequency (Eq. 16.2).	$\omega = 45\,\text{rad/s}$ $f = \frac{\omega}{2\pi} = \boxed{7.2\,\text{Hz}}$
e. Finally, use the wave speed equation (Eq. 17.7).	$v = \omega/k = \boxed{3.0\times10^{1}\ \text{m/s}}$

CHECK and THINK

In this case, the wave equation in the problem matches the form of Eq. 17.4. Therefore, we only need to match terms to determine the wave's properties.

17. A graph (profile) of a traveling wave at a moment in time is shown in Figure P17.17. The wave is traveling in the negative x direction with a propagation speed of 56.0 cm/s.

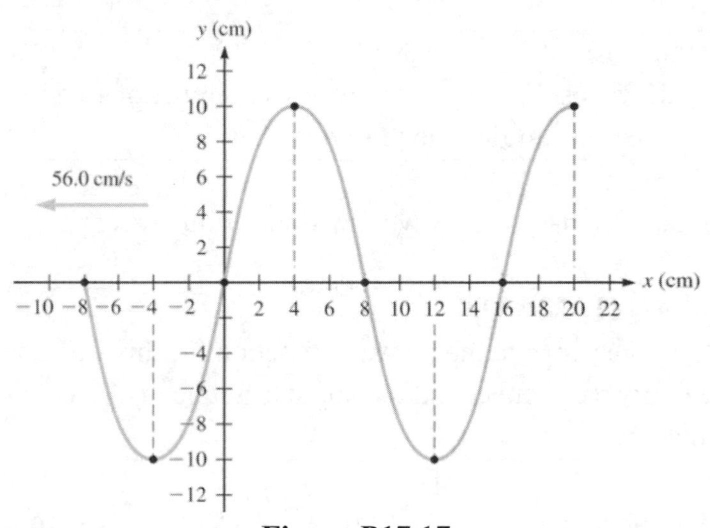

Figure P17.17

a. (N) What is the amplitude of the wave?

INTERPRET and ANTICIPATE

The amplitude is the height of the wave, measured from the axis to the peak height.

SOLVE

Examining the graph, the height of the wave is measured to be $\boxed{10\text{ cm}}$.

CHECK and THINK Information about the wave can often be found by examining a graph of the wave, or by comparing the equation to the form of the harmonic transverse wave (Eq. 17.4).

b. (N) What is the frequency of the wave?

INTERPRET and ANTICIPATE By examining the graph, the wavelength of the wave can be determined. Then, the frequency can be found using Eq. 17.8.	
SOLVE Examination of the graph allows for the measurement of the wavelength, the distance between the crests of the wave.	$\lambda = 16$ cm
Use the speed of the wave and the wavelength to calculate the frequency of the wave (Eq. 17.8).	$f = v/\lambda = 56.0 \text{ cm/s}/16 \text{ cm} = \boxed{3.5 \text{ Hz}}$
CHECK and THINK Eq. 17.8 applies to all waves. When you know any combination of a wave's frequency, wavelength, and speed, you can find the third quantity.	

c. (A) Write the transverse harmonic wave function for this wave.

INTERPRET and ANTICIPATE In order to write the transverse harmonic wave function for this wave, we must first determine the angular wave number and the angular frequency. Then, use Eq. 17.4 to write the wave function.	
SOLVE The angular wave number can be found using Eq. 17.5.	$k = 2\pi/\lambda = 2\pi/(16 \text{ cm}) = 0.39 \text{ cm}^{-1}$
The angular frequency can be found using Eq. 16.2.	$\omega = 2\pi f = 2\pi(3.5 \text{ Hz}) = 22 \text{ rad/s}$
Now, write the wave equation by using Eq. 17.4, and noting that the wave is traveling in the negative x direction.	$y(x,t) = y_{max}\sin(kx + \omega t)$ $\boxed{y(x,t) = (10 \text{ cm})\sin\left[(0.39 \text{ cm}^{-1})x + (22 \text{ rad/s})t\right]}$

CHECK and THINK
The harmonic transverse wave function describes the shape of the wave for any moment in time. It is important to be familiar with the construction of the wave function as well as interpreting values such as the angular frequency, or angular wave number, by inspection of the wave function.

25. A sinusoidal wave with amplitude 1.00×10^{-2} m and frequency 325 Hz is observed to be traveling with a speed of 55.0 m/s on a wire held under 195 N of tension.
a. (A) What is the wave function for this wave in SI units?

INTERPRET and ANTICIPATE	
We want an expression like Eq. 17.4, $y(x,t) = y_{max} \sin(kx - \omega t)$. The amplitude y_{max} is given, so we need only determine the wave number and angular velocity.	
SOLVE The angular frequency can be determined with the given frequency (Eq. 16.2).	$\omega = 2\pi f$ $\omega = 2\pi(325\text{ Hz})$ $\omega = 2.04 \times 10^3$ rad/s
The wave number is determined by the angular frequency and the speed of the wave (Eq. 17.7).	$k = \frac{\omega}{v}$ $k = \frac{2\pi(325\text{ Hz})}{55.0\text{ m/s}}$ $k = 37.1\text{ m}^{-1}$
Assuming all quantities are in metric units, insert values into Eq. 17.4.	$\boxed{y(x,t) = (1.00 \times 10^{-2}\text{ m})\sin(37.1x - 2.04 \times 10^3 t)}$
CHECK and THINK We are able to calculate wave properties with the information given and insert these into the general expression for a traveling wave.	

b. (N) What is the mass per unit length for the wire?

INTERPRET and ANTICIPATE
The speed of a wave on a wire depends on the tension and linear mass density. We can solve this expression for the mass density.

SOLVE Equation 17.11 relates the speed of the wave of the wire to the tension and mass density.	$v = \sqrt{\dfrac{F_T}{\mu}}$
Solve this expression for the mass density of the wire.	$\mu = \dfrac{F_T}{v^2}$
Insert numerical values.	$\mu = \dfrac{195\text{ N}}{(55.0\text{ m/s})^2}$ $\mu = \boxed{0.0645\text{ kg/m}}$

CHECK and THINK
The mass density and tension of the wire determine the wave speed. Given two of these (tension and speed in this case), we can calculate the third.

26. As in Figure 17.8A, a simple harmonic oscillator is attached to a rope, creating a transverse wave of wavelength 7.47 cm. At the other end of the rope is a hanging block of mass 6.30 kg. The oscillator has an angular frequency of 2340 rad/s and an amplitude of 34.6 cm.
a. (N) What is the linear mass density of the rope?

INTERPRET and ANTICIPATE
The speed of any harmonic wave equals its angular frequency and angular wave number (Eq. 17.7). Since this wave is on a rope, the speed also depends on the rope's linear mass density and tension in the rope (Eq. 17.11). We can combine these formulas to find the linear mass density of the rope.

SOLVE Using both Equation 17.7 and 17.11, we can solve for the linear mass density in terms of other quantities provided.	$v_x = \sqrt{\dfrac{F_T}{\mu}} \quad (17.11)$ $v_x = \dfrac{\omega}{k} \quad (17.7)$

Set the two equations equal to each other and then solve for linear mass density.	$\frac{\omega}{k}=\sqrt{\frac{F_T}{\mu}}$ $\frac{\omega^2}{k^2}=\frac{F_T}{\mu}$ $\mu=\frac{k^2}{\omega^2}F_T$
The tension is the weight of the hanging mass, $F_T=mg$.	$\mu=\frac{k^2}{\omega^2}mg$
Write this in terms of wavelength (Eq. 17.5).	$k=\frac{2\pi}{\lambda}$ $\mu=\frac{4\pi^2}{(\lambda\omega)^2}mg$
Now, substitute values.	$\mu=\frac{4\pi^2(6.30\text{ kg})(9.81\text{ m/s}^2)}{(7.47\times10^{-2}\text{ m}\cdot2340\text{ rad/s})^2}$ $\mu=\boxed{7.99\times10^{-2}\text{ kg/m}}$

CHECK and THINK

The linear mass density seems reasonable and has the correct units of kilograms per meter.

b. (C) If the angular frequency of the oscillator doubles, what properties (speed, angular wave number, amplitude) of the wave must change? Give the new values of these properties.

INTERPRET and ANTICIPATE

We would not expect the speed to change since it is the same rope and we are given no indication its tension is changing. Nor should the frequency affect the amplitude. We would expect the angular wave number to double since it and the angular frequency are related to the speed of the wave as in Eq. 17.7.

We start again with the relationship from part (a).	$\frac{\omega}{k}=\sqrt{\frac{F_T}{\mu}}$

We solve that expression for the angular wave number and write it given the new angular frequency. The tension and mass per unit length can then be expressed in terms of the original values using the same relationship.	$k_{\text{new}} = \omega_{\text{new}}\sqrt{\frac{\mu}{F_T}}$ $k_{\text{new}} = \omega_{\text{new}}\frac{k_{\text{old}}}{\omega_{\text{old}}}$
Using Eq. 17.5 to express the angular wave number in terms of the original wavelength, we get an expression for the new angular wave number in terms of the original wavelength	$k_{\text{new}} = \omega_{\text{new}}\frac{2\pi}{\lambda_{\text{old}}\omega_{\text{old}}}$ $k_{\text{new}} = 2\frac{2\pi}{\lambda_{\text{old}}}$ $k_{\text{new}} = \frac{4\pi}{\lambda_{\text{old}}}$
Use the original wavelength to find the new angular wave number.	$k_{\text{new}} = \frac{4\pi}{7.47\times10^{-2}\ \text{m}} = \boxed{168\ \text{rad/m}}$

CHECK and THINK

Given the ratio of angular frequency to angular wave number that we started with, we can see that the angular wave number must also double since the speed of the wave remains unchanged.

c. (C) If the frequency remains at its original value but the mass of the hanging block is doubled, what properties of the wave must change? Give the new values of these properties.

INTERPRET and ANTICIPATE

Since the mass will be increased, the tension will also increase. The linear mass density and frequency are not changing, and so we expect that the angular wave number must change.

We start again with the relationship from part (a).	$\frac{\omega}{k} = \sqrt{\frac{F_T}{\mu}}$
Solve the equation for k and express this for the new angular wave number, which will depend on the new tension. The tension is equal to the gravitational force on the new mass.	$k_{\text{new}} = \omega\sqrt{\frac{\mu}{m_{\text{new}}g}} = \omega\sqrt{\frac{\mu}{2m_{\text{old}}g}}$ $k_{\text{new}} = \frac{\omega}{\sqrt{2}}\sqrt{\frac{\mu}{m_{\text{old}}g}}$

Now, the square root involving the original mass is really just the inverse of the relationship with which we began this part. We can express it in terms of the angular frequency and the original angular wave number. Then, express the original angular wave number in terms of the original wavelength.	$k_{new} = \frac{\omega}{\sqrt{2}} \frac{k_{old}}{\omega}$ $k_{new} = \frac{k_{old}}{\sqrt{2}}$ $k_{new} = \frac{2\pi/\lambda_{old}}{\sqrt{2}}$ $k_{new} = \sqrt{2}\pi/\lambda_{old}$
Finally, plug in the original wavelength to find the new angular wave number.	$k_{new} = \sqrt{2}\pi/\left(7.47\times10^{-2}\ \text{m}\right) = \boxed{59.5\ \text{rad/m}}$

CHECK and THINK

The angular wave number has decreased which we might have suspected when looking at Eq. 17.7. Since the tension increased and the string's linear mass density stayed the same, the speed of the waves increased. With the angular frequency remaining constant, the angular wave number must have decreased.

30. (N) Dolphins have been trained to understand human voice commands. Imagine a trainer standing on a platform with her head and shoulders 5.20 m above the surface of a pool, with a dolphin waiting 10.4 m directly below the trainer (Fig. P17.30). Assume the speed of sound in air is 343 m/s and in water is 1497 m/s.

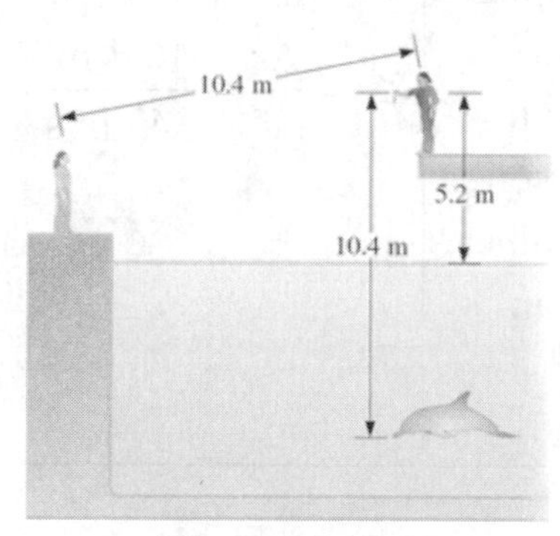

Figure P17.30

a. (N) If the trainer gives the command to jump, how long will it take the dolphin to hear the command?

b. (C) Would the dolphin hear the command before or after another person who is 10.4 m from the trainer? Justify your answer.

INTERPRET and ANTICIPATE

The distance to the person and the distance to the dolphin are the same. The difference is that the sound traveling toward the dolphin must pass through water. The speed of sound in water is faster than in air, so presumably the dolphin hears the sound before the person.

SOLVE	
a. Find the time it takes the sound to reach the surface of the water, which is 5.2 m away from the trainer using the definition of speed $v = \frac{\Delta r}{\Delta t}$. The sound is traveling in the air.	$\Delta t_{air} = \frac{\Delta r_{air}}{v_{air}}$
Find the time it takes the sound to travel through the water to the dolphin, which is 5.2 m below the surface.	$\Delta t_{water} = \frac{\Delta r_{water}}{v_{water}}$
Add to find the total time for the sound to reach the dolphin.	$\Delta t_{tot} = \frac{\Delta r_{air}}{v_{air}} + \frac{\Delta r_{water}}{v_{water}}$ $\Delta t_{tot} = \frac{5.20 \text{ m}}{343 \text{ m/s}} + \frac{5.20 \text{ m}}{1497 \text{ m/s}}$ $\Delta t_{tot} = \boxed{1.86 \times 10^{-2} \text{ s}}$
b. Since the speed of sound in water is higher, we expect that the dolphin will hear the sound first. We can confirm this by finding the time it takes the sound to reach the person who is 10.4 m away from the trainer to see that it is indeed longer than the answer to part (a).	$v = \frac{\Delta r}{\Delta t}$ $\Delta t = \frac{\Delta r}{v} = \frac{10.4 \text{ m}}{343 \text{ m/s}} = \boxed{3.03 \times 10^{-2} \text{ s}}$

CHECK and THINK

As expected the dolphin hears the command much sooner than the person. So, the dolphin hears the sound and can start to react, before you do.

33. (N) On July 17, 2006, an earthquake hit just south of Java, Indonesia (between Australia and Malaysia). The time of the earthquake was 08:19:25.02 UTC (Coordinated Universal Time). The strong earthquake was detected by more than 400 seismograph stations worldwide. Table P17.33 provides the arrival time of the P wave and the distance to three such stations. Assume the P wave traveled through the crust and the crust's average density is 3300 kg/m^3. Estimate the bulk modulus of the Earth's crust. *Hint*: See Problem 32.

Table P17.33 Java earthquake data.

Arrival time at seismograph (UTC)	Distance from epicenter to station (km)
08:21:55.58	1190
08:22:05.09	1230
08:23:29.89	1970

INTERPRET and ANTICIPATE

We can use the time and distance data to find the average speed of the P wave in the crust. A P-wave like sound is a longitudinal wave, so once we know its speed we can use Equation 17.13 to find the bulk modulus of the crust.

SOLVE

Find the travel time to each station by subtracting the time that the earthquake occurred from the arrival time of the P wave. To find the speed of the wave, divide the distance to the station by the travel time.

Travel time (s)	Speed of P wave (m/s)
150.6	7.90×10^3
160.1	7.68×10^3
244.9	8.04×10^3

Take the average of the speeds to the three stations, and assume this equals the speed of the P wave in the crust.

$$v_P = \frac{(7.90 + 7.68 + 8.04) \times 10^3 \text{ m/s}}{3}$$

$$v_P = 7.87 \times 10^3 \text{ m/s}$$

Solve Equation 17.13 for B.

$$v_P = \sqrt{\frac{B}{\rho}}$$

$$B = \rho v_P^2 = (3300 \text{ kg/m}^3)(7.87 \times 10^3 \text{ m/s})^2$$

$$B = \boxed{2.0 \times 10^{11} \text{ Pa}}$$

CHECK and THINK

This is a high bulk modulus similar to that of metals such as iron, steel and tungsten. (See Table 15.2) The bulk modulus of the crust is an important part of the models used to predict earthquakes.

36. (A) Use the result of Problem 35 to show that the potential energy in one wavelength of a harmonic transverse wave on a rope (Fig. 17.8A) is given by $U_\lambda = (1/4)\mu\omega^2 y_{max}^2 \lambda$.

INTERPRET and ANTICIPATE For this derivation we take a short cut. We are told to use the result of Problem 35, which is listed in the problem statement: $K_\lambda = (1/4)\mu\omega^2 y_{\max}^2 \lambda$. Therefore, we know the kinetic energy in one wavelength and we know the mechanic energy in one wavelength (Eq. 17.7). We can simply use conservation and subtract to find an expression for the potential energy.	
SOLVE The mechanical energy is the sum of the kinetic plus potential energy. Solve for potential energy.	$E_\lambda = K_\lambda + U_\lambda$ $U_\lambda = E_\lambda - K_\lambda$
Substitute Equation 17.17 for mechanical energy and the result of the previous problem for kinetic energy.	$U_\lambda = \frac{1}{2}\mu\omega^2 y_{\max}^2 \lambda - \frac{1}{4}\mu\omega^2 y_{\max}^2 \lambda$ $U_\lambda = \boxed{\frac{1}{4}\mu\omega^2 y_{\max}^2 \lambda}$
CHECK and THINK In this case, energy conservation allows us to quickly find the answer. It is consistent with our expectations that the potential energy is large if the displacement ($y_{\max}$) is large.	

37. (N) A 5.00-m rope with a mass of 0.650 kg is held under tension. If 1.00 kW of power is supplied to the rope, what is the amplitude of sinusoidal waves that will be generated with a wavelength of $\pi/4$ m and speed 50.0 m/s?

INTERPRET and ANTICIPATE The average power of the wave depends on the mass density, angular frequency, amplitude, and wave speed. We can solve for amplitude given the other variables.	
SOLVE The amplitude of the wave is related to its power by Equation 17.19.	$P_{\text{av}} = \frac{1}{2}\mu\omega^2 y_{\max}^2 v_x$
We can solve this expression for the maximum amplitude.	$y_{\max} = \frac{1}{\omega}\sqrt{\frac{2P}{\mu v_x}}$

With the given wave speed and wavelength, we can determine the frequency and angular frequency of the wave.	$f=\frac{v}{\lambda}=\frac{50.0}{\pi/4}$ Hz $\omega=2\pi f=400$ rad/s
Insert numerical values.	$y_{max}=\frac{1}{400 \text{ rad/s}}\sqrt{\frac{2(1000 \text{ W})}{(0.650 \text{ kg}/5.00 \text{ m})(50.0 \text{ m/s})}}$ $y_{max}=\boxed{0.0439 \text{ m}}$
CHECK and THINK The amplitude is around 4.4 cm compared to the wavelength of around 80 cm.	

41. (N) A sound produces a pressure amplitude of ΔP_{max} = 32 Pa. Determine the intensity of this sound wave. Assume the density of air is 1.3 kg/m^3 and the speed of sound is 343 m/s.

INTERPRET and ANTICIPATE The intensity of a sound wave depends on the pressure amplitude, density of air, and the sound speed. We are given each of these.	
SOLVE Equation 17.24 relates the intensity to the pressure amplitude.	$I=\frac{1}{2}\frac{\Delta P_{max}^2}{\rho v_s}$
Insert numerical values.	$I=\frac{1}{2}\frac{(32 \text{ N/m}^2)^2}{(1.3 \text{ kg/m}^3)(343 \text{ m/s})}$ $I=\boxed{1.1 \text{ W/m}^2}$
CHECK and THINK This would actually be a painfully large sound.	

48. (N) The speaker system at an open-air rock concert forms a ring around the entire circular stage and delivers 50,000 W of power output. Assume the sound radiates in all directions equally as if it were generated by an isotropic point source and assume the sound energy is not absorbed by air.

a. At what distance is the sound from the speakers barely audible? Note that your answer will be far too large since the model we are using for sound level ignores the power absorbed by the medium (air). How does your answer compare to the radius of the Earth?

INTERPRET and ANTICIPATE	
The sound intensity decreases with the square of the distance. We will find the distance at which the sound is barely audible, for which $\beta = 0$.	
SOLVE We use the formula for the sound intensity (Equation 17.25) to determine the intensity of the sound. For a barely audible sound, $\beta = 0$.	$\beta \equiv 10\log\left(\frac{I}{I_0}\right)$ $0\text{ dB} = (10\text{dB})\log\left(\frac{I}{10^{-12}\text{ W/m}^2}\right)$
Solve for *I*.	$10^0 = 10^{\log\left(\frac{I}{10^{-12}\text{ W/m}^2}\right)}$ $1 = \frac{I}{10^{-12}\text{ W/m}^2}$ $I = 1.00\times10^{-12}\text{ W/m}^2$
Now use Eq. 17.23 (i.e. the sound intensity decreases with the square of the distance) and solve for the distance at which the sound has the intensity we've determined.	$I = \frac{P_{av}}{4\pi r^2}$ $r = \sqrt{\frac{P}{4\pi I}}$
Insert numerical values.	$r = \sqrt{\frac{50000\text{ W}}{4\pi\left(1.00\times10^{-12}\text{ W/m}^2\right)}}$ $r = \boxed{6.31\times10^7\text{ m}}$
CHECK and THINK Over 63,000 km! Ok, clearly the assumption that the air does not absorb any of the power is not correct. This distance is more than 10 times the radius of the Earth!	

b. What is the closest distance audience members can be to the speakers if the sound is not to be painful to their ears?

INTERPRET and ANTICIPATE
We take painfully loud to be 120 dB and use the formulas from part (a) to solve for the distance.

SOLVE Follow the steps from part (a) for a sound of 120 dB.	$120\text{ dB} = (10\text{dB})\log\left[\frac{I}{10^{-12}\text{ W/m}^2}\right]$ $I = 1.00\text{ W/m}^2 = \frac{P}{4\pi r^2}$ $r = \sqrt{\frac{P}{4\pi I}} = \sqrt{\frac{50000\text{ W}}{4\pi(1.00\text{ W/m}^2)}} = \boxed{63.1\text{ m}}$
CHECK and THINK Again, this seems to be an overestimate. We have assumed no dissipation and that all of the sound emanates from a single source, which is the likely source of our overestimate.	

50. (N) A seismic P wave strikes a boundary between two types of material having the same bulk modulus. At the boundary, the density increases abruptly from 3400 kg/m^3 to 4100 kg/m^3 (Fig. P17.50). If the wave front of the P wave is planar and makes a 25° angle with the boundary, what is the angle of refraction θ?

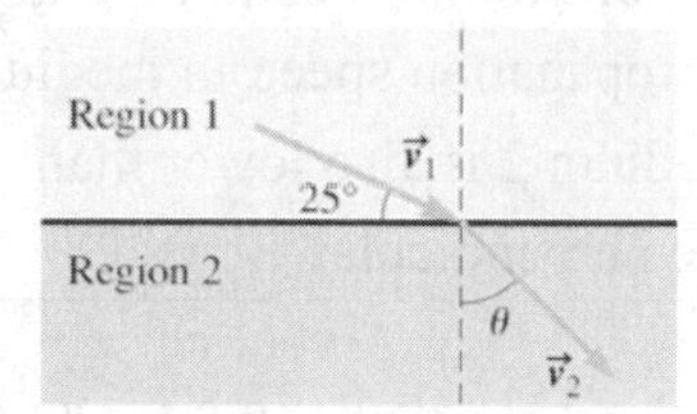

Figure P17.50

INTERPRET and ANTICIPATE The angle of refraction depends on the relative speed of the wave in the two media. We can find this from the densities of the two media and the fact that the bulk modulus is the same in the two media.	
SOLVE The angles of incidence and refraction are related by the wave speeds according to Eq. 17.26.	$\sin\theta_2 = \frac{v_2}{v_1}\sin\theta_1$
The wave speed depends on the bulk modulus and density (Equation 17.13). Using this equation, we can write an expression for the ratio of the speeds.	$v_s = \sqrt{\frac{B}{\rho}}$ $\frac{v_1}{v_2} = \sqrt{\frac{B}{\rho_1}}\sqrt{\frac{\rho_2}{B}}$

	$\frac{v_1}{v_2} = \sqrt{\frac{\rho_2}{\rho_1}}$
We now insert this relationship into Eq. 17.26.	$\sin\theta_2 = \sqrt{\frac{\rho_1}{\rho_2}}\sin\theta_1$
We must be careful, however, when we substitute values. The angles in this equation are measured with respect to the perpendicular while the angle give in the problem is measured with respect to the boundary. So $\theta_1 = 90°-25° = 65°$.	$\sin\theta_2 = \sqrt{\frac{3400\text{ kg/m}^3}{4100\text{ kg/m}^3}}\sin 65° = 0.825$ $\theta_2 = \sin^{-1}(0.825) = \boxed{56°}$
CHECK and THINK Our answer makes sense. Because the density is higher in the second medium (and the bulk modulus is the same), the speed in the second medium is lower than in the first medium. We expect that if the propagation speed in medium 2 is lower than in medium 1 (if $v_2 < v_1$) then the angle for medium 2 is also lower than the angle for medium 1 ($\theta_2 < \theta_1$) and the ray bends toward the perpendicular.	

57. (N) An ambulance traveling eastbound at 140.0 km/h with sirens blaring at a frequency of 7.00×10^2 Hz passes cars traveling in both the eastbound and westbound directions at 55.0 km/h.

What is the frequency observed by the eastbound drivers
a. as the ambulance approaches from behind and
b. after the ambulance passes them?

What is the frequency observed by the westbound drivers
c. as the ambulance approaches them and
d. after the ambulance passes them?

INTERPRET and ANTICIPATE
We must use the Dopplar formula to calculate the shifted frequency detected by the observer depending on the motion of the source and the observer. When the source and observer approach each other, the sound will shift to a higher frequency. When the source and observer recede from each other, the sound will shift to a lower frequency.

SOLVE In each case, we use the Doppler formula (Eq. 17.31). When the observer is moving towards the sound source, the observed frequency is larger and we use a plus sign in the numerator. As the observer moves away from the source, the frequency is lower and we use a minus sign in the numerator. When the source is moving towards the observer, the observed frequency is larger and we use a minus sign in the denominator. As the source moves away from the observer, the frequency is lower and we use a plus sign in the denominator.	$f_{\text{obs}} = \left(\dfrac{v_s \pm v_{\text{obs}}}{v_s \mp v_{\text{source}}}\right) f$
For parts (a) and (b), both the ambulance and car are moving in the same direction. **a.** The ambulance is moving towards the car (minus sign in denominator) and the car is moving away from the ambulance (minus sign in numerator).	$f_{\text{obs}} = \left(7.00\times10^2\right)\left(\dfrac{343-15.3}{343-38.9}\right)$ $f_{\text{obs}} = \boxed{754\text{ Hz}}$
b. The ambulance is moving away from the car (plus sign in denominator) and the car is moving towards the ambulance (plus sign in numerator).	$f_{\text{obs}} = \left(7.00\times10^2\right)\left(\dfrac{343+15.3}{343+38.9}\right)$ $f_{\text{obs}} = \boxed{657\text{ Hz}}$
For parts (c) and (d), the car and ambulance are moving in opposite directions. **c.** The ambulance is moving towards the car (minus sign in denominator) and the car is towards the ambulance (plus sign in numerator).	$f_{\text{obs}} = \left(7.00\times10^2\right)\dfrac{(343+15.3)}{(343-38.9)}$ $f_{\text{obs}} = \boxed{825\text{ Hz}}$
d. The ambulance is moving away from the car (plus sign in denominator) and the car is moving away from the ambulance (minus sign in numerator).	$f_{\text{obs}} = \left(7.00\times10^2\right)\dfrac{(343-15.3)}{(343+38.9)}$ $f_{\text{obs}} = \boxed{601\text{ Hz}}$

CHECK and THINK

When the ambulance and cars approach each other (c) the frequency is largest and when they move away from each other (d) the frequency is smallest.

58. (N) Joe, a skateboarder, wants to measure his downhill speed, thinking that he may reach 45 mph. While riding down the hill, he holds a panic alarm that operates at a frequency of 1250 Hz. Rochelle, who has perfect pitch, claims that she can identify any note on a piano from 27.5 Hz to 4186.0 Hz.

a. Rochelle listens from the top of the hill as Joe rolls away (Fig. P17.58). If Joe really reaches 45 mph, what frequency will Rochelle hear?

b. From the top of the hill, Rochelle hears a C6, which has a frequency of 1046.5 Hz. What is Joe's speed? In the **CHECK and THINK** step, describe whether or not your answer seems reasonable.

Figure P17.58

INTERPRET and ANTICIPATE

Joe is skateboarding away from Rochelle, who is stationary at the top of the hill. Since they are moving apart, we expect that the frequency is shifted lower (as we are told in part (b)) and that the Doppler equation will allow us to relate the speed with which Joe travels to the frequency that Rochelle hears.

SOLVE	
a. Choose the appropriate signs in the Doppler Equation 17.31. The observer is stationary and the source is moving *away*. So we choose the positive sign in the denominator. (We don't need to worry about the sign in the numerator since Rochelle is not moving.)	$f_{\text{obs}} = \left(\dfrac{v_s \pm v_{\text{obs}}}{v_s \mp v_{\text{source}}}\right) f$ $f_{\text{obs}} = \left(\dfrac{v_s \pm 0}{v_s + v_{\text{source}}}\right) f = \left(\dfrac{v_s}{v_s + v_{\text{source}}}\right) f \quad (1)$
Substitute values. Convert 45 mph to 20.1 m/s and keep three significant figures. She would hear a frequency of 1180 Hz.	$f_{\text{obs}} = \left(\dfrac{343\ \text{m/s}}{343\ \text{m/s} + 20.1\ \text{m/s}}\right) 1250\ \text{Hz}$ $f_{\text{obs}} = \boxed{1.18 \times 10^3\ \text{Hz}}$

b. Now rearrange equation (1) above to solve for the speed of the source as a function of the frequency.	$(v_s + v_{\text{source}}) f_{\text{obs}} = v_s f$ $v_{\text{source}} = v_s \left(\dfrac{f - f_{\text{obs}}}{f_{\text{obs}}} \right)$
Substitute numerical values. This is a speed of about 150 mph!	$v_{\text{source}} = 343 \text{ m/s} \left(\dfrac{1250 \text{ Hz} - 1046.5 \text{ Hz}}{1046.5 \text{ Hz}} \right)$ $v_{\text{source}} = \boxed{66.7 \text{ m/s}}$
CHECK and THINK With an estimated speed of 150 mph, our answer is not reasonable. We would imagine that he begins at rest at the top of the hill and picks up speed on his downward path, so she should hear a frequency close to 1250 Hz initially that decreases towards the 1180 Hz (part (a)) as he accelerates away and reaches the bottom of the hill.	

63. (N) A woman sees a supersonic plane directly overhead and then hears a sonic boom 3.7 s later. If the plane's altitude is 1.75×10^3 m, what is the Mach number of the plane? Assume the speed of sound is 331 m/s. Ignore any potential effects due to refraction of the sound.

INTERPRET and ANTICIPATE

It is best to start with a sketch so see the geometry involved. Use Figures 17.32 and 17.33 as guides. With this, we can relate the variables in the sketch.

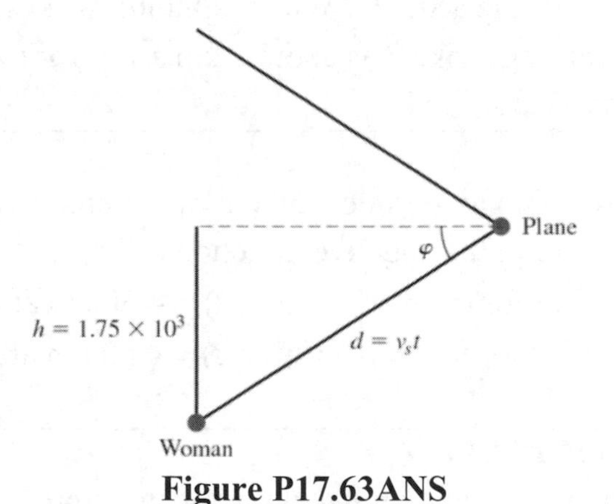

Figure P17.63ANS

SOLVE Use the figure to find the Mach angle.	$\cos\varphi = \dfrac{v_s t}{h}$ $\varphi = \cos^{-1} \dfrac{(331 \text{ m/s})(3.7 \text{ s})}{1.75 \times 10^3 \text{ m}} = 45.6°$

We can now determine the Mach number from the Mach angle with Eq. 17.32.	$\sin\varphi = \frac{v_s}{v_{source}} = \frac{1}{M}$ $M = \frac{1}{\sin 45.6°} = \boxed{1.4}$
CHECK and THINK The Mach number is larger than 1, as we would expect.	

65. How far are you located from a lightning strike if the thunderclap arrives 7.50 s after you see the lightning flash? Assume the speed of light in air is 3.0×10^8 m/s and the speed of sound at this location is 3.40×10^2 m/s.

INTERPRET and ANTICIPATE Since $v_{light} >> v_{sound}$, the light reaches you almost immediately compared to the relatively slow speed of sound. Therefore, the distance from the lightning is very nearly the distance that sound has traveled in that time interval.	
SOLVE Calculate the distance that the sound has traveled in 7.50 seconds.	$d = vt$ $d \approx (3.40 \times 10^2 \text{ m/s})(7.50 \text{ s}) = 2.55 \times 10^3 \text{ m}$ $d \approx \boxed{2.55 \text{ km}}$
CHECK and THINK We can use this as a general rule actually. When lightning strikes, start counting until you hear the thunder. The lightning strike was about 1 km away for every 3 seconds you count (or a mile for every five seconds).	

68. (A) A sinusoidal wave with amplitude 4.50 cm, wavelength 10.0 cm, and frequency 1.50 Hz is observed to travel in the negative direction.

a. What is the wave function for this wave if $y(x, t) = 4.50$ cm at $x = 0$ and $t = 0$?

b. What is the wave function for this wave if $y(x, t) = 4.50$ cm at $x = 5.00$ cm and $t = 0$?

INTERPRET and ANTICIPATE

We would like to write a wave equation like we've seen before, in a form like Eq. 17.9, though possibly with a non-zero phase:

$$y(x,t) = y_{max} \sin(kx + \omega t + \varphi)$$

We need to determine the wavenumber, angular frequency, and phase in each case.

SOLVE **a.** Considering Eq. 17.9, the fact that the *kx* and ω*t* terms have the same sign indicates that the wave is moving in the negative direction. The amplitude is given as 4.50 cm.	$y(x,t) = y_{max} \sin(kx + \omega t + \varphi)$ $y_{max} = 4.50 \text{ cm} = 0.0450 \text{ m}$
Calculate the wave number (Eq. 17.5).	$k = \frac{2\pi}{\lambda} = \frac{2\pi}{(0.100 \text{ m})} = 20.0\pi \text{ m}^{-1}$
Determine the angular frequency (Eq. 16.2).	$\omega = 2\pi f = 2\pi(1.50) = 3.00\pi \text{ rad/s}$
We also want $y(0, t) = 4.50$ cm at $t = 0$.	$4.50 \text{ cm} = (4.50 \text{ cm})\sin(0 + 0 + \varphi)$ $\sin\varphi = 1$ $\varphi = \pi/2$
Putting this all together and assuming metric units, we write the answer.	$\boxed{y = (0.0450)\sin\left(20.0\pi x + 3.00\pi t + \frac{\pi}{2}\right)}$
b. All quantities are the same as part (a) except for the phase. We require that $y(x, 0) = 4.50$ cm at $x = 0.0500$.	$0.0450 = 0.0450 \ \sin(20.0(0.0500)\pi + \phi)$ $1 = \sin(\pi + \phi)$ $\sin^{-1}(1) = \pi + \phi$ $\phi = \frac{\pi}{2} - \pi = -\frac{\pi}{2}$
We again write the wave equation assuming metric units.	$\boxed{y = 0.0450\sin\left(20.0\pi x + 3.00\pi t - \frac{\pi}{2}\right)}$

CHECK and THINK

In both cases, properties of the wave such as the wave number and angular frequency are the same. The only difference is where in the oscillation the wave is at a particular time, which is determined by the phase.

72. (A) The equation of a harmonic wave propagating along a stretched string is given as

$$y(x,t) = y_{max} \sin\left[2\pi\left(\frac{x}{a} - ft\right)\right]$$

where a is a constant and f is the frequency. If the magnitude of the maximum vertical speed

$$v_{y,\,\max} = \left.\frac{\partial y}{\partial t}\right|_{\max}$$

is equal to the three times the propagation speed, what is the value of a? Express your answer in terms of $y_{\max}$.

INTERPRET and ANTICIPATE

We are given the wave equation. We can take the derivative to find the maximum vertical speed and then require that this equals three times the propagation speed $v = f\lambda$.

SOLVE	
To find the maximum velocity, take the derivative. Since the derivative of the sine function produces a cosine, the maximum value occurs when the magnitude of the cosine function equals one (when the cosine term equals negative one in this case).	$v_{y,\max} = \left.\frac{\partial y}{\partial t}\right\|_{\max}$ $v_{y,\max} = \left(y_{\max}(-2\pi f)\cos\left[2\pi\left(\frac{x}{a} - ft\right)\right]\right)_{\max}$ $v_{y,\max} = 2\pi f y_{\max}$
This should be equal to three times the propagation speed, $v = f\lambda$. Comparing the wave function to the general wave equation for a harmonic wave propagating in the positive x direction, $y(x,t) = y_{\max}\sin(kx - \omega t)$, we compare with the given equation to determine the wavelength.	$k = \frac{2\pi}{\lambda} = \frac{2\pi}{a}$ $\lambda = a$
Now set the maximum velocity to three times the propagation speed.	$v_{y,\max} = 3v$ $2\pi f y_{\max} = 3f\lambda = 3fa$ $a = \boxed{\frac{2}{3}\pi y_{\max}}$

CHECK and THINK

We don't have much intuition about the value of the answer, but it is in terms of the maximum amplitude as expected.

76. (N) A stationary observer receives sound waves from two tuning forks, each oscillating at a frequency of 512 Hz. One of the tuning forks is approaching the observer, and the other tuning fork is receding from the stationary observer with the same speed as the first one. The observer hears a frequency difference (beat frequency; see Chapter 18) of 4 Hz when comparing the sound from each tuning fork. Find the speed of each tuning fork. Assume the speed of sound in air is 343 m/s.

INTERPRET and ANTICIPATE	
The frequency heard from the tuning fork approaching the observer will be larger than the tuning fork receding from the stationary observer. The difference in the two frequencies is the beat frequency.	
SOLVE Equation 17.31 can be used to determine the frequency heard by an observer when the source of the sound is moving. The observer's speed is zero. The frequency heard by the observer at rest from the tuning fork (source) coming towards the observer will be larger than the source frequency, so we use a negative sign in the denominator.	$f_{\text{towards}} = \dfrac{v_s}{v_s - v_{\text{source}}} f$
The frequency heard by the observer from the tuning fork (source) moving away the observer will be smaller than the source frequency, so we use a positive sign in the denominator.	$f_{\text{away}} = \dfrac{v_s}{v_s + v_{\text{source}}} f$
The frequency difference is the beat frequency of 4 Hz.	$\Delta f = f_{\text{towards}} - f_{\text{away}} = 4\,\text{Hz}$ $\left(\dfrac{v_s}{v_s - v_{\text{source}}} f\right) - \left(\dfrac{v_s}{v_s + v_{\text{source}}} f\right) = 4\,\text{Hz}$
Solve for the source velocity. All quantities are in metric units. We should retain all significant figures until we solve for the roots of the quadratic, but the final answer should have three significant figures.	$\left(\dfrac{2v_{\text{source}}}{v_s^{\,2} - v^2_{\text{source}}}\right) v_s f = 4$ $4v^2_{\text{source}} + 2v_s f v_{\text{source}} - 4v_s^{\,2} = 0$ $4v^2_{\text{source}} + 2(343)(512)v_{\text{source}} - 4(343)^2 = 0$ $v^2_{\text{source}} + 8.7808\times10^4 v_{\text{source}} - 1.17649\times10^5 = 0$

Finally, use the quadratic formula and take the positive root.

$$v_{\text{source}} = \frac{-8.781\times10^4 \pm\sqrt{\left(8.781\times10^4\right)^2 - 4\left(-1.176\times10^5\right)}}{2}$$

$$v_{\text{source}} = \boxed{1.34\text{ m/s}}$$

CHECK and THINK

The tuning forks are moving at a speed of 1.34 m/s.

79. (N) A careless child accidentally drops a tuning fork vibrating at 450 Hz from a window of a high-rise building. How far below the window is the tuning fork when the child hears sound waves with frequency 425 Hz? Remember to account for the time required for the sound to reach the child.

INTERPRET and ANTICIPATE

As the tuning fork falls away, the perceived frequency is Doppler shifted lower. We can determine how fast the tuning fork must be traveling for the child to hear the Doppler shifted frequency. Then, we can determine how long it takes for the tuning fork under free fall to reach that speed and how far the tuning fork has fallen. (To be precise, the tuning fork continues to fall as the sound travels back up to the window, so we consider this as well.)

SOLVE

The perceived frequency is Doppler shifted lower according to Eq. 17.31. The observer is not moving and the positive sign in the denominator leads to the decrease in frequency that the observer detects. The speed of the source increases as it accelerates due to gravity, so the observed frequency decreases in time. We can determine this time and then find how far the tuning fork has fallen.

$$f_{\text{obs}} = \left(\frac{v_s \pm v_{\text{obs}}}{v_s \mp v_{\text{source}}}\right) f$$

$$f_{\text{obs}} = \left(\frac{v_s}{v_s + v_{\text{source}}}\right) f$$

$$f_{\text{obs}} = \left(\frac{v_s}{v_s + gt}\right) f$$

We solve for the time at which the frequency is shifted by the amount specified.

$$t_f = \left(\frac{f}{f_{\text{obs}}} - 1\right)\frac{v_s}{g}$$

$$t_f = \left(\frac{450}{425} - 1\right)\frac{343}{9.81}$$

$$t_f = 2.06\text{ s}$$

During this time interval, the tuning fork falls a distance we can determine using kinematics.	$d_1 = \frac{1}{2}gt_f^2$ $d_1 = \frac{1}{2}(9.81\ \text{m/s}^2)(2.06\ \text{s})^2$ $d_1 = 20.8\ \text{m}$
This is a reasonable final answer for this question, but to be precise, the sound then must travel back to the window, during which the tuning fork continues to fall, so let's consider whether this matters. We first calculate the time for the sound to travel 20.8 meters back up to the window.	$t_{\text{return}} = \frac{20.8\ \text{m}}{343\ \text{m/s}} = 0.0606\ \text{s}$
So, the tuning fork falls for a total time of $t_{\text{total fall}} = t_f + t_{\text{return}} = 2.06\ \text{s} + 0.0606\ \text{s} = 2.119\ \text{s}$ until the child actually hears a frequency of 425 Hz. We now determine how far the tuning fork falls for this total time. The difference of 1.2 meters is about 6% larger, so the initial estimate was pretty good.	$d_{\text{total}} = \frac{1}{2}gt_{\text{total fall}}^2$ $d_{\text{total}} = \frac{1}{2}(9.81\ \text{m/s}^2)(2.119\ \text{s})^2$ $d_{\text{total}} = \boxed{22.0\ \text{m}}$

CHECK and THINK

In this case, we used the Doppler formula to relate the frequency heard to the speed of the source. As an object in free fall, we could then use kinematics to determine the position of the tuning fork when it reached this speed. The experiment could plausibly be done, though we might have trouble hearing a tuning fork falling 22 meters down from the window!

18

Superposition and Standing Waves

3. Two waves in the same medium are given by $y_1(x,t) = 2.5\sin(8x - 3.2t)$ and $y_2(x,t) = 2.5\sin(6x - 2.4t)$.

a. (C) In what ways do the two waves differ?

Wave 1 has an angular frequency of ω_1 = 3.2 rad/s and wave 2's angular frequency is ω_2 = 3.4 rad/s. Additionally, the angular wave number of wave 1 is 8 rad/m and wave 2's is 6 rad/m.

b. (N) Find the wave that results on this string. *Hint*: Recall that $\sin\alpha + \sin\beta = 2\sin\left[\frac{1}{2}(\alpha+\beta)\right]\cos\left[\frac{1}{2}(\alpha-\beta)\right]$.

INTERPRET and ANTICIPATE We need to add the two waves together and hope that the hint seems helpful!	
SOLVE We start by adding the two waves, $y_1(x,t) + y_2(x,t)$.	$y_1 + y_2 = 2.5\sin(8x - 3.2t) + 2.5\sin(6x - 2.4t)$ $y_1 + y_2 = 2.5\left[\sin(8x - 3.2t) + \sin(6x - 2.4t)\right]$
Comparing the hint to our expression, we define alpha and beta in order to carry out the sum of sines.	$\alpha = 8x - 3.2t$ $\beta = 6x - 2.4t$
To evaluate the sum, we need to determine $\alpha + \beta$ and $\alpha - \beta$.	$\alpha + \beta = 14x - 5.6t$ $\alpha - \beta = 2x - 0.8t$
Now, use the hint to evaluate the sum.	$y_1 + y_2 = 2.5\left[2\sin\left[\frac{1}{2}(14x - 5.6t)\right]\cos\left[\frac{1}{2}(2x - 0.8t)\right]\right]$ $y_1 + y_2 = \boxed{5.0\sin(7x - 2.8t)\cos(x - 0.4t)}$

CHECK and THINK
Conceptually, we literally needed only to add the two waves. The hint allows us to rewrite this sum in a more concise notation.

c. (C) What is the resultant wave's amplitude?

We expect a traveling wave to have a form that looks like $y_{max}\sin(kx - \omega t)$ or $y_{max}\cos(kx - \omega t)$ with the factor in front representing the amplitude. There is no simple way to describe the amplitude of this wave. If we use the remaining sine function to indicate the traveling wave's frequency and wavelength, the amplitude varies in time and space based on the remaining function, $\boxed{5.0\cos(x-0.4t)}$. This is one way to view the result.

9. CASE STUDY Sounds that reach your ear within 0.1 s of the initial sound are perceived as *reverberation* rather than as a distinct echo. In a music hall, a little reverberation can be a good thing. Too much can be distracting. Because sound reflects off the many surfaces in the hall several times, a listener may hear many reflections of the same sound. Each reflection absorbs some of the wave's energy, and the reflected waves get quieter. So, after a short time, the reflected waves are unheard. If the surfaces don't absorb enough energy, however, the reflected waves may be perceived for too long, and an echo is heard. Often, panels are placed in the interior of a music hall to control the reflected sounds. Let's consider a simple example. (We won't worry about speakers, reflections off of people, furniture, or other objects.) A single source of sound is on stage, and a listener sits directly in front of the source (Fig. P18.9). There are six reflecting panels set inside the hall. Suppose each panel absorbs sound so that each reflected wave has 10% of the intensity of the incident wave. The range of human hearing is quite large, so let's say that if the intensity is reduced by a factor of one million, the sound is just barely audible.
a. (N) How many reflections can the sound wave undergo before becoming inaudible?
b. (A) Use the distances shown in Figure P18.9 to write an expression for the duration Δt of a sound heard due to its many reflections in terms of the speed of sound v_s in the room.
c. (N) The duration can be longer than 0.1 s because reflected waves are quieter than the original sound and because the music being played overwhelms the quieter reflections. So, a duration Δt of about 1 s is quite good. Find the distance ℓ that gives a duration of 1 s.

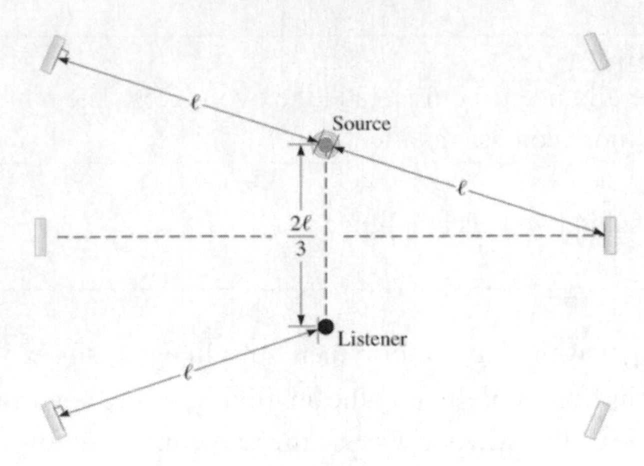

Figure P18.9

<table>
<tr><td colspan="2">INTERPRET and ANTICIPATE
There is a lot of information in this problem, but the major ideas are that each reflection is reduced to 10% of the original intensity and that when the sound is not audible once it's reduced to one millionth of the original intensity.</td></tr>
<tr><td>SOLVE
a. The question is "How many times can the intensity be reduced by a factor of 10 before it is reduced by 1,000,000 times?" It's possible to write a mathematical expression as shown on the right, but after 6 reflections, the intensity is reduced by 10 six times for a total reduction of a factor of 1,000,000.</td><td>after 1 reflection, reduced by 10
after 2 reflections, reduced by $10 \times 10 = 100$
after 3 reflections, reduced by $10 \times 10 \times 10 = 1000$
after N reflections, reduced by 10^N

Therefore, we want: $10^N = 1{,}000{,}000$
$N = \log(1{,}000{,}000) = \boxed{6}$</td></tr>
<tr><td>b. Each reflection surface is approximately a distance 1 from the sound/listener, so each reflection adds a distance (there and back) of around 2ℓ. We might estimate the difference between the longest path that is still audible after six reflections (12ℓ) and the direct path (around ℓ) as approximately 11ℓ, and determine the time for the sound to travel this distance.</td><td>$t \approx \boxed{\dfrac{11\ell}{v_s}}$</td></tr>
</table>

c. We want the time for the sound to travel a distance 11ℓ to be one second.	$\frac{11\ell}{v_s} \approx 1\text{ s}$ $\ell \approx \frac{1}{11} v_s \approx \frac{1}{11} 330 \approx \boxed{30\text{ m}}$
CHECK and THINK You've likely experienced this reverberation (or even an echo) in a large performance hall or cathedral. A distance of 30 meters seems like the right order of magnitude for a large room.	

12. Two speakers, facing each other and separated by a distance *d*, each emit a pure tone of the same amplitude *A* with frequency *f*. The speed of each of the sound waves is v_S. A listener stands between the speakers, a distance *x* from one of the speakers.
a. (A) What frequencies would cause a dead spot (complete destructive interference) at the listener's position?

INTERPRET and ANTICIPATE Since the speakers emit a pure tone, the difference in path length between the paths from each of the two speakers will determine whether the interference is constructive or destructive.	
SOLVE We can use Equation 18.3, which expresses the condition for destructive interference, that the difference in path length should be half a wavelength.	$\Delta d = \frac{n}{2}\lambda \quad (n = 1,3,5,...) \qquad (1)$
We know the distance from each speaker to the listener, x and $d - x$, and can express this path length difference. Equation 17.12 also allows us to relate the wavelength and the frequency of the wave ($v = \lambda f$).	$\Delta d = (d - x) - (x) = d - 2x \qquad (2)$ $f = \frac{v}{\lambda} \qquad (3)$
Plugging Equations 1 and 2 into Equation 3 leads to an expression for the frequencies that produce a dead spot.	$f_n = \frac{v}{\lambda} = \frac{nv}{2\Delta d} = \boxed{\frac{nv}{2(d - 2x)}}$
CHECK and THINK The expression is expressed in terms of the variables given. One limit that makes sense is that exactly between the speakers, when $x = d/2$, $f = \infty$. That is, there are no frequencies	

that lead to destructive interference when the path length from each speaker is the same.

b. (N) If the speakers are separated by 5.00 m with the listener 2.00 m from one of the speakers, what is the lowest frequency for which there is a dead spot? The speed of sound in air is 343 m/s.

INTERPRET and ANTICIPATE	
The result from part (a) can be used with the values given to find the answer.	
SOLVE The separation between speakers is d = 5.00 m and the listener is at x = 2.00 m. The smallest frequency corresponds to n = 1. We assume a speed of sound in air of 343 m/s.	$f_1 = \dfrac{(1)(343\text{ m/s})}{2(5.00\text{ m} - 2(2.00\text{ m}))} = \boxed{172\text{ Hz}}$
CHECK and THINK The values were used to find the lowest frequency leading to destructive interference. This frequency is a low frequency audible tone.	

16. (N) As in Figure P18.16, a simple harmonic oscillator is attached to a rope of linear mass density 5.4×10^{-2} kg/m, creating a standing transverse wave. There is a 3.6-kg block hanging from the other end of the rope over a pulley. The oscillator has an angular frequency of 43.2 rad/s and an amplitude of 24.6 cm.

a. What is the distance between adjacent nodes?

b. If the angular frequency of the oscillator doubles, what happens to the distance between adjacent nodes?

c. If the mass of the block is doubled instead, what happens to the distance between adjacent nodes?

d. If the amplitude of the oscillator is doubled, what happens to the distance between adjacent nodes?

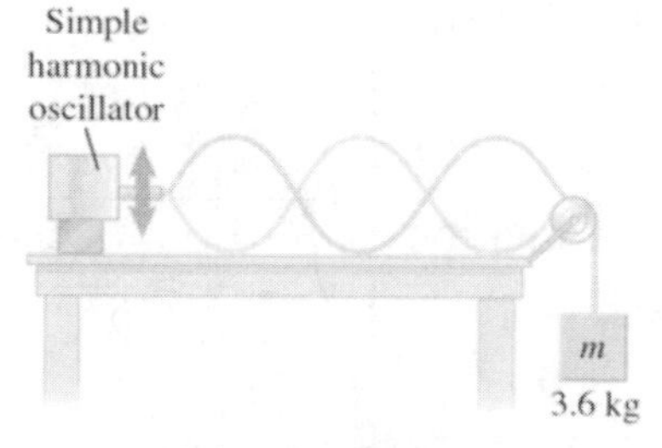

Figure P18.16

INTERPRET and ANTICIPATE The distance between adjacent nodes is half a wavelength. We can determine the wavelength from the frequency and speed, where the wave speed depends on the tension and linear mass density.	
SOLVE **a.** We can determine the wavelength from the speed and frequency using Equation 17.8.	$\lambda = \frac{v}{f}$
The wave speed depends on the tension and linear mass density according to Eq. 17.11.	$\lambda = \frac{1}{f}\sqrt{\frac{F_T}{\mu}} = \frac{2\pi}{\omega}\sqrt{\frac{F_T}{\mu}}$
Substitute numerical values.	$\lambda = \frac{2\pi}{43.2}\sqrt{\frac{(3.6\text{ kg})(9.81\text{ m/s}^2)}{5.4\times10^{-2}\text{ kg/m}}}$
The distance between adjacent nodes is half a wavelength.	$d_{\text{nodes}} = \left(\frac{2\pi}{43.2}\sqrt{\frac{(3.6\text{ kg})(9.81\text{ m/s}^2)}{5.4\times10^{-2}\text{ kg/m}}}\right)\Big/2 = \boxed{1.9\text{ m}}$
b. From part (a), the wavelength is inversely proportional to the angular frequency, so the distance between nodes would be halved.	$\lambda = \frac{2\pi}{\omega}\sqrt{\frac{F_T}{\mu}} \quad\rightarrow\quad \lambda \propto \frac{1}{\omega}$ $\lambda_{2\omega_0} = \frac{1}{2}\lambda_{\omega_0}$ $d_{2\omega_0} = \frac{\lambda_{\omega_0}/2}{2} = \left(\frac{2\pi}{43.2}\sqrt{\frac{(3.6\text{ kg})(9.81\text{ m/s}^2)}{5.4\times10^{-2}\text{ kg/m}}}\right)\Big/4$ $d_{2\omega_0} = \boxed{0.93\text{ m}}$
c. Using part (a), if the mass doubles instead, the tension force doubles, and the wavelength increases by a factor of $\sqrt{2}$.	$\lambda = \frac{2\pi}{\omega}\sqrt{\frac{F_T}{\mu}} \quad\rightarrow\quad \lambda \propto \sqrt{F_T}$ $\lambda_{2F_T} = \sqrt{2}\lambda_{\omega_0} = \sqrt{2}\left(\frac{2\pi}{43.2}\sqrt{\frac{(3.6\text{ kg})(9.81\text{ m/s}^2)}{5.4\times10^{-2}\text{ kg/m}}}\right)$ $d_{2F_T} = \frac{\lambda_{2F_T}}{2} = \sqrt{2}\left(\frac{2\pi}{43.2}\sqrt{\frac{(3.6\text{ kg})(9.81\text{ m/s}^2)}{5.4\times10^{-2}\text{ kg/m}}}\right)\Big/2$ $d_{2F_T} = \boxed{2.6\text{ m}}$

d. The wavelength is independent of amplitude, so the $\boxed{\text{distance between nodes remains unchanged}}$ compared to part (a).	$\lambda = \frac{2\pi}{\omega}\sqrt{\frac{F_T}{\mu}} \rightarrow \lambda =$ independent of amplitude $d_{2A} = \boxed{1.9 \text{ m}}$
CHECK and THINK Since the node distance is half the wavelength, we are ultimately asked to find how the wavelength depends on frequency, tension force, and amplitude. From the wave equation, wavelength depends inversely on the frequency. From the formula for wave speed on a wire, the wave speed depends on the square root of the tension, which in turn affects the wavelength. Our formulas are independent of amplitude.	

19. (N) A standing transverse wave on a string of length 60 cm is represented by the equation $y(x,t) = 4.0\sin(\pi x/15)\cos(96\pi t)$, where x and y are in centimeters and t is in seconds.

a. What is the maximum value of the standing wave at the point $x = 5.0$ cm?

b. Where are the nodes located along the string for this particular standing wave?

c. What is the vertical velocity v_y of the string at $x = 7.5$ cm when $t = 0.25$ s?

INTERPRET and ANTICIPATE We have an expression for the position of a standing wave on the string. Each part of the question asks us to find specific values of the wave properties that we can determine from our equation.	
SOLVE **a.** First, consider the general formula for a standing wave (Eq. 18.6). At any location on the string, the position of the string oscillates up and down with an amplitude $2y_{\max}\sin(kx)$. The oscillation in time is characterized by the $\cos(\omega t)$ term.	$y(x,t) = \left[2y_{\max}\sin(kx)\right]\cos(\omega t)$
The maximum y value at $x = 5$ cm occurs when the cosine term is at its maximum and equals one.	$y(x = 5 \text{ cm}, t) = 4.0\sin\left(\frac{\pi 5}{15}\right)\cos(\omega t)$ $y(x = 5 \text{ cm}, t)\Big\vert_{\max} = 4.0\sin\left(\frac{\pi 5}{15}\right) = \boxed{3.5 \text{ cm}}$

b. Nodes occur at locations that for all times are at zero amplitude. This occurs when the sine term is zero ($y = 0$ for any time t). That is, when the argument is a multiple of π.	Nodes when $\sin\left(\frac{\pi x}{15}\right) = 0$ $\frac{\pi x}{15} = n\pi \quad \rightarrow \quad x = 15n$ cm
For a string of length 60 cm, this occurs at multiples of 15 cm across the length of 0 to 60 cm.	Nodes at $x =$ $\boxed{0, 15, 30, 45, \text{and } 60 \text{ cm}}$
c. We take the time derivative of the displacement $y(x,t)$ to obtain the particle velocity.	$v_y(x,t) = \frac{\partial y(x,t)}{\partial t}$ $v_y(x,t) = 4.0(-96\pi)\sin\left(\frac{\pi x}{15}\right)\sin(96\pi t)$
Now we find the velocity at $x = 7.5$ cm and $t = 0.25$ s. The second sine term turns out to be zero, so the velocity is zero.	$v_y(x,t) = -(4.0)(96\pi)\sin\left(\frac{\pi(7.5)}{15}\right)\sin(96\pi(0.25))$ $v_y(x,t) = \boxed{0}$

CHECK and THINK

As we've seen with waves on a string in general, with a formula for the location of the string, we are able to identify kinematic properties, like position and velocity, at any point on the string at any time.

25. (N) Two successive harmonics on a string fixed at both ends are 66 Hz and 88 Hz. What is the fundamental frequency of the string?

INTERPRET and ANTICIPATE

The harmonic frequencies are multiples of the fundamental frequency. Therefore, we need a frequency for which 66 Hz and 88 Hz are sequential multiples.

SOLVE Harmonic frequencies for the string are multiples of the fundamental f_1 as given by Eq. 18.11.	$f_n = nf_1 \qquad (n = 1, 2, 3, \text{K})$
We need a lower frequency for which 66 Hz and 88 Hz are sequential multiples. A fundamental frequency of 22 Hz, with harmonics of 44 Hz, 66 Hz, 88 Hz, etc., satisfies this requirement. You	$f_n = nf_1 \qquad (1)$ $f_{n+1} = (n+1)f_1 \qquad (2)$

may realize that this is the largest common factor. Mathematically, we can determine this by first writing the two sequential frequencies as f_n and f_{n+1}.	
Subtract equation 2 from 1 to solve for the fundamental.	$88\text{ Hz} - 66\text{ Hz} = f_1$ $f_1 = \boxed{22\text{ Hz}}$
CHECK and THINK In this case the fundamental frequency of 22 Hz has harmonics at 44 Hz, 66 Hz, 88 Hz, … Two sequential harmonics are the 66 Hz and 88 Hz that we expect.	

28. (N) A 50.0-cm-long copper wire with radius 0.100 cm and density 8.96 g/cm^3 is placed under 45.0 N of tension.
a. What is the fundamental frequency of vibration for the wire?
b. What are the next two harmonic frequencies for standing waves on this wire?

INTERPRET and ANTICIPATE The fundamental frequency occurs for a standing wave in which half a wavelength fits across the wire, such that each end is a node. The frequency depends on this and the wave speed, which depends on the tension in the wire and its linear mass density. Harmonics occur at multiples of the fundamental frequency.	
SOLVE **a.** We are given the density of the wire (mass per volume), but we need the linear mass density (mass per length). They are related by the cross sectional area A. The cross-sectional area of the circular wire is $A = \pi r^2$.	$\mu = \rho A = \rho \pi r^2$ $\mu = \left(8.96 \frac{\text{g}}{\text{cm}^3}\right)\pi(0.100\text{ cm})^2$ $\mu = \left(8.96 \frac{\text{g}}{\text{cm}^3}\right)(3.14\times10^{-2}\text{cm}^2)$ $\mu = 0.281 \frac{\text{g}}{\text{cm}} = 2.81\times10^{-2} \frac{\text{kg}}{\text{cm}}$
The fundamental frequency for a standing wave on a wire (Eq. 18.10) depends on the wave speed on the wire and the length of the wire.	$f_1 = \frac{v}{2L}$

Use Eq. 17.11 to determine the wave speed, which depends on the tension and linear mass density.	$v=\left(\frac{F_T}{\mu}\right)^{1/2}$ $v=\left(\frac{45.0\ \text{N}}{2.81\times10^{-2}\ \text{kg/m}}\right)^{1/2}$ $v=40.0\ \text{m/s}$
Calculate the fundamental frequency with L = 0.500 m.	$f_1=\frac{v}{2L}=\frac{40.0\ \text{m/s}}{2(0.500\ \text{m})}$ $f_1=\boxed{40.0\ \text{Hz}}$
b. Harmonics are integer multiples of the fundamental frequency (Eq. 18.11).	$f_2=2f_1=\boxed{80.0\ \text{Hz}}$ $f_3=3f_1=\boxed{1.2\times10^2\ \text{Hz}}$

CHECK and THINK
As expected, we determine a fundamental frequency with harmonics that are integer multiples of the fundamental.

31. (A) Two strings fixed at both ends oscillate in the fourth harmonic. String A has twice the mass per unit length of string B. The length of the strings and the tension in each are identical. Find the ratio (A to B) of their

a. wavelengths and

b. frequencies.

INTERPRET and ANTICIPATE
The wavelengths supported by a string depend on its length while the frequency depends on the length and the wave speed. Given that the strings have the same length and different mass densities, we expect that they will have standing waves with the same wavelengths but different frequencies.

SOLVE	
a. The wavelength of the fundamental (and the harmonics) depends on the length of the string, which are the same.	$\frac{\lambda_A}{\lambda_B}=\boxed{1}$

b. Based on the dependence of the wave speed on tension and mass density (Eq. 17.11), we can determine the fundamental frequency for each string.	$f_1 = \frac{1}{2L}\sqrt{\frac{F_T}{\mu}}$
Now, take the ratio.	$\frac{f_A}{f_B} = \frac{\frac{1}{2L}\sqrt{\frac{F_T}{\mu_A}}}{\frac{1}{2L}\sqrt{\frac{F_T}{\mu_B}}} = \frac{\sqrt{\mu_B}}{\sqrt{\mu_A}} = \boxed{\frac{1}{\sqrt{2}}}$
CHECK and THINK String A is heavier and therefore has a lower fundamental frequency. (A familiar fact to anyone who plays a stringed instrument.) The lengths, and therefore the wavelengths supported, are the same.	

32. (N) A wire 1.50 m in length and with a mass of 25.0 g is placed under a tension of 33.0 N.

a. What are the first two harmonic frequencies of vibration for this wire?

INTERPRET and ANTICIPATE The fundamental frequency of a vibrating string occurs when half a wavelength is supported on the length of the string. The frequency depends on the wavelength and the wave speed on the string, which depends on the tension and linear mass density. Harmonics are integer multiples of the fundamental frequency.	
SOLVE The natural frequencies of vibration of a wire fixed at both ends are given by Equations 18.10 and 17.11.	$f_n = \frac{n}{2L}\sqrt{\frac{T}{\mu}}$ $f_n = \frac{n}{2(1.50\text{ m})}\sqrt{\frac{33.0\text{ N}}{\left(\frac{0.025\text{ kg}}{1.50\text{ m}}\right)}}$ where $n = 1, 2, 3, \ldots$
The first two harmonic frequencies correspond to $n = 1$ and $n = 2$.	$f_1 = \boxed{14.8\text{ Hz}}$ and $f_2 = \boxed{29.7\text{ Hz}}$
CHECK and THINK Given the tension, linear mass density, and length of the wire, we can determine the frequencies of the standing waves that can occur on a wire.	

b. The wire is observed to have a node at 30.0 cm from one end. What are the possible harmonic modes of vibration of the wire?

INTERPRET and ANTICIPATE

We know that there must be an integer number of half wavelengths between any two nodes of a standing wave. We know that the ends of the wire are nodes, which determine the possible standing waves on the wire. We are also told that there is an additional node, so we must find wavelengths supported by the wire for which this third point (30 cm from the end) is also a node.

SOLVE	
We first sketch the situation. We require that there is a node at each end as well as a point 30 cm from one end.	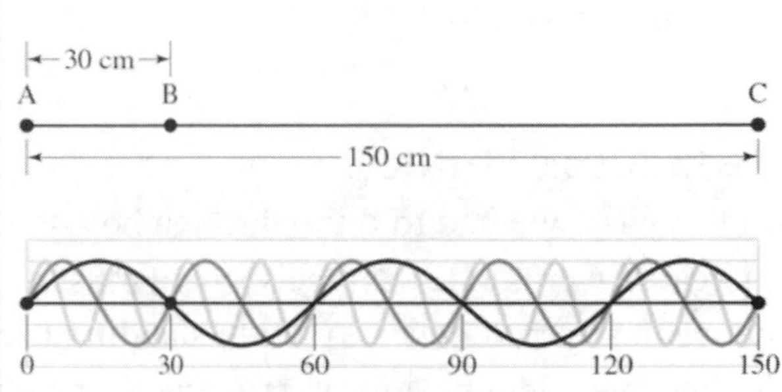 **Figure P18.32ANS**
If $D = 0.300$ m is the distance between adjacent nodes (points A and B), $D = \lambda/2$ is one possible standing wave that has a node at this point, so the wavelength of this standing wave would be 0.600 m. The wavelengths that are supported by the string are given by Equation 18.10 and we can find the mode n for which the wavelength 0.600 m occurs.	$\lambda = \dfrac{2L}{n}$
We verify that this wavelength we've determined is also one supported by the string. Since we find $n = 5$ (an integer), this means that the wavelength of 0.600 m is the fifth harmonic supported by this string. So, this mode corresponds to the harmonic.	$n = \dfrac{2L}{\lambda} = \dfrac{2L}{2D} = \dfrac{L}{D}$ $n = \dfrac{1.50\text{ m}}{0.300\text{ m}} = 5$ $f_5 = \dfrac{5}{2(1.50\text{ m})}\sqrt{\dfrac{33.0\text{ N}}{\left(\dfrac{0.025\text{ kg}}{1.50\text{ m}}\right)}} = 74\text{ Hz}$
This is actually the *longest* wavelength that satisfies this condition, but there could also be an *entire* wavelength fitting	$N\dfrac{\lambda}{2} = \dfrac{2L}{n}$

<table>
<tr><td>between the node at the end and at 0.3 m, (points A and B) in which case, $D = \lambda$. In fact, any half number of wavelengths could fit between these two points and still have nodes at both, so let $D = N\frac{\lambda}{2}$, where N is any integer number of half wavelengths between these two points on the string and repeat the above calculate to determine which harmonics of the string this corresponds to.</td><td>$n = N\frac{L}{D}$
$n = N\frac{1.50 \text{ m}}{0.300 \text{ m}}$
$n = 5N$

So the harmonics that satisfy this condition are the 5th, 10th, 15th, and so on.

The frequency could be the $n = 5$ harmonic at 74 Hz, $n = 10$ at 148 Hz, $n = 15$ at 222 Hz, etc.</td></tr>
<tr><td colspan="2">CHECK and THINK
In this case, we had to think through how to assure that both ends as well as the point indicated were nodes. Namely, there must be an integer number of half wavelengths between any two nodes. As with standing waves on a string in general, we find there is a longest wavelength (lowest frequency) possible, as well as higher multiples of this frequency. In the figure, we plot the first three possibilities that we've found and see that all three points (A, B, and C) are nodes as required.</td></tr>
</table>

36. (N) A sonometer is a single-stringed musical instrument on a wooden box with two movable bridges (Fig. P18.36). The string is fastened at one end, and the other end is attached to a hanging weight holder that is extended over a pulley. The distance between the end of the pulley and the fastened end define the length of the string. By adding weights to the weight holder, you can change the tension in the string. A particular sonometer wire has a length of 114 cm. Where should the two bridges be placed along this length so as to divide the wire into three segments whose fundamental frequencies are in the ratio 1:3:4? Assume the tension is the same throughout the wire.

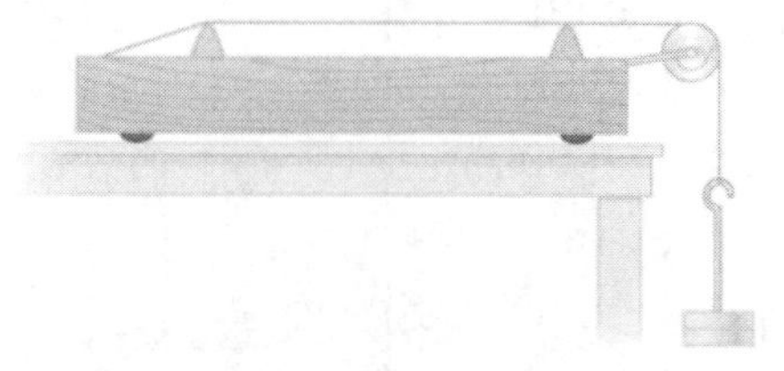

Figure P18.36

INTERPRET and ANTICIPATE
The two bridges on the wire will be the nodes for the standing wave. We want the fundamental frequencies of all three sections to be in the ratio indicated.

SOLVE Consider the lengths of three segments are L_a, L_b, and L_c, which must add to the 114 cm total length.	$L_a + L_b + L_c = 114\,\text{cm}$
Now for each segment we use the formula for the fundamental frequency of a standing wave fixed at both ends (Eq. 18.10). v is the speed of the wave in the wire.	$f_a = \frac{v}{2L_a} \quad f_b = \frac{v}{2L_b} \quad f_c = \frac{v}{2L_c}$

Now the ratio of the three frequencies is 1:3:4. We use that to solve for the lengths of the three segments. We express lengths L_a and L_c in terms of L_b.

$$\frac{f_a}{f_b} = \frac{1}{3} \quad \rightarrow \quad \frac{v/2L_a}{v/2L_b} = \frac{1}{3} \quad \rightarrow \quad \frac{L_b}{L_a} = \frac{1}{3} \quad \rightarrow \quad L_a = 3L_b$$

$$\frac{f_b}{f_c} = \frac{3}{4} \quad \rightarrow \quad \frac{v/2L_b}{v/2L_c} = \frac{3}{4} \quad \rightarrow \quad \frac{L_c}{L_b} = \frac{3}{4} \quad \rightarrow \quad L_c = \frac{3}{4}L_b$$

Now, we make sure the three lengths add to the total string length.	$L_a + L_b + L_c = 114\,\text{cm}$ $3L_b + L_b + \frac{3}{4}L_b = 114\,\text{cm}$
Solve for L_b and insert into the ratios above to find L_a and L_c. Then, quote the locations of the two bridges.	$L_a = 72\,\text{cm}$ $L_b = 24\,\text{cm}$ $L_c = 18\,\text{cm}$ One should be placed 72 cm from one end and the other should be 18 cm from the other end.

CHECK and THINK

The ratios of the lengths are related to the ratios of the frequencies ($L_a = 3L_b = 4L_c$).

39. (N) A very large organ may have pipes that are as short as about 1 ft and as long as 16 ft. Imagine such an organ kept in a church whose temperature in the middle of winter is 55°F and in the middle of summer is 90°F. Find the seasonal change in both the shortest and the longest pipe's fundamental frequency. Assume the pipes are open at both ends.

INTERPRET and ANTICIPATE

A change in temperature leads to a change in the speed of sound and therefore a change in the frequency.

SOLVE Convert temperatures to Celsius. Find the speed of sound in winter (T = 12.8°C) and in summer (T = 32.2°C) using Eq. 17.14.	$v_{\text{winter}} = 331\,\text{m/s}\sqrt{1+\dfrac{12.8°\text{C}}{273}}$ $v_{\text{summer}} = 331\,\text{m/s}\sqrt{1+\dfrac{32.2°\text{C}}{273}}$
The fundamental frequency for a pipe that is open at both ends is given by Eq. 18.13. The change in frequency depends on the difference in the speed of sound between summer and winter.	$f = \dfrac{v}{2L}$ $\Delta f = \dfrac{1}{2L}\left(v_{\text{summer}} - v_{\text{winter}}\right)$ $\Delta f = \dfrac{331\text{ m/s}}{2L}\left[\sqrt{1+\dfrac{32.2°\text{C}}{273}} - \sqrt{1+\dfrac{12.8°\text{C}}{273}}\right]$
Insert numerical values for both the short and long pipe.	$\Delta f_{\text{short}} = \dfrac{331\text{ m/s}}{2(0.305\text{ m})}\left[\sqrt{1+\dfrac{32.2°\text{C}}{273}} - \sqrt{1+\dfrac{12.8°\text{C}}{273}}\right]$ $\Delta f_{\text{short}} = \boxed{19\text{ Hz}}$ $\Delta f_{\text{long}} = \dfrac{331\text{ m/s}}{2(4.88\text{ m})}\left[\sqrt{1+\dfrac{32.2°\text{C}}{273}} - \sqrt{1+\dfrac{12.8°\text{C}}{273}}\right]$ $\Delta f_{\text{long}} = \boxed{1.2\text{ Hz}}$

CHECK and THINK

Both pipes have frequencies that are shifted significantly based on the change in temperature. It may be surprising that the short pipe is affected by the temperature change more than the long pipe.

47. (N) The third harmonic of an organ pipe that is closed at one end is in resonance with the first harmonic of an organ pipe that is open at both ends. Find the ratio of the lengths of the two pipes ($L_{\text{closed}}/L_{\text{open}}$).

INTERPRET and ANTICIPATE

We need to employ the formulas for the resonant frequencies of pipes that are open on both ends (Eq. 18.13) or closed on one end (Eq. 18.15).

SOLVE Consider an open pipe with fundamental frequency f_O and length L_O and a closed pipe with fundamental frequency f_C and length	$f_O = \dfrac{v}{2L_O}$ and $f_C = \dfrac{v}{4L_C}$

L_C. Apply Equations 18.13 and 18.15.	
We want the fundamental frequency of the open pipe to equal the third harmonic of the closed pipe.	$\frac{v}{2L_O} = 3\frac{v}{4L_C}$ $\frac{L_C}{L_O} = \boxed{\frac{3}{2}}$
CHECK and THINK The closed pipe is 1.5 times longer than the open pipe to satisfy the condition.	

50. A tube 1.00 m long that is closed at one end and open at the other is filled with room-temperature air, where the speed of sound is 343 m/s. Then water is used to fill the tube. The speed of sound in water is 1482 m/s

a. (N) Calculate the wavelength of the fundamental and the third harmonics both before and after the water is added.

b. (N) Calculate the fundamental and third harmonic frequencies both before and after the water is added.

INTERPRET and ANTICIPATE The harmonic wavelengths will be the same, regardless of whether the tube is filled with water or air. The harmonic frequencies will be different when the water is added to the tube. This is shown using Eq. 18.14 and Eq. 18.15 ($\lambda_n = 4L/n$ and $f_n = nv/4L$ where n = 1, 3, 5, …).	
SOLVE **a.** As stated, when examining Eq. 18.14 for the harmonic wavelengths in a tube with one open end, there is no effect from the medium. Thus, fundamental and third harmonic wavelength are unchanged when the water is added. Use, Eq.18.14 to find the wavelengths.	$\lambda_1 = 4(1.00 \text{ m})/1 = \boxed{4.00 \text{ m}}$ $\lambda_3 = 4(1.00 \text{ m})/3 = \boxed{1.33 \text{ m}}$
b. First, use Eq. 18.15 to find the fundamental and third harmonic frequencies when the tube is filled with air.	$f_{1,\text{ air}} = 1(343 \text{ m/s})/[4(1.00 \text{ m})] = \boxed{85.8 \text{ Hz}}$ $f_{3,\text{ air}} = 3(343 \text{ m/s})/[4(1.00 \text{ m})] = \boxed{257 \text{ Hz}}$

Then, use Eq. 18.15 to find the fundamental and third harmonic frequencies when the tube is filled with water.	$f_{1,\text{ water}} = 1(1482\text{ m/s})/[4(1.00\text{ m})] = \boxed{371\text{ Hz}}$ $f_{3,\text{ water}} = 3(1482\text{ m/s})/[4(1.00\text{ m})] = \boxed{1.11\times10^{3}\text{ Hz}}$

CHECK and THINK
The harmonic wavelengths are determined completely by the dimensions of the system (in this case, the tube), whereas the frequencies are medium-dependent because they depend on the speed with which waves travel in the medium. If the speed of the sound waves is greater in one medium, that medium will produce greater harmonic frequencies than a medium with slower sound waves.

c. (C) If someone were striking the tube and the sound inside were amplified through a speaker, would a listener hear a higher-pitched tone or a lower-pitched tone after the water is added? Explain your answer.

The listener would hear a higher-pitched tone after the water is added because the frequencies are greater. A greater frequency indicates a higher pitch, and the harmonic frequencies, including the fundamental became greater when the water replaced the air in the tube.

55. (N) A violinist and a pianist each play an A note. The piano was recently tuned to 440.0 Hz, but when the musicians play together, there are clear beats heard at a rate of about three per second. The violinist tightens the violin string, and the number of beats increases to about five beats per second. What is the frequency of the violin string after it is tightened?

INTERPRET and ANTICIPATE
The beat frequency depends on the difference in frequency of the two notes. This should allow us to determine the frequency of the violin assuming that the piano is in tune.

SOLVE	
We can use Equation 18.25 for the beat frequency. Initially, hearing three beats per second tells us that the violin must have a frequency of either 437 Hz or 443 Hz.	$f_{\text{beat}} = \frac{\omega_{\text{beat}}}{2\pi} = \left\lvert\frac{\omega_2}{2\pi} - \frac{\omega_1}{2\pi}\right\rvert = \lvert f_2 - f_1\rvert$ $3\text{ Hz} = \lvert f_{violin} - 440\text{ Hz}\rvert$ $f_{violin} = 437\text{ Hz or }443\text{ Hz}$

Tightening the violin string produces a higher frequency. A beat frequency of 5 Hz would correspond to a violin frequency of 335 Hz or 445 Hz. If it was tightened slightly, it suggests that the frequency changed from 443 Hz to 445 Hz. Therefore the frequency is now $\boxed{445\text{ Hz}}$.	$5\text{ Hz} = \lvert f_{violin} - 440\text{ Hz}\rvert$ $f_{violin} = 435\text{ Hz or } 445\text{ Hz}$
CHECK and THINK The beat frequency is due to a difference in frequency, allowing us to determine the frequency of the violin based on the known frequency of the piano.	

58. (N) In a classroom demonstration, two pipes of length 1.25 m open at both ends oscillate in their fundamental harmonic. One pipe has a collar allowing it to be lengthened by 15.0 cm. As one pipe is slowly lengthened, beats are heard.
a. Find the maximum beat frequency.
b. What is the length of the longer pipe when the beat frequency heard is half of its maximum?

INTERPRET and ANTICIPATE The beat frequency is equal to the differencc in frequencies of the two tones played at the same time. We need to determine the fundamental frequency of both open pipes to determine the beat frequency.	
SOLVE We use the fundamental frequency for a pipe open on both ends (Eq. 18.13) and calculate the difference to find the beat frequency (Eq. 18.20) for the largest difference in length.	$f_{beat} = \left\lvert \dfrac{v}{2L_1} - \dfrac{v}{2L_2} \right\rvert$ $f_{beat} = \dfrac{343\text{ m/s}}{2}\left\lvert \dfrac{1}{1.25\text{ m}} - \dfrac{1}{1.40\text{ m}} \right\rvert$ $f_{beat} = \boxed{14.7\text{ Hz}}$
To achieve half this beat frequency, use the formula from part (a) for a frequency of 14.7/2 = 7.35 Hz and determine the unknown length of the second pipe.	$7.35\text{ Hz} = \dfrac{v}{2}\left\lvert \dfrac{1}{L_1} - \dfrac{1}{L_2} \right\rvert$ $\dfrac{1}{L_2} = \dfrac{1}{L_1} - \dfrac{2f_{beat}}{v}$ $\dfrac{1}{L_2} = \dfrac{1}{1.25\text{ m}} - \dfrac{2(7.35\text{ Hz})}{343\text{ m/s}}$ $L_2 = \boxed{0.757\text{ m}}$

CHECK and THINK
The maximum beat frequency occurs when the two pipes have lengths that differ by the largest amount possible. To achieve half this beat frequency, the second pipe must be extended by about half the maximum possible extension.

61. (N) What is the amplitude of the resultant wave for two sinusoidal waves with equal amplitudes of 6.50 cm traveling in the same direction on a taut string, but 90.0° out of phase?

INTERPRET and ANTICIPATE	
With the information given, we can write expressions for each wave and add them together. If we can express the result in the usual form for a traveling wave, we should then be able to read off the amplitude as the prefactor.	
SOLVE First, write the sum of the two wave functions.	$y(x,t) = \left[(6.50\text{ cm})\sin(kx-\omega t)\right] + \left[(6.50\text{ cm})\sin(kx-\omega t+90.0°)\right]$ $y(x,t) = (6.50\text{ cm})\left[\sin(kx-\omega t)+\sin(kx-\omega t+90.0°)\right]$
One way to handle this is to use a trigonometric identity called the sum to product rule.	$\sin A + \sin B = 2\sin\left(\frac{A+B}{2}\right)\cos\left(\frac{A-B}{2}\right)$
Applying this, we get $\sin A + \sin B = 2\sin\left(\frac{A+B}{2}\right)\cos\left(\frac{A-B}{2}\right)$ $\sin(kx-\omega t+90.0°)+\sin(kx-\omega t) = 2\sin(kx-\omega t+45.0°)\cos(45.0°)$ $= \sqrt{2}\sin(kx-\omega t+45.0°)$	
Substitute this into the original expression. The amplitude of the resultant wave is **9.19 cm**.	$y(x,t) = (6.50\text{ cm})\sqrt{2}\sin(kx-\omega t+45.0°)$ $y(x,t) = (9.19\text{ cm})\sin(kx-\omega t+45.0°)$
CHECK and THINK A trigonometric identity allowed us to add two sine functions with different arguments. The result is in the form of a traveling wave though, so we can easily read off parameters. Though we weren't asked, notice also that in this case (in which the two initial waves have the same wavenumber and angular frequency), the resulting wave also has the same wavenumber and frequency, but the wave is shifted so that the peak value is between the peaks of the two incident waves.	

64. (N) The low E string and the high E string on a guitar are tuned to have fundamental frequencies of 82.4 Hz and 329.6 Hz, respectively. The difference in the fundamental frequency is primarily due to the different thickness, and therefore different mass per unit length, of the two strings. We could, however, try to design a guitar using the same type of strings for both notes by varying either the length of the string or its tension.
a. If the mass per unit length and the length were kept the same for both strings, what would be the ratio of the tension in the high E string to the tension in the low E string?

INTERPRET and ANTICIPATE The frequency increases with the square root of the tension. A larger frequency should result from a larger tension.	
SOLVE We can use the equation for the fundamental frequency of a string (using Eq. 17.11 and 18.10), expressing both the high and low frequencies with variables, assuming only the tension of the strings is different.	$f_{high} = \frac{1}{2L}\sqrt{\frac{F_{T,high}}{\mu}}$ $f_{low} = \frac{1}{2L}\sqrt{\frac{F_{T,low}}{\mu}}$
Taking the ratio of the two frequencies provides the dependence on the tension of the strings.	$\frac{f_{high}}{f_{low}} = \frac{\frac{1}{2L}\sqrt{\frac{F_{T,high}}{\mu}}}{\frac{1}{2L}\sqrt{\frac{F_{T,low}}{\mu}}} = \sqrt{\frac{F_{T,high}}{F_{T,low}}}$
This expression can then be written to find the ratio of the tensions in terms of the frequency ratio.	$\frac{F_{T,high}}{F_{T,low}} = \left(\frac{f_{high}}{f_{low}}\right)^2$
Plugging in the values of the frequency leads to the ratio of tension needed.	$\frac{F_{T,high}}{F_{T,low}} = \left(\frac{329.6\text{ Hz}}{82.4\text{ Hz}}\right)^2 = \boxed{16.0}$
CHECK and THINK The higher frequency note is produced if the same string is at a larger tension. While possible to make an instrument this way, the tension on the high-E string would need to be significantly higher than the low-E string due to the square root dependence of frequency on tension.	

b. If the mass per unit length and the tension were kept the same for both strings and we found a way to alter the length of the strings, what would be the ratio of the length of the high E string to the length of the low E string?

INTERPRET and ANTICIPATE The frequencies are inversely proportional to the wavelengths, which depend on the length of the strings. The higher pitched note should correspond to a smaller string.	
SOLVE As in part (a), we can express the high and low frequencies in terms of variables, assuming that they are the same in both cases except for the length of the string.	$f_{high} = \frac{1}{2L_{high}}\sqrt{\frac{F_T}{\mu}}$ $f_{low} = \frac{1}{2L_{low}}\sqrt{\frac{F_T}{\mu}}$
Take the ratio of the high frequency to the low frequency.	$\frac{f_{high}}{f_{low}} = \frac{\frac{1}{2L_{high}}\sqrt{\frac{F_T}{\mu}}}{\frac{1}{2L_{low}}\sqrt{\frac{F_T}{\mu}}} = \frac{L_{low}}{L_{high}}$
Substitute numerical values.	$\frac{L_{high}}{L_{low}} = \frac{f_{low}}{f_{high}} = \frac{82.4\text{ Hz}}{329.6\text{ Hz}} = \boxed{0.250}$
CHECK and THINK As expected, the high frequency note would need to be produced by a smaller string. It is four times smaller since the frequency is four times larger.	

70. (N) In Oshin's home theater, an oscillator operating at 425 Hz drives a pair of identical speakers placed 4.00 m apart and facing each other in phase. Where are the relative minima of sound pressure located in the line joining the two speakers?

INTERPRET and ANTICIPATE The facing speakers produce a standing wave in the space between them. We need to find locations where the waves from each speaker add destructively.	
SOLVE The requirement for destructive interference is that the path length from each speaker differs by an odd integer multiple of half a wavelength ($\lambda/2$, $3\lambda/2$, $5\lambda/2$...).	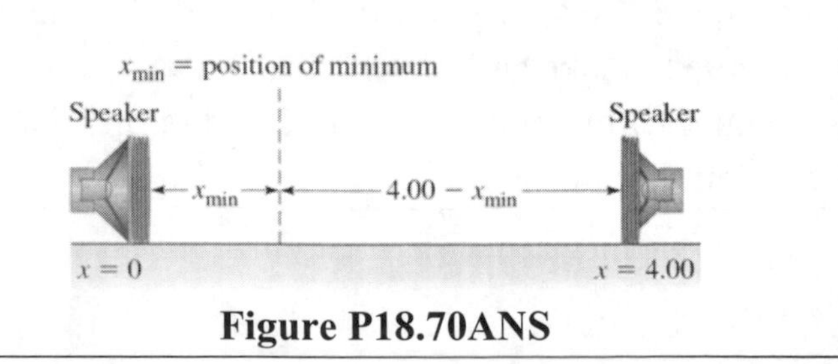 **Figure P18.70ANS**

For paths that differ by integer multiples of a wavelength, the waves add constructively. We first sketch the situation and determine the wavelength.	$\lambda = \frac{v}{f} = \frac{343 \text{ m/s}}{\left(425 \text{ s}^{-1}\right)} = 0.808 \text{ m}$
Take the position of a minimum in sound pressure to be x_{min}, where the speakers are at $x = 0.00$ and $x = 4.00$ (assume all distances in meters). The path length from the first speaker is x_{min} and the second is $(4.00 - x_{min})$, so we can determine the difference in path length, which must equal any odd multiple of half a wavelength. The integer can be positive or negative, simply meaning that the path from the speaker on the left might be longer or shorter than the path from the speaker on the right.	$\text{path difference} = x_{min} - \left(4.00 - x_{min}\right) = 2x_{min} - 4.00$ $2x_{min} - 4.00 = n\frac{\lambda}{2} \qquad \text{where } n = \ldots -3,\ -1,\ 1,\ 3\ldots$
Insert the wavelength and solve for values of x_{min} between 0 and 4.	$2x_{min} - 4.00 = n\frac{0.808}{2}$ $x_{min} = 0.202n + 2.00 \qquad \text{where } n = \ldots -3,\ -1,\ 1,\ 3\ldots$ $x_{min} = \boxed{0.182,\ 0.586,\ 0.990,\ 1.39,\ 1.80,\ 2.20,\ 2.61,\ 3.01,\ 3.41,\ 3.82 \text{ (meters)}}$

CHECK and THINK

There is a second way we can approach this as a double check. If the speakers vibrate in phase, the point halfway between them is an antinode of pressure, at $x = 0.200$. Then there is a node one-quarter of a wavelength away at:

$$2.00 \text{ m} - \frac{0.404 \text{ m}}{2} = 1.80 \text{ m}$$

There is a node every half-wavelength from this position (every 0.404 m), which results in the same set of point.

75. (N) A cylindrical glass tube is partially filled with water to create an air column with a length of 0.20 m. The air column is vibrating at its fundamental frequency. After the water column is slowly lowered, the same frequency is heard. What should be the new length of the air column so that the third harmonic has the same frequency as the previous fundamental frequency? Treat the air column as a case of resonance with one open end and one closed end.

INTERPRET and ANTICIPATE	
The glass tube with partially filled water has one side open and other side closed. So we use the expression for the frequency of standing waves in a tube open on one end (Eq. 18.14).	
SOLVE Start with the frequency of standing waves in a tube open on one end (Eq. 18.14).	$\lambda_n = 4\frac{L}{n} \quad (n = 1,3,5,\ldots)$
The lengths that lead to resonance are odd multiples of a quarter wavelength.	$L = n\frac{\lambda_n}{4} \quad (n = 1,3,5,\ldots)$
For the fundamental mode ($n = 1$) we can relate the length $L_1 = 0.2$ m to the wavelength.	$\lambda = 4L_1 = 0.80\text{ m}$
For the second resonance heard , we want the third harmonic ($n = 3$) at a new length L_2 to produce the same frequency (and therefore the same wavelength).	$\lambda = \frac{4L_2}{3}$ $L_2 = 3\frac{\lambda}{4} = 3\frac{0.80\text{ m}}{4} = \boxed{0.60\text{ m}}$ $L_1 = \frac{\lambda}{4}$ $L_2 = \frac{3\lambda}{4}$ A. B. **Figure P18.75ANS**
CHECK and THINK Given the type of tube (for instance open on one or both ends), we can determine the wavelengths of standing waves in the tube.	

80. (N) An accurate electronic tuner with an adjustable frequency is used to determine the actual frequency of a tuning fork marked as 512 Hz. When the tuner is set to 514 Hz and the tuning fork is sounded, a beat frequency of 2 Hz is heard. When the tuner is set to 510 Hz and the tuning fork is sounded, the beat frequency is 6 Hz. What is the frequency of the tuning fork?

Notice that the electronic tuner reading is correct and the frequency of the tuning fork is marked wrong. When the tuner reads 514 Hz and 2 beats are heard per second, the frequency of the tuning fork is $514 \pm 2 = 516$ Hz or 512 Hz. When the tuner reads 510 Hz and 6 beats are heard per second, the frequency of the tuning fork is $510 \pm 6 = 516$ Hz or 504 Hz. The common value is 516 Hz. So the frequency of the tuning fork must be $\boxed{516 \text{ Hz}}$.

19

Temperature, Thermal Expansion, and Gas Laws

3. (N) Stars (like the Sun) that are actively converting hydrogen into helium in their cores are known as main sequence stars. Their surface temperatures range from about 1700 K to 53,000 K. Express this temperature range on the Fahrenheit scale.

INTERPRET and ANTICIPATE	
Convert each of the temperatures to the Celsius scale first, and then the Fahrenheit scale. Only two significant figures are kept in the final answer because each of the temperatures only has two significant figures.	
SOLVE Use Eq. 19.1 to convert each of the temperatures to the Celsius scale. Because we have not performed the final calculation, we retain all digits at this point in the solution.	$T_1(°\text{C}) = (1{,}700\ \text{K} - 273.15\ \text{K})(1°\text{C}/1\ \text{K}) = 1{,}426.85°\text{C}$ $T_2(°\text{C}) = (53{,}000\ \text{K} - 273.15\ \text{K})(1°\text{C}/1\ \text{K}) = 52{,}726.85°\text{C}$
Now, use Eq. 19.3 to convert each temperature to the Fahrenheit scale.	$T_1(°\text{F}) = \frac{9}{5}T(°\text{C}) + 32 = \frac{9}{5}(1{,}426.85°\text{C}) + 32°\text{C}$ $T_1(°\text{F}) = \boxed{2.6 \times 10^3\ °\text{F}}$ $T_2(°\text{F}) = \frac{9}{5}T(°\text{C}) + 32 = \frac{9}{5}(52{,}726.85°\text{C}) + 32°\text{C}$ $T_2(°\text{F}) = \boxed{9.5 \times 10^4\ °\text{F}}$
CHECK and THINK Equations 19.1, 19.2, and 19.3 are used to express the temperature of an object in any of the three units. Note that the Fahrenheit temperature of an object will always be numerically greater than the Celsius temperature of the object.	

15. (N) A bridge spanning a river gorge is made up of concrete sections 15.24 m in length, poured and cured at 15.0°C. What is the minimum spacing between bridge sections required to prevent buckling if the summertime temperatures at the bridge reach 43.0°C?

INTERPRET and ANTICIPATE	
As the temperature rises from 15°C to 43°C, the concrete sections will expand by a length ΔL based on the coefficient of linear expansion, temperature change, and original length (Eq. 19.4). The minimum spacing is such that, at the high temperature, neighboring sections just make contact with each other. This occurs if there is a gap of length ΔL for each section to expand into. (Consider that each section is next to a gap that it expands into or that each section expands a distance $\frac{1}{2}\Delta L$ into the gap on either side.)	
SOLVE We consider one section expanding and apply Eq. 19.4.	$\Delta L = \alpha L_0 \Delta T$
The coefficient of linear expansion for concrete is listed in Table 19.1, $\alpha = 14.5\times10^{-6}$ °C^{-1}. The change in length calculated is the minimum width for each gap.	$\Delta L = (14.5\times10^{-6}\ ^\circ\text{C}^{-1})(15.24\ \text{m})(43.0^\circ\text{C} - 15.0^\circ\text{C})$ $\Delta L = 0.00619\ \text{m} = \boxed{6.19\times10^{-3}\ \text{m}}$
CHECK and THINK We need a gap of around 0.6 cm for this 15 m concrete section to expand into as the temperature rises nearly 30°C.	

17. (N) At 22.0°C, the radius of a solid aluminum sphere is 7.00 cm. Assume $\alpha = 22.2 \times 10^{-6}$ K^{-1} and β is 66.6×10^{-6} K^{-1} and

a. At what temperature will the volume of the sphere have increased by 3.00%?

INTERPRET and ANTICIPATE	
This is a case of thermal volume expansion. The change in volume depends on the original volume, coefficient of volume expansion, and the temperature change (Eq. 19.5).	
SOLVE Use Eq. 19.5 to relate the volume expansion to temperature change.	$\Delta V = \beta V_0 \Delta T$

The volume increases by 3.00% and the coefficient of volume expansion is listed in Table 19.2.	$\Delta V = 0.03\ V_0 = 3.00\times10^{-2}\ V_0$ $\beta = 66.6\times10^{-6}\ ^\circ\text{C}^{-1}$
Now, solve for the temperature change.	$\Delta T = \dfrac{\Delta V/V_0}{\beta}$ $\Delta T = \dfrac{3.00\times10^{-2}}{66.6\times10^{-6}\ ^\circ\text{C}^{-1}}$ $\Delta T = 450^\circ\text{C}$
Finally, find the final temperature (which is ΔT above the initial temperature of 22°C).	$T_f = T_i + \Delta T$ $T_f = 22 + 450 = \boxed{472^\circ\text{C}}$
CHECK and THINK A very large temperature change (over 400°C, or almost 800°F) is needed to change the volume of the aluminum sphere by a few percent.	

b. What is the increase in the sphere's radius if it is heated to 250°C?

INTERPRET and ANTICIPATE This is similar to part (a), but this is a *linear* distance and we need to use linear thermal expansion (Eq. 19.4).	
SOLVE We use Eq. 19.4 to determine the amount that the radius increases.	$\Delta r = \alpha r_0 \Delta T$
Insert numerical values. The radius expands by 0.035 cm (from 7.00 cm to 7.035 cm ≈ 7.04 cm).	$\Delta r = (7.00\text{ cm})(22.2\times10^{-6}\ ^\circ\text{C}^{-1})(250^\circ\text{C} - 22.0^\circ\text{C})$ $\Delta r = \boxed{0.035\text{ cm}}$
CHECK and THINK Similar to part (a), the change in temperature leads to a small change in radius. The main difference is that we used the coefficient of linear thermal expansion rather than the coefficient for volume expansion.	

20. An exterior wooden door fits comfortably in a steel frame on a day when the outdoor temperature is 50°F. The wood has been cut so that the grain runs vertically. The door is in a climate with a yearly temperature range from 0°F to 100°F. Keep four significant figures as you work and report final answers to one significant figure. Use values from Table 19.1.

a. (C) Would you expect the door to stick on a hot day or on a cold day? Explain.

The change in height or length of the door or gate depends on the change in temperature and coefficient of linear expansion (Eq. 19.14). We first look up the coefficients of linear expansion:

$$\alpha_{wood\perp} = 5.4\times10^{-6}\ °C^{-1}$$
$$\alpha_{wood\parallel} = 4.9\times10^{-6}\ °C^{-1}$$
$$\alpha_{steel} = 13.0\times10^{-6}\ °C^{-1}$$

On a hot day, the steel frame expands more than the wood and the door should be loose. The door is more likely to stick on a cold day.

b. (N) On a day when the temperature is 50°F, the door is 36 in. wide and 80.5 in. tall. What are its maximum and minimum width and height, given the range of temperatures?

INTERPRET and ANTICIPATE

We now apply the formula for linear thermal expansion to see how much the door expands. The grain runs vertically, so the wood has a slightly lower expansion coefficient in this direction than the horizontal direction.

SOLVE	
We use linear thermal expansion (Eq. 19.14). We looked up the coefficients for thermal expansion in part (a).	$\Delta L = \alpha L_0 \Delta T$ $\alpha_{wood\perp} = 5.4\times10^{-6}\ °C^{-1}$ $\alpha_{wood\parallel} = 4.9\times10^{-6}\ °C^{-1}$
Compared to 50°F, the coldest temperature is 50°F lower and the hottest temperature is 50°F hotter. We first convert this 50 degree difference to a difference in Celsius.	$\Delta T = 50°F\left(\frac{5}{9}\right) = 27.78°C$
Calculate the change in height of the wooden door (parallel to grain) from 80.5 inches (to either the coldest or hottest temperature).	$\Delta L_{\parallel} = (4.9\times10^{-6}\ °C^{-1})(80.5\ in)(27.78°C)$ $\Delta L_{\parallel} = 0.01096\ in$

Now calculate the change in width of the wooden door (perpendicular to grain) from 36 inches (to either the coldest or hottest temperature).	$\Delta L_{\perp} = (5.4\times10^{-6}\ ^{\circ}\text{C}^{-1})(36\ \text{in})(27.78^{\circ}\text{C})$ $\Delta L_{\perp} = 0.005400\ \text{in}$

CHECK and THINK
However, considering significant figures, $80.5 \pm 0.01\ \text{in} = 80.5\ \text{in}$, so the expansion or contraction is much smaller than the precision of our numbers. The same is true for the width of the door, $36 \pm 0.0054\ \text{in} = 36\ \text{in}$. Therefore, the door is essentially unchanged in size within the precision of our measurements.

c. (N) On a day when the temperature is 50°F, there is a 0.50-cm gap between the door and the frame. The steel frame has a uniform width of 2.5 in. What are the maximum and minimum gaps between the door and the frame, given the range of temperatures?

INTERPRET and ANTICIPATE
Similar to part (b), if we calculate the expansion of the 2.5 inch width of the frame, the expansion or contraction is negligible. The height of the steel beam of length 80.5 in = 205 cm will be the largest effect, so let's calculate that.

SOLVE Use linear expansion as in part (b).	$\Delta L = \alpha L_0 \Delta T$ $\Delta L = (13.0\times10^{-6}\ ^{\circ}\text{C}^{-1})(205\ \text{cm})(27.78^{\circ}\text{C})$ $\Delta L = 0.07403\ \text{cm} \approx 0.07\ \text{cm}$

CHECK and THINK
Therefore the gap changes by at most 0.07 cm, and the gap will vary between 0.43 cm and 0.57 cm. Since this is the longest dimension of the steel frame, it is the largest change in length as the temperature varies and the largest change in gap distance we would expect.

d. (C) Will the door ever get stuck in its frame? Explain.

No. There is always a gap, so we don't expect the gate to get stuck.

24. A clock with an iron pendulum is made so as to keep the correct time when at a temperature of 20°C. Assume the coefficient of linear expansion of iron is $12 \times 10^{-6}\ ^{\circ}\text{C}^{-1}$ and the pendulum is a simple pendulum.
a. (C) Will the clock be ahead or behind the correct time if the temperature rises to 30°C?
b. (N) What is the amount of time lost or gained each day?

INTERPRET and ANTICIPATE The length of the iron rod will increase with temperature. As the length of a pendulum increases, its period increases and its frequency decreases. Therefore, it swings fewer times in a day and the clock will lose time as fewer "seconds" will tick off the clock.	
SOLVE Start with the frequency or period of a simple pendulum (Eq. 16.29). The period is $T = 1/f$ (Eq. 16.2).	$\omega = 2\pi f = \frac{2\pi}{T} = \sqrt{\frac{g}{L}} \quad \rightarrow \quad T = 2\pi\sqrt{\frac{L}{g}}$
The pendulum gets longer as the temperature increases (Eq. 19.4).	$L = L_0(1+\alpha\Delta T)$
From this, we can write an expression for the ratio of the new period to the original one.	$\frac{T}{T_0} = \frac{2\pi\sqrt{\frac{L}{g}}}{2\pi\sqrt{\frac{L_0}{g}}} = \sqrt{\frac{L}{L_0}} = \sqrt{1+\alpha\Delta T}$
Substitute numerical values.	$\frac{T}{T_0} = \left(1+(12\times10^{-6}\ {}^\circ\text{C}^{-1})(30^\circ\text{C}-20^\circ\text{C})\right)^{1/2}$ $\frac{T}{T_0} = 1.00006$
The period is 0.006% slower, so in one day (86400 s), the clock will lose around $6\times10^{-5}(86400\,\text{s}) = \boxed{5\,\text{s}}$.	
CHECK and THINK Although 5 seconds does not sound like a lot of time, this could add up to a significant error over time.	

29. (N) An 80-ft^3 scuba tank is filled with air at room temperature (22°C) so that the gauge pressure is 2987 psi. The tank is forgotten on the shore in winter for hours, and the tank's temperature drops to –45°F. Ignore the slight decrease in the tank's size. What is the gauge pressure at this lower temperature?

INTERPRET and ANTICIPATE We will treat the air as an ideal gas. If the temperature decreases at constant volume, the pressure should also decrease. One thing to be careful about is that we need to use

<table>
<tr><td colspan="2">absolute pressure and temperature (kelvins) rather than the gauge pressure and temperatures in Celsius or Fahrenheit.</td></tr>
<tr><td>SOLVE
Convert the gauge pressure to pascals and calculate the initial absolute pressure. The gauge pressure (pressure above atmospheric, Eq. 15.15) is significantly higher than atmospheric pressure and therefore nearly identical to the absolute pressure.</td><td>$2987\ \text{psi}\left(\frac{6895\ \text{Pa}}{1\ \text{psi}}\right) = 2.060\times10^7\ \text{Pa}$
$P_i = P_{\text{atm}} + P_{\text{gauge}}$
$P_i = 1.01325\times10^5\ \text{Pa} + 2.060\times10^7\ \text{Pa}$
$P_i = 2.070\times10^7\ \text{Pa}$</td></tr>
<tr><td>Convert temperatures to kelvins.</td><td>$T_i = 22°\text{C} = 295.15\ \text{K}$
$T_f = -45°\text{F} = -42.7°\text{C} = 230.37\ \text{K}$</td></tr>
<tr><td>Use either the ideal gas law (Eq. 19.17), with constant volume and constant number of particles, or Gay-Lussac's law (Eq. 19.12) to calculate the final pressure.</td><td>$\frac{P_i V}{P_f V} = \frac{N k_B T_i}{N k_B T_f}$
$P_f = \frac{T_f}{T_i} P_i$</td></tr>
<tr><td>Insert numerical values.</td><td>$P_f = \frac{T_f}{T_i} P_i = \frac{230.37\ \text{K}}{295.15\ \text{K}}\left(2.070\times10^7\ \text{Pa}\right)$
$P_f = \boxed{1.62\times10^7\ \text{Pa}}$</td></tr>
<tr><td colspan="2">CHECK and THINK
The final pressure is indeed lower than the initial one as we expect. The absolute pressure decreases in proportion to the decrease in absolute temperature (kelvin).
$1.62\times10^7\ \text{Pa}\left(\frac{1\ \text{psi}}{6895\ \text{Pa}}\right) = 2350\ \text{psi}$</td></tr>
</table>

31. (N) A balloon is filled with air outdoors on a cold day when the temperature is –13.0°C so that it occupies a volume of 3.00 L. It is then brought inside where the temperature is 21.0°C. Assuming the pressure in the balloon remains constant, what is the volume of the balloon indoors?

INTERPRET and ANTICIPATE We will treat the gas in the bottle as an ideal gas at constant pressure, which obeys Charles's law. The increase in temperature should lead to an increase in volume for a gas at constant pressure.	
SOLVE Charles's Law is given by Equation 19.10. Equivalently, you can start with the ideal gas law ($PV = Nk_BT$, Eq. 19.17) at constant pressure, $\frac{PV_i}{PV_f} = \frac{Nk_BT_i}{Nk_BT_f} \rightarrow \frac{V_i}{V_f} = \frac{T_i}{T_f}.$	$\frac{V_f}{V_i} = \frac{T_f}{T_i}$ (if P is constant)
While we are given the initial temperatures, the laws are expressed in terms of absolute temperature in the Kelvin scale. Use Eq. 19.1.	$T_i = -13.0 + 273.15 = 260.2\text{ K}$ $T_f = 21.0 + 273.15 = 294.2\text{ K}$
Substitute values, with the initial volume of 3.00 liters.	$V_f = V_i \frac{T_f}{T_i} = (3.00\text{ L})\left(\frac{294.2\text{ K}}{260.2\text{ K}}\right) = \boxed{3.39\text{ L}}$
CHECK and THINK The volume increases as the temperature increases.	

38. (N) A very good laboratory vacuum can evacuate the gas in a chamber so that its pressure is about 10^{-12} Pa. Assuming the gas is at room temperature ($T \approx 20°\text{C}$), estimate the number density (number per unit volume) of gas molecules in the chamber.

INTERPRET and ANTICIPATE The number density is N/V, which we expect to be a small number for a vacuum. We can calculate this ratio by starting with the ideal gas law.	
SOLVE Treating this as an ideal gas (Eq.19.17), we can calculate the number density.	$PV = Nk_BT$ $\frac{N}{V} = \frac{P}{k_BT}$

Substitute values.	$\frac{N}{V}=\frac{10^{-12}\text{ Pa}}{(1.38\times10^{-23}\text{ J/K})(293\text{ K})}$ $\frac{N}{V}=\boxed{2.5\times10^{8}\text{ particles/m}^3}$

CHECK and THINK
This is a fairly straightforward ideal gas calculation. It might be surprising that the value of 250 million is a small number, but 250 million molecules per cubic meter is actually a *very small* value. A cubic meter of air on earth typically contains over 10^{23} molecules!

40. On a hot summer day, the density of air at atmospheric pressure at 35.0°C is 1.1455 kg/m^3.
a. (N) What is the number of moles contained in 1.00 m^3 of an ideal gas at this temperature and pressure?
b. (N) Avogadro's number of air molecules has a mass of 2.85×10^{-2} kg. What is the mass of 1.00 m^3 of air?
c. (C) Does the value calculated in part (b) agree with the stated density of air at this temperature?

INTERPRET and ANTICIPATE
Use the ideal gas law to find how many moles are contained in one cubic meter. We can then use the mass per mole given to find the total mass in this cubic meter.

SOLVE **a.** We can use the form of the ideal gas law in terms of moles (Eq. 19.21).	$n=\frac{PV}{RT}$
The volume is one cubic meter. The temperature must be converted to kelvins (Eq. 19.1).	$T=35.0+273.15=308.2\text{ K}$
Substitute values.	$n=\frac{(1.013\times10^{5}\text{ Pa})(1.00\text{ m}^3)}{(8.314\text{ J/mol}\cdot\text{K})(308.2\text{ K})}$ $n=\boxed{39.5\text{ moles}}$

b. An Avagadro's number of molecules is equivalent to a mole, so the problem statement is equal to "1 mole = 2.85 × 10^{-2} kg."	$m = (39.5\text{ moles})\left(\dfrac{2.85\times10^{-2}\text{ kg}}{\text{mole}}\right) = \boxed{1.13\text{ kg}}$
CHECK and THINK	
c. $\boxed{\text{This value }(1.13\text{ kg/m}^3)\text{ differs by about 1\% with the given density of }1.1455\text{ kg/m}^3.}$	

43. (N) Most of the Universe is made up of hydrogen, often found in huge clouds. The Orion Nebula is an example. The number density of hydrogen molecules (H_2) in such a cloud is estimated to be between 10^9 and 10^{12} molecules per cubic meter. The cloud temperature is typically between 10 K and 30 K. What is the range in such a cloud's gas pressure? Give your answer in pascals and atmospheres and comment on your results.

INTERPRET and ANTICIPATE	
Treat the hydrogen as an ideal gas. We are given ranges for the number density *N/V* and temperature *T* and calculate the range of possible pressures *P*.	
SOLVE Use the ideal gas law (Eq. 19.17) and solve for pressure.	$P = \dfrac{N}{V}k_B T$
We take the lowest and highest ends of the ranges given to get the lowest and highest pressures we would expect to find. So, to estimate the lowest pressure expected, use the smallest number density (10^9) and the lowest temperature (10 K). Values are already in absolute units, as we require.	$P_{low} = (10^9\text{ m}^{-3})(1.38\times10^{-23}\text{ J/K})(10\text{ K})$ $P_{low} \approx \boxed{1\times10^{-13}\text{ Pa}}$ $P_{low} \approx (10^{-13}\text{ Pa})\left(\dfrac{1\text{ atm}}{1.013\times10^5\text{ Pa}}\right) \approx \boxed{1\times10^{-18}\text{ atm}}$
Now calculate the highest expected pressure.	$P_{high} = (10^{12}\text{ m}^{-3})(1.38\times10^{-23}\text{ J/K})(30\text{ K})$ $P_{high} \approx \boxed{4\times10^{-10}\text{ Pa}}$ $P_{high} \approx (4\times10^{-10}\text{ Pa})\left(\dfrac{1\text{ atm}}{1.013\times10^5\text{ Pa}}\right) \approx \boxed{4\times10^{-15}\text{ atm}}$

CHECK and THINK

These "huge clouds" of hydrogen gas are actually around the best vacuum pressures we can produce on Earth.

49. (A) An ideal gas is trapped inside a tube of uniform cross-sectional area sealed at one end as shown in Figure P19.49. A column of mercury separates the gas from the outside. The tube can be turned in a vertical plane. In Figure P19.49A, the column of air in the tube has length L_1, whereas in Figure P19.49B, the column of air has length L_2. Find an expression (in terms of the parameters given) for the length L_3 of the column of air in Figure P19.49C, when the tube is inclined at an angle θ with respect to the vertical.

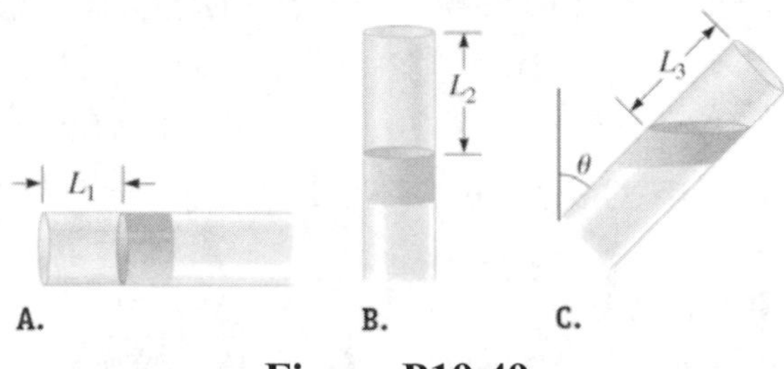

Figure P19.49

INTERPRET and ANTICIPATE

In each case, the mercury must be in static equilibrium. The forces include those due to the pressure in the tube and due to atmospheric pressure, as well as the component of the weight of the mercury along the tube length. We sketch these three cases and require that the pressures balance. The tube is at constant temperature, so we can use Boyle's law (Eq. 19.14) or the ideal gas law (Eq. 19.17) with constant T and N for the gas trapped in the tube.

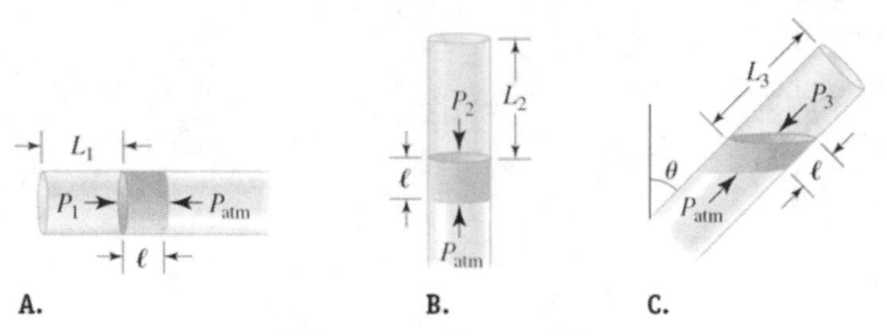

Figure P19.49ANS

SOLVE	
In the first position (A) the weight of the mercury is perpendicular to the orientation of the tube, so the mercury is in equilibrium (along the tube axis) when the pressure of gas in the tube is equal to the outside atmospheric pressure.	Position A: $P_1^{\text{gas}} = P_{\text{atm}}$

In the second position (B), the outside atmospheric pressure P_{atm} is balanced by the pressure of the trapped gas column of length L_2 plus the pressure due to the weight of the mercury divided by the cross-sectional area A of that tube. We also use the fact that the mass of mercury equals its density times volume and the volume is the length of the mercury in the column l times the cross-sectional area.	$P_2^{\text{mercury}} = \frac{mg}{A} = \frac{\rho_{\text{mercury}} Vg}{A} = \rho_{\text{mercury}} \ell g$ Position B: $P_2^{\text{gas}} + \rho_{\text{mercury}} \ell g = P_{\text{atm}}$
In the third position (C), atmospheric pressure is balanced by the pressure inside the tube and the vertical component of the mercury column.	$P_3^{\text{mercury}} = \rho_{\text{mercury}} g\, \ell \cos\theta$ $P_3^{\text{gas}} + \rho_{\text{mercury}} g\, \ell \cos\theta = P_{\text{atm}}$
Apply Boyle's law between cases A and B, and for cases A and C, and then solve for atmospheric pressure P_{atm}., which must be the same in all three cases.	Cases A & B: $P_1^{\text{gas}} V_1 = P_2^{\text{gas}} V_2$ $P_{\text{atm}} AL_1 = \left(P_A - \rho_{\text{mercury}} g\ell\right) L_2 A$ $P_{\text{atm}} = \left(\rho_{\text{mercury}} g\ell\right) \frac{L_2}{L_2 - L_1}$ (1)
	Cases A & C: $P_1^{\text{gas}} V_1 = P_3^{\text{gas}} V_3$ $P_{\text{atm}} AL_1 = \left(P_A - \rho_{\text{mercury}} g\Delta\, \ell \cos\theta\right) L_3 A$ $P_{\text{atm}} = \left(\rho_{\text{mercury}} g\ell \cos\theta\right) \frac{L_3}{L_3 - L_1}$ (2)
Solve for L_3 using Equations 1 and 2.	$\rho_{\text{mercury}} g\ell \frac{L_2}{L_2 - L_1} = \rho_{\text{mercury}} g\ell \frac{L_3 \cos\theta}{L_3 - L_1}$ $\frac{L_2}{L_2 - L_1} = \frac{L_3 \cos\theta}{L_3 - L_1}$

	$L_2L_3 - L_2L_1 = L_3(L_2 - L_1)\cos\theta$ $L_3\left(L_2 - (L_2 - L_1)\cos\theta\right) + L_1L_2$ $L_3 = \boxed{\dfrac{L_1L_2}{L_2 - (L_2 - L_1)\cos\theta}}$

CHECK and THINK

We can test for the couple cases that we know the answer to, when the tube is either horizontal or vertical (cases A and B).

If $\theta = 0°$, then $L_3 = L_2$, which is the second position of the tube.
If $\theta = 90°$, then $L_3 = L_1$, which is the first position of the tube.

53. (N) An air bubble starts rising from the bottom of a lake. Its diameter is 3.6 mm at the bottom and 4.0 mm at the surface. The depth of the lake is 2.50 m, and the temperature at the surface is 40°C. What is the temperature at the bottom of the lake? Consider the atmospheric pressure to be 1.01×10^5 Pa and the density of water to be 1000 kg/m^3. Model the air as an ideal gas.

INTERPRET and ANTICIPATE

Use ideal gas law for the bottom (point 1) and the surface (point 2) of the lake. At the surface, the pressure is atmospheric pressure. However, at the bottom it is equal to the sum of the atmospheric pressure and the pressure due to 2.50 m column of water.

SOLVE Use ideal gas law (Eq. 19.17) for the gas contained in the bubble. The total amount of gas (N molecules) is the same both at the bottom (1) and top (2) of the lake.	$\dfrac{P_1V_1}{P_2V_2} = \dfrac{Nk_BT_1}{Nk_BT_2}$ $T_1 = \dfrac{P_1V_1}{P_2V_2}T_2$
At the bottom of the lake, the pressure is equal to the sum of the atmospheric pressure and the pressure due to 2.50 m column of water (Eq. 15.6).	$P_2 = 1.01 \times 10^5 \text{ Pa}$ $P_1 = P_2 + \rho_W g h_W$ $P_1 = 1.01 \times 10^5 \text{ Pa} + (1.00 \times 10^3 \text{ kg/m}^3)(9.81 \text{ m/s}^2)(2.50 \text{ m})$

The volume ratio at the bottom and top of the lake can be calculated with the diameters given. The radii of the bubble at the bottom and top of the lake are half the diameter, 1.8 mm and 2.0 mm.	$\frac{V_1}{V_2}=\frac{\frac{4}{3}\pi r_1^3}{\frac{4}{3}\pi r_2^3}=\left(\frac{r_1}{r_2}\right)^3=\left(\frac{1.8}{2.0}\right)^3$
Now substitute values.	$T_1=\frac{P_1}{P_2}\left(\frac{V_1}{V_2}\right)T_2$ $T_1=\left[\frac{1.01\times10^5\text{ Pa}+(1.00\times10^3\text{ kg/m}^3)(9.81\text{ m/s}^2)(2.50\text{ m})}{1.01\times10^5\text{ Pa}}\right]$ $\times\left[\left(\frac{1.80}{2.00}\right)^3(40.0+273.15\text{ K})\right]$ $T_1=\boxed{284\text{ K}}$

CHECK and THINK

The temperature is lower at the bottom of the lake than the top. It makes sense that the volume of the bubble is smaller at the bottom of the lake, both because the colder temperature and higher pressure would lead to a decrease in volume.

56. (N) Jerome's physics lab on heat expansion involves a copper ring with a diameter of 3.000 cm and a stainless steel ball with a diameter of 3.015 cm. The linear expansion coefficient for copper is $17\times10^{-6}\ ^\circ\text{C}^{-1}$ and that for stainless steel is $11\times10^{-6}\ ^\circ\text{C}^{-1}$, and both the copper ring and the steel ball are at 22.0°C.

a. In the first phase of the experiment, Jerome warms the copper ring so that the stainless steel ball will just slip through. How hot is the copper ring when that occurs?

INTERPRET and ANTICIPATE

In order for the ball to fit, its diameter must be smaller than the inner diameter of the ring. We will use linear thermal expansion to find out the change in temperature necessary to heat the ring sufficiently that its inner diameter is just large enough to fit the ball (3.015 cm).

SOLVE	
Apply Eq. 19.4 to find the temperature at which the ring expands just enough to fit the L = 3.015 cm ball.	$L=L_0(1+\alpha\Delta T)=L_0(1+\alpha(T_f-T_i))$

The ring's original diameter is $L_0 = 3.000$ cm and the coefficient of linear expansion is given.	$3.015\text{ cm} = 3.000\text{ cm}\left[1+\left(17.0\times10^{-6}\,{}^\circ\text{C}^{-1}\right)\left(T-22.0^\circ\text{C}\right)\right]$ $T = \dfrac{\dfrac{3.015}{3.000}-1}{17.0\times10^{-6}}+22.0 = \boxed{316^\circ\text{C}}$
CHECK and THINK The ring must be heated significantly to fit the ball.	

b. In the second phase of the experiment, both the copper ring and the stainless steel ball are warmed together. What is the temperature at which the ball slips through the ring?

INTERPRET and ANTICIPATE This is similar to part (a) with the exception that the ring *and* the ball expand as the temperature rises. The coefficient of expansion is larger for the copper ring than the stainless steel ball, so the ring will expand more than the ball and will eventually be large enough for the ball to slip through. Since the ring must expand even more than part (a) to fit the now larger ball, we expect the necessary temperature to be higher than in part (a).	
SOLVE We must find when $L_{\text{Cu}} = L_{\text{Steel}}$ for some ΔT.	$L_{0,\text{Cu}}\left(1+\alpha_{\text{Cu}}\Delta T\right) = L_{0,\text{Steel}}\left(1+\alpha_{\text{Steel}}\Delta T\right)$
Insert numerical values. $3.000\text{ cm}\left[1+\left(17.0\times10^{-6}\,{}^\circ\text{C}^{-1}\right)\Delta T\right] = 3.015\text{ cm}\left[1+\left(11.0\times10^{-6}\,{}^\circ\text{C}^{-1}\right)\Delta T\right]$	
Solve for ΔT. This gives the temperature increase beyond the initial temperature.	$\Delta T = 841^\circ\text{C}$ $T_f = T_f + \Delta T =$ $T_f = 841^\circ\text{C} + 22^\circ\text{C} = \boxed{863^\circ\text{C}}$
CHECK and THINK Note that if the steel ball was just slightly larger, then the copper ring would actually melt before expanding sufficiently to allow the ball to pass through.	

62. (N) In December 2006, two Missouri representatives filed legislation requiring that the price of motor fuel be adjusted for volume changes caused by temperature fluctuations. The industry standard assumes the temperature is 60°F. In the summer, when the temperature easily exceeds 60°F in Missouri, gasoline expands and its density decreases. Because gasoline is sold by volume, you buy fewer gasoline molecules (less energy) per gallon during the hot months. This change in volume is good for consumers in the winter months, when cold temperatures increase gasoline density and consumers get more molecules of gasoline per gallon. The industry has accepted temperature-based price adjustments in cold places like Canada, but not in warm places. In 2006, it was estimated that Missouri consumers paid $12 million extra due to the expansion of hot fuel.

a. The density of gasoline at 60°F is 737.22 kg/m^3. What is the density of gasoline on a very hot day ($T = 95$°F)?

INTERPRET and ANTICIPATE	
We will need to consider the expansion of volume equation. Based on the problem statement, we would expect that on a hot day, the gasoline will be less dense and we will be getting less mass per tank than on a cooler day.	
SOLVE We will need to calculate the change in volume using Equation 19.5.	$\Delta V \approx \beta V_0 \Delta T$ (1)
First, convert temperatures to Celsius using Eq. 19.2. The temperature difference is $\Delta T =$ 19.4°C.	$T(°\text{C}) = \frac{5}{9}[T(°\text{F}) - 32]$ $T_{60°\text{F}} = \frac{5}{9}[60° - 32] = 15.6°\text{C}$ $T_{95°\text{F}} = \frac{5}{9}[95° - 32] = 35°\text{C}$ $\Delta T = 35°\text{C} - 15.6°\text{C} = 19.4°\text{C}$
The density decreases as the temperature and volume of the gas increase. The original density at 60°F depends on the mass of gas and the original volume.	$\rho_0 = \frac{m}{V_0}$ (2)
The new density is lower since the same mass occupies a larger volume. Write a formula for the new density. Then use equation (2) to rewrite the mass as $m = \rho_0 V_0$ and write the new volume as $V = V_0 + \Delta V$.	$\rho = \frac{m}{V} = \frac{\rho_0 V_0}{V_0 + \Delta V}$

Insert equation (1).	$\rho = \dfrac{\rho_0 V_0}{V_0 + V_0 \beta \Delta T}$ $\rho = \dfrac{\rho_0}{1 + \beta \Delta T}$
Finally, insert values. The coefficient of volume expansion can be found in Table 19.2.	$\rho = \dfrac{737.22 \text{ kg/m}^3}{1 + (950 \times 10^{-6} \text{ °C}^{-1})(19.4\text{°C})}$ $\rho = \boxed{723.88 \text{ kg/m}^3}$
CHECK and THINK The density is lower at this higher temperature, just as we expected.	

b. If you buy 12.0 gal of gasoline at 60°F, what mass of gasoline have you purchased?
c. If you buy 12.0 gal of gasoline at 95°F, what mass of gasoline have you purchased?
d. If gasoline costs $2.50 per gallon (in 2006), how much money did a consumer lose by buying gasoline on a very hot day?

INTERPRET and ANTICIPATE We have the density of gas at both temperatures from part (a), so we can find out the mass in each case. Since we're getting less mass on the hot day, we can figure out how much it would cost us to purchase a little more to replace the "missing" gasoline.	
SOLVE We need to convert 12 gallons to metric units. 1 gallon = 0.00379 m^3.	$12.0 \text{ gal}\left(\dfrac{0.00379 \text{ m}^3}{1 \text{ gal}}\right) = 0.0454 \text{ m}^3$
b. Calculate the mass given the density at 60 degrees using the density formula (equation (2) in part (a)).	$m_{60\text{°F}} = \rho_0 V$ $m_{60\text{°F}} = (737.22 \text{ kg/m}^3)(0.0454 \text{ m}^3)$ $m_{60\text{°F}} = \boxed{33.5 \text{ kg}}$
c. Repeat to find the mass at 95 degrees.	$m_{95\text{°F}} = \rho V$ $m_{95\text{°F}} = (723.88 \text{ kg/m}^3)(0.0454 \text{ m}^3)$ $m_{95\text{°F}} = \boxed{32.9 \text{ kg}}$

d. In either case, the consumer spent \$30 for the 12 gallons of fuel, but on the hot day, received 0.6 kg less compared to the 33.5 kg expected from part (b), or 1.8% less. Therefore, they lost about 1.8% of the \$30 gas bill, or around 54 cents. Equivalently, we can use the density to convert 0.6 kg to gallons and find the cost for this volume, as shown, reaching the same answer.	$\Delta m = m_{60°\text{F}} - m_{95°\text{F}}$ $\Delta m = 33.5\text{ kg} - 32.9\text{ kg} = 0.6\text{ kg}$ $V_{\text{missing}} = \frac{m}{\rho} = \frac{0.6\text{ kg}}{737.22\text{ kg/m}^3} \cdot \frac{1\text{ gal}}{0.00379\text{ m}^3} = 0.215\text{ gal}$ $\text{cost of missing gasoline} = 0.215\text{ gal}\left(\frac{\$2.50}{\text{gal}}\right) = \boxed{\$0.54}$

CHECK and THINK

In a tankful of gas, this is about 50 cents out of 30 dollars. Not a lot of money, but it could add up. Of course, gasoline prices typically fluctuate much more than this difference, so it's likely that this seasonal variation in temperature is one of many factors that contribute to the price of gasoline at the pump.

63. (A) A glass container of volume V_0 is completely filled with a liquid, and its temperature is then increased by ΔT. Find an expression for the volume of the liquid that will overflow. Assume the coefficient of linear expansion of glass is α_g and the coefficient of volume expansion of the liquid is β_ℓ, where $\beta_\ell > 3\alpha_g$.

INTERPRET and ANTICIPATE

As the temperature increases, the liquid expands, but the glass expands too such that more liquid can be held. Since the liquid expands more than the glass, some liquid spills out. We need to find the difference between how much the liquid expands and how much the glass expands.

SOLVE

Both the liquid and the interior space of the container have the same initial volume V_0. The increase in volume of liquid due to thermal expansion is given by Eq. 19.5.	$\Delta V_\ell = \beta_\ell V_0 \Delta T$

In general, the coefficients of volume and linear expansion are related as $\beta \approx 3\alpha$, so the volume held by the glass also increases.	$\Delta V_g = \beta_g V_0 \Delta T$ $\Delta V_g = 3\alpha_g V_0 \Delta T$
Therefore, the amount of liquid that overflows is the difference in these volumes.	$\Delta V = \Delta V_\ell - \Delta V_g$ $\Delta V = \boxed{V_0\, \Delta T \left(\beta_\ell - 3\alpha_g\right)}$
CHECK and THINK Since we are told that $\beta_\ell > 3\alpha_g$., liquid does in fact overflow.	

67. (N) A tank contains gas at 15.0°C pressurized to 14.0 atm. The temperature of the gas is increased to 90.0°C, and half the gas is removed from the tank. What is the pressure of the remaining gas in the tank?

INTERPRET and ANTICIPATE This is a straight-up ideal gas problem. We simply need to identify the variables in the initial and final situations and use the ideal gas formula.	
SOLVE We will use the ideal gas law (Eq. 19.17). The volume of the tank is assumed to be unchanged, or $V_f = V_i$.	$\frac{P_f V_f}{P_i V_i} = \frac{N_f k_B T_f}{N_i k_B T_i} \quad \Rightarrow \quad P_f = \left(\frac{N_f}{N_i}\right)\left(\frac{T_f}{T_i}\right) P_i$
We must use absolute temperatures for the ideal gas formula (kelvins), so we first use Eq. 19.1 to convert temperatures.	$T_i = 15.0 + 273.15 = 288 \text{ K}$ $T_f = 90.0 + 273.15 = 363 \text{ K}$
Since half of the gas is withdrawn, $N_f = N_i / 2$. Substitute numerical values.	$P_f = \left(\frac{1}{2}\right)\left(\frac{363 \text{ K}}{288 \text{ K}}\right)(14.0 \text{ atm})$ $P_f = \boxed{8.82 \text{ atm}}$
CHECK and THINK The temperature has increased by about 25% (which would increase pressure) but the number of molecules has been cut in half (which would decrease pressure). The net effect is that the pressure goes down.	

74. (A) A gas is in a container of volume V_0 at pressure P_0. It is being pumped out of the container by a piston pump. Each stroke of the piston removes a volume V_s through valve A and then pushes the air out through valve B as shown in Figure P19.74. Derive an expression that relates the pressure P_n of the remaining gas to the number of strokes n that have been applied to the container.

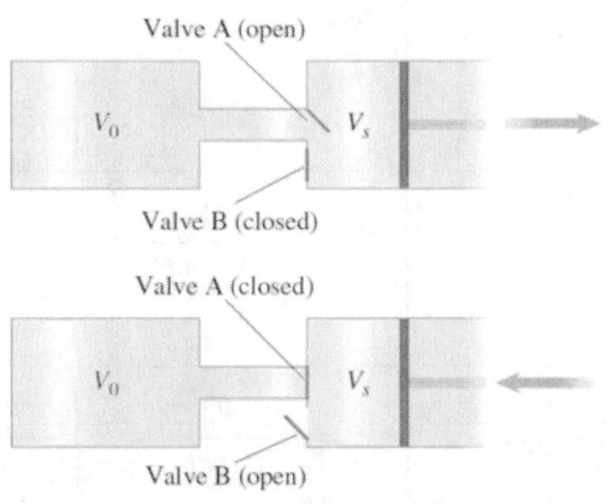

Figure P19.74

<table>
<tr><td colspan="2">INTERPRET and ANTICIPATE
As the container is pumped, each stroke removes a constant volume, but this is a decreasing number of molecules removed per stroke as the pressure decreases. We can think about the first few strokes, one at a time, using the ideal gas law to relate the pressure before the stroke to the pressure after the stroke. We'll continue this process to see what would happen after n strokes.</td></tr>
<tr><td>SOLVE
Since the temperature is constant, when valve A opens and B closes, the gas expands, reaching a new density (N is constant during this stage). Boyle's law applies (Eq. 19.14), or equivalently, the ideal gas law (Eq. 19.17) with constant temperature. So, for the first stroke the original volume V_0 expands to $V_0 + V_s$.</td><td>$P_0V_0 = P_1(V_0 + V_s)$
$P_1 = P_0\frac{V_0}{V_0 + V_s}$ (1)</td></tr>
<tr><td>In the next step, valve A closes and B opens, and a volume V_s of the gas is expelled. The gas in the sealed container is still at the new pressure</td><td>V_s expelled
V_0 remaining at pressure P_1</td></tr>
</table>

P_1. (We're not using the ideal gas law in this step, so we don't need to worry about N changing.)	
Now, we repeat this process. A volume V_0, now at pressure P_1, expands to $V_0 + V_s$ and the pressure decreases from P_1 to P_2.	$P_1V_0 = P_2(V_0 + V_s)$ $P_2 = P_1 \dfrac{V_0}{V_0 + V_s}$ (2)
Now plug equation (1) into (2).	$P_2 = P_0 \left(\dfrac{V_0}{V_0 + V_s}\right)^2$ (3)
Let's do one more to see the pattern. The gas is expelled so that there is a volume V_0 of gas remaining at pressure P_2, which expands again to a volume $V_0 + V_s$ and the pressure decreases from P_2 to P_3.	$P_3 = P_2 \dfrac{V_0}{V_0 + V_s} = P_0 \left(\dfrac{V_0}{V_0 + V_s}\right)^3$
Consider the P_n to be the pressure after stroke n. We can extrapolate this pattern.	$P_n = \boxed{P_0 \left(\dfrac{V_0}{V_0 + V_s}\right)^n}$

CHECK and THINK

Consider the following example: Assume that each stroke expels 5% of the volume of the gas, which starts at atmospheric pressure. After 20 strokes, rather than removing all 100% of the gas and reaching zero pressure (which we'd expect if we assumed that 5% of the *molecules* were removed at each stroke), about 2/3 of the gas molecules have been removed and the pressure is around $P_{atm}/3$.

75. (N) CASE STUDY A constant-volume thermometer uses mercury and oxygen. In the thermometer, 0.750 g of oxygen is kept at a volume of 1.000 L. When the thermometer is placed in contact with a triple-point cell, the mercury rises to 1.674 cm. When the thermometer is placed in contact with an unknown substance, the mercury rises to 7.896 cm. What is the temperature (in kelvins) of the unknown substance?

INTERPRET and ANTICIPATE

Refer to Example 19.7. The temperature change indicated by the thermometer is related to the height that the mercury rises, which is derived in that example. As the temperature

<table>
<tr><td colspan="2">of the gas is increased, its pressure increases. The flexible tube (for instance in Figure 19.22) is used to raise the arm with the mercury, until the increased hydrostatic pressure causes the gas to return to its original volume. The height of the mercury column is then related to the temperature of the gas.</td></tr>
<tr><td>SOLVE
Example 19.7 derives a relationship between the temperature change of the gas and the change in height of the mercury.</td><td>$\Delta T = \rho g \Delta h \left(\frac{V}{N k_B} \right)$</td></tr>
<tr><td>We are provided the height increase and the density of mercury can be found in Table 15.1 (1.36×10^4 kg/m^3). One liter is equal to 10^{-3} m^3.</td><td>$\Delta h = 7.896 \text{ cm} - 1.674 \text{ cm} = 6.222 \text{ cm}$</td></tr>
<tr><td>The number of oxygen gas molecules (O_2) in the volume given can be calculated using its mass, formula weight, and Avagadro's number.</td><td>$N = nN_A$
$N = \left(\frac{0.750 \text{ g}}{32.0 \text{ g/mol}} \right) \left(\frac{6.022 \times 10^{23} \text{ molecules}}{1 \text{ mole}} \right)$
$N = 1.41 \times 10^{22} \text{ molecules}$</td></tr>
<tr><td colspan="2">Now, apply the formula from Example 19.7.
$\Delta T = \rho g \Delta h \left(\frac{V}{N k_B} \right)$
$\Delta T = (1.36 \times 10^4 \text{ kg/m}^3)(9.81 \text{ m/s}^2)(0.06222 \text{ m}) \left(\frac{1.000 \times 10^{-3} \text{ m}^2}{(1.41 \times 10^{22})(1.38 \times 10^{-23} \text{ J/K})} \right)$
$\Delta T = 42.6 \text{ K}$</td></tr>
<tr><td>Therefore, the temperature of the unknown substance is 42.6 K above the triple point of water (273.15 K).</td><td>$T_{\text{unknown}} = 42.6 \text{ K} + 273.15$
$T_{\text{unknown}} = \boxed{315.8 \text{ K}}$</td></tr>
<tr><td colspan="2">CHECK and THINK
The height of the mercury is related to the temperature. We are not surprised then that the mercury rising corresponded to the temperature increasing above the reference point (the triple point of water).</td></tr>
</table>

78. (N) Children playing in the neighborhood park on a cold day find a 50.0-cm-long section of PVC pipe open at both ends, and they begin blowing through it to produce trumpet-like sounds. The pipe is initially at 2.00°C and warms to 25.0°C when warm air is blown through it. What is the change in the fundamental frequency of the pipe after it warms? The coefficient of thermal expansion for PVC is 54×10^{-6} °C^{-1}.

INTERPRET and ANTICIPATE We assume the change in the speed of sound as a function of temperature is negligible but that the length of the tube increases due to thermal expansion. When the instrument warms up, it will become longer and the pitch will decrease.	
SOLVE The fundamental frequency played by the cold-walled pipe depends on the speed of sound and the length of the pipe.	$f_i = \dfrac{v}{2L_i}$ (1)
We assume the change in the speed of sound as a function of temperature is negligible but that the length of the tube increases due to thermal expansion (Eq. 19.4). When the instrument warms up, the length increases.	$L_f = L_i + L_i\alpha\Delta T$ $L_f = L_i(1+\alpha\Delta T)$ (2)
Write an expression for the final frequency, using equation (2) for the final length and equation (1) to substitute the initial frequency.	$f_f = \dfrac{v}{2L_f}$ $f_f = \dfrac{v}{2L_i(1+\alpha\Delta T)}$ $f_f = \dfrac{f_i}{1+\alpha\Delta T}$
The final frequency is lower. The change in frequency can be calculated.	$\Delta f = f_f - f_i = f_i\left(\dfrac{1}{1+\alpha\Delta T} - 1\right)$
Substitute numerical values.	$\Delta f = \dfrac{(343\ \text{m/s})}{2(0.500\,\text{m})}\left(\dfrac{1}{1+(54.0\times10^{-6}/\text{C°})(23.0°\text{C})} - 1\right)$ $\Delta f = \boxed{-0.425\ \text{Hz}}$

CHECK and THINK
This change in frequency is very small (0.425 Hz) compared to the original frequency of 343 Hz, though it could be enough to be out of tune if this were a musical instrument being played in a band.

81. (N) Two glass bulbs of volumes 500 cm^3 and 200 cm^3 are connected by a narrow tube of negligible volume. The apparatus is filled with air and sealed. The initial pressure of the air is 1.0×10^5 Pa, and the initial temperature is 17°C. The smaller bulb is immersed in a large amount of ice at 0°C, and the larger bulb is immersed in a large pot of boiling water at a temperature of 100°C for a long period of time. Determine the final pressure of the air in the bulbs. Ignore any change in the volume of the glass. Model the air as an ideal gas.

INTERPRET and ANTICIPATE
The key is that the pressure on each side must be equal for the gas to stay in equilibrium (if it was different, gas would be pushed from the high pressure to low pressure side and pressure would even out). We can treat each side as an ideal gas, keeping the pressure equal between the two bulbs and the total number of gas molecules in the system constant.

SOLVE
We consider the initial and final situations for both bulbs:

The total number of molecules N of the gas in the two bulbs remains constant and $N = N_1 + N_2$ for bulb 1 and bulb 2.

The pressure on each side also must be equal to each other at equilibrium, but might change from an initial value $P_i = 1.0 \times 10^5$ Pa to an unknown final value P_f.

The temperature is initially the same in each bulb $T_0 =$ 17°C = 290 K, but in the final situation, one side is cooled $T_1 =$ 0°C = 273 K and the other side heated $T_2 =$ 100°C = 373 K.

Using the ideal gas law (Eq. 19.17, $N = \frac{PV}{k_B T}$), we can determine the total number of molecules in both bulbs, both initially and in the final situation. The total number of molecules is fixed at $N = N_1 + N_2$.	initial: $N = \frac{P_i V_1}{k_B T_0} + \frac{P_i V_2}{k_B T_0}$ final: $N = \frac{P_f V_1}{k_B T_1} + \frac{P_f V_2}{k_B T_2}$
Now we equate the two equations above to find out the final pressure P_f.	$\frac{P_1 V_1}{k_B T_0} + \frac{P_1 V_2}{k_B T_0} = \frac{P_f V_1}{k_B T_1} + \frac{P_f V_2}{k_B T_2}$ $\frac{P_1}{T_0}(V_1 + V_2) = P_f\left(\frac{V_1}{T_1} + \frac{V_2}{T_2}\right)$

Solve for the final pressure.

$$\frac{1.0\times10^5\ \text{Pa}}{290\ \text{K}}(5.0\times10^{-4}\ \text{m}^3 + 2.0\times10^{-4}\ \text{m}^3) = P_f\left(\frac{5.0\times10^{-4}\ \text{m}^3}{373\ \text{K}} + \frac{2.0\times10^{-4}\ \text{m}^3}{273\ \text{K}}\right)$$

$$P_f = \frac{0.2414\ \text{Pa}}{2.073\times10^{-6}} = \boxed{1.2\times10^5\ \text{Pa}}$$

CHECK and THINK

The average temperature of the gas molecules increased, as does the pressure.

20

Kinetic Theory of Gases

4. (A) The speeds of three molecules are v, $4v$, and $8v$, respectively. What are the average and rms speeds for the system consisting of these molecules?

INTERPRET and ANTICIPATE

We can calculate the average or rms (root mean square) value for any series of numbers. The rms is always larger than or equal to the average value.

SOLVE	
To find the average, add up all of the values and divide by the number of values (in this case three).	$v_{av} = \dfrac{\sum_{j=1}^{n} v_j}{N}$ $v_{av} = \dfrac{v+4v+8v}{3} = \boxed{13v/3}$
Now, find the rms speed using Equation 20.3. The rms is "root mean square": square the numbers, take the mean (average), and then the square root of the result.	$v_{\text{rms}} = \sqrt{\dfrac{1}{N}\sum_{j=1}^{N} v_j^2}$ $v_{\text{rms}} = \sqrt{\dfrac{1}{3}[v^2 + (4v)^2 + (8v)^2]}$ $v_{\text{rms}} = \sqrt{\dfrac{81}{3}v^2}$ $v_{\text{rms}} = \boxed{3\sqrt{3}v}$

CHECK and THINK

The average value is about $4.3v$ and the rms value is around $5.2v$. The rms value is larger as we expected.

9. (N) Particles in an ideal gas of molecular oxygen (O_2) have an average momentum in the x direction of 2.726×10^{-23} kg·m/s. What is **a.** their root mean square speed and **b.** the root mean square of their x component of velocity?

INTERPRET and ANTICIPATE By "average momentum," the question must be referring to the rms average momentum. An actual average momentum (a vector quantity) in a sample of an ideal gas would be essentially zero, since there are, on average, just as many molecules traveling in the $+x$ direction as in the $-x$ direction. With the rms average momentum and the mass, we can determine an rms velocity.	
SOLVE **a.** We take the question to refer to the rms value of the x momentum in Equation 20.8. The rms average x velocity is equal to 1/3 of the rms average of total velocity. (The x, y, and z components each contribute one third to the total average v^2.)	$p_{av,x}^2 = m^2 v_{av,x}^2 = \frac{1}{3} m^2 v_{av}^2$
We solve this expression for the rms speed as in Equation 20.3.	$v_{rms} = \sqrt{v_{av}^2} = \sqrt{\frac{3p_{av,x}^2}{m^2}} \sqrt{3}\frac{p_{av,x}}{m}$
Finally, insert numerical values.	$v_{rms} = \sqrt{3}\frac{2.726\times10^{-23}\ \text{kg}\cdot\text{m/s}}{32\left(1.66\times10^{-27}\ \text{kg}\right)}$ $v_{rms} = \boxed{889\ \text{m/s}}$
b. Given the average x momentum, we can determine the average x velocity simply by using the definition of momentum: $p = mv$.	$v_{rms,x} = \frac{p_{rms,x}}{m}$ $v_{rms,x} = \frac{2.726\times10^{-23}\ \text{kg}\cdot\text{m/s}}{32\left(1.66\times10^{-27}\ \text{kg}\right)}$ $v_{rms,x} = \boxed{513\ \text{m/s}}$
As a way to double-check this result, based on what we said in part (a), we also expect the rms x component of the velocity to be related to the rms value of the total velocity, $v_{av,x}^2 = \frac{1}{3}v_{av}^2$. Take the square root and use Eq. 20.3 ($v_{rms} = \sqrt{v_{av}^2}$) to get $v_{rms,x} = \frac{1}{\sqrt{3}}v_{rms}$. Substitute the rms velocity from part (a) to get the average x velocity. The answer agrees with what we just found in the last step.	$v_{rms,x} = \frac{889\ \text{m/s}}{\sqrt{3}} = 513\ \text{m/s}$

CHECK and THINK

We've calculated the rms average velocity to be almost 900 m/s. This is the right order of magnitude for speeds of molecules in a gas. Each component (x, y, and z) contributes 1/3 to the average v^2 (as in Eq. 20.6), so $v_{\text{rms},\,x} = v_{\text{rms},\,y} = v_{\text{rms},\,z} = \frac{1}{\sqrt{3}} v_{\text{rms}}$.

12. (N) Find the rms speed of hydrogen molecules (H_2) at a temperature of 27°C.

INTERPRET and ANTICIPATE

Since the temperature is the average random kinetic energy of the molecules (Eq. 20.17, $K_{\text{av}} = \frac{1}{2} m v_{\text{av}}^2 = \frac{3}{2} k_B T$), the rms speed depends on the temperature and mass of the hydrogen molecule. For molecules of a gas at room temperature, we expect to find molecular speeds of hundreds to thousands of meters per second.

SOLVE We can use Equation 20.18 to relate the rms velocity to the temperature and mass of the molecules.	$v_{\text{rms}} = \sqrt{\frac{3k_B T}{m}}$
We take room temperature to be around 300 K and the mass of an H_2 molecule is 2 u, where 1 u = 1.66×10^{-27} kg.	$v_{\text{rms}} = \sqrt{\frac{3(1.38\times10^{-23}\text{ J/K})(300\text{ K})}{2(1.66\times10^{-27}\text{ kg})}}$ $v_{\text{rms}} = \boxed{1.93\times10^3\text{ m/s}}$

CHECK and THINK

The rms speed of nearly 2000 m/s is on the right order of magnitude. (Less massive molecules, like hydrogen, have higher average velocities, so molecules like oxygen and nitrogen might have speeds around 500 m/s at this temperature.)

15. (N) The mass of a single hydrogen molecule is approximately 3.32×10^{-27} kg. There are 5.64×10^{23} hydrogen molecules in a box with square walls of area 49.0 cm^2. If the rms speed of the molecules is 2.72×10^3 m/s, calculate the pressure exerted by the gas.

INTERPRET and ANTICIPATE

There are a few ways to approach this problem. With the rms speed of the molecules, we could find the change in momentum at the wall of the box, with which we could determine the average force on the wall, and then the average pressure. We can also think of this in terms of the ideal gas law. The rms speed is related to the temperature, so with

the number of molecules given and the volume of the box, which we can calculate, we can directly calculate the pressure.	
SOLVE We'll take the second approach mentioned above and use Eq. 20.18 to relate the temperature to the rms velocity.	$v_{rms} = \sqrt{\frac{3k_B T}{m}}$ $k_B T = \frac{1}{3} m v_{rms}^2$ (1)
We can then use the ideal gas law (Eq. 19.17).	$PV = Nk_B T$ (2)
Solve equation (2) for pressure and insert equation (1).	$P = \frac{Nk_B T}{V} = \frac{1}{3}\frac{Nmv_{rms}^2}{V}$ (3)
Now, plug in numerical values into equation (3).	$P = \frac{1}{3}\frac{Nmv_{rms}^2}{V}$ $P = \frac{1}{3}\frac{(5.64\times10^{23})(3.32\times10^{-27}\text{ kg})(2720\text{ m/s})^2}{(3.43\times10^{-4}\text{ m}^3)}$ $P = \boxed{1.35\times10^{7}\text{ Pa}}$
CHECK and THINK The pressure calculated is around 10 atmospheres.	

17. (N) The noble gases neon (atomic mass 20.1797 u) and krypton (atomic mass 83.798 u) are accidentally mixed in a vessel that has a temperature of 90.0°C. What are the
a. average kinetic energies and
b. the rms speeds of neon and krypton molecules in the vessel?

INTERPRET and ANTICIPATE Temperature is the average random kinetic energy of the molecules in the gas. In other words, the temperature determines the average translational kinetic energy of the molecules and will be the same for each type of molecule. Since the kinetic energy $K = \frac{1}{2}mv^2$ is the same for each molecule, the more massive the molecule, the slower the average speed will be.	
SOLVE **a.** The average kinetic energy of	

both neon and krypton depends on the temperature of the gas according to Equation 20.17. Notice that it depends *only* on temperature and is therefore the same for both neon and krypton.	$K_{av} = \frac{3}{2} k_B T$
Insert numbers and calculate the resulting energy in Joules.	$K_{av} = \frac{3}{2}(1.38\times10^{-23}\ \text{J/K})(90.0+273.15\ \text{K})$ $K_{av} = \boxed{7.52\times10^{-21}\ \text{J}}$
b. The rms speed can be found by equating the average kinetic energy found in part (a) to $\frac{1}{2}mv_{rms}^2$ or by using Equation 20.18.	$K_{av} = \frac{1}{2} m_0 v_{rms}^2 = 7.52\times10^{-21}\ \text{J}$ $v_{rms} = \sqrt{\frac{1.50\times10^{-20}\ \text{J}}{m_0}}$
The mass of each molecule is given in atomic mass units (1 u = 1.66 × 10^{-27} kg), so we can plug those in and solve for the rms velocity.	$v_{rms,\ Ne} = \sqrt{\frac{1.50\times10^{-20}\ \text{J}}{20.1797(1.66\times10^{-27}\ \text{kg})}}$ $v_{rms,\ Ne} = \sqrt{\frac{1.50\times10^{-20}\ \text{J}}{3.35\times10^{-26}\ \text{kg}}}$ $v_{rms,\ Ne} = \boxed{670\ \text{m/s}}$
	$v_{rms,\ Kr} = \sqrt{\frac{1.50\times10^{-20}\ \text{J}}{83.798(1.66\times10^{-27}\ \text{kg})}}$ $v_{rms,\ Kr} = \sqrt{\frac{1.50\times10^{-20}\ \text{J}}{1.38\times10^{-25}\ \text{kg}}}$ $v_{rms,\ Kr} = \boxed{329\ \text{m/s}}$

CHECK and THINK

The average kinetic energy is the same for both molecules since they are at the same temperature. Krypton is more massive than neon, so the rms speed of these molecules is lower. Specifically, it's about 4 times more massive, so the speed is about $\sqrt{4}$ or 2 times smaller.

25. (N) CASE STUDY Because the Moon is about the same distance from the Sun as is the Earth, assume the Moon is at the same temperature as the Earth. Find the mass (in kilograms and atomic mass units) of the lightest molecule or atom that can be retained in the Moon's atmosphere. Comment on your results.

<table>
<tr><td colspan="2">INTERPRET and ANTICIPATE
This question is very similar to Example 20.2, which you might want to refer to, but we need to compare the rms speed of the gas molecules to the escape velocity. The Moon has much lower gravity, so the escape velocity is lower and lighter molecules are more likely to escape into space. The lightest molecule retained therefore should be higher than on Earth. Molecules need to be heavier to have sufficiently low rms speeds to not escape the atmosphere.</td></tr>
<tr><td>SOLVE
The escape velocity of a particle trapped in a gravitational field (for instance of the Moon in this case) was covered in Chapter 8 (Eq. 8.17).</td><td>$v_{esc} = \sqrt{\frac{2GM}{R}}$</td></tr>
<tr><td>The rms velocity of molecules in a gas depends on their mass and the temperature of the gas (Eq. 20.18). More massive molecules move slower at a given temperature.</td><td>$v_{rms} = \sqrt{\frac{3k_B T}{m}}$</td></tr>
<tr><td>In Example 20.2, we're told that "The condition for retaining a certain gas in the atmosphere for several billion years (the age of the solar system) is $10v_{rms} \le v_{esc}$," so we'll use that same criterion here. Equivalently, the rms speed should be less than 10% of the escape velocity to make sure that the molecules are retained in the atmosphere. The mass m_{min} refers to the minimum mass molecule that we expect will be retained. Heavier molecules are likely to be kept in the atmosphere, but</td><td>$10v_{rms} = v_{esc}$
$10\sqrt{\frac{3k_B T}{m_{min}}} = \sqrt{\frac{2GM_{Moon}}{R_{Moon}}}$</td></tr>
</table>

lighter, faster molecules are likely to escape into space over time.	
Square each side and solve for $m_{\min}$.	$100\left(\frac{3k_B T}{m_{\min}}\right)=\frac{2GM_{\text{Moon}}}{R_{\text{Moon}}}$ $m_{\min}=\frac{150k_B R_{\text{Moon}} T_{\text{Moon}}}{GM_{\text{Moon}}}$
We first list the values needed for the moon.	$M_{\text{moon}}=7.36\times10^{22}\text{ kg}$ $R_{\text{moon}}=1737.4\text{ km}=1.7374\times10^{6}\text{ m}$ $T=277\text{ K}$
Plug in values.	$m_{\min}=\frac{150\left(1.38\times10^{-23}\text{ J/K}\right)\left(1.7374\times10^{6}\text{ m}\right)\left(277\text{ K}\right)}{\left(6.67\times10^{-11}\text{ N}\cdot\text{m}^2/\text{kg}^2\right)\left(7.36\times10^{22}\text{ kg}\right)}$ $m_{\min}=\boxed{2.03\times10^{-25}\text{ kg}}$
We also convert to amu.	$m_{\min}=2.03\times10^{-25}\text{ kg}\left(\frac{1\text{ u}}{1.66\times10^{-27}\text{ kg}}\right)$ $m_{\min}=\boxed{122\text{ u}}$

CHECK and THINK

The Moon has weaker gravity and therefore lighter molecules escape the atmosphere more easily. The minimum mass molecule is significantly larger than that for Earth (5.5 u). The major components of the Earth's atmosphere, nitrogen gas (N_2, 28 u) and oxygen gas (O_2, 32 u), would certainly not remain in the Moon's atmosphere.

29. (N) Consider the Maxwell-Boltzmann distribution function graphed in Problem 28. For those parameters, determine the rms velocity and the most probable speed, as well as the values of $f(v)$ for each of these values. Compare these values with the graph in Problem 28.

INTERPRET and ANTICIPATE

The Maxwell-Boltzmann distribution captures the probability for a molecule in a gas to have a particular speed. We'll plot the distribution as done in Problem 29 and determine the rms and most probable speeds. The most probable speed should be at the peak of the distribution (the value with the highest probability) and the rms for the Maxwell-Boltzmann distribution is to the right of this peak.

SOLVE	
Recapping Problem 28, The	

Maxwell-Boltzmann speed distribution is given by Equation 20.19.	$f(v)=4\pi\left(\dfrac{m}{2\pi k_B T}\right)^{3/2} v^2 e^{-\left(mv^2/2k_B T\right)}$
We can evaluate the terms. $4\pi\left(\dfrac{m}{2\pi k_B T}\right)^{3/2}=4\pi\left(\dfrac{4.68\times10^{-26}\text{ kg}}{2\pi(1.38\times10^{-23}\text{ J/K})(295\text{ K})}\right)^{3/2}=3.11\times10^{-8}\ \dfrac{\text{s}}{\text{m}}$ $\dfrac{m}{2k_B T}=\dfrac{4.68\times10^{-26}\text{ kg}}{2(1.38\times10^{-23}\text{ J/K})(295\text{ K})}=5.75\times10^{-6}\ \dfrac{\text{s}^2}{\text{m}^2}$	
And then plot the function:	$f(v)=\left(3.11\times10^{-8}\right)v^2 e^{-\left(5.75\times10^{-6}v^2\right)}$ (1)
Now to find the values asked for. Equation 20.18 can be used to find the rms velocity given the temperature, Boltzmann's constant, and the mass of the atom or molecule.	$v_{\text{rms}}=\sqrt{\dfrac{3k_B T}{m}}$
The mass of a nitrogen molecule is 28 u = 28(1.66 × 10^{-27} kg) = 4.68 × 10^{-26} kg.	$v_{\text{rms}}=\sqrt{\dfrac{3(1.38\times10^{-23}\text{ J/K})(295\text{ K})}{4.68\times10^{-26}\text{ kg}}}$ $v_{\text{rms}}=\boxed{511\text{ m/s}}$
Using equation (1) above and the rms velocity, we can calculate the value of $f(v)$.	$f(v_{rms})=\left(3.11\times10^{-8}\right)(511)^2 e^{-\left(5.75\times10^{-6}(511)^2\right)}$ $f(v_{rms})=\boxed{0.00181}$
The most probable speed, for which this function has its maximum value, is given by Equation 20.20.	$v_{\text{mp}}=\sqrt{\dfrac{2k_B T}{m}}$
Insert values.	$v_{\text{mp}}=\sqrt{\dfrac{2(1.38\times10^{-23}\text{ J/K})(295\text{ K})}{4.68\times10^{-26}\text{ kg}}}$ $v_{\text{mp}}=\boxed{417\text{ m/s}}$

Again, use equation (1) to find $f(v)$.	$f\left(v_{\text{mp}}\right)=\left(3.11\times10^{-8}\right)(417)^2 e^{-\left(5.75\times10^{-6}(417)^2\right)}$ $f\left(v_{\text{mp}}\right)=\boxed{0.00199}$
We plot these points on the speed distribution. They lie on the distribution, which is a good sign!	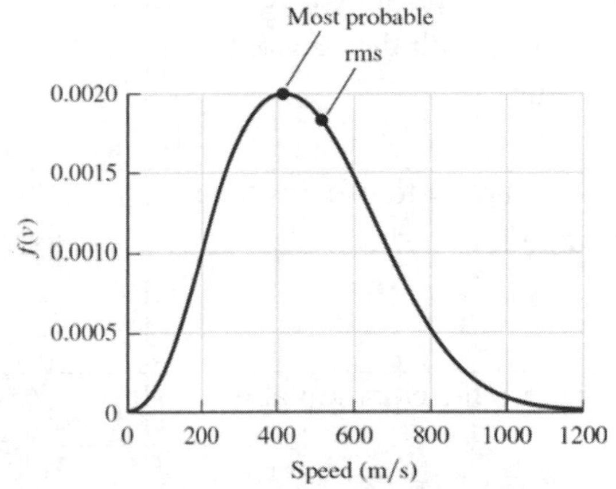 **Figure P20.29ANS**

CHECK and THINK

The most probable speed is indeed at the peak of the distribution function. Because the function is not symmetric, the rms velocity is somewhat higher than the most probable speed.

30. A container of helium is at 20°C.

a. (N) If you increase the temperature to 25°C, by what amount does the most probable speed increase?

b. (N) If you increase the temperature to 100°C, by what amount does the most probable speed increase?

c. (C) Compare your results and comment.

INTERPRET and ANTICIPATE

The most probable speed can be calculated using the temperature and the mass of the molecule with Equation 20.20, $v_{\text{mp}}=\sqrt{\dfrac{2k_{\text{B}}T}{m}}$. Since we only need to calculate the amount that it increases, we can take a ratio of the final to initial speeds, being careful to use absolute temperature units.

SOLVE **a.** First, calculate the ratio of final to initial most probable speeds. Only temperature changes, so the mass and Boltzmann constant cancel. The speed increases with the square root of temperature.	$\dfrac{(v_{\text{mp}})_f}{(v_{\text{mp}})_i} = \dfrac{\sqrt{\dfrac{2k_B T_f}{m}}}{\sqrt{\dfrac{2k_B T_i}{m}}} = \sqrt{\dfrac{T_f}{T_i}}$
We must be careful to use absolute temperature units.	$T_i = 20°\text{C} = (20 + 273)\ \text{K} = 293\ \text{K}$ $T_f = 25°\text{C} = (25 + 273)\ \text{K} = 298\ \text{K}$
Now, plug into the equation above.	$\dfrac{(v_{\text{mp}})_f}{(v_{\text{mp}})_i} = \sqrt{\dfrac{T_f}{T_i}} = \sqrt{\dfrac{298\ \text{K}}{293\ \text{K}}} = \boxed{1.01}$
b. Repeat the calculation for an increase of 100°C.	$T_i = 20°\text{C} = (20 + 273)\ \text{K} = 293\ \text{K}$ $T_f = 120°\text{C} = (120 + 273)\ \text{K} = 393\ \text{K}$
	$\dfrac{(v_{\text{mp}})_f}{(v_{\text{mp}})_i} = \sqrt{\dfrac{T_f}{T_i}} = \sqrt{\dfrac{393\ \text{K}}{293\ \text{K}}} = \boxed{1.16}$
CHECK and THINK **c.** The most probable speed increases as a square root of temperature. From room temperature, an increase in temperature of 5°C has very little effect (1% increase in speed of the molecules). An increase of 100°C has a larger effect (16% increase in speed).	

34. (N) Galaxies are usually found in clusters. There are so many galaxies that it is common for them to pass through one another, but the stars rarely collide. A typical star has a radius of $0.63R_\odot$. The density of stars is about 0.098 star per cubic parsec (1 pc = 3.086×10^{16} m). **a.** What fraction of the galaxy's volume is occupied by stars? Use this result to help you think about your answer to part (c). **b.** Find the mean free path of a star through a galaxy. **c.** If an intruder star travels 1000 pc through a galaxy, what is the probability of a collision?

INTERPRET and ANTICIPATE Even though this is at a much larger scale than an ideal gas, the principles are the same. Given the density of objects, whether stars in a galaxy or atoms in a gas, and the speed of travel, we can estimate how far a particle will typically travel before it might collide with another particle.	
SOLVE **a.** We calculate the ratio of the volume of the stars V_{stars} to the total volume of the galaxy V, taking the volume of stars to be the number of stars times the volume of each one.	$V_{stars} = NV_{star} = N\left(\frac{4}{3}\pi R_{star}^3\right)$ $\frac{V_{stars}}{V} = \frac{N}{V}\left(\frac{4}{3}\pi R_{star}^3\right)$
N/V is the number density (number of stars per cubic meter), which we can calculate with the information given: "0.098 star per cubic parsec"	$\frac{N}{V} = \frac{0.098}{\text{pc}^3}$ $\frac{N}{V} = \frac{0.098}{\left(3.086\times10^{16}\text{ m}\right)^3}$ $\frac{N}{V} = 3.33\times10^{-51}\text{m}^{-3}$
The radius is given as $0.63R_\odot$.	$R_{star} = 0.63R_\odot$ $R_{star} = 0.63\left(6.96\times10^8\text{ m}\right)$ $R_{star} = 4.38\times10^8\text{ m}$
Now, put the pieces together. The fraction of space occupied is *very* low, as we would expect.	$\frac{V_{stars}}{V} = \left(3.33\times10^{-51}\text{m}^{-3}\right)\left(\frac{4}{3}\pi\left(4.38\times10^8\text{m}\right)^3\right)$ $\frac{V_{stars}}{V} = \boxed{1.2\times10^{-24}}$
b. The mean free path depends on the radius and number density of the particles (determined in part (a)) according to Eq. 20.24.	$\lambda = \frac{1}{4\sqrt{2}\pi r^2 (N/V)}$
Insert values. The mean free path to travel through this space (with the low density of stars) is *very* high. A star will travel *very* far before it's likely to collide with another star.	$\lambda = \frac{1}{4\sqrt{2}\pi\left(4.38\times10^8\text{ m}\right)^2\left(3.33\times10^{-51}\text{ m}^{-3}\right)}$ $\lambda = \boxed{8.81\times10^{31}\text{ m}}$

c. The mean free path is an estimate of how far the star will travel before it is likely to hit another star. So, the probability of a collision increases as the star travels further. After it travels a distance equal to λ, we might say that there's around 100% chance that it will hit another star. An estimate for the probability of a collision is therefore the distance traveled (1000 pc or about 10^{19} meters) divided by the mean free path (from part (b)). The answer is indeed a *very* small probability.	$\frac{1000\left(3.086\times10^{16}\ \text{m}\right)}{8.81\times10^{31}\ \text{m}} = \boxed{3.5\times10^{-13} = 3.5\times10^{-11}\ \%}$

CHECK and THINK

It might be surprising to see how low the probability estimate actually is, but the results seem consistent with what we expect from the problem statement: stars occupy a *very* small fraction of space and therefore another star will travel a *very* long distance before it is likely to collide with another star, so the probability of a collision, even after a significant distance of 1000 pc, or about 10^{19} meters, the probability for a collision is *very* small.

39. Monica and Kennedy are going to have a race of sorts. They measure the distance across a room as 10.0 m, declaring one end the starting line and the other the finish line. They each have a container filled with a different gas and plan to release the gases simultaneously, timing the travel of the gases across the room. (Don't worry about the details such as how they detect the gases and don't worry about chemical reactions.)
a. (C) If Monica has a container of hydrogen and Kennedy has a container of oxygen, who wins the race? Explain your answer.

The average speed of the gases would be given by $v_{\text{av}} = \sqrt{(8k_{\text{B}}T)/(\pi m)}$. Given that the gases are at the same temperature, and that the oxygen molecules are more massive than the hydrogen molecules, then the average speed of the hydrogen molecules should be greater. $\boxed{\text{Because the average speed is greater Monica (or the hydrogen) will win the race.}}$

b. (N) Assume the room's temperature is 0°C. By what factor is the one gas faster than the other?

INTERPRET and ANTICIPATE	
We can use the formula for the average speed of a gas, $v_{av} = \sqrt{(8k_BT)/(\pi m)}$, and calculate the ratio of the average speed of the hydrogen versus the average speed of the oxygen to determine the factor by which the hydrogen's speed is greater. Because the difference is the mass of the molecules, H_2 and O_2, we will need to use the periodic table to find these masses.	
SOLVE The mass of a hydrogen atom is approximately 1.0 u and the mass of an oxygen atom is approximately 16.0 u. The mass of each molecule can then be found by multiplying each atomic mass by two.	$m_{H_2} = 2(1.0\text{ u}) = 2.0\text{ u}$ $m_{O_2} = 2(16.0\text{ u}) = 32.0\text{ u}$
Then, we can compute the ratio of the average speeds.	$\frac{v_{av_{H_2}}}{v_{av_{O_2}}} = \frac{\sqrt{(8k_BT)/(\pi m_{H_2})}}{\sqrt{(8k_BT)/(\pi m_{O_2})}}$ $\frac{v_{av_{H_2}}}{v_{av_{O_2}}} = \sqrt{\frac{m_{O_2}}{m_{H_2}}} = \sqrt{\frac{32\text{ u}}{2\text{ u}}} = \sqrt{16}$ $\frac{v_{av_{H_2}}}{v_{av_{O_2}}} = \boxed{4}$
CHECK and THINK When comparing two gases with the same temperature, the gas with the least massive molecules will have a greater average speed.	

42. (N) There are 1789 moles of carbon dioxide in a container of volume 15.75 m^3 at a temperature of 45.6°C. Can you model this situation as an ideal gas? Show your work and explain your conclusion.

INTERPRET and ANTICIPATE
To compare, we calculate the pressure using the ideal gas law and then use the van der Waals equation of state, which takes into account deviations from ideal behavior. The degree to which they differ is the degree to which the model of the gas as ideal breaks down.

SOLVE First, use the ideal gas law (Eq. 19.21) to calculate the pressure. We have all the information we need, so the calculation is straightforward.	$P_{\text{ideal}} = \frac{nRT}{V}$ $P_{\text{ideal}} = \frac{(1789\text{ mol})(8.31\frac{\text{J}}{\text{K·mol}})(318.75\text{ K})}{15.75\text{ m}^3}$ $P_{\text{ideal}} = 3.01\times10^5\text{Pa}$
Now use the van der Waals equation of state (Eq. 20.31). The parameters a and b for carbon dioxide can be found in Table 20.3: $a = 0.359\text{ Pa}\cdot\text{m}^6/\text{mol}^2$ $b = 4.27\times10^{-5}\text{ m}^3/\text{mol}$	$P_{vdW} = \frac{nRT}{V-nb} - a\left(\frac{n}{V}\right)^2$
Plug in values and calculate. $P_{vdW} = \frac{(1789\text{ mol})(8.31\frac{\text{J}}{\text{K·mol}})(318.75\text{ K})}{(15.75\text{ m}^3)-((1789\text{ mol})(4.27\times10^{-5}\frac{\text{m}^3}{\text{mol}}))} - 0.359\left(\frac{1789\text{ mol}}{V}\right)^2 = 2.98\times10^5\text{ Pa}$	
We can determine the relative difference between our ideal and van der Waals estimates. The difference is only 1%.	$\frac{\Delta P}{P} = \frac{P_{\text{ideal}} - P_{vdW}}{P_{vdW}}$ $\frac{\Delta P}{P} = \frac{3.01\times10^5\text{ Pa} - 2.98\times10^5\text{ Pa}}{2.98\times10^5\text{ Pa}}$ $\frac{\Delta P}{P} = 0.010$ $\frac{\Delta P}{P} = 1.0\%$
CHECK and THINK The pressures actually agree within about a percent, so the ideal gas approximation is pretty good in this case. For most purposes, 1% accuracy is sufficient and the calculation using the ideal gas law is simpler and quicker to perform. In retrospect, it's maybe not a huge surprise… we typically so use the ideal gas law and we probably wouldn't get away with this if it generally produced large errors.	

51. (N) In a pressure cooker, water boils at 120°C. What is the pressure inside the cooker?

<table>
<tr><td colspan="2">INTERPRET and ANTICIPATE
Water boils when the partial pressure of the water vapor (i.e. the gas in the pressure cooker) equals the saturated vapor pressure. Using Table 20.5, we can find the saturated vapor pressure at 120°C, which must be the pressure of the gaseous water vapor when the water boils.</td></tr>
<tr><td>SOLVE
Refer to Table 20.5 to see that the saturated vapor pressure is 1.013 × 10^5 at 100°C and 2.320 × 10^5 at 125°C.</td><td>$P_{sat,\ 100°C} = 1.013\times10^5\ \text{Pa}$
$P_{sat,\ 125°C} = 2.320\times10^5\ \text{Pa}$</td></tr>
<tr><td>We need to determine the saturated vapor pressure at 120°C, which we'll do by interpolating between the two points that we know. Basically, draw a straight line between the known values at 100°C and 125°C to determine what the value would be at 120°C.</td><td>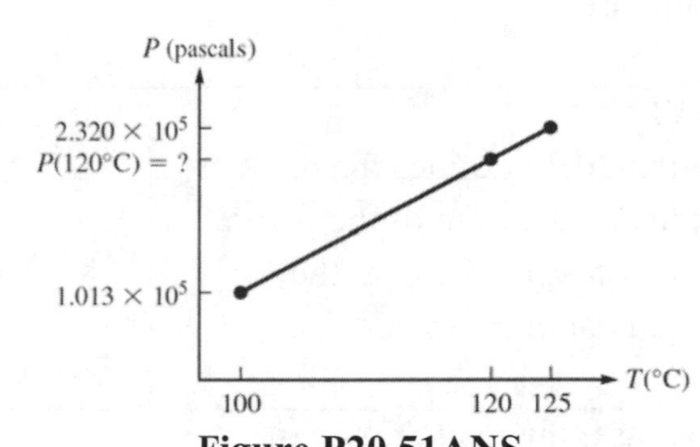

Figure P20.51ANS</td></tr>
<tr><td>Using the equation of the line: $(y_2 - y_1) = m(x_2 - x_1)$, or in this case, $(P_2 - P_1) = m(T_2 - T_1)$, we can first determine the slope between the two know points.</td><td>$m = \dfrac{\Delta P}{\Delta T}$
$m = \dfrac{2.320\times10^5\ \text{Pa} - 1.013\times10^5\ \text{Pa}}{25°\text{C}}$
$m = 5.23\times10^3\ \dfrac{\text{Pa}}{°\text{C}}$</td></tr>
<tr><td>Now, use this to find the pressure at 120°C, which is 20°C above the known value at 100°C.</td><td>$P_{\text{svp},\ 120°C} = P_{\text{svp},\ 100°C} + m(120°\text{C} - 100°\text{C})$
$P_{\text{svp},\ 120°C} = 1.013\times10^5\ \text{Pa} + \left(5.23\times10^3\ \dfrac{\text{Pa}}{°\text{C}}\right)(20°\text{C})$
$P_{\text{svp},\ 120°C} = 2.1\times10^5\ \text{Pa}$</td></tr>
<tr><td>At boiling, the pressure in the cooker must be equal to the saturated vapor pressure at that temperature.</td><td>$P_{\text{ext}} = P_{\text{svp}} = \boxed{2.1\times10^5\ \text{Pa}}$</td></tr>
</table>

CHECK and THINK
The key is that at the boiling point, the gas pressure must equal the saturated vapor pressure at that temperature, which we could find by interpolating between the two known points. The pressure we found is between the values at 100°C and 125°C, as we would expect.

54. (N) On a summer day, the temperature is 34.5°C, and the partial pressure of water vapor in air is 2.75×10^3 Pa. What is the relative humidity?

INTERPRET and ANTICIPATE	
The relative humidity is the ratio of the partial pressure of water P_{water} to the saturated vapor pressure P_{svp}. We need to use Table 20.5 to find the saturated vapor pressure at this temperature.	
SOLVE Equation 20.33 defines the relative humidity as the ratio of the partial pressure of water P_{water} to the saturated vapor pressure P_{svp}.	
In this case, the temperature of 34.5°C is very close to 35°C, which is listed in Table 20.5, so we will use this value for the saturated vapor pressure. (In Problem 51 above, we interpolated between two known points, which we could also do here, but in this case the desired temperature is nearly the same as one listed on the table.)	At 35°C, from Table 20.5: $P_{\text{svp}} = 5.621 \times 10^3\,\text{Pa}$
Now, plug in values.	$RH = \frac{2.75 \times 10^3\ \text{Pa}}{5.621 \times 10^3\ \text{Pa}} \times 100\%$ $RH = \boxed{49\%}$
CHECK and THINK The partial pressure of water is around half the saturated vapor pressure so the relative humidity is around 50%.	

57. (A) A box with volume V confines 18 particles with identical mass m and the following speeds: three particles with speed v, one particle with speed $2v$, four particles with speed $3v$, three particles with speed $4v$, five particles with speed $5v$, and two particles with speed $6v$.
a. What is the average pressure exerted by the particles on the walls of the box?

INTERPRET and ANTICIPATE
The average pressure is related by the ideal gas law to the temperature, which is in turn related to the rms speed of the gas molecules. So, we should be able to get from the list of speeds to temperature and pressure.

SOLVE The rms speed of the particles depends on the square root of temperature (Equation 20.18).	$v_{rms} = \sqrt{\frac{3k_BT}{m}}$ $k_BT = \frac{1}{3}mv_{rms}^2$ (1)
With the ideal gas law (Eq. 19.17), we can determine pressure in terms of temperature.	$PV = Nk_BT$ (2)
Solve equation (2) for pressure and plug in equation (1).	$P = \frac{N}{V}k_BT = \frac{N}{V}\left(\frac{1}{3}mv_{rms}^2\right)$ $P = \frac{1}{3}\frac{Nmv_{rms}^2}{V}$ (3)
Calculate the square of the rms speed directly with Equation 20.3.	$v_{rms}^2 = \frac{1}{N}\sum_{j=1}^{N} v_j^2$

We need to add the square of all 18 values and divide by 18. In cases such as "four particles with speed $3v$," this contributes 4 times $(3v)^2$.

$$v_{rms}^2 = \frac{1}{18}\left[3(v)^2 + 1(2v)^2 + 4(3v)^2 + 3(4v)^2 + 5(5v)^2 + 2(6v)^2\right]$$
$$v_{rms}^2 = 16v^2$$

Now, plug this into equation (3).	$P = \frac{18}{3}\frac{\left[m(16)v^2\right]}{V}$ $P = \boxed{96\left(\frac{mv^2}{V}\right)}$
CHECK and THINK It's difficult to know what the factor in front should look like, but it has a form like equation (3), which looks right.	

b. What is the average kinetic energy per particle in this box?

INTERPRET and ANTICIPATE The average kinetic energy for each particle is just $\frac{1}{2}mv_{\text{av}}^2$. The "average squared speed" is the same as the square of the rms speed that we calculated in part (a), as we can see in Equation 20.3: $v_{\text{rms}} = \sqrt{v_{\text{av}}^2} \quad \rightarrow \quad v_{\text{rms}}^2 = v_{\text{av}}^2$.	
SOLVE Calculate the average kinetic energy of the molecules using Equation 20.1. This depends on the average *squared* speed, which is the same as the average squared rms speed.	$K_{\text{av}} = \frac{1}{2}mv_{\text{av}}^2 = \frac{1}{2}mv_{\text{rms}}^2$
Insert the expression from part (a).	$K_{\text{av}} = \frac{1}{2}m\left(16v^2\right)$ $K_{\text{av}} = \boxed{8mv^2}$
CHECK and THINK Again, though it's not obvious that there should be an 8.00 in front, the mv^2 definitely looks like a kinetic energy, which seems like a good sign!	

64. (N) A spherical balloon with a radius of 10.0 cm is filled with oxygen gas at atmospheric pressure and at a temperature of 15.0°C. What are the **a.** number of molecules, **b.** average kinetic energy, and **c.** rms speed of oxygen in the balloon?

INTERPRET and ANTICIPATE The problem provides the size of the balloon (a way to calculate volume), pressure, and temperature. The number of molecules can be found using the ideal gas law and should be a very large number on the order of Avagadro's number. The kinetic energy and rms speed can be found from temperature. Molecular speeds of hundreds or thousands of m/s are typical at standard temperature and pressure.	
SOLVE **a.** Using the radius, we can calculate the volume of a spherical balloon. With *P, V,* and *T* given, we can determine the number of molecules *N* using the ideal gas law (Eq. 19.17). First calculate the volume.	$V=\frac{4}{3}\pi(0.100\text{ m})^3=4.19\times10^{-3}\text{m}^3$
Insert values into the ideal gas law.	$N=\frac{PV}{k_B T}$ $N=\frac{(1.013\times10^5\text{ Pa})(4.19\times10^{-3}\text{ m}^3)}{(1.38\times10^{-23}\text{ J/K})(288.15\text{ K})}$ $N=\boxed{1.07\times10^{23}\text{ molecules}}$
b. The average kinetic energy depends on the temperature of the gas (Eq. 20.17).	$K_{av}=\frac{3}{2}k_B T$ $K_{av}=\frac{3}{2}(1.38\times10^{-23})(288.15)\text{ J}$ $K_{av}=\boxed{5.96\times10^{-21}\text{ J}}$
c. The rms speed can be found from the average translational kinetic energy (Eq. 20.18). The mass of an O_2 molecule is 32 u where 1 u = 1.66×10^{-27} kg.	$v_{rms}=\sqrt{\frac{3k_B T}{m}}$ $v_{rms}=\sqrt{\frac{3(1.38\times10^{-23}\text{ J/K})(288.15\text{ K})}{32(1.66\times10^{-27}\text{ kg})}}$ $v_{rms}=\boxed{474\text{ m/s}}$
CHECK and THINK As expected, there are on the order of 10^{23} molecules in the balloon traveling at hundreds of meters per second on average.	

66. (N) At what temperature will the mean free path of the molecules of an ideal gas be twice that at 27°C, if the pressure is kept constant but the volume changes?

INTERPRET and ANTICIPATE	
The mean free path depends on the number density if the particles, which depends on the pressure and temperature according to the ideal gas law. At higher temperatures, the number density should decrease, which would lead to a larger mean free path.	
SOLVE Start with the relationship for the mean free path, Equation 20.24.	$\lambda = \frac{1}{4\sqrt{2}\pi r^2 (N/V)}$
We are not given the number density (N/V), but we can determine this quantity using the ideal gas law (Eq. 19.17).	$PV = Nk_BT$ $\frac{N}{V} = \frac{P}{k_BT}$
Insert this into Eq. 20.24.	$\lambda = \frac{1}{4\sqrt{2}\pi r^2 (P/k_BT)} = \frac{k_BT}{4\sqrt{2}\pi r^2 P}$
The mean free path is directly proportional to temperature, so for the mean free path to double, the temperature must double.	$\frac{\lambda_f}{\lambda_i} = \frac{T_f}{T_i} \quad \rightarrow \quad T_f = \frac{\lambda_f}{\lambda_i} T_i = 2T_i$
Before inserting numerical values, we must be sure to use absolute temperature units.	$T_i = 27°\text{C} = (27 + 273)\ \text{K} = 300\ \text{K}$ $T_f = 2T_i = \boxed{600\ \text{K} = 327°\text{C}}$
CHECK and THINK It turns out that the mean free path is proportional to temperature, so we need to double the absolute temperature from 300 K to 600 K.	

69. Consider the situation described in Problem 68 and focus on the right chamber shown in Figure P20.68. Suppose the gas filling the chambers is O_2. Assume the initial average speed is 900.0 m/s.

a. (A) Derive an expression for the rate of change of temperature in terms of the rate of change of the average speed.

b. (N) If the temperature rises at a rate of 2.00 K/s, what is the rate of change of the average speed?

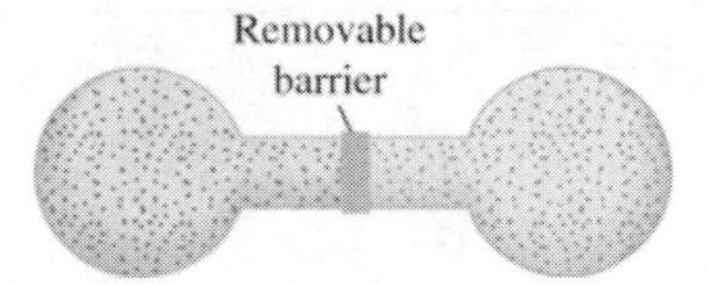

Figure P20.68

INTERPRET and ANTICIPATE	
We can take the derivative of Eq. 20.21, $v_{av} = \sqrt{\frac{8k_B T}{\pi m}}$, with respect to time in order to find an expression for the rate of change of temperature in terms of the rate of change of the average speed of the gas molecules. We can then use the rate given in part (b), along with the initial average speed, to find the rate of change of the average speed.	
SOLVE **a.** We first take the derivative of both sides of Eq. 20.21 with respect to time.	$\left(\frac{d}{dt}\right)v_{av} = \left(\frac{d}{dt}\right)\sqrt{\frac{8k_B T}{\pi m}}$ $\frac{dv_{av}}{dt} = \sqrt{\frac{8k_B}{\pi m}}\left(\frac{1}{2}\right)\frac{1}{\sqrt{T}}\frac{dT}{dt}$
Substituting for the temperature, T, using Eq. 20.21, our expression is now in terms of the rate of change of the average speed and the average speed.	$\frac{dv_{av}}{dt} = \sqrt{\frac{8k_B}{\pi m}}\left(\frac{1}{2}\right)\sqrt{\frac{8k_B}{v_{av}^2 \pi m}}\frac{dT}{dt}$
Now, solve for dT/dt.	$\frac{dv_{av}}{dt} = \frac{1}{2}\frac{8k_B}{v_{av}\pi m}\frac{dT}{dt}$ $\boxed{\frac{dT}{dt} = \frac{v_{av}\pi m}{4k_B}\frac{dv_{av}}{dt}}$
b. Using the periodic table, the mass of an O_2 molecule is 32.0 u, which can be converted to kilograms.	$m = (32.0\text{ u}) \times \left(\frac{1.66 \times 10^{-27}\text{ kg}}{1\text{ u}}\right) = 5.31 \times 10^{-26}\text{ kg}$

Given the initial average speed and the rate of change of the temperature, the rate of change of the average speed can be found using the result of part (a).	$\frac{dT}{dt} = \frac{v_{av}\pi m}{4k_B}\frac{dv_{av}}{dt}$ $\frac{dv_{av}}{dt} = \frac{4k_B}{v_{av}\pi m}\frac{dT}{dt}$ $\frac{dv_{av}}{dt} = \frac{4\left(1.38\times10^{-23}\text{ J/K}\right)}{\left(900.0\text{ m/s}\right)\left(\pi\right)\left(5.31\times10^{-26}\text{ kg}\right)}\left(2.00\text{ K/s}\right)$ $\frac{dv_{av}}{dt} = \boxed{0.735\text{ m/s}}$
CHECK and THINK By taking the derivative of an equation with respect to time, we are able to derive a relationship between the rates of change of a dependent and independent variable. In this case, we can see from the expression in (a) that the rate at which the average speed increases depends linearly on the rate at which the temperature increases.	

73. (N) CASE STUDY Although the nebula from which our solar system formed was comprised mostly of hydrogen and helium, the terrestrial planets including the Earth and Mars quickly lost these gases from their atmospheres. The mass of a helium atom is 6.64 $\times 10^{-27}$ kg. What is the temperature at which the average speed of helium atoms exceeds the escape speed from **a.** the Earth (11.2 km/s) **b.** Mars (5.00 km/s)?

INTERPRET and ANTICIPATE We are given the escape speed of Earth and Mars. We can calculate the average speed of molecules in a gas (this time using the average speed in a Maxwell-Boltzmann distribution, though using the rms speed should result in a similar answer). The escape speed is lower on Mars, which means that it's easier for gas molecules to escape and a lower temperature is required to excite the molecules to the escape speed.	
SOLVE **a.** The average speed of molecules in a gas depends on temperature and is given by Eq. 20.21.	$v_{av} = \sqrt{\frac{8k_BT}{\pi m}}$
Solve the expression for temperature.	$T = \frac{\pi m v_{av}^2}{8k_B}$

Insert numerical values.	$T=\dfrac{\pi\left(6.64\times10^{-27}\ \text{kg}\right)\left(1.12\times10^{4}\ \text{m/s}\right)^2}{8\left(1.38\times10^{-23}\ \text{J/mol}\cdot\text{K}\right)}$ $T=\boxed{2.37\times10^{4}\ \text{K}}$
b. Now, repeat the calculation for Mars. The only difference is the escape speed.	$T=\dfrac{\pi\left(6.64\times10^{-27}\ \text{kg}\right)\left(5.00\times10^{3}\ \text{m/s}\right)^2}{8\left(1.38\times10^{-23}\ \text{J/mol}\cdot\text{K}\right)}$ $T=\boxed{4.72\times10^{3}\ \text{K}}$

CHECK and THINK
The temperature required for the average helium molecule to reach the escape speed is indeed lower on Mars, since it has a lower escape speed.

c. If the average temperature of air at sea level is 20.0°C, how is it possible for helium to escape from the Earth's atmosphere?

Although the sea-level temperature of the atmosphere is low, solar energy heats the upper atmosphere so that the thermosphere, at an altitude of 180 km, has a temperature exceeding 2200 K. This is still smaller than the escape speed, but this is an average — a small fraction of helium in the upper atmosphere always has a velocity above the escape speed and is lost to space. Over a long time, the helium slowly leaks out of the atmosphere.

21

Heat and the First Law of Thermodynamics

7. (N) Late for his morning workout, a 75.0-kg man eats a 200.0-Calorie energy bar for breakfast.
a. What is the energy equivalent of this energy bar in joules?
b. On a stair-climbing exercise machine, each "step" can be thought of as increasing the gravitational potential energy of the man–Earth system. If the equivalent height of each step is 12.0 cm, how many steps must the man take to work off the energy of the energy bar? Assume the efficiency of the human body in converting chemical energy to mechanical energy is 20.0%.

INTERPRET and ANTICIPATE	
The first part is a straightforward unit conversion as long as we remember that 1 Calorie (a nutritional Calorie with a big C) is 1000 calories (metric unit for heat with a little c). In the second part, the man needs to expend enough energy (climbing to a higher potential energy on each step) to add up to the energy in the bar. Two hundred Calories is a lot of energy (compared to his daily intake of maybe 2500 Calories), so it should take a lot of steps to work that off.	
SOLVE **a.** 1 kcal or 1000 calories, where 1 calorie = 4.190 J. (This is an unfortunate historical fact that we need to keep in mind. In many countries, the energy in food is listed as kilojoules.)	$Q = 200.0\ \cancel{\text{Cal}}\left(\frac{10^3\ \cancel{\text{cal}}}{1\ \cancel{\text{Cal}}}\right)\left(\frac{4.190\ \text{J}}{1\ \cancel{\text{cal}}}\right)$ $Q = \boxed{8.380\times10^5\ \text{J}}$
b. The work done lifting his weight (mg) up one step of height h is $W = mgh$. The total work needed for him to move up N steps is just N times this.	$W_{1\ \text{step}} = mgh$ $W_{N\ \text{steps}} = Nmgh$

If only 20% of the energy from the energy bar goes into mechanical energy, that means that if he uses an amount of food energy Q, he gets only 20% of that energy out as useful work W. (Or, another way to think about it, if he wants to do a certain amount of work W, he needs to consume 5 times as much energy in the form of food. The rest of the energy is lost to heat, friction, etc.	$W = 0.200\,Q$ or $Q = 5\,W$
Now, put this all together.	$N = \dfrac{0.200Q}{mgh}$
Fill in numbers.	$N = \dfrac{0.200\left(8.38\times10^{5}\text{ J}\right)}{\left(75.0\text{ kg}\right)\left(9.81\text{ m/s}^2\right)\left(0.120\text{ m}\right)}$ $N \simeq \boxed{1.90\times10^{3}\text{ steps}}$

CHECK and THINK

This is about 2000 steps! That's a lot, but 200 Calories is also potentially close to 10% of the man's daily energy intake. It would definitely take walking up a lot of steps to burn this off.

13. (N) A system's thermal energy increases by 15.0 J when the environment does 15.0 J of work on it. According to the first law of thermodynamics, how much heat is transferred to the system? Does energy enter or leave the system by work? Does energy enter or leave the system by heat?

INTERPRET and ANTICIPATE

The thermal energy of a system increases if work is done on the system or if heat enters the system. 15 J of work are done on the system. In the absence of any heat flow, the thermal energy should increase by 15 J, which is exactly what happens. There must be no heat flow.

SOLVE	
The thermal energy of a system depends on work done on the system	$\Delta E_{\text{th}} = W_{\text{tot}} + Q$

W_{tot} and heat flow into the system Q according to Eq. 21.1.	
We're told that thermal energy increases by 15.0 J and the environment does 15.0 J of work on the system.	$\Delta E_{th} = 15.0\text{ J}$ $W = 15.0\text{ J}$
Solve for the heat Q. We can draw an energy diagram corresponding to this situation as well.	$Q = \Delta E_{th} - W_{tot} = \boxed{0}$ zero $E_{th,i} + W_{tot} + Q = E_{th,f}$ **Figure P21.13ANS**

CHECK and THINK We have to be careful about signs, but this particular process turns out to be pretty simple. 15 J of work are done on the system leading to an increase of 15 J in its thermal energy (and no heat flow).

16. (N) A block of metal of mass 0.250 kg is heated to 150.0°C and dropped in a copper calorimeter of mass 0.250 kg that contains 0.160 kg of water at 30°C. The calorimeter and its contents are insulated from the environment and have a final temperature of 40.0°C upon reaching thermal equilibrium. Find the specific heat of the metal. Assume the specific heat of water is 4.190×10^3 J/(kg · K) and the specific heat of copper is 386 J/(kg · K).

INTERPRET and ANTICIPATE Because the container is insulated, any energy lost by the sample must be gained by the water and the container. Mathematically, we can write that the change in thermal energy of the metal, water, and container must add up to zero (with the metal losing energy and the water and container gaining energy): $\Delta E_{metal} + \Delta E_w + \Delta E_c = 0$. It's also possible to set the magnitude of heat lost by the metal equal to that gained by the water and container to confused about signs: $\lvert\Delta E_{metal}\rvert = \lvert\Delta E_w\rvert + \lvert\Delta E_c\rvert$.

SOLVE Use energy conservation and then	

Equation 21.5 to express changes in thermal energy of each component in terms of mass, specific heat capacity, and change in temperature. The water is in thermal equilibrium with its container, so they both have the same change in temperature ΔT_w.	$\Delta E_{metal} + \Delta E_w + \Delta E_c = 0$ $-\Delta E_{metal} = \Delta E_w + \Delta E_c$ $-m_{metal} c_{metal} \Delta T_{metal} = m_w c_w \Delta T_w + m_c c_c \Delta T_w$
Solve for c_{metal}.	$c_{metal} = -\dfrac{(m_w c_w + m_c c_c)\Delta T_w}{m_{metal} \Delta T_{metal}}$

Substitute values into the equation. Each change in temperature (ΔT) equals the final minus initial temperature for that object.

$$c_{metal} = -\frac{\left[(0.160\ \text{kg})\left(4190\ \frac{\text{J}}{\text{kg}\cdot\text{K}}\right) + (0.250\ \text{kg})\left(386\ \frac{\text{J}}{\text{kg}\cdot\text{K}}\right)\right]\left[313\ \text{K} - 303\ \text{K}\right]}{(0.250\ \text{kg})(313\ \text{K} - 423\ \text{K})}$$

$$c_{metal} = \boxed{2.79 \times 10^2\ \frac{\text{J}}{\text{kg}\cdot\text{K}}}$$

CHECK and THINK

We don't really have a guess as to the magnitude of the final answer, but the final answer of 279 J/kg·K looks like it's in line with a lot of materials in Table 21.1, so it sounds reasonable.

20. (N) From Table 21.1, the specific heat capacity of milk is 3.93×10^3 J/(kg · K), and the specific heat capacity of water is 4.19×10^3 J/(kg · K). Suppose you wish to make a large mug (0.500 L) of hot chocolate. Each liquid is initially at 5.00°C, and you need to raise their temperature to 80.0°C. The density of milk is about 1.03×10^3 kg/m^3, and the density of water is 1.00×10^3 kg/m^3.

a. How much heat must be transferred in each case?

INTERPRET and ANTICIPATE

The amount of heat needed to increase the temperature of a material depends on the mass, specific heat capacity, and temperature change $(Q = mc\Delta T)$. For each liquid, we are given a specific heat capacity, the temperature change required, and the volume, which can be used to determine the mass. The density and specific heats are close to each other, so the heat needed in each case should be similar.

SOLVE Heat added changes the temperature of a substance according to Eq. 21.5. We are given the specific heat capacities and temperature change (75.0°C for each).	$Q = mc\Delta T$
We also need the mass of milk and water, which can be determined from the density and volume. One liter is 10^{-3} m^3.	$m_m = \rho_m V_m$ $m_m = (1.03 \times 10^3 \text{ kg/m}^3)(0.500 \times 10^{-3} \text{ m}^3)$ $m_m = 0.515 \text{ kg}$ $m_w = \rho_w V_w$ $m_w = (1.00 \times 10^3 \text{ kg/m}^3)(0.500 \times 10^{-3} \text{ m}^3)$ $m_w = 0.500 \text{ kg}$
Calculate the heat required for each case.	$Q_m = (0.515 \text{ kg})(3.93 \times 10^3 \text{ J/kg} \cdot \text{K})(75.0°\text{C})$ $Q_m = \boxed{1.52 \times 10^5 \text{ J}}$ $Q_w = (0.500 \text{ kg})(4.19 \times 10^3 \text{ J/kg} \cdot \text{K})(75.0°\text{C})$ $Q_w = \boxed{1.57 \times 10^5 \text{ J}}$

CHECK and THINK

As expected, the amount of heat needed to heat either milk or water is very similar, differing by only a few percent.

b. If you use a small electric hot plate that puts out 455 W, how long would it take to heat each liquid?

INTERPRET and ANTICIPATE

Power is energy over time. Given the power of the hot plate and the energy needed from part (a), we can calculate the time. Again, they should be quite similar, though the water requires a little more energy and should take a little longer.

SOLVE Power is energy (in this case, heat added to the liquid) per time. Solve	

this relationship for the time.	$P=\frac{Q}{\Delta t} \quad \rightarrow \quad \Delta t=\frac{Q}{P}$
Insert the heat needed from part (a) and the 455 W (or 455 J/s) power of the hot plate.	$\Delta t_m=\frac{Q_m}{P}=\frac{1.52\times10^5\text{ J}}{455\text{ J/s}}$ $\Delta t_m=3.34\times10^2\text{s}=\boxed{5.57\text{ min}}$ $\Delta t_w=\frac{Q_w}{P}=\frac{1.57\times10^5\text{ J}}{455\text{ J/s}}$ $\Delta t_w=3.45\times10^2\text{s}=\boxed{5.75\text{ min}}$
CHECK and THINK The water does take a little bit longer to heat up by 75 degrees.	

23. (N) An ideal gas is confined to a cylindrical container with a movable piston on one end. The 3.57 mol of gas undergo a temperature change from 300.0 K to 350.0 K. If the total work done on the gas during this process is 1.00×10^4 J, what is the energy transferred as heat during this process? Is the heat flow into or out of the system?

INTERPRET and ANTICIPATE The work and heat flow in a thermodynamic process are related by the first law of thermodynamics, $\Delta E_{\text{th}}=W_{\text{tot}}+Q$. In order to make use of this, we must first find the change in thermal energy. We expect the change in thermal energy to be positive because the temperature increases during the process. The sign of the heat flow will depend on the relative values of the change in thermal energy and the work. If Q turns out to be negative, it will mean the heat flow must have been out of the system, and vice-versa if it is positive.	
SOLVE Using Equation 21.2, we can write the change in internal energy for an ideal monatomic gas.	$\Delta E_{\text{th}}=3/2\,nR\Delta T$
Then, using the initial and final temperatures, we find the change in thermal energy for the gas.	$\Delta E_{\text{th}}=3/2(3.57\text{ mol})(8.315\text{ J/mol}\cdot\text{K})(350.0\text{ K}-300.0\text{ K})$ $\Delta E_{\text{th}}=2.23\times10^3\text{ J}$
Writing the first law of thermodynamics, and solving for Q, we can use	$Q=\Delta E_{\text{th}}-W_{\text{tot}}$ $Q=2.23\times10^3\text{ J}-1.00\times10^4\text{ J}=\boxed{-7.77\times10^3\text{ J}}$

the work to find the heat flow.	
CHECK and THINK The heat flow for this process is negative, indicating that the heat flow must be out of the system.	

26. (N) A 25-g ice cube at 0.0°C is heated. After it first melts, the temperature increases to the boiling point of water (100.0°C), and the water then boils to form 25 g of water vapor at 100.0°C. How much energy in total is added to the ice/water? Which process (melting, increasing temperature, or boiling) requires the most energy? Water has a latent heat of vaporization of 2.256×10^6 J/kg, a latent heat of fusion of 3.33×10^5 J/kg, and specific heat of 4190 J/(kg · K).

INTERPRET and ANTICIPATE For phase changes, the energy needed to melt or boil the water depends on the latent heats of fusion and vaporization respectively. The heat needed to increase the temperature of a substance can be found using the specific heat capacity.	
SOLVE The heat needed to melt a substance is given by Equation 21.9, which depends on the mass and latent heat of fusion.	$Q_{\text{melt}} = mL_F$ $Q_{\text{melt}} = (0.025\text{ kg})\left(3.33\times10^5 \dfrac{\text{J}}{\text{kg}}\right) = 8325\text{ J}$
The heat required to increase the temperature of the water is given by Equation 21.5. The change in temperature from the freezing point to the boiling point of water is 100K. The mass of water is the same as the initial mass of ice.	$Q_{0\to100} = mc\Delta T$ $Q_{0\to100} = (0.025\text{ kg})\left(4190\dfrac{\text{J}}{\text{kg}\cdot\text{K}}\right)(100\text{ K}) = 10{,}475\text{ J}$
The heat needed to boil a substance is given by Equation 21.10, where L_V is the latent heat of vaporization. The energy needed to boil the water is higher than that needed to melt the ice or to increase the temperature to the boiling point.	$Q_{\text{boil}} = mL_V = (0.025\text{ kg})\left(2.256\times10^6 \dfrac{\text{J}}{\text{kg}}\right) = 56{,}400\text{ J}$

The total energy needed for all three processes is the sum of these three values.	$Q_{tot} = Q_{melt} + Q_{0\to100} + Q_{boil}$ $Q_{tot} = 8325\text{ J} + 10{,}475\text{ J} + 56{,}400\text{ J} = \boxed{75{,}200\text{ J}}$

CHECK and THINK
The total energy (in the form of heat) required has been calculated. The energy needed to boil the water exceeds the energy required to melt an equivalent mass of ice or to raise the temperature of the water from 0°C to 100°C.

31. (N) Consider the latent heat of fusion and the latent heat of vaporization for H_2O, 3.33×10^5 J/kg and 2.256×10^6 J/kg, respectively. How much heat is needed to **a.** melt 2.00 kg of ice and **b.** vaporize 2.00 kg of water? Assume the temperatures of the ice and steam are at the melting point and vaporization point, respectively.

INTERPRET and ANTICIPATE
For phase changes, the energy needed to melt or boil the water depends on the latent heats of fusion and vaporization respectively (Equations 21.9 and 21.10). Knowing the mass of the water (or ice), we use the values for the latent heats from Table 21.2 in each case. Both heat flows should be positive to change the phase from solid to liquid, and then liquid to vapor.

SOLVE

a. In order to melt the ice, energy must flow into the system, given by Equation 21.9.	$Q = mL_F$ $Q = (2.00\text{ kg})(3.33\times10^5\text{ J/kg}) = \boxed{6.66\times10^5\text{ J}}$
b. In order to vaporize the water, energy must flow into the system, given by Equation 21.10.	$Q = mL_V$ $Q = (2.00\text{ kg})(2.256\times10^6\text{ J/kg}) = \boxed{4.51\times10^6\text{ J}}$

CHECK and THINK
Note that if we were condensing, or freezing the water, the heat flows must be negative with respect to the water, or out of the system. When using Equations 21.9 and 21.10, we should always evaluate the direction of the heat flow and determine whether it is appropriate to make the numerical result negative.

34. (N) A thermodynamic cycle is shown in Figure P21.34 for a gas in a piston. The system changes states along the path *ABCA*.
a. What is the total work done by the gas during this cycle?
b. How much heat is transferred? Does heat flow into or out of the system?

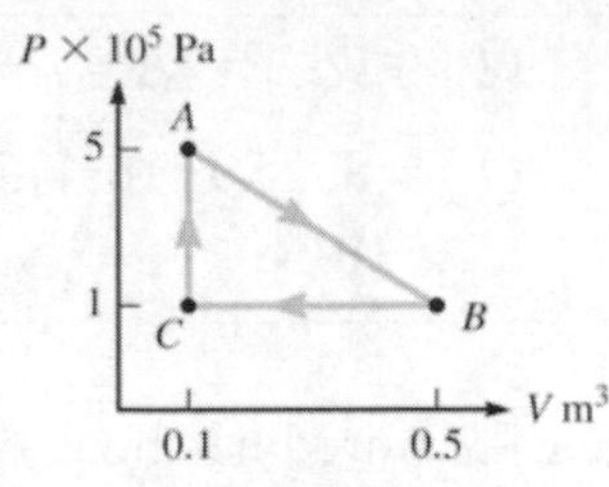

Figure P21.34

<table>
<tr><td colspan="2">INTERPRET and ANTICIPATE
The total work done in a cycle is the sum of the work done in each part of the cycle. The magnitude of the total work is equal to the area of the cycle on the PV diagram. The gas does more work as it expands (AB) than is done on it when it is compressed (BC), so the gas does net work or, equivalently, the total work by the environment on the gas is negative. For a closed cycle, the net change in thermal energy is zero and $Q = -W_{tot}$.</td></tr>
<tr><td>SOLVE
a. The work done during a thermodynamic process is $W = \int PdV$ or equivalently the area under the curve in a PV diagram. For a cycle, the total (net) work done is the area enclosed by the cycle on a PV diagram, so we calculate the area of the triangle. Since the gas does more work on the environment as it expands than is done on the gas while it is compressed (i.e. there is a larger area under line AB than BC), the gas does net work on the environment (piston). For a constant volume process (CA), no work is done.</td><td>$W_{\text{by gas}} = \frac{1}{2}bh$
$W_{\text{by gas}} = \frac{1}{2}(0.5\text{ m}^3 - 0.1\text{ m}^3)(5\times10^5\text{ Pa} - 1\times10^5\text{ Pa})$
$W_{\text{by gas}} = \boxed{80{,}000\text{ J}}$</td></tr>
<tr><td>The total work done by the environment on the gas is equal and opposite the work done by the gas.</td><td>$W_{\text{tot}} = -W_{\text{by gas}} = -80{,}000\text{ J}$</td></tr>
</table>

b. Using Equation 21.1, we set the change in thermal energy for a cycle to zero since the system returns to the same state after one cycle.	$W_{tot} + Q_{tot} = \Delta E_{th} = 0$
Solve for the heat Q. The net heat for the cycle is positive, meaning that heat is added to the gas from the environment.	$Q_{tot} = -W_{tot} = \boxed{+80,000\ \text{J}}$

CHECK and THINK
In a cycle, the net change in thermal energy is zero and the area of the cycle on the *PV* diagram determines the total work and heat during the cycle. In this case, for one cycle heat is added to the gas from the environment and this energy is used to do work by the gas on the environment.

42. An ideal gas at $P = 2.50 \times 10^5$ Pa and $T = 295$ K expands isothermally from 1.25 m^3 to 2.75 m^3. The gas then returns to its original state through a two-part process: constant-pressure followed by constant-volume.
a. (G) Draw a *PV* diagram for this gas.

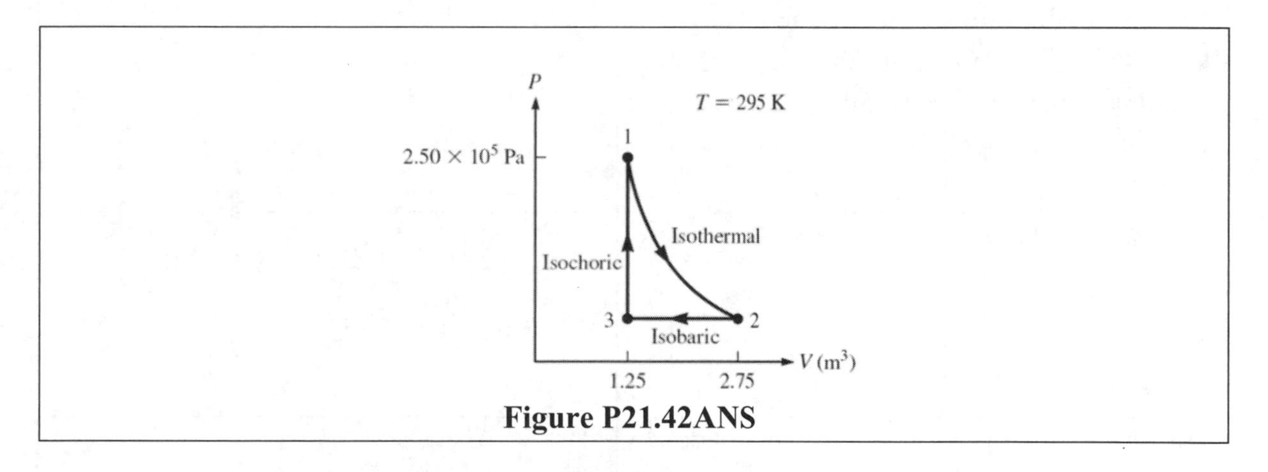

Figure P21.42ANS

b. (C) What is the change in thermal energy?
c. (N) Find the work done by the environment on the gas.
d. (N) Find the heat that flows into the gas.

INTERPRET and ANTICIPATE
For a cycle, the change in thermal energy is zero and the total work and heat are equal and opposite. The gas does more work on the environment between points 1 and 2 than

the environment does on the gas from points 2 and 3. So, in total, the gas does work on the environment and the total work by the environment on the gas is negative.	
SOLVE **b.** For a cyclic process, the net change in thermal energy is zero.	$\Delta E_{th} = \boxed{0}$
c. The net work is the area inside the curve. Since this shape is not something simple, like a triangle, for which we could easily calculate the area, we determine the work for the isothermal part (1 → 2) and the constant pressure part (2 → 3) and add them to get the total. For a constant volume process (3 → 1), there is no work done (no area under that vertical line!).	$W_{tot} = W_{1\to 2} + W_{2\to 3}$
For the isothermal part, work is done by the gas on the environment (Eq. 21.15).	$W_{isothermal} = Nk_B T \ln \frac{V_1}{V_2}$
We don't know the number of molecules N, but we can determine it using the ideal gas law $PV = Nk_BT$. We apply the ideal gas law at point 1 since we know this pressure and volume ($P_1V_1 = Nk_BT$).	$W_{1\to 2} = P_1V_1 \ln \frac{V_1}{V_2}$
Now plug in values.	$W_{1\to 2} = \left(2.50\times 10^5 \text{ Pa}\right)\left(1.25 \text{ m}^3\right)\ln\frac{1.25}{2.75}$ $W_{1\to 2} = -2.46\times 10^5 \text{ J}$
For the constant pressure part, work is done on the gas (Eq. 21.16).	$W = -P\Delta V$
We need the pressure at point 2, which we don't have yet, but we can use the ideal gas law and the information for the isothermal process.	$PV = \text{constant}$ (isothermal) $P_1V_1 = P_2V_2$

	$P_2 = \frac{V_1}{V_2} P_1$ $P_2 = \frac{1.25}{2.75}\left(2.50\times10^5\ \text{Pa}\right)$ $P_2 = 1.14\times10^5\ \text{Pa}$
Now determine the work for the constant pressure process.	$W_{2\to3} = P_2\left(V_2 - V_3\right)$ $W_{2\to3} = \left(1.14\times10^5\ \text{Pa}\right)\left(2.75\ \text{m}^3 - 1.25\ \text{m}^3\right)$ $W_{2\to3} = 1.70\times10^5\ \text{J}$
There is no work done in a constant volume process, so add these two for the entire cycle.	$W_{tot} = -2.46\times10^5\ \text{J} + 1.70\times10^5\ \text{J}$ $W_{tot} = \boxed{-7.56\times10^4\ \text{J}}$
d. Since $\Delta E_{th} = 0$, according to Eq. 21.1, the heat added to the gas is equal and opposite the work done on the gas.	$Q = -W_{tot} = \boxed{7.56\times10^4\ \text{J}}$

CHECK and THINK

For the cycle, the net change in thermal energy is zero. The total work done on the gas in this cycle is negative as we expected and the heat added to the gas is positive. So, in this cycle, heat is added to the gas and that energy is used by the gas to perform work, resulting in no net change of the thermal energy of the gas.

43. (N) Suppose 3.67 mol of a monatomic gas is at a temperature of 300.0 K and undergoes an isobaric (constant-pressure) expansion from an initial volume of 0.025 m^3 to a final volume of 0.065 m^3.

a. What is the final temperature of the gas?

INTERPRET and ANTICIPATE

Since the initial volume and temperature are known, we can use the ideal gas law to find the pressure during the isobaric process. The pressure stays constant in an isobaric process. Then, knowing the pressure and final volume, the ideal gas law can again be applied to find the final temperature.

SOLVE Solving the ideal gas law, $PV = nRT$, for P, we use the initial	$P_i = nRT_i/V_i$ $P_i = \left[(3.67\ \text{mol})(8.315\ \text{J/mol}\cdot\text{K})(300.0\ \text{K})\right]/\left(0.025\ \text{m}^3\right)$ $P_i = 3.7\times10^5\ \text{Pa}$

volume and temperature to get the pressure.	
Applying the ideal gas law again in the final state, we get the final temperature.	$T_f = P_f V_f / nR$ $T_f = \left[(3.7\times10^5\ \text{Pa})(0.065\ \text{m}^3)\right] / \left[(3.67\ \text{mol})(8.315\ \text{J/mol}\cdot\text{K})\right]$ $T_f = \boxed{7.8\times10^2\ \text{K}}$

CHECK and THINK

The ideal gas law describes the state of an ideal gas at any moment and can be applied at the beginning, or end of a thermodynamics process, as well as at any moment between.

b. What is the change in thermal energy of the gas as it undergoes this process?

INTERPRET and ANTICIPATE

A monatomic ideal gas has 3 degrees of freedom. Using Equation 21.2, the thermal energy in a monatomic gas would be, $E_{\text{th}} = 3/2\, nRT$. Knowing the initial and final temperatures, this may be used to find the change in thermal energy.

SOLVE Using Equation 21.2 and the temperatures, we solve for the change in thermal energy.	$\Delta E_{\text{th}} = 3/2\, nR\Delta T$ $\Delta E_{\text{th}} = 3/2(3.67\ \text{mol})(8.315\ \text{J/mol}\cdot\text{K})(7.8\times10^2\ \text{K} - 300.0\ \text{K})$ $\Delta E_{\text{th}} = \boxed{2.2\times10^4\ \text{J}}$

CHECK and THINK

A gas with more degrees of freedom than another at the same temperature will necessarily have a greater thermal energy.

c. What is the work done on the gas?

INTERPRET and ANTICIPATE

The work done on the gas can be found using Equations 21.16 since this is an isobaric process. Because the gas expands, we expect this work to be negative.

SOLVE Using Equation 21.16 along with the pressure and volumes, we find the work done on the gas.	$W = -P\Delta V$ $W = -(3.7\times10^5\ \text{Pa})(0.065\ \text{m}^3 - 0.025\ \text{m}^3)$ $W = \boxed{-1.5\times10^4\ \text{J}}$

CHECK and THINK
The work performed on the gas is negative when the gas expands and positive when the gas is contracted.

50. (N) A sample of a monatomic gas is in a container with a movable piston such that it is maintained at constant pressure. If 25.0 J of heat were transferred into the gas, the temperature would increase by 75.0°C.
a. If, instead, the piston was initially locked in place such that the *volume* remains fixed, by how much would the temperature increase if 25.0 J of heat were transferred into the container?

INTERPRET and ANTICIPATE	
A change in temperature depends on the heat added, the specific heat capacity of the substance, and the amount of the substance. The specific heat can be expressed as either a specific heat at constant volume (C_V) or a specific heat at constant pressure (C_P). We ultimately need to apply $Q_V \equiv nC_V\Delta T$ to find the heat required in the constant volume process. We must first use the information given about the constant pressure process to determine the number of moles *n* and then calculate the answer. As you add heat at constant pressure, some of the energy goes into work done by the gas as it expands. So, the final temperature of the gas at constant volume should be higher since *all* of the heat goes into raising the temperature.	
SOLVE The initial situation is a constant pressure process, which is described by Equation 21.19. We need to use this information to determine the number of moles of gas in order to apply a similar formula for the constant volume process.	$Q_P \equiv nC_P\Delta T$
The molar specific heat at constant volume for a monatomic gas is given by Equation 21.27. This is based on Equation 21.29 and the fact that a monatomic gas has three degrees of freedom.	$C_V = \frac{3}{2}R$ monatomic
For an ideal gas, the specific heat at constant pressure is related to that at constant volume by Equation 21.24. We can now calculate C_P.	$C_P - C_V = R$

	$C_P = \frac{5}{2}R$ monatomic
We now use the given information to determine how many moles of gas are in the piston.	$n = \frac{Q_P}{C_P \Delta T}$ $n = \frac{25.0\text{ J}}{\frac{5}{2}\left(8.315\ \frac{\text{J}}{\text{mol}\cdot\text{K}}\right)(75\text{ K})}$ $n = 0.0160\text{ mol}$
Using Equation 21.18, we can calculate a temperature change assuming the volume is kept constant.	$Q_V \equiv nC_V \Delta T$ $\Delta T = \frac{Q}{nC_V}$ $\Delta T = \frac{25.0\text{ J}}{(0.0160\text{ mol})\left(\frac{3}{2}\left(8.315\frac{\text{J}}{\text{mol}\cdot\text{K}}\right)\right)}$ $\Delta T = \boxed{125\text{ K}}$

CHECK and THINK

The gas reaches a higher temperature when it is kept at constant volume. At constant pressure (as opposed to constant volume), the piston moves as the gas expands, therefore some of the energy added to the gas is used as work in increasing the volume of the gas.

b. If, instead, the same number of moles of a *diatomic* gas were contained in the box and 25.0 J of heat were transferred in at constant pressure, by how much would the temperature increase? (Assume there are five active degrees of freedom for the diatomic gas.)

INTERPRET and ANTICIPATE

The heat capacities are larger for a diatomic gas because there are more degrees of freedom than a monatomic gas. Some of the energy goes into rotational motion, so less of the energy goes into the translational degrees of freedom that determine temperature. The final temperature should therefore be lower.

SOLVE	
Using Equation 21.27 and 21.29 again, we can write the heat capacities for the diatomic gas.	$C_V = \frac{5}{2}R$ diatomic $C_P = \frac{7}{2}R$ diatomic

Assuming we have the same number of moles of gas, we can determine the temperature change when 25J of heat is added (as in part (a)).	$\Delta T = \frac{Q}{nC_P}$ $\Delta T = \frac{25.0\text{ J}}{(0.0160\text{ mol})\left(\frac{7}{2}\left(8.315\frac{\text{J}}{\text{mol}\cdot\text{K}}\right)\right)}$ $\Delta T = \boxed{53.7\text{ K}}$

CHECK and THINK

Since the diatomic gas has more degrees of freedom and a higher heat capacity, more heat is needed to cause the same increase in temperature (related to translational degrees of freedom). Some of the energy goes into rotational degrees of freedom. That is, with a given amount of heat added, the temperature change is smaller.

55. A diatomic ideal gas at room temperature undergoes an adiabatic process such that its final pressure is 2.75 times its initial pressure.

a. (C) Did the gas expand or contract?

As pressure increases on a gas, the gas **contracts**. The precise *amount* by which it contracts and the final temperature depend on details like whether the gas is monatomic or diatomic and whether the compression is adiabatic or isothermal (or something else), but with a fixed amount of gas, compressing the sample leads to an increase in pressure and decrease in volume.

b. (N) What is the ratio of its final volume to its initial volume?

INTERPRET and ANTICIPATE

Now, to determine exactly what the final volume is, we need to know that this is adiabatic compression and that it's a diatomic gas. We will look for a relationship between pressure and volume for an adiabatic process and go from there. The final volume should be less than the initial volume based on part (a).

SOLVE An adiabatic process has a pressure and volume that depend on Eq. 21.30, where γ is a number that depends on whether the gas is monatomic, diatomic, etc. We can rearrange this to determine the ratio of final to initial volumes.	$P_f V_f^\gamma = P_i V_i^\gamma$ $\frac{V_f}{V_i} = \left(\frac{P_i}{P_f}\right)^{1/\gamma}$

We're told that $P_f = 2.75\,P_i$. γ is the ratio of specific heats at constant pressure and constant volume (Eq. 21.31). From Table 21.3, for a diatomic gas, $\gamma \approx \frac{7}{5}$.	$\frac{V_f}{V_i} = \left(\frac{1}{2.75}\right)^{5/7} = \boxed{0.486}$
CHECK and THINK The final volume is around half of the initial volume. Though we weren't asked, the temperature also increases during this process.	

59. (N) A lake is covered with ice that is 2.0 cm thick. The temperature of the ambient air is −20°C. Find the rate of thickening of ice. Assume the thermal conductivity of ice is 200.0 W/(m · K), the density of ice is 9.0×10^2 kg/m^3, and the latent heat of fusion is 3.33 $\times 10^5$ J/kg.

INTERPRET and ANTICIPATE The air temperature is cold enough to freeze more water, but the water is *below* the ice. We can use the temperature difference to determine the rate at which heat is conducted out through the ice and then the rate at which new ice is formed. We know from experience that this must be a pretty slow process, so the rate should be very small.	
SOLVE The rate at which heat is drawn out through the ice layer depends on the temperature difference and thermal conductivity (Equation 21.33).	$\frac{Q}{\Delta t} = kA\frac{\Delta T}{\Delta x}$
The heat Q that's absorbed by the air comes from the water and freezes more ice. We need Eq. 21.9 to determine how much ice freezes based on how much heat comes out of the lake.	$Q = L_f \Delta m$
This will tell us how much *mass* of water is frozen, but we want to know how the *thickness*	$\Delta m = \rho \Delta V = \rho A \Delta y$

changes. The density of ice can be used to relate a certain new mass of ice Δm with the new volume of ice ΔV, which is equal to the new thickness Δy of ice over some area A. (We have to be a little careful keeping in mind that Δy is the *new* thickness of ice and Δx in the equation above is the *total* thickness of the ice, 2 cm. Let's sketch this just to be clear.)	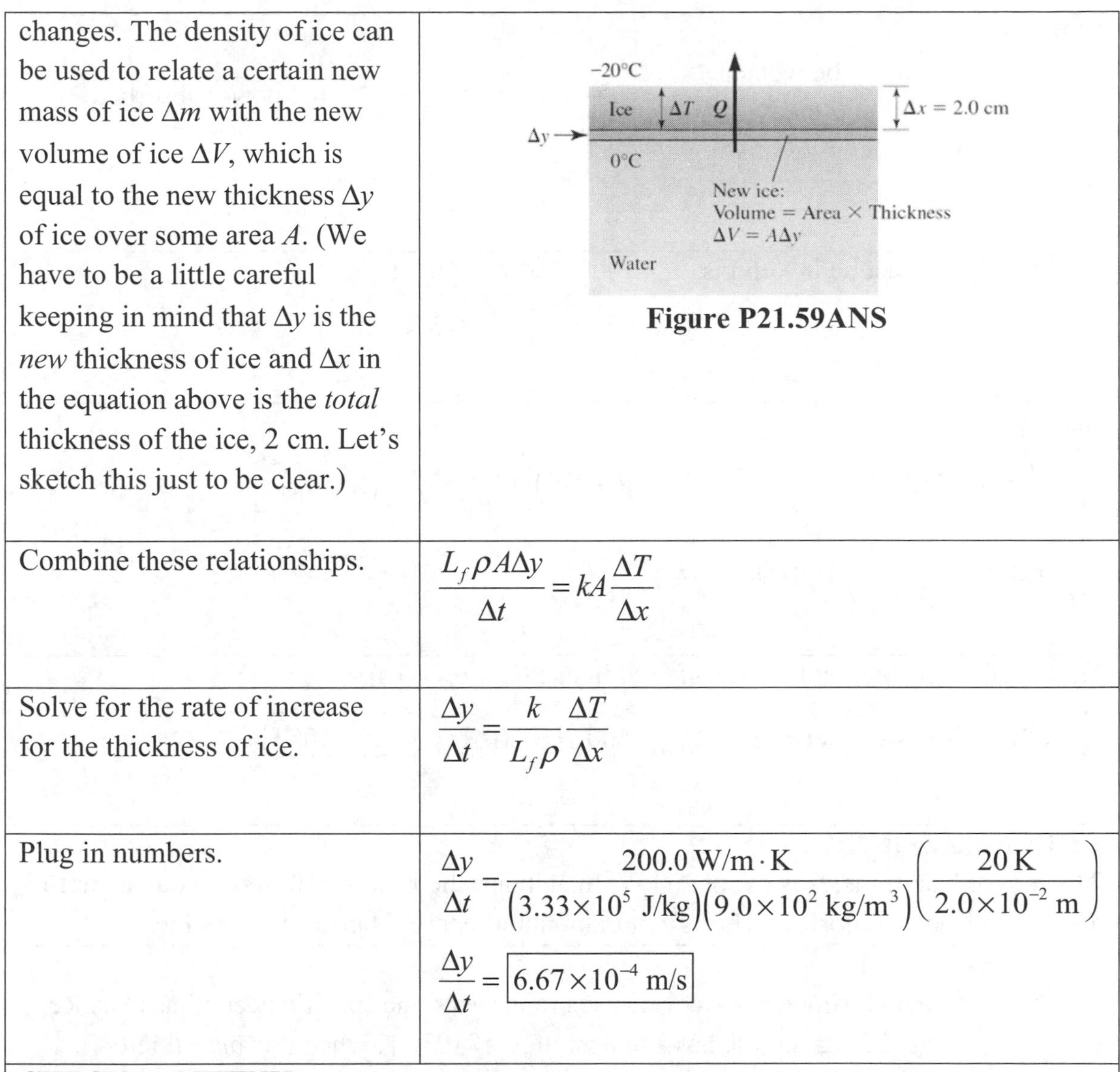 **Figure P21.59ANS**
Combine these relationships.	$\frac{L_f \rho A \Delta y}{\Delta t} = kA\frac{\Delta T}{\Delta x}$
Solve for the rate of increase for the thickness of ice.	$\frac{\Delta y}{\Delta t} = \frac{k}{L_f \rho}\frac{\Delta T}{\Delta x}$
Plug in numbers.	$\frac{\Delta y}{\Delta t} = \frac{200.0\text{ W/m}\cdot\text{K}}{\left(3.33\times10^5\text{ J/kg}\right)\left(9.0\times10^2\text{ kg/m}^3\right)}\left(\frac{20\text{ K}}{2.0\times10^{-2}\text{ m}}\right)$ $\frac{\Delta y}{\Delta t} = \boxed{6.67\times10^{-4}\text{ m/s}}$

CHECK and THINK

Note that the rate depends on $1/dx$, so as the thickness grows the rate decreases.

60. (N) A concerned mother is dressing her child for play in the snow. The child's skin temperature is 36.0°C, and the outside air temperature is 2.00°C. If the emissivity of the child's skin is 0.790 and he has 1.10×10^{-2} m^2 of exposed skin area, what is the amount of energy transferred from his body to the surroundings in 1.00 h?

INTERPRET and ANTICIPATE

Presumably the mother is concerned about dressing the child and not about calculating the radiative heat loss, but let's see if we can help her out. Based on the question, we are only considering the heat loss due to thermal radiation (not including any conduction through the air, convection effects due to the wind, or evaporation). The Stefan-Boltzmann equation allows us to calculate heat transfer through radiation.

SOLVE The net heat transfer between a system (the child) and the environment is given by Equation 21.36.	$\left.\frac{Q}{\Delta t}\right\|_{net} = \sigma\varepsilon A\left(T_{env}^4 - T_{sys}^4\right)$ net power absorbed
Temperatures must be in kelvins.	$T_{sys} = 36°C = 309\text{ K}$ $T_{sys} = 2°C = 275\text{ K}$

Insert numbers.

$$\frac{Q}{\Delta t} = \left(5.6703\times10^{-8}\ \frac{\text{W}}{\text{m}^2\cdot\text{K}^4}\right)\left(1.10\text{ m}^2\right)(0.970)\left[(309\text{ K})^4 - (275\text{ K})^4\right]$$

$$\frac{Q}{\Delta t} = 206\text{ W}$$

Now, determine the net heat transferred in one hour (3600 s).

$$\frac{Q}{\Delta t} = 206\text{ W} \quad\rightarrow\quad Q = (206\text{ J/s})(3600\text{ s}) = 740\text{ kJ} = \boxed{7.40\times10^5\text{ J}}$$

CHECK and THINK

The rate of heat transfer is about 200 W. In an hour, the heat transferred is equivalent to around 177 kcal (Calories). That's a fair amount of energy, but seems possible.

65. (N) A water calorimeter is a device used to measure the specific heat of a substance (Fig. 21.6). A 0.125-kg sample has a temperature of 1030 K when it is placed into 1.00 kg of water at a temperature of 280 K. The container is made of steel and has a mass of 0.250 kg. After the sample, the water, and its container reach thermal equilibrium, their common temperature is 293 K. What is the specific heat of the substance?

INTERPRET and ANTICIPATE

This is similar to Example 21.2, so we can use that as a model. The heat leaving the substance must be equal to the heat entering the water plus the heat entering the container.

SOLVE We first express energy conservation, with the change in energy of each constituent (substance, water, and container) adding up to zero. Equivalently, the heat leaving the substance	$\Delta E_s + \Delta E_w + \Delta E_c = 0$ $-\Delta E_s = \Delta E_w + \Delta E_c$

<table>
<tr><td>equals the heat gained by the water and container. (To avoid confusion with signs, it's also possible to set the magnitude of each side equal to each other. For instance, translating the phrase "the heat leaving the substance equals the heat gained by the water and container" into an equation $\left|\Delta E_s\right| = \left|\Delta E_w\right| + \left|\Delta E_c\right|$.)</td><td></td></tr>
<tr><td>The change in thermal energy of each depends on the temperature change according to Eq. 21.7.</td><td>$\Delta E_{\text{th}} = mc\Delta T$</td></tr>
<tr><td>This leads to equation (1) from Example 21.2.</td><td>$-m_s c_s \Delta T_s = m_w c_w \Delta T_w + m_c c_c \Delta T_w$

$c_s = -\dfrac{\left(m_w c_w + m_c c_c\right)\Delta T_w}{m_s \Delta T_s}$</td></tr>
<tr><td>The specific heats of water and steel can be found from Table 21.1.</td><td>$c_w = 4190\ \dfrac{\text{J}}{\text{kg}\cdot\text{K}}$
$c_c = 450\ \dfrac{\text{J}}{\text{kg}\cdot\text{K}}$</td></tr>
<tr><td colspan="2">Substitute the masses, specific heats, and temperature changes.

$c_s = -\dfrac{\left[(1.00\ \text{kg})\left(4190\ \dfrac{\text{J}}{\text{kg}\cdot\text{K}}\right) + (0.250\ \text{kg})\left(450\ \dfrac{\text{J}}{\text{kg}\cdot\text{K}}\right)\right]\left[293\ \text{K} - 280\ \text{K}\right]}{(0.125\ \text{kg})(293\ \text{K} - 1030\ \text{K})}$

$c_s = \boxed{6.07\times10^2\ \dfrac{\text{J}}{\text{kg}\cdot\text{K}}}$</td></tr>
<tr><td colspan="2">CHECK and THINK
We don't know what the substance is, but the specific heat of 607 J/kg·K looks like a reasonable value based on the range of values in Table 21.1.</td></tr>
</table>

69. (N) Three 100.0-g ice cubes initially at 0°C are added to 0.850 kg of water initially at 22.0°C in an insulated container.

a. What is the equilibrium temperature of the system?

b. What is the mass of unmelted ice, if any, when the system is at equilibrium?

INTERPRET and ANTICIPATE Based on the way the question is phrased, it sounds like there might be ice remaining in the final situation, which would mean that the final mixture is an ice/water mixture at 0°C. We'll first check for this possibility. If there's ice remaining, we can calculate how much of the original ice melts as the water reaches 0°C.	
SOLVE **a.** Let's first determine how much energy is needed to melt *all* the ice using Eq. 21.9. The total mass for the three ice cubes is 300 g = 0.300 kg.	$Q_{\text{melt all ice}} = mL_f$ $Q_{\text{melt all ice}} = (0.300\text{ kg})(3.33\times10^5\text{ J/kg})$ $Q_{\text{melt all ice}} = 1.00\times10^5\text{ J}$
Now, we figure out how much energy must be removed from the water in order to bring it down to 0°C using Eq. 21.5, with the specific heat for water from Table 21.1.	$Q_{22\to0°\text{C}} = mc\Delta T$ $Q_{22\to0°\text{C}} = (0.850\text{ kg})\left(4190\dfrac{\text{J}}{\text{kg}\cdot\text{K}}\right)(22.0\text{ K})$ $Q_{22\to0°\text{C}} = 7.84\times10^4\text{ J}$
Since the heat required to melt 300 g of ice at 0°C *exceeds* the heat required to cool 850 g of water from 22.0°C to 0°C, the final temperature of the system (water + ice) must be 0°C. That is, all of the water cools to 0°C and there is still more energy needed to melt the rest of the ice.	$T_f = \boxed{0°\text{C}}$
b. The 7.84×10^4 J energy needed to cool the water goes into melting a certain amount of ice (mass M_{melted}). Use Eq. 21.9 to determine how much ice melts.	
The amount of ice remaining is the original mass minus the mass that has melted.	$M_{\text{remaining}} = M_{\text{original}} - M_{\text{melted}}$ $M_{\text{remaining}} = 0.300\text{ kg} - 0.235\text{ kg}$ $M_{\text{remaining}} = \boxed{0.065\text{ kg}}$

CHECK and THINK
In this case, the energy removed to cool the water is less than the energy needed to melt all the ice, so at the end, there is an ice/water mixture. The energy to cool the water down can be used to find out *how much* of the ice melts, as we've done here.

74. (A) In Figure 21.27, we saw that the gas initially has a large volume, low pressure, and low temperature and finally it has smaller volume, higher pressure, and higher temperature. Check our derivation of the adiabatic path by making sure these changes are correctly predicted by the equations $P_f V_f^\gamma = P_i V_i^\gamma = \text{constant}$ and $T_f / T_i = \left(V_f / V_i\right)^{(1-\gamma)}$.

INTERPRET and ANTICIPATE	
Let's imagine a situation where the gas volume decreases. We expect that the pressure and temperature will both go up. We can use these relationships to verify that the predicted behavior occurs.	
SOLVE Assume the gas's volume decreases.	$\frac{V_f}{V_i} < 1$
To determine how pressure changes in this process, rewrite Equation 21.30.	$\frac{P_i}{P_f} = \left(\frac{V_f}{V_i}\right)^\gamma$
γ is a positive number so if the volume ratio (final over initial volume, in parentheses) is less than one, the ratio on the left side of the equality is also less than one. Notice that this is *initial* over *final* pressure though, so the initial pressure is smaller than the final and pressure increases as we expected.	If $\frac{V_f}{V_i} < 1$, then $\frac{P_i}{P_f} < 1$, therefore $P_i < P_f$
Now let's confirm that the temperature goes up. Let's consider the relationship above (Eq. 21.32).	$\frac{T_f}{T_i} = \left(\frac{V_f}{V_i}\right)^{(1-\gamma)}$
The ratio of specific heats γ is a positive number greater than 1, therefore the exponent is negative. A negative exponent is the same as taking one over the argument and raising it to a positive exponent. So, if the volume ratio (*final*	If $\frac{V_f}{V_i} < 1$, then $\frac{T_f}{T_i} > 1$, therefore $T_f > T_i$

over *initial*, in parentheses) is less than one, the temperature ratio (also *final* over *initial*) is greater than 1 and the final temperature is larger than the initial, as we were expecting to find.	
CHECK and THINK The pressure and temperature both increase as the volume decreases, just as we expected.	

75. (A) A container contains equal moles of two ideal gases A and B at temperature T. Gas A is monatomic, and gas B is diatomic. Find the average thermal energy per molecule at temperature T. Assume the diatomic molecule has five active degrees of freedom.

For the monatomic gas (A), the number of degrees of freedom per molecule is 3. The average thermal energy of the molecules in gas A is therefore $\frac{3}{2}k_B T$ (Eq. 21.25 or 21.28). For the diatomic gas (B), the number of degrees of freedom is 5, so the average thermal energy of gas molecules B is therefore $\frac{5}{2}k_B T$ (Eq. 21.28). As the mixture contains equal moles of A and B, the average thermal energy is simply the average of the two values:

$$\bar{E}_{\text{th, av}} = \frac{1}{2}\left(\frac{3+5}{2}\right)k_B T = \boxed{2k_B T}$$

79. (N) How much faster does a cup of tea cool by 1°C when at 373 K than when at 303 K? Consider the tea to be a blackbody (an ideal object whose emissivity is 1) and assume the room temperature is 293 K.

INTERPRET and ANTICIPATE The net heat flow due to radiation can be determined with the Stefan-Boltzmann equation. The energy radiated depends on the difference in T^4 for the system and the environment. We expect that the tea will cool at a faster rate at 373 K than at 303 K. (This intuition that a larger temperature difference leads to faster cooling is also true for conduction.)	
SOLVE Assuming only blackbody radiation, the net rate of energy transfer out of the tea is given by the Stefan-Boltzmann equation (Eq. 21.36).	$\left.\frac{Q}{\Delta t}\right\|_{\text{net}} = \sigma\varepsilon A\left(T_{\text{sys}}^4 - T_{\text{env}}^4\right)$

We consider the two cases for the system temperature $T_1 = 373$ K and $T_2 = 303$ K. The temperature of the environment is 293 K in both cases.	$\left.\frac{Q}{\Delta t}\right\|_{\text{net, 373 K}} = \sigma\varepsilon A\left((373\text{ K})^4 - (293\text{ K})^4\right)$ $\left.\frac{Q}{\Delta t}\right\|_{\text{net, 303 K}} = \sigma\varepsilon A\left((303\text{ K})^4 - (293\text{ K})^4\right)$
Now, let's take the ratio.	$\frac{\left.\frac{Q}{\Delta t}\right\|_{\text{net, 373 K}}}{\left.\frac{Q}{\Delta t}\right\|_{\text{net, 303 K}}} = \frac{(373)^4 - (293)^4}{(303)^4 - (293)^4} = \boxed{11.3}$

CHECK and THINK

The energy radiated from the tea is about 11 times larger at the higher temperature.

82. (A) In an adiabatic process, the relationship between pressure and volume is $PV^\gamma = \text{constant}$, where γ is the adiabatic exponent that depends on the properties of the gas. Two different adiabatic processes *AD* and *BC* intersect two isothermal processes *AB* at temperature T_1 and *DC* at temperature T_2 as shown in Figure P21.82. Is the ratio V_A/V_D greater than, smaller than, or equal to the ratio V_B/V_C?

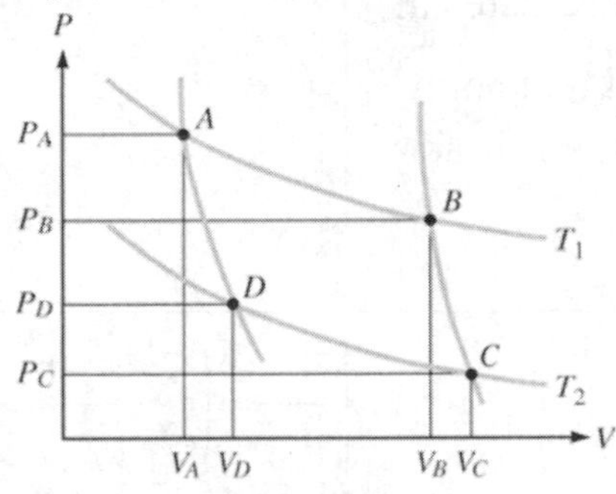

Figure P21.82

INTERPRET and ANTICIPATE

Just by looking at the figure, it's not clear how these ratios are related. We'll have to write each ratio and use equations for adiabatic processes, isothermal processes, and the ideal gas law to express them in terms of the same variables. Then, we should be able to conclude which one is larger or if they are equal.

SOLVE

We want to write expressions for V_A/V_D and V_B/V_C. Points *A* and *D* are on the same adiabat, so we can use the relationship $PV^\gamma = \text{constant}$ to relate their pressures and volumes.	$P_A V_A^\gamma = P_D V_D^\gamma$ $\left(\frac{V_A}{V_D}\right)^\gamma = \frac{P_D}{P_A}$

	$\dfrac{V_A}{V_D} = \left(\dfrac{P_D}{P_A}\right)^{1/\gamma}$ (1)
Points *B* and *C* are on the same adiabat, so we can write a similar relationship.	$\dfrac{V_B}{V_C} = \left(\dfrac{P_C}{P_B}\right)^{1/\gamma}$ (2)
We want to compare equations (1) and (2), but we need to get them into the same variables. Since points *A* and *B* and points *C* and *D* are each at the same temperature, perhaps it will be easier to express the volume ratio as a ratio of temperatures rather than a ratio of pressures. We can use Equation 21.32 to do this, but let's run through the derivation of that equation. We can use the ideal gas law $(PV = Nk_\text{B}T)$ to rewrite the pressures in equation (1), keeping in mind that point *D* is at temperature T_2 and point *A* is at T_1.	$P_D V_D = Nk_\text{B}T_2$ and $P_A V_A = Nk_\text{B}T_1$ $\dfrac{P_D}{P_A} = \dfrac{T_2}{T_1}\dfrac{V_A}{V_D}$ (3)
Now Plug equation (3) into (1).	$\dfrac{V_A}{V_D} = \left(\dfrac{T_2}{T_1}\dfrac{V_A}{V_D}\right)^{1/\gamma}$ $\left(\dfrac{V_A}{V_D}\right)^{\gamma} = \left(\dfrac{T_2}{T_1}\dfrac{V_A}{V_D}\right)$ $\left(\dfrac{V_D}{V_A}\right)\left(\dfrac{V_A}{V_D}\right)^{\gamma} = \left(\dfrac{T_2}{T_1}\right)$ $\left(\dfrac{V_A}{V_D}\right)^{\gamma-1} = \left(\dfrac{T_2}{T_1}\right)$ $\dfrac{V_A}{V_D} = \left(\dfrac{T_2}{T_1}\right)^{1/(\gamma-1)}$ (4)

It's possible to use the ideal gas law in the same way to rewrite equation (2) in terms of volume and temperature.	$P_C V_C = Nk_B T_2$ and $P_B V_B = Nk_B T_1$ $\frac{P_C}{P_B} = \frac{T_2}{T_1}\frac{V_B}{V_C}$ $\frac{V_B}{V_C} = \left(\frac{T_2}{T_1}\right)^{1/(\gamma-1)}$ (5)
Finally, we compare equations (4) and (5). While the expressions are not immediately transparent, we see that they are actually identical. Therefore the ratios of the volumes are equal to each other!	$\boxed{\frac{V_A}{V_D} = \frac{V_B}{V_C} = \left(\frac{T_2}{T_1}\right)^{1/(\gamma-1)}}$

CHECK and THINK

The problem involved a lot of algebraic manipulation, but the essence of it was: write the ratios of the volumes and use our adiabatic, isothermal, and ideal gas laws to put them in terms of the same quantities (in this case, as a ratio of the temperatures). After a lot of work, we find that they are actually the same.

22

Entropy and the Second Law of Thermodynamics

4. (N) During each cycle, a heat engine does 50.0 J of work and absorbs 425 J of energy from a hot reservoir.
a. What is the efficiency of this engine?
b. What is the amount of energy that is exhausted to the cold reservoir during one cycle?

INTERPRET and ANTICIPATE
For a heat engine, energy is absorbed from the hot reservoir. Some of this energy can be converted to mechanical work and the rest is output to the cold reservoir as waste heat. The efficiency must be between 0 and 1 (or between 0% and 100%).

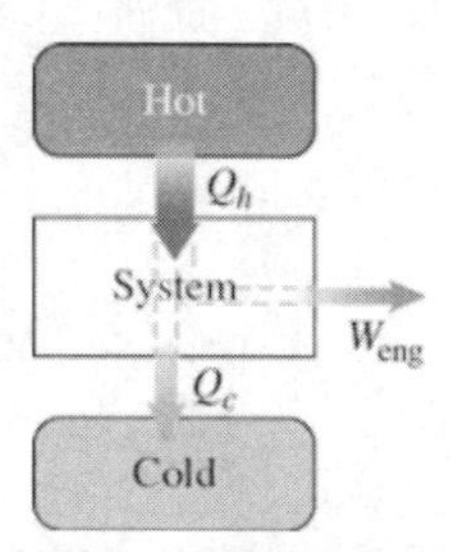

Figure P22.4ANS

SOLVE	
a. The efficiency is how much work you get out compared to the amount of heat drawn from the hot reservoir. This is expressed by Eq. 22.4 and we are given both the work generated and the heat input.	$e = \dfrac{W_{eng}}{Q_h}$ $e = \dfrac{50.0\text{ J}}{425\text{ J}}$ $e = 0.118 = \boxed{11.8\%}$
b. Eq. 22.3 expresses energy conservation. Namely, the heat from the hot reservoir must add up to the work done and the heat output to the cold reservoir, as shown in the	$\lvert Q_c \rvert = Q_h - W_{eng} = 425\text{ J} - 50.0\text{ J} = \boxed{375\text{ J}}$

figure. We have everything we need from part (a).	

CHECK and THINK

The efficiency is between 0 and 1 as expected. It's fairly low. Of the 425 J drawn from the hot reservoir, about 11.8% is converted to work (50 J) and the other 88.2% is dumped out as waste heat (375 J).

10. (N) A heat engine with 40.0% efficiency produces 2.50×10^4 W of power while exhausting 1.00×10^4 J to the cold reservoir during each cycle.
a. How much energy does the engine absorb from the hot reservoir during each cycle?
b. How long does each cycle take?

INTERPRET and ANTICIPATE

A heat engine draws heat from the hot reservoir, converts some of this energy to mechanical work, and then outputs the rest of the energy as waste heat to the cold reservoir (see the answer to Problem 4 for a sketch). The efficiency is the fraction of incoming heat that's converted to useful work. Since each cycle generates a certain amount of energy, the faster the cycle, the larger the power generated.

SOLVE	
a. The efficiency is the fraction of Q_h that is converted to work. With energy conservation, the work can be expressed as $Q_h - \|Q_c\|$ to produce Eq. 22.5.	$e = \frac{W_{eng}}{Q_h} = 1 - \frac{\|Q_c\|}{Q_h}$
Rearrange this to solve for Q_h.	$Q_h = \frac{\|Q_c\|}{1-e}$
Substitute values, using $\|Q_c\| = 1.00\times10^4$ J and $e = 0.400$.	$Q_h = \frac{1.00\times10^4\text{ J}}{1-0.400}$ $Q_h = 1.67\times10^4\text{ J} = \boxed{16.7\text{ kJ}}$
b. Power is work over time. Solve for the time for one cycle in terms of the power and work per cycle.	$P = \frac{W_{eng}}{\Delta t} \quad \rightarrow \quad \Delta t = \frac{W_{eng}}{P}$

We're given the power and we can use the definition of efficiency from part (a) to determine the work.	$P = 2.50\times10^4$ W $W_{\text{eng}} = eQ_h$
Substitute into the equation for time.	$\Delta t = \frac{W_{\text{eng}}}{P} = \frac{eQ_h}{P}$ $\Delta t = \frac{0.400\left(1.67\times10^4 \text{ J}\right)}{\left(2.50\times10^4 \text{ J/s}\right)}$ $\Delta t = \boxed{0.267 \text{ s}}$
CHECK and THINK A heat engine harnesses a fraction of heat flowing from hot to cold to perform work. We are able to relate the heat and work as well as determine the rate at which work is performed (power).	

13. (N) An internal combustion engine ignites a mixture of fuel and air that reaches a temperature of 1800 K and exhausts gas at 440 K. In the process, the volume of the gas expands such that its maximum volume is nine times larger than the minimum volume. The adiabatic coefficient for air is $\gamma = 1.4$.

a. What is the maximum efficiency possible for the engine, assuming gas in the piston follows the Otto cycle?

INTERPRET and ANTICIPATE The Otto cycle, shown in Figure 22.12, is described in Example 22.2. The efficiency of the cycle depends on the compression ratio, the ratio of the maximum to the minimum volume of the gas during the cycle. We know the efficiency should be a number between 0 and 1.	
SOLVE The efficiency of the Otto cycle was derived in Example 22.2. The result is Equation 22.14.	$e_{\text{Otto}} = 1 - \left(\frac{V_{\text{max}}}{V_{\text{min}}}\right)^{1-\gamma}$
The volume ration ($V_{\text{max}}/V_{\text{min}}$) is 9.	$e_{\text{Otto}} = 1 - (9)^{1-1.4}$ $e_{\text{Otto}} = \boxed{0.58}$
CHECK and THINK The efficiency of the Otto cycle depends on the compression ratio, or ratio of the	

maximum to the minimum volume of the gas in the piston. It is between 0 and 1, as expected.

b. What is the ideal efficiency for a Carnot cycle operating between the same high and low temperatures?

INTERPRET and ANTICIPATE The efficiency of the Carnot cycle depends on the temperatures of the hot and cold reservoir. It represents the most efficient cycle possible, so we expect the efficiency to be at least as large as the Otto cycle efficiency, but less than 1.	
SOLVE The Carnot efficiency is given by Equation 22.6.	$e_{\text{Carnot}} = 1 - \frac{T_c}{T_h}$
Substitute the temperatures of the hot and cold reservoirs.	$e_{\text{Carnot}} = 1 - \frac{440}{1800} = \boxed{0.76}$
CHECK and THINK The efficiency is higher than that of the Otto cycle. The Carnot efficiency represents the ideal cycle and the highest efficiency possible given T_c and T_h.	

17. (N) A Carnot heat engine operates in contact with a cold reservoir and a hot reservoir such that 4.310×10^5 J of heat is transferred into the cold reservoir and 9.678×10^5 J of heat is absorbed from the hot reservoir.
a. If the temperature of the cold reservoir is 315.0 K, what is the temperature of the hot reservoir?

INTERPRET and ANTICIPATE The efficiency of the engine can be found from the heat transferred out of the hot reservoir and the amount transferred into the cold reservoir by Eq. 22.5. This must be equal to the Carnot efficiency as described in Eq. 22.6. Knowing the cold reservoir temperature, we can then find the temperature of the hot reservoir, which should be greater.	
SOLVE Equating Eq. 22.5 and Eq. 22.6, we can solve for the temperature of the hot reservoir.	$e = 1 - \lvert Q_c \rvert / Q_h = 1 - T_c / T_h$

	$\lvert Q_c\rvert/Q_h = T_c/T_h$ $T_h\left(\lvert Q_c\rvert/Q_h\right) = T_c$ $T_h = (T_c Q_h)/\lvert Q_c\rvert$ $T_h = \left[(315.0\text{ K})(9.678\times10^5\text{ J})\right]/4.310\times10^5\text{ J}$ $T_h = \boxed{707.3\text{ K}}$

CHECK and THINK
The temperature is greater than the temperature of the cold reservoir, as we would expect. Note that Eq. 22.5 can be applied to any engine, while Eq. 22.6 is only for the ideal Carnot engine.

b. What ratio of volumes for an Otto cycle would cause it to have the same efficiency as the above engine? Assume $\gamma = 1.4$.

INTERPRET and ANTICIPATE
The efficiency for the Otto cycle (Eq. 22.14) can be set equal to the efficiency for a Carnot cycle (Eq. 22.6), since we are trying to compare it to the engine described in part (a). Knowing the temperatures of the hot and cold reservoirs, we can solve for the ratio of the volumes. If we solve for $V_{\max}/V_{\min}$, we should expect a number that is positive and greater than 1.

SOLVE	
Writing Eq. 22.14 and equating it to Eq. 22.6, we find an expression that can be manipulated to solve for the ratio of volumes. Then, use the answer from part (a) and the given information to find the ratio.	$e_{\text{Otto}} = 1 - \left(\dfrac{V_{\max}}{V_{\min}}\right)^{1-\gamma} = 1 - T_c/T_h$ $\left(V_{\max}/V_{\min}\right)^{1-\gamma} = T_c/T_h$ $V_{\max}/V_{\min} = \left(T_c/T_h\right)^{\frac{1}{1-\gamma}}$ $V_{\max}/V_{\min} = (315.0\text{ K}/707.3\text{ K})^{\frac{1}{1-1.4}}$ $V_{\max}/V_{\min} = (315.0\text{ K}/707.3\text{ K})^{-\frac{1}{0.4}}$ $V_{\max}/V_{\min} = \boxed{7.556}$

CHECK and THINK
The ratio is bigger than 1, as expected. The efficiency of the Otto cycle decreases as the ratio of the volumes increases.

21. (N) A car engine has an efficiency of 22%. If it produces 1.20×10^4 J of work, how much heat does it expel?

INTERPRET and ANTICIPATE	
We are asked to calculate Q_c, the heat output to the cold reservoir. The efficiency tells us how much of the incoming heat from the hot reservoir is converted to work. At 22%, most of the energy is lost as heat to the cold reservoir, so we expect Q_c to be larger than the work produced.	
SOLVE Using the definition of efficiency (Eq. 22.4), first determine the heat input from the hot reservoir. With energy conservation, we'll be able to determine the heat that leaves to the cold reservoir.	$e \equiv \frac{W_{eng}}{Q_h} \quad \rightarrow \quad Q_h = \frac{W_{eng}}{e}$
Substitute the efficiency and work.	$Q_h = \frac{1.20\times10^4 \text{ J}}{0.22}$ $Q_h = 5.5\times10^4 \text{ J}$
Now use energy conservation, Eq. 22.3, to solve for Q_c.	$\lvert Q_c\rvert = Q_h - W_{eng}$ $\lvert Q_c\rvert = 5.5\times10^4 \text{ J} - 1.20\times10^4 \text{ J}$ $\lvert Q_c\rvert = \boxed{4.3\times10^4 \text{ J}}$
CHECK and THINK With an efficiency of 22%, of the 55 kJ entering from the hot reservoir, 22% of it (12 kJ) is converted to work and the rest (78%, or 43 kJ) is lost as waste heat to the cold reservoir.	

25. A diesel engine (Problem 24) has the volume ratios $r = 4.57$ and $R = 21.0$. When air is taken into the cylinder, the pressure is 1.00 atm, and the temperature is 22.5°C.

a. (N) What is the temperature of the air when the fuel is injected?

b. (C) The air–fuel mixture must be at least 483 K for ignition to take place. Is ignition achieved? What minimum intake temperature would achieve ignition?

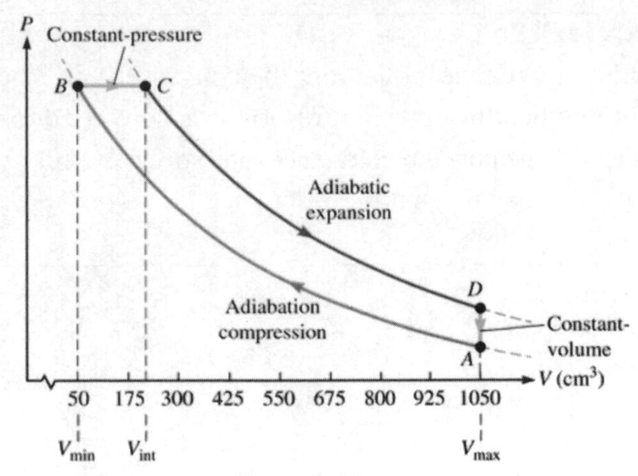

Figure P22.24

INTERPRET and ANTICIPATE

We are given quantities for point *A*, where air is taken into the cylinder at a temperature $T_A = 22.5°\text{C} = 295.65$ K. The fuel is injected at point *B*, so we want to determine the temperature at this point. Process *AB* is adiabatic compression, so we can use material from Chapter 21 to determine the temperature at point *B*.

SOLVE	
a. For an adiabatic process, Eq. 21.32 relates temperature and volume in an adiabatic process. (It's also possible to use Eq. 21.30, $PV^{\gamma} = \text{constant}$, and the ideal gas law to derive this relationship.)	$\frac{T_f}{T_i} = \left(\frac{V_f}{V_i}\right)^{(1-\gamma)}$
Looking at the Figure provided, $V_f = V_B = V_{\text{min}}$ and $V_i = V_A = V_{\text{max}}$. The ratio is given in Problem 24 as $R = (V_{\text{max}}/V_{\text{min}})$.	$\frac{T_B}{T_A} = \left(\frac{V_{\text{min}}}{V_{\text{max}}}\right)^{(1-\gamma)} = \left(\frac{1}{R}\right)^{(1-\gamma)}$
Insert numerical values. It's also possible to take a slightly different approach using the fact that $\left(\frac{1}{x}\right)^{a} = x^{-a}$ to reach the same answer: $T_B = T_A\left(\frac{1}{R}\right)^{(1-\gamma)} = T_A R^{(\gamma-1)}$ $T_B = (295.65\text{ K})(21.0)^{(1.4-1)} = 999\text{ K}$	$\frac{T_B}{T_A} = \left(\frac{V_{\text{min}}}{V_{\text{max}}}\right)^{(1-\gamma)} = \left(\frac{1}{R}\right)^{(1-\gamma)}$ $T_B = T_A\left(\frac{1}{R}\right)^{(1-\gamma)}$ $T_B = (295.65\text{ K})\left(\frac{1}{21.0}\right)^{(1-1.4)}$ $T_B = \boxed{999\text{ K}}$

b. The answer from part (a) is well above the 483 K necessary to achieve ignition. We want the minimum temperature T_A to achieve the necessary final temperature T_B = 483 K. Use the same relationship as in part (a).	$T_A = T_B\left(\frac{V_{max}}{V_{min}}\right)^{(1-\gamma)} = T_B R^{1-\gamma}$ $T_A = (483\text{ K})(21.0)^{1-1.4} = \boxed{143\text{ K}}$

CHECK and THINK

The problem turned out to be practice with an adiabatic process as in Chapter 21. In this case, the gas is compressed and the temperature rises dramatically based on the compression ratio to achieve a sufficiently high temperature to ignite the fuel.

31. (N) The coefficient of performance of a refrigerator that takes 210.0 J of energy from the cold reservoir during each cycle is $\kappa = 4.00$.

a. What is the work done during each cycle?

b. How much energy is exhausted into the hot reservoir during each cycle?

INTERPRET and ANTICIPATE

Similar to efficiency for a heat engine, the coefficient of performance for a refrigerator is the ratio of "*what you get*" (in this case, how much heat is drawn from the cold reservoir) to "*what you put in*" (the work). We're given the first two and asked for the third. Energy conservation can be used to determine the heat output to the hot reservoir.

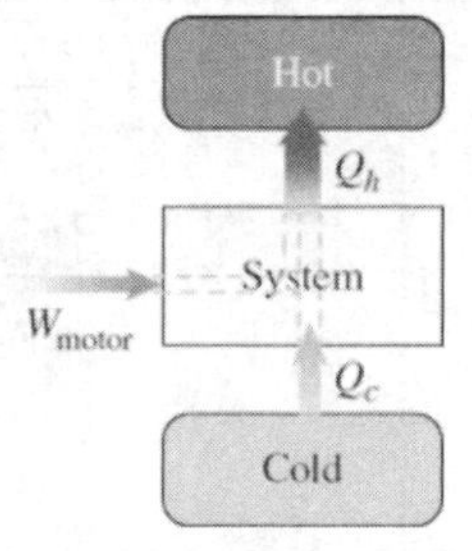

Figure P22.31ANS

SOLVE	
a. For a refrigerator, the coefficient of performance is given by Eq. 22.15.	$\kappa = \frac{\lvert Q_c \rvert}{\lvert W_{motor} \rvert}$
If $\lvert Q_c \rvert = 210.0$ J and $\kappa = 4.00$, then we can calculate the work.	$W_{motor} = \frac{210.0\text{ J}}{4.00} = \boxed{52.5\text{ J}}$

b. Referring to the figure, we can use conservation of energy (Eq. 22.16).	$\lvert Q_h\rvert = \lvert Q_c\rvert + \lvert W_{\text{motor}}\rvert$ $\lvert Q_h\rvert = 210\text{ J} + 52.5\text{ J}$ $\lvert Q_h\rvert = \boxed{262.5\text{ J}}$
CHECK and THINK The coefficient of performance is 4, which means that by expending 52.5 J of work, you can move four times that (210 J) out of the cold reservoir.	

35. (N) What is the change in entropy of 100.0 g of water when heated so that its temperature changes from 20.0°C to 100.0°C? The specific heat capacity of water is 4186 J/(kg K).

INTERPRET and ANTICIPATE As heat is added to the water its entropy increases. The heat added depends on the specific heat capacity.	
SOLVE In this case, the temperature is changing, so we must integrate from the initial to final temperature using Eq. 22.22.	$\Delta S = \int_i^f \frac{dQ}{T}$
The integration variable dQ depends on temperature. They are related by the heat capacity, $dQ = mc\ dT$. Using this relationship, we can integrate to find the total change in entropy.	$\Delta S = mc\int_i^f \frac{dT}{T}$ $\Delta S = mc\ln\left(\frac{T_f}{T_i}\right)$
Plugging in values, using $c = 4190$ J/(kg K) and $T_f = 373$ K, $T_i = 293$ K.	$\Delta S = (0.100\text{ kg})\left(4190\frac{\text{J}}{\text{kg}\cdot\text{K}}\right)\ln\left(\frac{373\text{ K}}{293\text{ K}}\right)$ $\Delta S = \boxed{101\text{ J/K}}$
CHECK and THINK The change in entropy as we add heat is positive as expected.	

40. (N) An ideal gas undergoes a free expansion, doubling in volume. If there are 15 moles of the gas, what is the change in the gas's entropy? Use Example 22.6 to check your results.

INTERPRET and ANTICIPATE	
We have a formula for the change in entropy of an expanding gas. As the gas occupies more volume, there are (roughly speaking) more ways for the gas molecules to arrange themselves and more microstates/disorder, so the entropy increases.	
SOLVE The entropy change due to free expansion is given by Eq. 22.24. The volume doubles, so $V_f = 2V_i$.	$\Delta S = nR\ln\left(\frac{V_f}{V_i}\right)$ $\Delta S = (15\text{ mol})\left(8.314\frac{\text{J}}{\text{K}\cdot\text{mol}}\right)\ln(2)$ $\Delta S = \boxed{86.4\text{ J/K}}$
CHECK and THINK This is 15 times greater than Example 22.6 since there are 15 moles instead of 1.	

42. (N) A cold reservoir at $T_c = 15.0°\text{C}$ and a hot reservoir at $T_h = 450.0°\text{C}$ are connected via a steel bar. The energy transferred from the hot reservoir to the cold reservoir in this irreversible process is 1.85×10^4 J. What is the change in the entropy of

a. the cold reservoir and

b. the hot reservoir during this process?

c. Ignoring the change in entropy of the steel bar, what is the change in the entropy of the Universe during this process?

INTERPRET and ANTICIPATE	
In general, the change of entropy as heat is transferred into a system is $dS = \frac{dQ}{T}$. Since the reservoirs are assumed to be isothermal, we can simply divide the total heat transfer by the temperature of the reservoir. When heat enters the cold reservoir, its entropy goes up, and when heat leaves the hot reservoir, its entropy goes down. For an irreversible process, the entropy of the Universe should increase.	
SOLVE The change in entropy of a reservoir is $\Delta S = Q/T$ (Eq. 22.19), where Q is the energy absorbed ($Q > 0$) or expelled ($Q < 0$) by the reservoir, and T is the absolute temperature of the reservoir.	$\Delta S = \frac{Q}{T}$

a. The temperatures of the cold reservoir is T_c = 15.0°C + 273 = 288 K. Energy enters the cold reservoir, so Q is positive.	$\Delta S_c = \dfrac{+1.85\times10^4\ \text{J}}{288\ \text{K}}$ $\Delta S_c = \boxed{+64.2\ \text{J/K}}$
b. The temperatures of the cold reservoir is T_h = 450°C + 273 = 723 K. Energy leaves the hot reservoir, so Q is negative.	$\Delta S_h = \dfrac{-1.85\times10^4\ \text{J}}{723\ \text{K}}$ $\Delta S_h = \boxed{-25.6\ \text{J/K}}$
c. For the total entropy change of the Universe, add both together. It's positive, as we expected.	$\Delta S_U = \Delta S_h + \Delta S_c$ $\Delta S_U = 64.2\ \text{J/K} - 25.6\ \text{J/K}$ $\Delta S_U = \boxed{+38.6\ \text{J/K}}$

CHECK and THINK

The entropy change for the Universe increases in this irreversible process, as we would expect.

45. (N) A nonporous membrane divides a vessel in the form of two identical 5.00-L chambers that are welded together (Fig. P22.45). Chamber A is filled with 1.22 mol of argon gas, and chamber B is filled with 1.22 mol of krypton gas. What is the change in entropy of the system if the membrane is removed and the gases mix together?

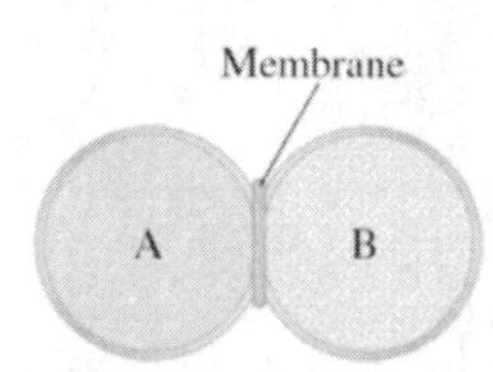

Figure P22.45

INTERPRET and ANTICIPATE

Both gases expand irreversibly to fill a larger volume, so the entropy increases. Thinking of entropy as disorder, the system goes from having one type of gas on each side to a mixed (more disordered) system, which again suggests that the entropy increases.

SOLVE	
Each gas quickly diffuses into the other half of the vessel, expanding to fill the entire volume as though the other gas were not there; therefore, consider each gas to undergo a free	$\Delta S = nR\ln\left(\dfrac{V_f}{V_i}\right)$

expansion process in which its volume doubles (Eq. 22.24).	
The total entropy change for both gases is twice that for a single gas.	$\Delta S = 2\left[nR\ln\left(\frac{V_f}{V_i}\right)\right]$ $\Delta S = 2\left[(1.22)\left(8.314\ \frac{\text{J}}{\text{mol}\cdot\text{K}}\right)\ln(2)\right]$ $\Delta S = \boxed{14.1\ \text{J/K}}$
CHECK and THINK The entropy goes up as expected.	

47. A spring obeys Hooke's law and has a spring constant of 125 N/m. You stretch the spring from 1.60 m to 1.85 m at 22.2°C.
a. (N) What is the change in the spring's entropy?

INTERPRET and ANTICIPATE The entropy is related to heat flow, which is not something we think about with springs typically. But if we stretch a spring and let it go, it will snap back. We might think of this as its entropy going up (since isolated systems tend to maximize entropy, the spring "naturally" snaps back to a higher entropy configuration), so we might guess that the stretched spring has a lower entropy. Let's see…	
SOLVE **a.** We know that we have to do work to stretch the spring. And work and heat are related to the change in thermal energy. Assuming that this all happens in a room at a constant temperature, the change in thermal energy of the spring is zero.	$\Delta E_{\text{th}} = 0$
From the first law of thermodynamics (Eq. 21.1), that means that the heat transferred into the spring is the negative of work done on the spring.	$W_{\text{tot}} + Q = \Delta E_{\text{th}} = 0$ $Q = -W_{\text{tot}}$
Now use the definition of entropy for an isothermal process (Eq. 22.19).	$\Delta S_{\text{spring}} = \frac{Q}{T} = \frac{-W}{T}$

The work done on the spring is equal to the change in its potential energy.	$W = \Delta U$ $W = \frac{1}{2}kx_f^2 - \frac{1}{2}kx_i^2$ $W = \frac{1}{2}\left(125\,\frac{\text{N}}{\text{m}}\right)\left[(1.85\text{ m})^2 - (1.60\text{ m})^2\right]$ $W = 53.9\text{ J}$
Insert the work into the equation for entropy change at a temperature of 22.2°C = 295.35 K.	$\Delta S_{\text{spring}} = -\frac{W}{T}$ $\Delta S_{\text{spring}} = -\frac{53.9\text{ J}}{295.35\text{ K}}$ $\Delta S_{\text{spring}} = \boxed{-0.183\text{ J/K}}$

CHECK and THINK
The change in entropy of the spring is negative, so the stretched spring has lower entropy as we predicted. While we don't normally think of heat flow with something like a spring, you can actually try this: Find a large rubber band, stretch it quickly, and hold it against your forehead… you should find that it feels a little warm as it releases heat in the process (negative Q and decrease in entropy).

b. (C) Is the process reversible? Explain.

Yes, it's easily possible to imagine the reverse process occurring. (As opposed to dye diffusing in water for instance, where you couldn't possibly imagine diluted dye in water diffusing all to the same place and forming a little drop of concentrated dye in the water.)

c. (N) What is the change in the Universe's entropy?

If the process is reversible, then the total change in entropy is zero.

$$\Delta S_{\text{universe}} + \Delta S_{\text{spring}} = 0$$

$$\Delta S_{\text{universe}} = -\Delta S_{\text{spring}} = \boxed{+0.183\text{ J/K}}$$

57. (N) What is the entropy of a freshly shuffled deck of 52 playing cards?

INTERPRET and ANTICIPATE
Entropy depends on the natural log of the number of microstates, or the number of different configurations for the system. We need to count how many ways there are to arrange the 52 cards and then apply the formula for entropy.

SOLVE The entropy depends on the log of the number of microstates according to Eq. 22.25.	$S = k_B \ln \mathcal{W}$
To count the number of microstates, imagine laying out all 52 cards randomly in a line. For the first card, you have 52 options for where to put it in the line. For the next card, you can put it in any of the 51 remaining spots. The next one can go into any of the 50 remaining spots… and so on. Therefore, the total number of ways to lay down all the cards is $52 \times 51 \times 50 \times 49 \times \ldots 2 \times 1 = 52!$ This turns out to be a *huge* number of possibilities at around 10^{67}.	$\mathcal{W} = 52! = 8.066 \times 10^{67}$
Now, determine the entropy.	$S = k_B \ln \mathcal{W}$ $S = k_B \ln(52!) = 156 k_B$ $S = 156(1.38 \times 10^{-23} \text{ J/K})$ $S = \boxed{2.16 \times 10^{-21} \text{ J/K}}$

CHECK and THINK
It's not clear what the value of this entropy means, but the technique is to count all the possible microstates of the system and use $S = k_B \ln \mathcal{W}$. The more possible states, the larger the entropy.

58. Feeling lucky, George is rolling a pair of dice at a craps table.
a. (N) His first roll of the dice is a 2. How many microscopic states are there for this roll?
b. (N) His second roll of the dice is an 8. How many microscopic states are there for this roll?
c. (C) Which of the two rolls corresponds to a state with lower entropy?

INTERPRET and ANTICIPATE
In this case, we can count the number of microstates for each value rolled. The number of microstates is the number of different ways to achieve the given total. The entropy is larger for the value with more microstates.

SOLVE **a.** There is only one way to roll a 2 — if each die	$\mathcal{W} = \boxed{1}$

shows a one (1 + 1).	
b. There are multiple ways to roll an 8 with two dice — 2 + 6, 3 + 5, 4 + 4, 5 + 3, and 6 + 2… five ways in total.	$\mathcal{W} = \boxed{5}$
c. The entropy depends on the natural log of the number of microstates according to Eq. 22.25. Since ln 2 < ln 8, the 2 has a lower entropy than the 8.	$S = k_B \ln \mathcal{W}$
CHECK and THINK The entropy depends on the log of the number of microstates, which we can count up.	

64. a. (N) What is the entropy change for 5.0 g of ice that melts at 0.0°C to form water? The heat of fusion of water is 333 J/g.
b. (N) What is the entropy change for 5.0 g of water that boils at 100°C to form steam? The heat of vaporization for water is 2260 J/g.

INTERPRET and ANTICIPATE The entropy change for a process at constant temperature equals the heat added divided by the temperature. The heat added depends on the latent heat of fusion and mass of ice.	
SOLVE **a.** The entropy change for this process can be calculated using Equation 22.19. The heat absorbed depends on the heat of fusion, characterized by Equation 21.9.	$\Delta S \equiv \frac{Q}{T}$ $Q = mL_F$
Substitute the values for each variable. We are given the mass and heat of fusion. The temperature in absolute units is 0.0°C = 273 K.	$\Delta S = \frac{mL_F}{T} = \frac{(5.0\text{ g})(333\text{ J/g})}{273\text{ K}} = \boxed{6.1\text{ J/K}}$
b. We use the same approach, but the heat absorbed now depends on the heat of vaporization (Equation 21.10).	$Q = mL_V$

Substitute values. Converting to absolute units, 100°C = 373 K.	$\Delta S = \frac{mL_V}{T} = \frac{(5.0\text{ g})(2260\text{ J/g})}{373\text{ K}} = \boxed{3.0\times10^1\text{ J/K}}$

CHECK and THINK
The entropy change is larger for boiling than melting for the same mass because the heat of vaporization is significantly higher than the heat of fusion. In terms of disorder, we might imagine that water molecules that separate into rapidly moving gas molecules gain more entropy than ice molecules that are now able to flow past each other in a liquid.

67. (N) A 1.55-kg sample of liquid water at 50°C is heated and boiled away, and eventually it is converted entirely to water vapor. The resulting water vapor is then heated until it reaches a final temperature of 115°C. Assume the specific heat capacity of the liquid water and the water vapor are 4190 J/(kg · K) and 2010 J/(kg · K), respectively, and the latent heat of vaporization is 2.256×10^6 J/kg. No work is performed on or by the water or the water vapor.
a. What is the change in entropy of the liquid water as it is heated to the vaporization point?

INTERPRET and ANTICIPATE
The change in entropy for a thermodynamic process is defined by Eq. 22.22, $\Delta S = \int_i^f \frac{dQ}{T}$. By using Eq. 21.5, we can derive a formula for the change in entropy for a substance with specific heat capacity, c, and mass, m, that undergoes a change in temperature ΔT. After converting the temperatures to Kelvins, they can be used with the other given info to find the change in entropy. The water's boiling point is 100 °C. Since the temperature is increasing, we expect that the change in entropy will be positive.

SOLVE First, we convert the temperatures to Kelvins.	$T_i = 50°\text{C} + 273.15 = 323.15\text{ K}$ $T_f = 100 \cong °\text{C} + 273.15 = 373.15\text{ K}$
Then, writing Eq. 22.22 and substituting Eq. 21.5, we can find the change in entropy.	

	$\Delta S = \int_i^f \frac{dQ}{T}$ $\Delta S = \int_i^f \frac{mcdT}{T}$ $\Delta S = mc\int_i^f \frac{dT}{T}$ $\Delta S = mc\ln\left(\frac{T_f}{T_i}\right)$ $\Delta S = (1.55\text{ kg})\left(4190\frac{\text{J}}{\text{kg}\cdot\text{K}}\right)\ln\left(\frac{373.15\text{ K}}{323.15\text{ K}}\right)$ $\Delta S = \boxed{934\text{ J/K}}$
CHECK and THINK The change in entropy is positive, as expected. It is interesting to note that if another substance with the same mass, but with a lower specific heat capacity underwent the same change in temperature, it would not experience as great a change in entropy.	

b. What is the change in entropy of the water vapor as it is heated from the vaporization point to the final temperature of 115°C?

INTERPRET and ANTICIPATE We can apply the general result from part (a) to the water vapor, using its specific heat capacity. We expect the change in entropy to be positive again, because the temperature increases (from 100 to 115°C).	
SOLVE First, we convert the temperatures to Kelvins.	$T_i = 100°\text{C} + 273.15 = 373.15\text{ K}$ $T_f = 115°\text{C} + 273.15 = 388.15\text{ K}$
Then, using the result from part (a), we can find the change in entropy.	$\Delta S = mc\ln\left(\frac{T_f}{T_i}\right)$ $\Delta S = (1.55\text{ kg})\left(2010\frac{\text{J}}{\text{kg}\cdot\text{K}}\right)\ln\left(\frac{388.15\text{ K}}{373.15\text{ K}}\right)$ $\Delta S = \boxed{123\text{ J/K}}$
CHECK and THINK The change in entropy is positive.	

c. What is the change in entropy of the water as it undergoes the phase transformation?

<table>
<tr><td colspan="2">INTERPRET and ANTICIPATE
The temperature stays constant during a phase change (373.15 K for the vaporization of water). The change in entropy is given by Eq. 22.19. This again should be positive because the heat must flow into the water in order to break bonds between molecules, resulting in water vapor. The heat flow would be given by $Q = mL_V$.</td></tr>
<tr><td>SOLVE
Beginning with Eq. 22.19 and substituting for Q, we find the change in entropy.</td><td>$$\Delta S = \frac{Q}{T}$$ $$\Delta S = \frac{mL_V}{T}$$ $$\Delta S = \frac{(1.55\ \text{kg})(2.256\times10^6\ \text{J/kg})}{373.15\ \text{K}} = \boxed{9.37\times10^3\ \text{J/K}}$$</td></tr>
<tr><td colspan="2">CHECK and THINK
If the water was condensing from a gaseous state, the heat flow would necessarily be negative and the entropy would decrease, as the molecules bond, forming a liquid state.</td></tr>
</table>

71. (N) A system consisting of 10.0 g of water at a temperature of 20.0°C is converted into ice at −10.0°C at constant atmospheric pressure. Calculate the total change in entropy of the system. Assume the specific heat of water is 4.19×10^3 J/(kg · K), the specific heat of ice is 2.10×10^3 J/(kg · K), and the latent heat of fusion is 3.33×10^5 J/kg.

<table>
<tr><td colspan="2">INTERPRET and ANTICIPATE
We break the problem in three parts, (i) cooling the water to 0°C, (ii) freezing water to ice, and (iii) cooling ice to –10°C and find the change in entropy for each step. Since the water is cooling/freezing, heat is removed and the entropy decreases.</td></tr>
<tr><td>SOLVE
Break up the process into (i) cooling the water to 0°C, (ii) freezing water to ice, and (iii) cooling ice to –0°C.

For the first and third steps, the temperature is changing as heat is added to either water or ice (a single phase), so we consider a small amount of heat $dQ = mc_P dT$ (from Eq. 21.5, in terms of mass and specific heat</td><td>For processes (i) and (iii):
$$\Delta S = mc_P \ln\left(\frac{T_f}{T_i}\right) \qquad \text{(Eq. 22.23)}$$
For process (ii):
$$\Delta S = \frac{Q}{T} = \frac{-mL_f}{T} \qquad \text{(Eqs. 22.19 and 21.9)}$$</td></tr>
</table>

capacity) and integrate the entropy $dS = \frac{dQ}{T} = mc_P \frac{dT}{T}$ (from Eq. 22.22) to find the change in entropy. This results in Eq. 22.23. For the second step, which occurs at constant temperature, the heat removed from the ice (which is therefore negative) is due to the latent heat of fusion (Eq. 21.9). So, using Eq. 22.19, we can write the entropy change.	
(i) Apply Eq. 22.23, with 10 g = 0.010 kg and 20°C = 293 K and 0°C = 273 K.	$\Delta S_i = mc_P \ln\left(\frac{T_f}{T_i}\right)$ $\Delta S_i = (0.010\ \text{kg})\left(4.19\times10^3\ \frac{\text{J}}{\text{kg}\cdot\text{K}}\right)\ln\left(\frac{273\,\text{K}}{293\,\text{K}}\right)$ $\Delta S_i = -2.96\ \text{J/K}$
(ii) Use Eq. 22.19, with temperature of 273 K and heat $Q = -mL_F$.	$\Delta S_{ii} = \frac{-mL_F}{T}$ $\Delta S_{ii} = \frac{-(1.0\times10^{-2}\ \text{kg})(3.33\times10^5\ \text{J/kg})}{273\ \text{K}}$ $\Delta S_{ii} = -12.2\ \text{J/K}$
(iii) Repeat the steps for process (i) with the specific heat capacity of ice.	$\Delta S_{iii} = mc_P \ln\left(\frac{T_f}{T_i}\right)$ $\Delta S_{iii} = (0.010\,\text{kg})\left(2.1\times10^3\ \frac{\text{J}}{\text{kg}\cdot\text{K}}\right)\ln\left(\frac{263\,\text{K}}{273\,\text{K}}\right)$ $\Delta S_{iii} = -0.78\ \text{J/K}$
Finally, to get the total, add all three entropy changes together.	$\Delta S_{\text{tot}} = \Delta S_i + \Delta S_{ii} + \Delta S_{iii}$ $\Delta S_{\text{tot}} = (-2.96 - 12.20 - 0.78)\ \text{J/K}$ $\Delta S_{\text{tot}} = \boxed{-15.9\ \text{J/K}}$

CHECK and THINK
The entropy change is negative as the water cools and freezes. In fact, in each step, heat is removed, so the entropy change is negative in each process.

73. (N) Figure P22.73 illustrates the cycle *ABCA* for a 2.00-mol sample of an ideal diatomic gas, where the process *CA* is a reversible isothermal expansion. What is
a. the net work done by the gas during one cycle?

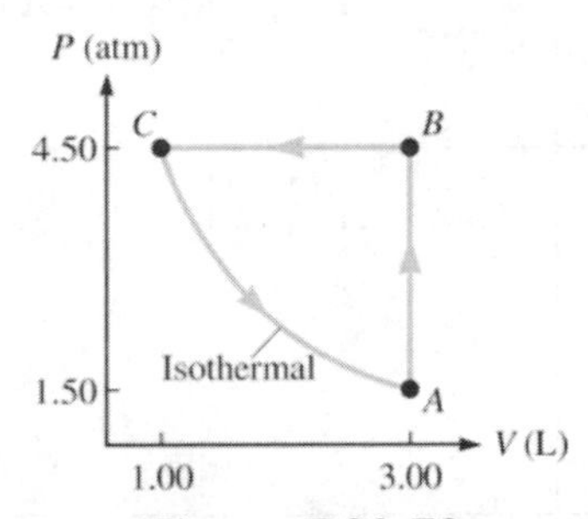

Figure P22.73

INTERPRET and ANTICIPATE	
The work done on the gas for each process can be determined using material from Chapter 21 and added together to get the total work. The work done *by the gas* is then the negative of the work done *on the gas*.	
SOLVE The net work done is the total of all three processes. There is no work done in the isochoric process *AB*.	$W_{tot} = W_{AB} + W_{BC} + W_{CA}$ $W_{BC} = 0$ (constant volume)
In the isobaric process *BC*, work is done on the gas as it is compressed (Eq. 21.16). Use the facts that 1.00 atm = 1.013×10^5 Pa and 1.00 L = 1.00×10^{-3} m^3.	$W_{BC} = -P_B \Delta V$ (constant pressure) $W_{BC} = -4.50(1.013\times10^5 \text{ Pa})\left[(1.00-3.00)\times10^{-3}\right]\text{m}^3$ $W_{BC} = 912 \text{ J}$

For the isothermal process *CA*, the work done on the gas can be calculated using Eq. 21.15.	$W_{CA} = P_C V_C \ln\left(\frac{V_C}{V_A}\right)$ (isothermal) $W_{CA} = 4.50\left(1.013\times10^5 \text{ Pa}\right)\left(1.00\times10^{-3} \text{ m}^3\right)\ln\left(\frac{1.00 \text{ L}}{3.00 \text{ L}}\right)$ $W_{CA} = -501 \text{ J}$
The total work done *on the gas* is then:	$W_{\text{gas}} = -W_{eng} = W_{CA} + W_{BC}$ $W_{\text{gas}} = 912 \text{ J} - 501 \text{ J} = 411 \text{ J}$
The work done *by the gas* (or by the engine or heat pump) has the same magnitude but opposite sign.	$W_{eng} = \boxed{-411 \text{ J}}$

CHECK and THINK

The net work on the gas is positive, meaning that the net work done by the gas on the surroundings is negative. This is a heat pump… work is done on the system and heat is transported from cold to hot.

b. How much energy is added to the gas by heat during one cycle?

INTERPRET and ANTICIPATE

Similar to part (a), we can find the heat entering the gas for each process using material from Chapter 21. Heat is added to the gas during processes *CA* and *AB* and it is exhausted from the gas during process *BC*. We can confirm this with the signs for each after we calculate them.

SOLVE	
Since *CA* is an isothermal process, $\Delta E_{\text{th, }CA} = 0$. From the conservation of energy (Eq. 21.1), the work and heat must have opposite signs. Q is positive meaning heat is added to the gas.	$Q_{CA} = -W_{CA,\ gas} = 501$ J.
The heat transferred into the gas during process *AB* (constant volume) can be found using the molar heat capacity and change in temperature (Eq. 21.6). For an ideal diatomic gas, $C_V = \frac{5R}{2}$.	$Q_{AB} = nC_V \Delta T$

To calculate the temperature difference, we need to calculate the temperatures at points *A* and *B* using the ideal gas law.	$T_A = T_C = \frac{P_A V_A}{nR}$ $T_A = \frac{1.50\left(1.013\times10^5 \text{ Pa}\right)\left(3.00\times10^{-3} \text{ m}^3\right)}{(2.00 \text{ mol})(8.314 \text{ J/K}\cdot\text{mol})}$ $T_A = 27.4 \text{ K}$ $T_B = \frac{P_B V_B}{nR}$ $T_B = \frac{4.50\left(1.013\times10^5 \text{ Pa}\right)\left(3.00\times10^{-3} \text{ m}^3\right)}{(2.00 \text{ mol})(8.314 \text{ J/K}\cdot\text{mol})}$ $T_B = 82.2 \text{ K}$
Now determine the heat. It is again positive and heat is added to the gas.	$Q_{AB} = nC_V\Delta T$ $Q_{AB} = 2.00\left(\frac{5}{2}R\right)(82.2 \text{ K} - 27.4 \text{ K})$ $Q_{AB} = 2280 \text{ J}$
The total energy absorbed by the gas as heat is the sum of these two.	$Q_{AB} + Q_{CA} = 2280 \text{ J} + 501 \text{ J} = \boxed{2.78\times10^3 \text{ J}}$

CHECK and THINK
The heat for processes *CA* and *AB* are positive, meaning that heat enters the gas.

c. How much energy is exhausted from the gas by heat during one cycle?

INTERPRET and ANTICIPATE
Heat is exhausted from the gas during process *BC*. We can confirm this with the sign after we calculate this quantity.

SOLVE We follow the same procedure as with process *AB* except that process *BC* is at constant pressure. For an ideal diatomic gas, $C_P = \frac{7R}{2}$. We calculated the temperatures of points	$Q_{BC} = nC_P\Delta T$

B and C in part (b).	$Q_{BC}=\frac{7}{2}\left(nR\left(T_C-T_B\right)\right)$ $Q_{BC}=\frac{7}{2}\left((2.00)(8.314)(27.4-82.2)\right)$ $Q_{BC}=-3190\text{ J}$ $\lvert Q_{BC}\rvert=\boxed{3.19\times10^3\text{ J}}$

CHECK and THINK

The heat calculated is indeed negative, meaning that heat is leaving the gas. Its magnitude is 3190 J.

d. What is the efficiency of the cycle?

e. What would be the efficiency of a Carnot engine operated between the temperatures at points A and B during each cycle?

INTERPRET and ANTICIPATE

By comparing the work needed to transfer an amount of energy through heat, we can determine the efficiency of the cycle, which must be between 0 and 1. Carnot efficiency, which depends only on the high and low temperatures, is the most efficient cycle theoretically possible, so it should be higher than the efficiency of this cycle.

SOLVE	
d. If we run this cycle (which is a heat pump) as an engine, we actually need to run this cycle in reverse. The efficiency is given by Eq. 22.4.	$e=\frac{W_{\text{eng}}}{Q_h}$
We must be careful when substituting values though. The work done by the engine running the cycle in reverse is +411 J (from part (a), but all the signs are reversed as well). The heat that enters the gas from the hot reservoir, for the cycle in reverse, is actually during process BC, +3190 J (from part (c), again positive means entering the gas and the sign is reversed when we reverse the cycle).	$e=\frac{W_{\text{eng}}}{Q_h}=\frac{W_{\text{eng}}}{Q_{BC}}$ $e=\frac{411\text{ J}}{3190\text{ J}}$ $e=\boxed{0.129}$ or $\boxed{12.9\%}$

e. A Carnot engine operating between $T_{hot} = T_B =$ 82.3 K and $T_{cold} = T_A = 27.4$ K has an efficiency given by Eq. 22.6.	$e_{Carnot} = 1 - \frac{T_c}{T_h}$ $e_{Carnot} = 1 - \frac{1}{3}$ $e_{Carnot} = \boxed{0.667} = \boxed{66.7\%}$
CHECK and THINK The efficiencies are between 0 and 1 with the Carnot efficiency being much higher than the cycle given in this problem.	

78. (N) The coefficient of performance of a household heater that consumes 850.0 W of power is $\kappa = 2.95$, where $\kappa = |Q_h|/W$ represents the coefficient of performance for a heater.

a. If the heater is in operation for 6.00 h each night, how much energy does it deliver?

b. What is the energy extracted by the heater from the outside air (the cold reservoir) during this time?

INTERPRET and ANTICIPATE A heat pump transfers heat from a colder space (outdoors) to a warmer space (indoors). We're given a way to calculate the work done by the heater, so we can calculate the amount of heat transferred with the coefficient of performance.	
SOLVE **a.** The coefficient of performance of a heat pump is $\kappa = |Q_h|/W$, where $|Q_h|$ is the thermal energy delivered to the warm space (hot reservoir) and W is the work input required to operate the heat pump, which equals the power of the heater times time.	$|Q_h| = \kappa W = \kappa(P\Delta t)$
Insert values.	$|Q_h| = \left[\left(850\frac{\text{J}}{\text{s}}\right)(6.00\text{ h})\left(\frac{3600\text{ s}}{1\text{ h}}\right)\right]2.95$ $|Q_h| = \boxed{5.42\times10^7\text{ J}}$

b. The energy extracted from the colder space (the cold reservoir, outside air) can be found using energy conservation, Eq. 22.16 (see figure in solution to Problem 31 above).	$\lvert Q_c\rvert = \lvert Q_h\rvert - \lvert W\rvert$ $\lvert Q_c\rvert = \lvert Q_h\rvert - \frac{\lvert Q_h\rvert}{K}$ $\lvert Q_c\rvert = \lvert Q_h\rvert\left(1-\frac{1}{K}\right)$
Plug in numbers.	$\lvert Q_c\rvert = (5.42\times10^7\text{ J})\left(1-\frac{1}{2.95}\right)$ $\lvert Q_c\rvert = \boxed{3.58\times10^7\text{ J}}$

CHECK and THINK

The coefficient of performance tells us how much heat is transferred to the hot reservoir relative to the amount of work we need to put in. The coefficient of performance of around three is consistent with the fact that $\lvert Q_h\rvert$ is around three times the value of W.

81. (N) Consider a system A with two subsystems A_1 and A_2. The total number of microstates of A_1 and A_2, respectively, are 10^{10} and 2×10^{10}.

a. What is the number of microstates available to the combined system, A?

b. What are the entropies of the system A and the subsystems A_1 and A_2?

INTERPRET and ANTICIPATE

The number of microstates available to system A will be much more that the individual subsystems. The entropy will therefore also be larger.

SOLVE

a. For every individual state in the 10^{10} possible states in subsystem A_1, there are 2×10^{10} separate possible states in subsystem A_2. The total number of possible states is then the number of A_1 states times the number of A_2 states.	$w = w_1 w_2 = 10^{10}\times(2\times10^{10}) = \boxed{2\times10^{20}}$

b. We calculate the entropies of the subsystems A_1 and A_2 using Eq. 22.25 ($S = k_B \ln \mathcal{W}$).	$S_{A_1} = k_B \ln(10^{10})$ $S_{A_1} = 10k_B \ln 10$ $S_{A_1} = \boxed{3.18 \times 10^{-22} \text{ J/K}}$ $S_{A_2} = k_B \ln(2 \times 10^{10})$ $S_{A_2} = k_B(10 \ln 10 + \ln 2)$ $S_{A_2} = \boxed{3.27 \times 10^{-22} \text{ J/K}}$
Now calculate the entropy of the system A. The entropy of the system A is equal to the sum of the entropies of subsystems A_1 and A_2.	$S_A = k_B \ln(2 \times 10^{20})$ $S_A = k_B(20 \ln 10 + \ln 2)$ $S_A = \boxed{6.45 \times 10^{-22} \text{ J/K}}$

CHECK and THINK

The number of microstates in the combined system is much larger than in either individually. The entropies of the individual systems add up to the entropy of the combined system.